CANCER AND PRE-CANCER OF THE CERVIX

CANCER AND PRE-CANCER OF THE CERVIX

Edited by

D.M. Luesley

Professor of Gynaecological Oncology
Directorate of Obstetrics and Gynaecology
City Hospital
Birmingham, UK

R. Barrasso

Director of the Colposcopy Unit
Department of Obstetrics and Gynaecology
Bichat University Hospital
Paris, France

CHAPMAN & HALL MEDICAL
London · Weinheim · New York · Tokyo · Melbourne · Madras

Published by Chapman & Hall, an imprint of Lippincott-Raven Publishers Inc, 2–6 Boundary Row, London SE1 8HN, UK

Lippincott-Raven Publishers Inc., 227 East Washington Square, Philadelphia, PA 19106-3780, USA

First edition 1998

© 1998 Lippincott-Raven Publishers

Typeset in 10/12 pt Palatino and produced by Gray Publishing, Tunbridge Wells, Kent
Printed at the University Press, Cambridge

ISBN 0 412 56600 1

A catalogue record for this book is available from the British Library

Library of Congress Catalog Card Number: 98–70533

∞ Printed on acid-free text paper, manufactured in accordance with ANSI/NISO Z39.48-1992 (Permanence of Paper)

CONTENTS

CONTRIBUTORS

R. BARRASSO
Colposcopy Unit
Department of Obstetrics and Gynaecology
Bichat University Hospital
Paris, France

T. BEYNON
Consultant/Senior Lecturer
Sainsbury's Department of
Palliative Medicine
UMDS
St Thomas's Hospital
London, UK

C. BERGERON
Laboratoire Cerba
95066 Cergy Pontoise Cedex 9
France

A.D. BLACKETT
University of Sheffield
Department of Obstetrics and Gynaecology
Northern General Hospital
Sheffield, UK

P.I. BLOMFIELD
Consultant Gynaecological Oncologist
Department of Gynaecological Oncology
Mercy Hospital for Women
Clarendon Street
Melbourne, Victoria
Australia

F.X. BOSCH
Servei d'Epidemiologia i Registre
del Càncer
Institut Català d'Oncologia
Barcelona, Spain

M. BYRNE
Consultant Physician in Genitourinary
Medicine
Department of Genitourinary Medicine
St Mary's Hospital
London, UK

X. CASTELLSAGUÉ
Servei d'Epidemiologia I Registre del Càncer
Institut Català d'Oncologia, Spain

R. CRAWFORD
Consultant Gynaecological Oncologist
Addenbrook's Hospital
Cambridge, UK

J. CUZICK
Head, Department of Mathematics Statistics
and Epidemiology
ICRF
London, UK

D. DARGENT
Gynaecologie Obstetricque
Hôpital Edouard Herriot
Lyon Cedex, France

M. DOMEIKA
Institute of Clinical Bacteriology
Uppsala University
Uppsala, Sweden

A.S. EVANS
Consultant Gynaecologist
Department of Obstetrics and Gynaecology
University Hospital of Wales
Cardiff, UK

H. FOX
Department of Pathological Sciences
University of Manchester
Manchester, UK

S. GARLAND
Department Head
Microbiology and Infectious Diseases
The Royal Women's Hospital
Carlton, Victoria, Australia

K.M. GREVEN
Department of Radiation Oncology
Wake Forest University School of Medicine
Winston Salem
NC, USA

A.P.M. HEINTZ
Professor of Obstetrics and Gynaecology
Gynaecological Oncology Centre
Utrecht, The Netherlands

S.J. HOUGHTON
Well Being Research Fellow
Department of Obstetrics and Gynaecology
City Hospital NHS Trust
Birmingham, UK

S. JABLONSKA
Department of Dermatology
Warsaw School of Medicine
Warsaw, Poland

G. KENTER
Department of Gynaecology
Leiden University Hospital
Leiden, The Netherlands

R. LANCIANO
Department of Radiation Oncology
Delaware County Memorial Hospital
Drexel Hill, PA, USA

D.M. LUESLEY
Professor of Gynaecological Oncology
Directorate of Obstetrics and Gynaecology
City Hospial
Birmingham, UK

P.N. MAINWARING
Medical Oncology Registrar
Royal Prince Alfred Hospital
Department of Medical Oncology
Sydney, NSW, Australia

S. MAJEWSKI
Department of Dermatology
Warsaw School of Medicine
Warsaw, Poland

J. MALEJCZYK
Department of Dermatology
Warsaw School of Medicine
Warsaw, Poland

P.-A. MÅRDH
Uppsala University
Centre for STD Research
Uppsala, Sweden

N. MUÑOZ
Field and Intervention Studies
International Agency for Research
on Cancer
Lyon Cedex, France

J. NEVIN
Department of Obstetrics and Gynaecology
University of Cape Town
Groote Schuur Hospital
Cape Town, South Africa

C.J. POOLE
MacMillan Senior Lecturer in Medical
Oncology and Palliative Care
CRC Institute for Cancer Studies
University of Birmingham
Birmingham, UK

T. NATASHA POSNER
Lecturer in Medical Sociology
Department of Social and Preventive
Medicine
University of Queensland Medical School
Brisbane
Queensland, Australia

K.S. RAJU
Consultant
Department of Gynaecological Oncology
Guy's and St Thomas's Hospital Trust
St Thomas's Hospital
London, UK

T.P. ROLLASON
Consultant Pathologist
Department of Pathology
Birmingham Women's Hospital
Birmingham, UK

M.I. SHAFI
Consultant Gynaecological Oncologist
Birmingham Women's Hospital
Birmingham, UK

F. SHARP
University of Sheffield
Department of Obstetrics and Gynaecology
Northern General Hospital
Sheffield, UK

J.H. SHEPHERD
Department of Gynaecological Oncology
Directorate of Women's Services
St Bartholomew's Hospital
London, UK

The late P. SKRABANEK
Trinity College
Dublin, Ireland

M.H.N. TATTERSALL
Professor of Cancer Medicine
University of Sydney
Department of Cancer Medicine
NSW, Australia

J.A. TIDY
Senior Lecturer on Gynaecological Oncology
University of Sheffield
Northern General Hospital
Sheffield, UK

M.A. VAN EIJKEREN
Department of Obstetrics and Gynaecology
Gynaecological Oncology Centre
Utrecht, The Netherlands

P. VAN GEENE
Senior Registrar/Fellow in Gynaecological
Oncology
Northern General Hospital
Sheffield, UK

C.D.A. WOLFE
Senior Lecturer
Division of Public Health Medicine
UMDS
St Thomas' Hospital
London, UK

A.M. WRIGHT
Senior Registrar
Department of Obstetrics and Gynaecology
University Hospital of Wales
Heath Park
Cardiff, UK

PREFACE

As the millennium approaches we might be forgiven for congratulating ourselves on our continued acquisition of new knowledge and a greater understanding of the genesis of cervical cancer. On the other hand, this disease continues to ravage womankind as the second most common cause of cancer death. The majority of women who will acquire and die from this disease will do so in the least economically developed areas of the globe. Perhaps the most obvious goal to which we should now aspire is to translate the advances in biological understanding into real health benefits for these hundreds of thousands of women world-wide.

The objectives of this particular book were to be comprehensive yet focus more attention on those areas where we feel there has been real change in the recent past. We make no excuses for the more detailed appraisal of human papillomaviruses and their role in cervical carcinogenesis as this has been an area where molecular biology has made a huge contribution to our understanding of this disease and where change has been rapid and extensive. Few would now doubt that these oncogenic viruses are central to the carcinogenic concept and as this field expands we will undoubtedly uncover more information with regard to host susceptibility, host response, vaccination and eventually real prevention. Despite the advances we have yet to see, many would argue that this disease is largely preventable through detection and treatment of pre-invasive precursors. For this reason, we believe that pre-invasive disease of the lower genital tract deserves inclusion here. There would appear to be ample evidence in those countries with well established and economically sound medical infrastructures that screening with Papanicolaou smears is effective. However, this area too is not without its critics and there is certainly more to come in the screening arena.

There has also been considerable change with regard to the management of early and late established disease. There is certainly scope for improvement in both areas. Risk prediction with a view to minimizing morbidity would appear to be central to the modern approach to management of early invasive cancer. Endoscopic techniques may have clinical utility in this role and certainly deserve thorough evaluation in clinical trials. In advanced or bulky disease a consensus is emerging whereby multidisciplinary management may hold at least some hope for improved outcome. For these reasons these areas also receive additional emphasis here.

This book has had a long gestation and we would wish to thank all of the contributors for their time, patience and effort. The pace of change in medical care is now almost too rapid to allow the production of contemporaneous literature. Topics, which at the copy-editing stage seem obvious may not have been anticipated at the outset. On reflection we feel that the major omission here is in the field of the economies of this disease. Most would not consider the inclusion of health economics in a largely scientific and clinical text as an omission yet it is obvious to all practising clinicians that it is of growing importance. Another effect of time is changing views and opinions as new information becomes synthesized and assimilated into the common consensus and thus clinical practice. All but one of our contributors has had the opportunity of updating their contributions. The exception was Dr Petr Skrabenek who tragically died during the production of this book. Petr's views were often outspoken and not necessarily conformist. His views on cervical screening are seen by many as no longer tenable, given the weight of circumstantial evidence that is now available. Although his views may not reflect our own, we feel that his intellect has much to offer the ongoing debate on the values of screening programmes and as such deserves a place in this text.

As alluded to in the introductory paragraph of this preface, the major challenge that we will face in the next millennium is that of translational research. The economics of our current and future efforts will only make sense when transposed to health environments where they will make the greatest impact. We both hope that this small contribution will provide some of the impetus to set the global medical community in the right direction.

D.M. Luesley and R. Barrasso
Birmingham and Paris

T.P. Rollason

GROSS ANATOMY

The cervix is the most caudal portion of the uterus and protrudes into the upper vagina. It measures 2.5–3 cm in length in the adult multigravida and makes up one-third to one-half of the length of the uterus. This ratio is not constant throughout life. At birth the uterus is approximately 3 cm long. It grows with the cervix in the first 2 years of life but after this time no changes occur in the uterus until approximately 9 years of age, by which time the cervix is larger than the corpus. By 13 years the two components are approximately the same length due to more rapid growth of the corpus [1]. The cervix is demarcated from the uterine corpus by a fibromuscular junction termed the internal os and the endocervical canal opens into the vaginal vault at the external os. The vagina is fused circumferentially to the cervix dividing it into an upper, supravaginal and a lower, vaginal portion. These portions are of approximately the same length. As the uterus is normally anteverted the cervix is usually angulated downward and backward. The shape of the cervix is highly variable. The nulliparous cervix has a circular external os and a diameter of approximately 2.0–2.5 cm. The multiparous cervix is larger and more protruding and has a transverse, slit-like external os. This altered appearance is brought about by size increase, stretching and laceration at the time of delivery and later repair. The reflections of the vaginal epithelium around the sides of the cervix constitute the vaginal fornices. The vaginal portion of the cervix (portio vaginalis) is divided into anterior and posterior lips, the anterior is shorter and projects lower than the posterior. Both cervical lips are normally in contact with the posterior vaginal wall (as is the anterior vaginal wall).

The cervical canal connects the uterine isthmus (internal os) with the external os. This is an elliptical cavity showing longitudinal ridges (plicae palmatae) composed of epithelium and connective tissue. The canal has a maximum diameter of approximately 7–8 mm. It is approximately 3 cm long and is flattened antero-posteriorly. The use of the terms anatomical and histological internal os relates to the difference seen in the position of the macroscopic os and the histological point at which the epithelium changes from isthmic endometrial to endocervical. These terms have little clinical relevance and the point of epithelial alteration is highly variable and often poorly defined.

The supravaginal portion of the cervix is separated anteriorly from the bladder by the parametrial connective tissues, which also extend to the sides of the cervix and between the layers of the broad ligaments. Posteriorly

Cancer and Pre-cancer of the Cervix
Edited by D.M. Luesley and R. Barrasso. Published in 1998 by Chapman & Hall, London. ISBN 0 412 56600 1.

the supravaginal cervix is covered with peritoneum, which continues down over the upper vaginal wall and is reflected onto the rectum forming the Pouch of Douglas (recto-uterine pouch).

The cervical stroma is made up of fibrous, muscular and elastic tissue. Fibrous tissue predominates, with smooth muscle located mainly in the endocervix [2] and increasing in relative proportion as the internal os is approached. At the isthmus smooth muscle and fibrous tissue are present in approximately equal proportions and in concentric arrangement, making up a functional sphincter. Damage to this sphincter, or congenital abnormality, may lead to cervical incompetence during pregnancy.

The arterial supply of the cervix is derived from descending branches of the uterine arteries [3], which pass to the lateral walls along the superior margins of the paracervical ligaments (transverse cervical ligaments or cardinal ligament of Mackenrodt). The terminal branches of these arteries anastomose with the azygous vaginal arteries, producing a complex vascular bed around the cervix. The transverse cervical or cardinal ligaments together with the utero-sacral ligaments, which attach the supra-vaginal portion of the cervix to the second to fourth sacral vertebrae, are the main source of fixation and support. The ligaments consist largely of fibrous tissue with smaller amounts of smooth muscle. The uterosacral ligaments appear to be the main ligaments holding the uterus in an anteverted position. The lateral ligaments appear to provide most general support. The ureter runs downwards and forwards within the parametrium some 2 cm from the cervix.

The venous drainage parallels the arterial system but communications exist between the cervical vessels and venous supply of the neck of the urinary bladder. There are two sets of cervical lymphatics, one superficial, beneath the cervical epithelium, the other deep within the stroma [4]. The superficial channels anastomose to produce collecting channels, originally perpendicular to the canal and radiating outward, and then parallel to the canal [5]. These vessels fuse with the deep lymphatic set and efferent lymphatics then lie alongside the vessels of the corpus until the regional nodes are reached.

There are three major lymphatic trunks draining the cervix [5]: lateral, posterior and anterior. The anterior trunk passes to the inter-iliac nodes and the posterior to the sub-aortic, common iliac and para-aortic nodes. The lateral trunks divide into three branches: upper, middle and lower. The upper branches pass to the internal iliac (highest interiliac) nodes; the middle pass to the external iliac, lower and deeper interiliac and common iliac nodes. The lower branches drain into the inferior and superior gluteal, sacral and sub-aortic nodes. Variations on this pattern are common, including direct drainage to the para-aortic and common iliac nodes [6].

The cervical nerves derive from the superior, middle and inferior hypogastric plexuses of the pelvic autonomic system. The nerve supply is largely limited to the endocervix and deep ectocervix, accounting for the relative insensitivity of the portio vaginalis.

DEVELOPMENTAL ABNORMALITIES OF THE CERVIX

Congenital abnormalities of the cervix usually result from abnormal development and fusion of the Müllerian ducts and are usually therefore seen in association with uterine body maldevelopment. Infrequently abnormalities of the cervix may relate to transitions between Müllerian and urogenital sinus epithelium. The cause of such abnormalities is usually unknown. Almost half of all females with trisomy D have a duplex or septate uterus and thalidomide exposure *in utero* was associated with uterus didelphys and pseudo-didelphys [7].

Agenesis or absence of the corpus and cervix is due to the early and bilateral agenesis or aplasia of the urogenital primordium or Müllerian ducts. Abnormal development of the cervical segment of the Müllerian ducts results in an absence of the cervix (uterus acollis), small corpus or deformed cervix. If the mesonephric duct development is prevented then the ipsilateral duct will be absent at the same level and often the kidney on the same side will be absent.

Müllerian duct fusion abnormalities result in uterus didelphys with cervical and vaginal duplication (complete defect) or, more commonly, with fusion of the vagina (pseudodidelphys) or lower uterus. Bicornuate uterus shows two horns with fusion of the lower corpus. The cervix may be duplicated (bicollis) or single (unicollis).

If the fused Müllerian ducts do not undergo normal resorption then a uterine septum may persist, which may extend the full length of the corpus and cervix or be partial, involving only corpus or only cervix.

In women exposed to diethylstilboestrol *in utero* cervical changes may accompany the well-known vaginal abnormalities. Between 22 and 58% of exposed women have gross cervical abnormalities in the form of hypoplasia, ridges and/or pseudopolyps [8, 9]. There may also be protrusion of the anterior cervical lip with a smooth or jagged outline. This is caused by excessive stromal proliferation and is often called a cock's comb cervix [9, 10]. Ridges may exist in several forms, the most frequent is a circular or transverse depression dividing the ectocervix into inner and outer zones. Pseudopolyps are broad-based protruberances arising from the ectocervix near the os and usually covered by endocervical epithelium. They are highly vascular. Abnormal extension of endocervical epithelium is also commonly seen with the other DES-associated abnormalities described above, as is an abnormally extensive congenital transformation zone (see later) and proliferation of surface and subepithelial glandular elements, similar to those seen in vaginal adenosis, are frequent.

HISTOLOGY

'ORIGINAL' (NATIVE) SQUAMOUS EPITHELIUM

The vaginal portion of the cervix (ectocervix) is lined by stratified squamous epithelium, which in its normal state is not keratinized on light microscopy (Figure 1.1). This epithelium is replenished by proliferation of basal cells every 4–5 days during reproductive life. Maturation may be accelerated by oestrogens and inhibited, at the mid-zone of the epithelium, by progestagens [2, 11]. In adult life this epithelium is fully mature and glycogen-laden due to oestrogenic stimulation. A similar pattern is seen in early postnatal life under the influence of maternal oestrogen. In postmenopausal women the epithelium undergoes atrophy, with thinning, loss of differentiation and loss of glycogen; the whole epithelium then appears to consist of basal and parabasal-type cells. In the pre-menopausal female the epithelium is of similar appearance to that of the post-menopausal woman. During pregnancy superficial maturation is lost under the influence of elevated progesterone.

In the past the cervical squamous epithelium was often divided histologically into five zones [12, 13] but this is an unnecessary complication. It is now usual to divide the ectocervical epithelium into three zones – basal, mid-zone and superficial. The basal zone is composed of

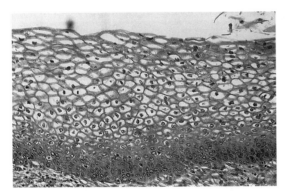

Figure 1.1 Mature ectocervical squamous epithelium. The surface is not keratinized and cellular vacuolation due to glycogen is prominent. The basal zone is clearly seen but the mid-zone and superficial zone blend imperceptibly.

one or two layers of cylindrical or elliptical cells approximately 10 mm in diameter. These have scant cytoplasm and nuclei orientated perpendicular to the underlying basal lamina (basal membrane), which is well demonstrated by electron microscopy or immunohistochemical staining with anti-laminin. The cells of this layer are actively dividing. On electron microscopy these cells contain abundant ribosomes but only occasional tonofilaments. Hemidesmosomes anchor these cells to the basal lamina.

The lower few layers of the mid-zone contain larger cells than the basal layer with more cytoplasm, often termed parabasal cells. Here the cells ultrastructurally show increased tonofilaments and numerous desmosomes at regular intervals. Many microvillous processes project into the intercellular spaces. In normal epithelium mitoses are seen in these cells as well as the basal layer but with less frequency. Glycogen synthesis occurs in this layer.

The upper mid-zone or intermediate cell zone is composed of non-dividing, glycogen-rich cells that show a gradual increase in cytoplasm with increasing height. The overall pattern of this zone is often termed 'basket-weave'. Nuclear size is constant from this level to the surface. On electron microscopy tonofilaments and desmosomes reach their maximal concentration here.

The cells of the superficial zone show flattening and an overall cell diameter of approximately 50 μm. The nuclei are small and pyknotic and the cytoplasm glycogen-rich and eosinophilic. Keratinosomes are evident on electron microscopy and nuclei and desmosomes undergo degeneration. The epithelial surface is cornified and on electron microscopy a complex surface pattern of microridges is present; these are believed together to help prevent trauma to the underlying layers and stop infective agents entering the deep epithelium. Under some exogenous stimuli keratinization occurs above the superficial cells; this is represented by a dense, eosinophilic layer of variable thickness (Figure 1.2). It is due

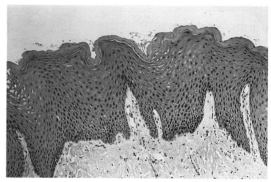

Figure 1.2   The ectocervical epithelium is hyperplastic and shows a thick surface layer of keratinization related in this case to simple procidentia.

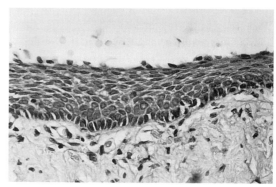

Figure 1.3   Atrophic post-menopausal epithelium. There is reduced cellular maturation and the regular dark cells give a superficial similarity to cervical intra-epithelial neoplasia.

to a relative increase in cell cytoplasmic keratin production. The keratinized cells are most commonly anucleate.

The atrophic epithelium of post-menopausal women (Figure 1.3) shows little or no surface epithelial maturation and absent or sparse stromal papillae (rete pegs are not normally seen even in the mature cervix). These papillae are finger-like extensions of vessels and stroma into the epithelium [14]; they may be more prominent when the epithelium is altered, e.g. by inflammation or papillomavirus effect. Occasional squamous epithelial cells showing a large solitary vacuole, leading to a 'signet-ring'

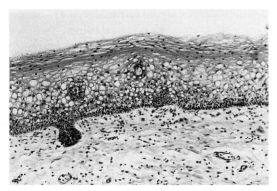

Figure 1.4 Glycogen vacuolation of the ectocervical squamous epithelium has produced a 'signet-ring' appearance to many cells. This is a normal variant, although similar vacuolation can be seen in cervical intra-epithelial neoplasia.

cellular appearance, are common in the ectocervical epithelium (Figure 1.4). Occasionally this cell pattern becomes dominant [15]. The cause of this change is uncertain but it appears entirely benign and no specific colposcopic appearance has been described in association with it. Similar cellular changes may, however, be seen in cervical intra-epithelial neoplasia.

Dendritic Langerhans' cells are seen in the cervical epithelium as in the vulva. Langerhans' cells are bone-marrow derived cells with extensive elongated cell processes (dendritic cells). They are intimately concerned with the body's immune system and act as antigen-presenting cells [6, 16] and in activating T-cells. They express a range of surface antigens including Fc and C3 receptors. They occur throughout the epithelial thickness but are usually evident as high level clear cells on haemaloxylin and eosin (H & E) staining. They can be demonstrated by impregnation with gold chloride or by immunostaining with anti-OKT6. They do not react to ultraviolet light. Their role in the development of malignancy is unclear but they appear reduced in numbers in smokers [17].

Rarely, apparently at any age, there may be sebaceous glands beneath the ectocervical epithelium. This has been suggested to be due to trauma, surgery or inflammation [18]. Hair follicles have also been reported. Whether these elements are formed by mesodermal metaplasia or represent heterotopias is unclear.

'ORIGINAL' (NATIVE) COLUMNAR EPITHELIUM

The endocervical columnar epithelium is composed of a single layer of mucin secreting, columnar cells. These cells have basally placed, round or oval nuclei and uniform, slightly granular cytoplasm filled with mucin droplets (Müllerian mucinous epithelium). The secretory cells appear to undergo cyclical synthesis, secretion and exhaustion under β-adrenergic control [19]. Acidic and neutral mucins are produced, with acidic predominating. The relative proportions of different mucins vary with the menstrual cycle. Sialomucins increase relative to sulphomucins in the pre-ovulatory phase of the cycle, but fall after ovulation and postmenopausally, when sulphomucins predominate [20, 21]. Secretion appears to be both apocrine and merocrine in type. The mucin secreting cells are 20–35 μm in height and 5–9 μm width in the sexually mature woman but decrease in size after the menopause. These changes in the histochemical composition of the mucins are reflected in the actual physical consistency of the mucus: at mid-cycle it is more watery, less viscous and more abundant than at other times in the cycle and it shows the capacity for 'ferning' in smears. This arborizing pattern is due to parallel alignment of longitudinal chains of micelles (the individual macromolecules align to produce filaments, aggregates of filaments make up micelles). Such macromolecular alignment is believed to facilitate sperm transport.

Occasional non-secretory cells with cilia are present, resembling tubal or endometrial ciliated cells; these probably play a role in mucin movement. Solitary neuroendocrine cells may also be identified on specific staining [22] but appear more frequent in some cases of intra-epithelial glandular neoplasia and some tumours.

On two-dimensional sections the endocervix appears to show surface epithelium and variably spaced, underlying tubular elements. Whilst the endocervical surface epithelium is often referred to as a mucosa and the tubular elements as glands, neither is actually true. The surface epithelium, and epithelium of the underlying structures, has no associated submucosa and is a simple epithelium, not a mucosal surface. Fluhmann [13] demonstrated conclusively that the endocervical 'glands' are

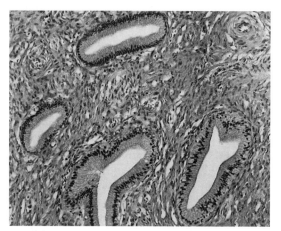

Figure 1.5 The endocervical crypts, although actually branching infoldings of the surface epithelium, appear here as isolated tubular structures.

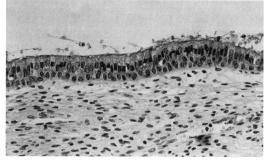

Figure 1.6 A well defined layer of subcolumnar reserve cells is seen in this epithelium which originated near to the squamo-columnar junction.

actually deep, cleft-like infoldings of the surface epithelium with numerous blind, secondary outpouchings. The complex pattern of these crypts leads to their histological appearance as isolated tubular glandular units (Figure 1.5). The epithelium of the crypts is identical to that of the surface whereas true glands have different epithelium in their ductal elements to that in their secretory portions. Although occasional crypts have been stated to occasionally be more than 1 cm deep, the maximum depth in well orientated sections is closer to 8 mm (mean 3.4 mm) [23].

It is usually stated that the endocervical epithelium has an origin in the subcolumnar reserve cells [24] (Figure 1.6) and that mitoses are not seen in the columnar epithelium in normal conditions. Whilst some experimental work supports a subcolumnar cell origin (see later), such reserve cells are, in practice, difficult to clearly define, even at the ultrastructural level, in endocervical epithelium well within the endocervical canal and in the experience of the author mitoses are seen, although infrequently, in normal endocervical epithelial cells and may be quite frequent in circumstances of increased cell turnover, e.g. active inflammation. Radiolabelling studies have shown no evidence of a crypt proliferation zone (as seen in the intestine).

The endocervical epithelium, as well as crypt infolding, also shows coarse mounds or cushions called rugae, which are present on both lips of the cervix [2]. This rugal pattern fuses with the longitudual 'arbor vitae' in the canal (the plicae palmatae previously referred to). There is a further fine grouping of folds to produce pendulous areas resembling bunches of grapes [25]. The basic single surface subunit is the villus which is usually ovoid and between 0.15 and 1.5 mm diameter.

SQUAMOUS METAPLASIA AND THE TRANSFORMATION ZONE

The squamo-columnar junction (SCJ) of the cervix is the point at which the endocervical

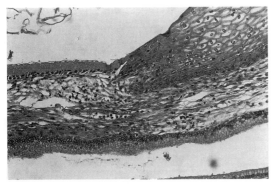

Figure 1.7 The squamo-columnar junction is clearly seen where the multilayered squamous epithelium changes abruptly to a monolayer of endocervical columnar cells. An endocervical crypt is present beneath the squamo-columnar junction.

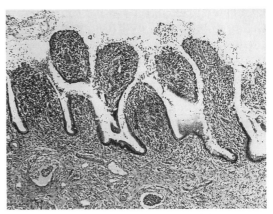

Figure 1.8 Everted endocervical epithelium is present which has taken on a papillary outline with an inflammatory cell infiltrate in the stromal cores of the papillae.

columnar epithelium meets the ectocervical stratified squamous epithelium (Figure 1.7). This junction is not at a fixed point on the cervix throughout life. The understanding of the changes that occur at the SCJ throughout life is fundamental to an understanding of the processes leading up to tumour formation in the cervix.

Before puberty, the SCJ is usually accepted to be located at, or close to, the external os of the cervix. This point is often called the 'original' SCJ. The junction is a sharp one. However, the term 'original' SCJ in these circumstances is usually a misnomer, as in 75% of infants there is evidence that there is squamous metaplastic epithelium rather than true 'original' squamous epithelium at the SCJ [26] (this phenomenon is discussed later in relation to the congenital transformation zone) and perhaps 45% of children have endocervical epithelium colposcopically visible on the ectocervical aspect of the cervix [27]. The changes produced by the use of a speculum make assessment of minor degrees of eversion difficult and arbitrary.

Under the influence of increasing ovarian hormones at puberty there is an increase in the size of both the corpus and cervix. This leads to eversion of the cervix, which is more

marked anteriorly and posteriorly than laterally. The endocervical epithelium then comes to lie on the vaginal portion of the cervix. This endocervical epithelium appears red and rough and is often clinically termed an erosion (incorrect as no ulceration is present) or an ectopy (ectropion). This zone of eversion is most extensive in women under 20 years of age and following the first pregnancy. It is usually more extensive on the anterior lip of the cervix. The everted zone, particularly when extensive, commonly takes on a papillary pattern with a chronic inflammatory cell infiltrate in the stromal cores of the papillae (Figure 1.8). This pattern is often termed papillary cervicitis but is a physiological change, not a true cervicitis. At what point the intensity of the inflammatory cell infiltrate merits a diagnosis of cervicitis proper is unclear, but a very intense lymphocyte and plasma cell infiltrate or the presence of lymphoid follicles with germinal centres have been suggested as markers of true inflammation [28].

The zone of eversion is exposed to the acidic environment of the vagina and it appears to be predominantly this stimulus which leads to the series of changes that follow and culminate in replacement of the everted endocervical epithelium by more resilient squamous epithelium. Two major mechanisms have in the past been favoured, the first is direct ingrowth of the adjacent squamous epithelium of the portio. Tongues of squamous epithelium grow beneath the adjacent columnar epithelium and expand between the endocervical mucinous cells and the basement membrane [29]. The endocervical cells are gradually displaced upwards, degenerate and are sloughed. It is unclear how important a role this mechanism has, some arguing that it is the more important [29, 30] others that it is of minor or no relevance [25]. Certainly this process cannot explain the occasional presence of isolated foci of squamous metaplasia within the endocervical canal.

The second process is usually called squamous metaplasia, but the process is not a truly metaplastic one, metaplasia being the replacement of one adult, differentiated epithelium by another of different type. Fluhmann [13] termed the process in the cervix squamous prosoplasia; this term has not achieved common usage but has much to recommend it. In the first part of the process small, non-differentiated, cuboidal reserve cells, with a high nucleo-cytoplasmic ratio, appear beneath the columnar epithelium. These usually appear first on the upper, more exposed parts of the villous outgrowths and superficial crypts. The origin of these reserve cells has been contentious. It has been argued that they arise from a pre-existing population of inconspicuous reserve cells beneath the columnar cells (first proposed by Mayer [31]), from the columnar cells themselves [13], or from stromal [32, 33], or circulating mononuclear [34] cells. The latter now seems certainly not the case [35]. Recent immunohistochemical and gel electrophoretic studies provide evidence both for the differentiation of reserve cells to squamous epithelium and for derivation of endocervical columnar epithelium from simple subcolumnar reserve cells [36, 37]. Older studies by electron microscopy, autoradiography and organ culture tend to favour direct transformation of endocervical cells to reserve cells or even dif-

Figure 1.9   Immature squamous metaplasia. There is a virtually complete layer of persisting columnar mucinous cells overlying immature squamous metaplastic epithelium.

ferentiated squamous cells [38–40]. Whatever their origin, the reserve cells proliferate to produce a layer several cells thick (reserve cell hyperplasia). At this stage the columnar cells remain as a complete or incomplete surface layer. Electron microscopically, these reserve cells resemble the parabasal cells of the ectocervical epithelium [24, 29]. These multilayered reserve cells then begin differentiation to clearly squamous cells with increasing amounts of eosinophilic cytoplasm, but without surface maturation and with little intracellular glycogen and a now incomplete persisting surface columnar cell layer (incomplete or immature squamous metaplasia) (Figure 1.9). The proliferation of reserve cells and later squamous differentiation tends to obliterate the spaces between the 'villi' of the zone of eversion. Finally all of the surface columnar cells are shed or degenerate and the squamous epithelium fully matures. There is therefore now a new squamo-columnar junction (the 'physiological' or 'functional' SCJ). The zone where columnar epithelium has been converted to squamous is termed the 'transformation zone' (Figure 1.10). Viewed at the end of the process of metaplasia, the transformation zone epithelium may be indistinguishable from the native ectocervical epithelium.

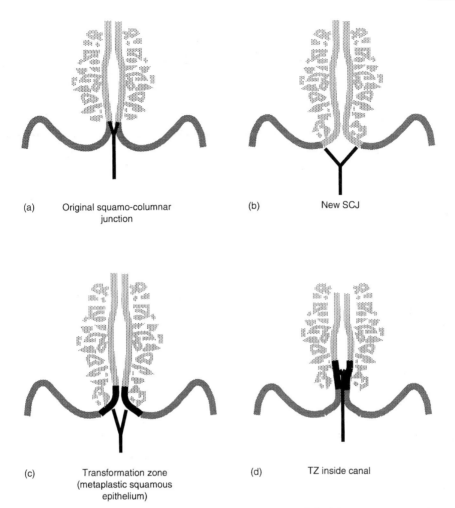

(a)   Original squamo-columnar junction

(b)   New SCJ

(c)   Transformation zone (metaplastic squamous epithelium)

(d)   TZ inside canal

Figure 1.10 Transformations in the transformation zone in adult life. (a) Prior to the menarche. (b) Eversion of the endocervix under the influence of ovarian hormones. (c) Squamous metaplasia converts the everted columnar epithelium to more resilient squamous. (d) With the fall in ovarian hormonal stimulation at the menopause the eversion is reversed and the SCJ passes into the lower canal.

Apart from hormonal effects and vaginal acidity, other possible causes of metaplasia, and accelerants of the process, include inflammatory damage, chronic irritation, coitus (prostaglandin exposure) and direct trauma.

The process of metaplasia may extend into the shallower underlying crypts for their full depth and eventually obliterate them, but usually the crypts either persist, lined by endocervical epithelium, or are partly lined by squamous cells. That surface epithelium is metaplastic may therefore be deduced from the presence of underlying crypts, as there is very little overlapping of crypts by 'original' squamous surface cells. The openings of the crypts may still be evident on the cervical surface of the transformation zone but the squamous proliferation may lead to their blockage; this produces the very common 'Nabothian follicles'. These are in reality mucus retention

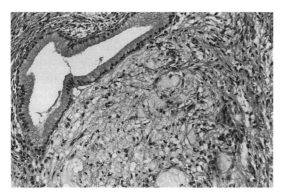

Figure 1.11   In the middle of the photograph the stroma is altered by the presence of extravated mucin from the endocervical crypt adjacent. The actual site of crypt rupture is not seen.

cysts of the crypts due to continued mucin production, with cystic dilatation related to lack of mucin drainage. The cysts may rupture leading to a local macrophage response, sometimes with associated inflammation and fibrosis (Figure 1.11). If the crypts become completely separated from the surface epithelium after the crypt epithelium has undergone replacement by squamous cells then a squamous inclusion cyst may develop.

It has been recently recognized that simple 'Nabothian cysts' may extend through most of the cervical wall. In these cases the cervical wall may appear replaced by mucin-filled cysts up to 1 cm in diameter [41]. These retention cysts are lined by benign endocervical cells, often flattened and without mitoses. There does not appear to be any premalignant potential, although histological confusion with minimum deviation adenocarcinoma (adenoma malignum) is possible.

Whilst, as previously indicated, cervical eversion and thus squamous metaplasia are most marked during adolescence and pregnancy the process continues throughout adult life and all stages of the processes described above are commonly seen in cervical biopsy specimens. Interestingly, there is evidence that the transformation zone of virgins is larger than that of sexually active young women of the same age

[25, 42]. After the menopause the shrinkage of the cervical stroma causes 'retraction' of the SCJ into the endocervical canal. The process of squamous metaplasia is not a reversible one and the canal is then lined in its lower portion by squamous epithelium. Reserve cell hyperplasia in the canal is often extensive in the post-menopausal woman, particularly when atrophic changes are prominent in the stratified squamous epithelium. The reasons for this are obscure. It may in fact represent atrophy in immature metaplasia, mimicking the appearance of reserve cell overgrowth.

Another simple variant pattern of metaplasia that may be seen in atrophic cervices particularly is 'transitional metaplasia', so called because it leads to an appearance resembling transitional urothelial epithelium. The condition in fact appears to be a replacement of the epithelium by basal and parabasal cells and is essentially a variant of basal cell hyperplasia. The epithelium lacks the 'umbrella' cells of true transitional epithelium and has a different immunohistochemical profile [43]. Because dark, small cells replace the full thickness of the epithelium it may be misdiagnosed as CIN III; there is, however, no mitotic activity or nuclear atypia.

The condition termed atypical immature metaplasia appears to be due to human papillomavirus infection in a zone of immature metaplasia [44]. It also may closely mimic CIN III but shows less nuclear pleomorphism, cellular crowding and disorganization than that condition and mitoses are more frequent in high-grade CIN.

THE CONGENITAL TRANSFORMATION ZONE (CTZ) AND CERVICAL CHANGES IN CHILDHOOD

The CTZ is essentially a zone where endocervical epithelium has undergone squamous metaplasia in late intrauterine or early extrauterine life. It appears to be related to metaplasia in a zone of endocervical epithelium which passed onto the portio under the influence of

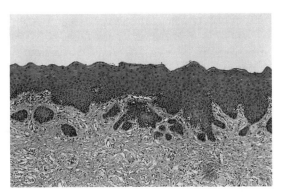

Figure 1.12 One variant of congenital transformation zone showing fine, blunt-ended downgrowths of epithelium, epithelial thickening (thinning is more usual) and fine surface keratinization.

maternal oestrogen and was replaced by squamous epithelium when the oestrogenic stimulus declined. Alternatively, it may be that the CTZ is formed in a similar manner to essentially identical zones seen in DES-exposed women [45], i.e. due to incomplete conversion of the early cuboidal epithelium of the vaginal anlage at the upper (uterine) end to squamous epithelium, followed by gradual squamous replacement in late intra-uterine and extra-uterine life. This second theory does not explain, however, the fine downgrowths of epithelium that must relate to preexisting shallow crypts, these can really only be explained on the basis of a previous endocervical eversion.

The histological features of the CTZ in a typical case are: a thinned epithelium with shallow, fine (but blunt-ended) epithelial downgrowths, sometimes with squamous 'eddies' at the base, low or absent epithelial glycogen and a very fine layer of surface keratinization. The epithelium gives the impression of being immature in its lower half, but maturing abnormally rapidly to keratinization over a few cell layers. No nuclear pleomorphism is seen, few mitoses are present and the adjacent stroma appears normal. Parakeratosis may be seen and variant patterns can be identified, which show some

epithelial thickening rather than thinning (Figure 1.12). The junction with the 'normal' ectocervical squamous epithelium is usually tangential but sharp and, when seen in the adult, the CTZ is usually separated from the SCJ by a zone of more typical 'adult' type squamous metaplasia. The low glycogen, thin epithelium, etc., may lead to a mistaken colposcopic impression of CIN; it must, however, be borne in mind that both CIN and papillomavirus effect commonly involve the CTZ, indeed warty changes are sometimes particularly prominent in this zone.

Pixley [26] has shown that, even *in utero*, after 36 weeks gestation, 75% of foetuses actually show a zone of metaplastic squamous epithelium adjacent to the truly 'original' endocervical surface epithelium (he termed this zone the 'original transformation zone'). The admixture of mature and immature metaplastic patterns with endocervical epithelium in the zone was highly variable and seen at all gestational ages. Before 30 weeks gestation the 'original' transformation zone was flat without intra-epithelial vascular projections. This pattern could be identified through infancy and childhood in 40–50% of cases. Whilst it is still not clear how this relates to the CTZ seen in the adult it seems highly likely that the CTZ represents the end result of the earliest stages of metaplasia, which Pixley suggests are an integral part of organogenesis at this site. He demonstrated that the original transformation zone could pass in a small proportion of patients well onto the vaginal surface, and this is certainly also true of the CTZ.

There is evidence that the position of the SCJ at or near the external os in pre-menarchal females is due in part to the elongation of the upper cervix around 1 year of age [46]. This is suggested to draw up the SCJ from the vaginal aspect as the enlargement occurs only in the stromal component As indicated previously, however, some studies have shown colposcopic evidence of columnar epithelium on the ectocervix in more than 40% of children between the ages of 1 and 13 [27].

CERVICAL CHANGES DURING PREGNANCY

Under the stimulus of gestational hormones the cervix softens and enlarges. This is due to increased vascularity and stromal oedema. Acute inflammatory changes are also commonly seen in the superficial stroma. In late pregnancy there is accumulation of large amounts of extracellular glycoprotein and collagen disruption leading to further softening, facilitating dilatation, etc., in labour. Decidualization of the stroma under progestational effects is common in the superficial stroma. It may be patchy or diffuse and affects both endo- and ectocervix. Some degree of decidualization occurs in more than one-third of pregnant women and takes some weeks to disappear after delivery [47].

Macroscopically, decidual foci appear as raised, vascular nodules and colposcopically they may closely resemble invasive carcinoma. Histologically the stromal cells are indistinguishable from decidualized endometrial cells, being large and pale with clearly defined cytoplasmic margins and large, round, monomorphic nuclei. Very occasionally foci of decidualization may be seen in the absence of pregnancy or obvious endometriosis; usually in association with progestagen therapy.

As indicated previously, very extensive zones of cervical 'erosion' are classically seen in pregnancy and immature metaplasia and reserve cell hyperplasia are extensive. This is most striking in primigravidae. There are probably two major processes underlying the epithelial changes seen. The first is the eversion of the endocervical canal epithelium, and the second is gaping of the os [48]. Both allow exposure of the columnar epithelium to vaginal acidity leading to metaplasia. These changes are more marked in first pregnancy and tend to occur later in pregnancy in multiparous women.

An Arias–Stella reaction is well described in the cervix in gravid hysterectomy specimens [49] but is usually very focal and limited in extent. The nuclear atypia may be worrying

and care must be taken histologically to avoid overdiagnosis as clear cell adenocarcinoma. True erosions may be seen also, usually immediately adjacent to the portio squamous epithelium [47]. Microglandular hyperplasia (see later) is often very florid and may produce visible protrusions into the endocervical canal.

Immediately after delivery lacerations, true areas of ulceration, bruising and tissue necrosis may be seen in a high percentage of cases [50], but appear to heal rapidly within a few months. The anterior lip is more commonly and severely affected.

SIMPLE GLANDULAR HYPERPLASIA (TUNNEL CLUSTERS, ADENOMATOID PROLIFERATION)

This very common condition, seen in perhaps 8–10% of cervices [13, 51, 52], consists simply of a localized proliferation of small and cystic, crowded endocervical crypts (Figure 1.13). It usually occurs some distance into the endocervical canal rather than around the SCJ. Two variants have been described [13], but this is an unnecessary complication. The glandular

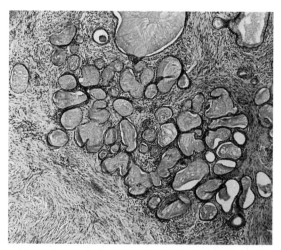

Figure 1.13 Simple glandular hyperplasia in the form of tunnel clusters. Small and cystic crypts are seen which are crowded but show no epithelial atypia.

clusters are usually discrete and rounded but multiple foci are common and cluster size is highly variable. The mean diameter of the foci is approximately 2.5 mm and they are composed of 20–50 closely packed, mucin-containing tubules. The tubules are oval or rounded and the cellular cytoplasm is usually pale. The epithelium is often attenuated or flattened. Squamous metaplasia is often seen above the proliferated crypts. These foci appear to be more common on the posterior cervical lip and those with cystic dilatation (40% of the total) lead to cervical distortion in one-third of cases [52]. The condition is entirely benign and the only danger lies in histological overdiagnosis as malignancy. Its cause is unknown. It may be a reaction to blockage of the superficial crypt lumen [45] or a regressive phenomenon. The term 'tunnel clusters' has been commonly used for these foci and is preferable to 'adenomatoid' or 'adenomatous' hyperplasia, which suggest a tumorous process.

DIFFUSE LAMINAR ENDOCERVICAL GLANDULAR HYPERPLASIA

This condition is simply a hyperplasia of the superficial (inner third) endocervical crypts producing a discrete layer sharply demarcated from the underlying cervical stroma [53]. An inflammatory cell reaction may be seen and focal oedema is well described but no desmoplastic stromal response is evident. The condition is completely benign, the cause is unknown and no clear association with hormonal stimulation is evident, though most cases are seen in pre-menopausal women.

There is a further type of simple glandular hyperplasia that does not show the classical appearances of tunnel clusters or a laminar pattern. In this poorly defined type the crypts just seem to be irregularly proliferated and crowded, without deep extension, atypia or a high mitotic rate. The cause of this simple hyperplasia is unknown but it is not uncommonly the cause of some histological concern [54].

MESONEPHRIC REMNANTS AND HYPERPLASIA

Mesonephric remnants are common in the lateral walls of the cervix (Figure 1.14). Whilst stated to be found in 8% of cervices [55], personal experience suggests that this is an underestimate. Occasionally these remnants appear unduly prominent and extensive, a condition termed mesonephric hyperplasia. Recently such simple hyperplasia has been divided into three patterns – lobular, diffuse and pure ductal types [56]. Lobular hyperplasia is the commonest pattern and shows rounded tubules in poorly defined lobular aggregates. The diffuse type may look very worrying and be mistaken for adenocarcinoma as the tubules are irregularly distributed and deep in the cervical wall. The ductal type shows large ducts with epithelium which often forms micropapillae. This form may be misdiagnosed as cervical intra-epithelial glandular neoplasia. All of these patterns are entirely benign and have only been associated with mesonephric carcinoma in extremely rare cases. The tubules contain pink hyaline material and have regular cuboidal epithelium, usually without atypia.

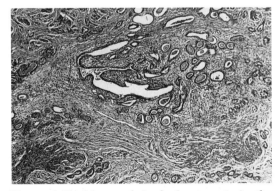

Figure 1.14 Mesonephric duct remnants in the lateral wall of the cervix. The duct itself is seen in the centre of these pictures with small proliferated tubules adjacent containing eosinophilic secretions.

MICROGLANDULAR ENDOCERVICAL HYPERPLASIA (MEH)

Whilst common in women with no excess hormonal stimulation, this condition appears to be related, particularly in its more florid forms, predominantly to progesterone stimulation. Such stimulation is most commonly due to progestagens in the combined-type oral contraceptive pill and less frequently to pregnancy, where it may be very extensive [57]. On some occasions it appears to be stimulated by oestrogen alone. It is typically seen in young women but does occur post-menopausally. In some cone biopsy series it has been seen in more than one-quarter of all cases [57] and in more than 40% of patients using the oral contraceptive pill. The lesion may occasionally produce the macroscopic appearances of a polyp, erosion or even very rarely a carcinoma [58].

Microscopically MEH may be sessile or polypoid. It may be seen both in otherwise normal areas of the cervix or within true, pre-existing endocervical polyps. It consists of aggregated, closely packed, predominantly small, round glands (Figure 1.15). Typically the stroma is hyalinized and contains a marked acute and chronic inflammatory cell infiltrate. Polymorphs are also seen in the gland lumina. Solid and reticular variants are described [59].

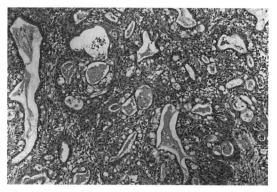

Figure 1.15 Microglandular endocervical hyperplasia. Crowded small glands are seen with cellular vacuolation and an acute inflammatory cell infiltrate. No significant cellular atypia is seen.

The lining cells are usually cuboidal and contain small, regular nuclei, sometimes with subnuclear vacuoles, which are filled with mucin, not glycogen. Mitoses are uncommon. Cellular vacuolation may be prominent, occasionally mimicking signet ring adenocarcinoma [59]. The condition may also be confused with clear cell or other adenocarcinomas, particularly those MEH variants with nuclear pleomorphism [60] or 'hobnail' cells, but it is itself benign [61]. Immunohistochemistry utilizing anti-CEA can help in differentiation of atypical MEH variants and carcinoma in the rare cases where difficulties arise but interpretation needs caution as clear cell and mesonephric carcinoma may not stain and by no means all other carcinomas are strongly positive [54].

EPITHELIAL METAPLASIAS AND ENDOMETRIOSIS

Tubal metaplasia

This is a common condition, present in almost a third of all cervices in cone biopsy and hysterectomy specimens [62]. Essentially it consists of replacement of the endocervical epithelium by ciliated, non-ciliated and peg cells, as seen in the tubal epithelium (Figure 1.16). It may occur within variant glandular patterns such as MEH. No atypia is seen and mitoses are very sparse. There is an undoubted association of tubal and endometrial metaplasia (see below) with previous cervical surgery, particularly cone biopsy [63]. In nontraumatized cervices it occurs most commonly in deep crypts of the upper endocervix but also in superficial crypts and surface epithelium. As normal crypts approach the isthmus the proportion with a tubo-endometrial cellular lining increases; in this area tubal 'metaplasia' is therefore entirely physiological and the cut-off between normal and abnormal entirely arbitrary. In postconization specimens the changes tend to occur at, or around, the SCJ.

Whilst tubal metaplasia is entirely benign it may cause problems in histological diagnosis.

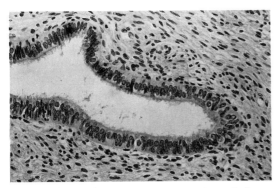

Figure 1.16 This crypt from the mid-canal shows replacement of the normal endocervical mucinous epithelium by tubal-type cells, i.e. tubal metaplasia.

Particularly when inflammatory changes are superadded it may be mistaken for cervical intra-epithelial glandular neoplasia [64]. Immunohistochemical staining with HMFG1 and CEA may help in the differentiation [65, 66]. Similar diagnostic problems may be encountered cytologically and are experienced more commonly [67, 68].

Endometrial (endometrioid) metaplasia

This is usually seen in association with tubal metaplasia and is difficult to differentiate from it, hence the common usage of the term 'tubo-endometrial metaplasia'. It is essentially the replacement of the endocervical mucinous epithelium by endometrial-type cells. As with tubal metaplasia considerable problems in cytological, and to a lesser extent histological, intepretation and differentiation from glandular neoplasia may occur. Endometrial metaplasia differs from superficial endometriosis in the absence of an endometrioid stromal component. Earlier suggestions that it was a form of endometrial glandular ectopy [69] now appear unfounded.

Intestinal (goblet cell) metaplasia

This is very rare [70] and, as goblet cells are common in high-grade CIGN, it is a diagnosis that should be made with caution. Argentaffin cells may be seen in association with the goblet cells but are also occasionally seen in normal endocervical epithelium when the term intestinal metaplasia should not be used.

Endometriosis

Cervical endometriosis (i.e. organized endometrioid glands and stroma) probably occurs in around 4% of cervices [55] but is undoubtedly more common at the SCJ after previous surgery or trauma [63]. It may be divided into superficial and deep variants. The superficial variant is rarely associated with endometriosis elsewhere in the pelvis and is the type seen after trauma. On the ectocervix it appears as small blue or red nodules, usually a few millimetres in diameter. Exfoliated cells in cytological preparations, as with tubo-endometrial metaplasia, may lead to an erroneous diagnosis of glandular neoplasia. Decidualization of the stroma and Arias–Stella reaction in glands in pregnancy also may lead to diagnostic difficulties. Superficial endometriosis has sometimes been referred to as 'primary' and deep as 'secondary'. These terms are misleading as the superficial type often relates to a specific cause and the deep does not.

REFERENCES

1. Valdes-Dapena, M.A. (1973) The development of the uterus in late fetal life, infancy and childhood. In *The Uterus* (eds H.J. Norris, A.T. Hertig and M.R. Abell), Williams & Wilkins, Baltimore, MD, p. 40.
2. Krantz, K.E. (1973) The anatomy of the human cervix, gross and microscopic. In *The Biology of the Cervix* (eds R.J. Blandau and K. Moghissi), University of Chicago Press, Chicago, IL, p. 57.
3. Gustafson, R. (1976) The vascular, lymphatic and neural anatomy of the cervix. In *The Cervix* (eds J. Jordan and A. Singer), W.B. Saunders, London, p. 50.
4. Reiffenstuhl, G. (ed) (1964) *The Lymphatics of the Female Genital Organs*, 1st edn, J.B. Lippincott, Philadelphia, PA.
5. Plentl, A.A. and Friedman, E. (1971) *Lymphatic System of the Female Genitalia*. W.B. Saunders, Philadelphia, PA.

6. Fu, Y.S. and Reagan, J.W. (1989) Development, anatomy and histology of the lower female genital tract. In *Pathology of the Uterine Cervix, Vagina and Vulva*, W.B. Saunders, Philadelphia, PA, p. 21.

7. Benirschke, K. (1973) Congenital anomalies of the uterus with emphasis on genetic causes. In *The Uterus* (eds A.T. Hertig, H.J. Norris and M.R. Abell), Williams & Wilkins, Baltimore, MD, p. 68.

8. Herbst, A.L., Poskanzer, D.C., Robboy, S.J., *et al.* (1975) Prenatal exposure to stilbestrol: a prospective comparison of exposed female offspring with unexposed controls. *New England Journal of Medicine*, **292**, 334–9.

9. Sandberg, E.C. (1976) Benign cervical and vaginal changes associated with exposure to stilbestrol *in utero*. *American Journal of Obstetrics and Gynecology*, **125**, 777–89.

10. Pomerance, W. (1973) Post-stilbestrol secondary syndrome. *Obstetrics and Gynecology*, **42**, 12–18.

11. Koss, L.G. (1979) *Diagnostic Cytology and its Histopathologic Bases*, 3rd edn, Lippincott, Philadelphia, PA.

12. Papanicolaou, G.N., Traut, H.F. and Marchetti, A.A. (1948) *The Epithelia of Woman's Reproductive Organs*, Commonwealth Fund, New York.

13. Fluhmann, C.F. (1961) *The Cervix Uteri and its Diseases*, 1st edn, W.B. Saunders, Philadelphia, PA.

14. Kolstad, P. and Stafl, A. (1977) *Atlas of Colposcopy*, 2nd edn University Park Press, Baltimore, MD.

15. Kupryjanczykj, J. and Kujawa, M. (1992) Signet ring cells in squamous cell carcinoma of the cervix and in non-neoplastic ectocervical epithelium. *International Journal of Gynecological Cancer*, **2**, 152–6.

16. Lever, W.F. and Schaumberg-Lever, G. (1983) Histology of the skin. *Histopathology of the Skin*, 3rd edn, Lippincott, Philadelphia, PA, p. 8.

17. Barton, S.E., Maddox, P.H., Jenkins, D., Edwards, R., Cuzick, J. and Singer, A. (1988) Effect of cigarette smoking on cervical epithelial immunity: a mechanism for neoplastic change? *Lancet*, **2**, 652–4.

18. Robledo, M.C., Vasquez, J.J., Contreras-Mejuto, F. and Lopez-Garcia, G. (1992) Sebaceous glands and hair follicles in the cervix uteri. *Histopathology*, **21**, 278–80.

19. Whitaker, E.M., Nimmo, A.J., Morrison, J.F.B., Griffin, N.R. and Wells, M. (1989) The distribution of b-adrenoreceptors in the human cervix. *Quarterly Journal of Experimental Physiology*, **74**, 573–6.

20. Wakefield, E.A. and Wells, M. (1985) Histochemical study of endocervical glycoproteins throughout the normal menstrual cycle and adjacent to cervical intraepithelial neoplasia. *International Journal of Gynecological Pathology*, **4**, 230–9.

21. Gilks, C.B., Reid, P.E., Clement, P.B. and Owen, D.A. (1989) Histochemical changes in cervical mucus-secreting epithelium during the normal menstrual cycle. *Fertility and Sterility*, **51**, 286–91.

22. Fetissof, F., Berger, G., Dubois, M.P., Arbeille-Brassart, B., Lansac, J., Sam-Giao, M. and Jobard, P. (1985) Endocrine cells in the female genital tract. *Histopathology*, **9**, 133–46.

23. Anderson, M.C. and Hartley, R.B. (1980) Cervical crypt involvement by intraepithelial neoplasia. *Obstetrics and Gynecology*, **55**, 546–50.

24. Gould, P.R., Barter, R.A. and Papadimitriou, J.M. (1979) An ultrastructural, cytodynamical and autoradiographic study of the mucous membrane of the human cervical canal with reference to subcolumnar cells. *American Journal of Pathology*, **95**, 1–16.

25. Singer, A. and Jordan. J.A. (1976) The anatomy of the cervix. In *The Cervix* (eds J.A. Jordan and A. Singer), W.B. Saunders, London, p. 13.

26. Pixley, E. (1976) Morphology of the fetal and prepubertal cervicovaginal epithelium. In *The cervix*, (eds J.A. Jordan and A. Singer), W.B. Saunders, London.

27. Linhartova, A. (1978) Extent of columnar epithelium on the ectocervix between the ages of 1 and 13 years. *Obstetrics and Gynecology*, **52**, 451–6.

28. Kurman, R.J., Norris, H.J. and Wilkinson, E. (1992) Anatomy of the lower female genital tract. In *Tumors of the Cervix, Vagina and Vulva. Atlas of Tumor Pathology*, 3rd series, fascicle 4, Armed Forces Institute of Pathology, Bethesda, MD, p. 10.

29. Feldman, D., Romney, S.L., Edgcomb, J. and Valentine, T. (1984) Ultrastructure of normal, metaplastic and abnormal uterine cervix: use of montages to study the topographical relationship of epithelial cells. *American Journal of Obstetrics and Gynecology*, **150**, 573–685.

30. Ferenczy, A. and Winkler, B. (1987) Anatomy and histology of the cervix. In *Blaustein's Pathology of the Female Genital Tract*, 3rd edn (ed. R. Kurman), Springer-Verlag, New York, p. 141.

31. Mayer, R. (1910) Die Epithelentwicklung der cervix und Portio vaginalis uteri. *Archive Gynaekologica*, **91**, 579–86.

32. Song, J. (1964) *The Human Uterus: Morphogenesis and Embryological Basis for Cancer*, C.C. Thomas, Springfield, IL.

33. Lawrence, W.D. and Shingleton, H.M. (1980) Early physiologic squamous metaplasia of the cervix, light and electron microscopic

observations. *American Journal of Obstetrics and Gynecology*, **137**, 661–71.

34. Singer, A., Reid, B.L. and Coppleson, M. (1968) The role of peritoneal mononuclear cells in the regeneration of the uterine epithelium of the rat. *Australia and New Zealand Journal of Obstetrics and Gynaecology*, **8**, 163–70.

35. Morris, H.H.B., Galter, K.G., Sykes, G., Casemore, V. and Reason, D.Y. (1983) Langerhans cells in human cervical epithelium: an immunohistochemical study. *British Journal of Obstetrics and Gynaecology*, **90**, 400–11.

36. Gigi-Leitner, O., Geiger, B., Levy, R. and Czernobilsky, B. (1986) Cytokeratin expression in squamous metaplasia of the human uterine cervix. *Differentiation*, **31**, 191–205.

37. Weikel, W., Wagner, R. and Moll, R. (1987) Characterization of subcolumnar reserve cells and other epithelia of the human uterine cervix. Demonstration of diverse cytokeratin polypeptides in reserve cells. *Virchows Arch [Cell Pathology]*, **54**, 98–110.

38. Lawrence, W.D. and Shingleton, H.M. (1980) Early physiologic squamous metaplasia of the cervix: light and electron microscopic observations. *American Journal of Obstetrics and Gynecology*, **137**, 661–71.

39. Gould, P.R., Barter, R.A. and Papadimetriou, J.M. (1979) An ultrastructural, cytochemical and autoradiographic study of the mucous membrane of the human cervical canal with reference to subcolumnar basal cells. *American Journal of Pathology*, **95**, 1–16.

40. Schurch, W., McDowell, E.M. and Trump, B.F. (1978) Long term organ culture of human uterine endocervix. *Cancer Research*, **38**, 3723–33.

41. Clement, P.B. and Young, R.H. (1989) Deep Nabothian cysts of the endocervix. A possible source of confusion with minimal-deviation adenocarcinoma (adenoma malignum). *International Journal of Gynecology and Pathology*, **8**, 340–8.

42. Coppleson, M., Reid, B., Singer, A. and Sullivan, J. (1976) Quoted in: *The Cervix* (eds J. Jordan and A. Singer), W.B. Saunders, London, p. 88.

43. Fetissof, F., Serres, G., Arbeille, B., de Muret, A., San Giao, M. and Lansac, J. (1991) Argyrophilic cells and ectocervical epithelium. *International Journal of Gynecology and Pathology*, **10**, 177–90.

44. Crum, C.P., Egawa, K., Fu, Y.S., *et al.* (1983) Atypical immature metaplasia (AIM). A subset of human papilloma virus infection of the cervix. *Cancer*, **51**, 2214–19.

45. Anderson, M.C. (1991) The cervix, excluding cancer. In *Female Reproductive System*, 3rd edn, Churchill Livingstone, London, p. 47.

46. Graham, C.E. (1967) Uterine cervical epithelium of fetal and immature human females in relation to estrogenic stimulation. *American Journal of Obstetrics and Gynecology*, **97**, 1033–40.

47. Johnson, L.D. (1973) Dysplasia and carcinoma in-situ in pregnancy. In *The Uterus* (eds H.I. Norris, A.T. Hertig and M.R. Abell), International Academy of Pathology Monograph. Williams & Wilkins, Baltimore, MD, p. 382.

48. Coppleson, M. and Reid, B. (1966) A colposcopic study of the cervix during pregnancy and the puerperium. *Journal of Obstetrics and Gynaecology of the British Commonwealth*, **73**, 575–85.

49. Schneider, V. (1981) Arias–Stella reaction of the endocervix. Frequency and location. *Acta Cytologica*, **25**, 224.

50. Singer, A. (1975) The uterine cervix from adolescence to the menopause. *British Journal of Obstetrics and Gynaecology*, **82**, 81–99.

51. Fluhmann, C.F. (1961) Focal hyperplasia (tunnel clusters) of the cervix uteri. *Obstetrics and Gynecology*, **17**, 206–14.

52. Segal, G.H. and Hart, W.R. (1990) Cystic endocervical tunnel clusters. A clinicopathologic study of 29 cases of so-called adenomatous hyperplasia. *American Journal of Surgical Pathology*, **14**, 895–903.

53. Jones, M.A., Young, R.H. and Scully, R.E. (1991) Diffuse laminar endocervical glandular hyperplasia: a report of seven cases. *American Journal of Surgical Pathology*, **15**, 1123–9.

54. Young, R.H. and Clement, P.B. (1993) Tumorlike lesions of the uterine cervix. In *Tumors and Tumorlike Lesions of the Uterine Corpus and Cervix* (eds P.B. Clement and R.H. Young), Churchill Livingstone, New York, p. 15.

55. Brown, L.J.R. and Wells, M. (1986) Cervical glandular atypia associated with squamous intraepithelial neoplasia: a premalignant lesion. *Journal of Clinical Pathology*, **39**, 22–8.

56. Ferry, J.A. and Scully, R.E. (1990) Mesonephric remnants, hyperplasia and neoplasia in the uterine cervix: a study of 49 cases. *American Journal of Surgical Pathology*, **14**, 1100–11.

57. Nichols, T.M. and Fidler, H.I.C. (1971) Microglandular hyperplasia in cervical cone biopsies taken for suspicious and positive cytology. *American Journal of Clinical Pathology*, **56**, 424–9.

58. Wilkinson, E. and Dufour, D.R. (1976) Pathogenesis of microglandular hyperplasia of the cervix uteri. *Obstetrics and Gynecology*, **47**, 189–95.

59. Young, R.H. and Scully, R.E. (1989) Atypical forms of microglandular hyperplasia of the cervix simulating adenocarcinoma: a report of five cases and review of the literature. *American Journal of Surgical Pathology*, **13**, 50–6.

60. Young, R.H. and Scully, R.E. (1992) Uterine carcinomas simulating microglandular hyperplasia. A report of six cases. *American Journal of Surgical Pathology*, **16**, 1092–7.

61. Jones, M.W. and Silverberg, S.G. (1989) Cervical adenocarcinoma in young women. Possible relationship to microglandular hyperplasia and use of oral contraceptives. *Obstetrics and Gynecology*, **73**, 984–9.

62. Jonasson, J.G., Wang, H.H., Antonioli, D.A. and Ducatman, B.S. (1992) Tubal metaplasia of the uterine cervix: a prevalence study in patients with gynecologic pathologic findings. *International Journal of Gynecology and Pathology*, **11**, 89–95.

63. Ismail, S.M. (1991) Cone biopsy causes cervical endometriosis and tubo-endometrial metaplasia. *Histopathology*, **18**, 107–14.

64. Jaworski, R.C., Pacey, N.F., Greenberg, M.C. and Osborne, R.A. (1988) The histologic diagnosis of adenocarcinoma-in-situ and related lesions of the cervix uteri. *Cancer*, **61**, 1171–81.

65. Brown, L.J.R., Griffin, N.R. and Wells, M. (1987) Cytoplasmic reactivity with the monoclonal antibody HMFG1 as a marker of cervical glandular atypia. *Journal of Pathology*, **151**, 203–8.

66. Rollason, T.P., Byrne, P., Williams, A. and Brown, G. (1988) Expression of epithelial membrane and 3-fucosyl-N-acetyllactosamine antigens in cervix uteri with particular reference to adenocarcinoma-in-situ. *Journal of Clinical Pathology*, **41**, 547–52.

67. Pacey, F., Ayer, B. and Greenberg, M. (1988) The cytologic diagnosis of adenocarcinoma in situ of the cervix uteri and related lesions. III. Pitfalls in diagnosis. *Acta Cytologica*, **32**, 325–30.

68. Novotny, D.B., Maygarden, J.J., Johnson, D.E. and Frable, W.J. (1992) Tubal metaplasia. A frequent potential pitfall in the cytologic diagnosis of endocervical glandular dysplasia on cervical smears. *Acta Cytologica*, **36**, 1–10.

69. Noda, K., Kimura, K., Ikeda, M. and Teshima, K. (1983) Studies on the histogenesis of cervical adenocarcinoma. *International Journal of Gynecology and Pathology*, **1**, 336–46.

70. Trowell, J.E. (1985) Intestinal metaplasia with argentaffin cells in the uterine cervix. *Histopathology*, **9**, 551–9.

THE TRANSFORMATION ZONE FROM NORMAL TO INTRA-EPITHELIAL LESIONS: COLPOSCOPIC AND HISTOLOGIC FEATURES

Renzo Barrasso and Christine Bergeron

THE NORMAL TRANSFORMATION ZONE

EMBRYOLOGY

The cervix and vagina are initially lined by Müllerian-type columnar epithelium but, after 6 weeks of embryonic life, this Müllerian epithelium is replaced by squamous epithelium growing up from the urogenital sinus. The columnar epithelium that remains on the cervix and forms the so-called native columnar epithelium is the remnant of the original Müllerian epithelium that once covered the whole cervix and vagina. The junction between the squamous epithelium of the urogenital sinus origin and the columnar epithelium is referred to as the squamo-columnar junction, and the line at which these two epithelia meet at the completion of embryonic development is referred to as the original squamo-columnar junction.

DEFINITION

The transformation zone is made up of a variable portion of the ectocervix whose covering epithelium undergoes a transformation from columnar to squamous during a woman's life. The presence of the endocervical ectropion is a physiologic finding. The position of the original squamo-columnar junction and the surface area of the endocervical ectropion are determined embryologically where inward migration of squamous epithelium from the lower third of the vagina stops. The endocervical ectropion is at its largest in the younger woman and following the first pregnancy, becoming reduced in size as remodelling of the transformation zone occurs with time. In DES daughters, in whom normal migration of the squamous epithelium is prematurely halted, the original squamo-columnar junction is often located in the vagina rather than on the exocervix.

The formation of a transformation zone is considered to be physiological but the original extent of the columnar epithelium on the ectocervix is variable. Some women do not present a transformation zone, because they have had progression of urogenital epithelium up to the external os. The transformation may occur at any time during a woman's life, but mostly occurs during foetal life, after puberty and during pregnancy. Particularly after puberty, with the changes in vaginal pH and the beginning of sexual life, the one-cell layer columnar epithelium is not adapted any more to protect the cervix. This epithelium is replaced by a multilayered squamous epithelium, which is

Cancer and Pre-cancer of the Cervix
Edited by D.M. Luesley and R. Barrasso. Published in 1998 by Chapman & Hall, London. ISBN 0 412 56600 1.

thicker and more adapted to the vaginal cavity. The speed of the transformation is extremely variable and may probably go from a few months to decades. The process seems to progress by steps, followed by static phases. It has been reported that 60% of European women at first pregnancy still present some degree of columnar ectropion [1]. By the age of 40, the process will have ended in the vast majority of cases. For most women, the transformation results in the development of a squamous tissue, first immature and then gradually reaching maturity. The new squamous epithelium becomes thus indistinguishable from native squamous epithelium, except for the vascular architecture and the persistence of few glandular crypts within the stroma. About 10% of women of childbearing age present, at the time of examination, a stable transformation zone with a squamo-columnar junction at the external os. In a few cases, for unknown reasons, some women of childbearing age will have a squamo-columnar junction in the canal. After the menopause, the squamo-columnar junction will be found in the canal in most women. This process seems to be linked to hormonal status.

Colposcopically, the glandular and vascular features allow the identification of the normal transformation zone in most cases [2]. A part or the whole process may slow down and even stop its development before the tissue reaches full maturity. Areas of immature or partially mature (arrested) squamous epithelium will thus be present. Those areas mostly react to the acetic acid test, colposcopically producing abnormal findings within the transformation zone.

LIMITS OF THE TRANSFORMATION ZONE

The caudal limit of the transformation zone is defined by the outer border of the pre-existing columnar epithelium. Squamous epithelium caudal to this limit is defined as native epithelium, since it has always been and always will be squamous. The border between the native

epithelium and the newly formed squamous epithelium of the transformation zone is defined as the squamo-squamous junction, and it never moves during life [2]. The cephalic limit of the transformation zone is composed of the last squamous cells, in contact with the first columnar cells. This limit is defined as physiologic or anatomic squamo-columnar junction. It moves cranially toward the external os, which constitutes the ideal end point of the process (Plates 1a and b).

THE ORIGIN OF THE NORMAL TRANSFORMATION ZONE

There are two proposed histogenetic mechanisms by which the cervical epithelium is replaced by squamous epithelium.

EPITHELIALIZATION

The first and most important mechanism consists of the direct ingrowth from the native portio epithelium bordering the columnar epithelium. Histologic, colposcopic and electron microscopic observations have shown that the tongues of native squamous epithelium of the portio grow beneath the adjacent columnar epithelium and expand in between the mucinous epithelium and its basement membrane (Plate 2). As the squamous cells expand and mature, the endocervical cells are gradually displaced upward, degenerate and are eventually sloughed away. A similar process is observed in the re-epithelialization of true pathologic erosion of the endocervix. The progression of squamous transformation of the endocervical ectropion is primarily dependent on local environmental factors with the initial stimulus being the low pH of the vagina after puberty. Trauma, chronic irritation or cervical infection also facilitate the development and maturation of the transformation zone because they stimulate repair and remodelling; rapid squamous re-epithelialization of the columnar ectropion may also be produced iatrogenically by electrocautery, cryosurgery or laser surgery. The process of squamous epithelialization is thought to be responsible for the obliteration of the outer two-thirds of endocervical ectopy.

SQUAMOUS METAPLASIA

The second mechanism involved in the genesis of squamous epithelium of the transformation zone is most commonly termed squamous metaplasia. Squamous metaplasia involves proliferation of the differentiated subcolumnar reserve cells of the endocervical epithelium and their gradual transformation into a fully mature squamous epithelium (Plate 3). The basic feature of reserve cells is their bipotency and ability to evolve into either columnar epithelium, including ciliated cells or squamous epithelium. The first state of metaplasia is manifested by the appearance of small cuboidal cells beneath the columnar mucinous epithelium, the so-called subcolumnar reserve cells. The subsequent steps are characterized by a progressive growth and stratification of reserve cells (subcolumnar reserve cells hyperplasia) followed by differentiation into immature squamous (Plates 4a and 4b) and ultimately, fully mature squamous epithelium indistinguishable from the native portio epithelium. Histologically, only the sinusoid line of the basal layers and the finding of gland crypts within the stroma will allow the identification of the external limits of the transformation zone (Plate 5).

Unlike squamous epithelialization, which involves the peripheral regions of the endocervical tissue, squamous metaplasia has a random distribution within the ectropion. The process is supposed to start at the tip of the villi, suggesting a possible reaction of the contact surface to changing vaginal pH. This also implies that metaplasia may uniformly proceed as a wave but it may also exist as an assembly of patches, independently originating on top of several villi expanding centripetally and developing delicate epithelial bridges that fuse with neighbouring epithelial

proliferations reflecting its asynchronous development. Prior to the transformation, each columnar villus is observed with its stromal core and terminal vascular structures covered by a single layer of epithelial mucus-secreting cells. Reserve cells will rapidly divide, capping the villus and extending into the space between two adjacent villi. The earliest detectable colposcopic change of metaplasia is a loss of translucency of the tip of the villus, the vascular structures becoming distinct [3]. Individual villi will appear fused, but at the beginning the original structure may still be imagined, since minute humps are present. During the formation of the full-thickness undifferentiated epithelium, the metaplastic cells are, by definition, very active and do not contain glycogen. This explains how immature metaplasia will colposcopically appear as a white, iodine-negative area. In the second phase, this undifferentiated epithelium will undergo differentiation and maturation. Usually, maturation occurs fast. Metaplastic, well-differentiated epithelium appears darker than normal epithelium (because it is thinner) but not acetowhite. Only a tiny ring of acetowhite epithelium is observed by the colposcopist at the squamocolumnar junction (SCJ) in most cases, representing the few cephalic immature epithelial layers (Plate 6). After the Schiller test, only the cranial portion of metaplasia remains iodine-negative, and the transition with iodine-positive tissue is gradual, since maturation is progressive. During the process, a few columnar villi may not be involved by the metaplastic progression, giving origin to glandular openings. In these cases, the feature is due to partial re-epithelialization of the gland. Metaplastic epithelium colonizes the walls, but does not penetrate into the os, the glandular structure being preserved. At high magnification, gland openings are mostly surrounded by a white ring, witness of the immaturity of the 'micro-SCJ' between squamous and columnar tissue. Occlusion instead of replacement of a gland will result in the formation of mucus cysts, the Nabothian cysts. Columnar cells may keep producing mucus, these cysts becoming very prominent.

The spatial endpoint of metaplasia is considered to be the external os. When the process is over and metaplastic tissue has reached full maturity, only the angioarchitecture and eventually glandular patterns will allow the colposcopist to know the existence and the extent of the transformation zone.

ABNORMAL FINDINGS WITHIN THE TRANSFORMATION ZONE

IMMATURE METAPLASIA

In some cases, the maturation of metaplastic tissue is not as fast as described. In this case, large parts and even the entire metaplastic tissue remains immature long enough to be detected by cytology and colposcopy. On the same cervix, immature areas may be intermingled with more mature portions of the process.

The colposcopist can detect abnormal findings within the transformation zone such as a red area (thinner epithelium and stromal hyperaemia), a strong reaction to the acetic acid test (high cellular and nuclear activity) with the possible presence of punctation and mosaic (stromal projections centred by capillaries within the epithelium); uniformly iodine-negative (lack of glycogen) [4]. Therefore, colposcopical features of immature metaplasia recall those of intra-epithelial lesions (Plate 7). However, the experienced colposcopist can differentiate immature metaplasia from cervical intra-epithelial neoplasia because the atypical findings detected in the case of immature metaplasia present as a smooth or papillary, markedly but homogeneously white area, punctation and mosaic are regularly distributed and the size and shape of capillaries are uniform [2]. Finally, the area is homogeneously iodine-negative, with no partial Lugol's uptake. However, some early intra-epithelial lesions will present

with the same features, rendering criteria not fully specific for differential diagnosis. Thus, the colposcopist will not be surprised or disappointed if the biopsy performed on markedly acetowhite epithelium shows either immature metaplasia or CIN.

ARRESTED METAPLASIA

The maturation of the metaplastic process may also be arrested at almost terminal stages. If the finding of immature metaplasia is occasional, arrested metaplasia is frequently encountered and it is at the origin of the most controversial subject in modern colposcopy [5]. In these cases, histology shows a well-differentiated but not fully mature metaplastic epithelium, without cytologic atypia but lacking glycogen. Its main morphologic character is the sharp and totally flat caudal border with normal squamous epithelium [4]. The explanation for arrested metaplasia is unknown. Arrested metaplasia may be infected by human papillomavirus (HPV) but it may also be detected in the absence of HPV infection, confirming that the phenomenon is not related to papillomaviruses [4].

Colposcopical consecutive observations have shown that the process may originate at any time during the metaplastic process and that the arrested metaplasia progresses toward the os, as in the normal case. This is why the caudal border is always sharp, while the cephalic border is indistinguishable from that of normal metaplasia, until, at the terminal stages, it shows lack of glycogen. The cervix looks normal at direct observation (normal height of the epithelium, normal amount of stromal vascularization). A flat acetowhite area, mostly tiny (mild degrees of cellular and nuclear activity) will appear after the acetic acid test. It may contain various degrees of punctation and mosaic, since due to the stromal arrangement during metaplasia, capillaries are regular in size and shape and the intercapillary distance is uniform. Those areas remain homogeneously iodine-negative (lack of glycogen) after the Schiller test. As in histology, the most characteristic colposcopical feature is the very sharp and flat border between the caudal part of this area and the normal squamous epithelium. The stereoscopic vision of the colposcope clearly allows the differential diagnosis with the elevated border of intra-epithelial lesions as well as with the progressively iodine-positive external border of normal metaplasia (Plates 5, 8a and b).

The internal border of the area may be located at the original junction, showing that the maturation of the entire metaplastic epithelium may stop at this stage. Arrested metaplasia may also originate during the metaplastic process. In those cases, a caudal area of normal transformation will be observed. Therefore, arrested metaplasia may appear as a circumferential area or it may co-exist with normal metaplasia, covering only one or more sections of the T-zone [3]. In this case, it will have a geographic map-like appearance. Since areas of normal metaplasia progress faster toward the os, arrested metaplasia may be found isolated on the ectocervical part of the transformation zone.

INTRA-EPITHELIAL LESIONS

Until recently, pathologists and clinicians have used the cervical intra-epithelial neoplasia or dysplasia/carcinoma *in situ* classifications for identifying pre-cancerous lesions of the cervix. The cervical intra-epithelial neoplasia (CIN) concept and the terminology were proposed originally to acknowledge the continuous process beginning with CIN I and progressing through CIN II and CIN III to invasive carcinoma as opposed to discrete steps in the disease process [6]. More recently, the Bethesda cytologic classification has proposed to condense the myriad terms for dysplasia or CIN to just two: low-grade and high-grade squamous intra-epithelial lesions [7]. The chief advantage of this approach is the elimination of terminology that has little bearing on the therapy of cervical disease. Accordingly, many investigators

have adopted this classification for histology as well [8, 9]. At the moment, HPV oncogenic types appear as the most important factors in the causation of cervical carcinoma. It is also possible to distinguish virologically two categories of lesions in the CIN spectrum of disease. One, designated CIN I or low-grade CIN, is associated with a heterogeneous group of HPV types including low, intermediate and high oncogenic risk types that produce a lesion whose outcome is then unpredictable. Any one of the viral types that infect the female genital tract may be found in CIN I lesions but they all produce changes that are indistinguishable histologically from one another [10, 11]. It is impossible by light microscopy to identify the lesions that are infected by high oncogenic risk types. The infection is productive and induces HPV-related cytopathogenic effect or koilocytosis. Cytologic atypia and rare abnormal mitotic figures are confined to the basal cell layer of the squamous epithelium (Plate 9a). The other type of lesions that contains mainly the high-risk HPV types and corresponds to CIN II and III or high-grade CIN would be expected to behave as a precursor lesion. This type of lesion contains disorganization, cytologic atypia and an increased mitotic rate with abnormal mitotic figures spread in at least half of the squamous epithelium [10, 11] (Plate 9b).

Opinions vary on the unicellular, as opposed to the multicellular, field origin of intra-epithelial neoplasia. Proponents of the unicellular theory maintain that intra-epithelial neoplasia begins in one cell or at most a single clone of cells. The metaplastic cell and more specifically the basal squamous cell is the target for infection by oncogenic HPV types [8, 9]. Furthermore, most CIN originates at the squamo-columnar junction (SCJ) of the transformation zone (Plate 9c). The reason may be mechanical (the SCJ represents the thinner squamous area) or biological (reserve cells might be particularly receptive to HPV infection). For the same reason, the origin of an intra-epithelial lesion at the micro-SCJ of a gland opening is also possible. The development of an intra-epithelial lesion may take place at any time during the metaplastic process, thus originating anywhere on the cervix, in relation to the location of the SCJ. Cervical intra-epithelial neoplasia develops in the same way as does squamous metaplasia. One can suppose that extension of such epithelium takes place by horizontal spread with lifting off and replacement of adjacent normal epithelia. The destruction of the entire transformation zone at the same time explains the rarity of recurrences of new CIN in treated patients and the overall treatment results strongly support the unifocal cervical carcinogenesis concept [12].

Followers of the multicellular theory believe that neoplasia arises in predetermined fields containing abnormal cell populations [13, 14]. The lesions enlarge by recruitment and coalescence of newly formed neighbouring fields. An increase in size of the abnormal area is effected by coalescence of neighbouring fields already predestined to be the origin of different forms of intra-epithelial neoplasia. Even if the fully mature original squamous epithelium is less vulnerable to oncogenic HPV types, intra-epithelial neoplasia, however, has been shown to occur in 5–15% of cases outside the last gland in the native squamous epithelium [13]. Squamous metaplasia produces multiple islands of squamous epithelium, not only in the transformation zone but also in the external os and endocervical canal. These isolated foci of squamous metaplasia in the endocervical canal could be the starting point of a precursor lesion explaining the recurrence of CIN which had been treated with free margins.

HPV-associated lesions may develop within the transformation zone at different biological phases, i.e. normal, immature, arrested metaplasia or when only mature squamous epithelium is present. The colposcopic features of HPV-asssociated lesions will depend not only on HPV types but also on the biological status of the transformation zone. At colposcopic examination, intra-epithelial lesions will appear as abnormal findings within the transformation zone [15, 16]: a normally pink or, most often, red area (stromal hyperaemia) reacting to the acetic acid test (increased cellular and nuclear activity) and mostly showing areas of punctation and mosaic (stromal fingerlike projections or ridges within the epithelium and/or angiogenesis), homogeneously iodine-negative (absence of glycogen) or showing irregular uptake of the iodine solution (focal maturation) (Plates 10a and b). Areas of keratosis may also be present. The line of progression toward the os (SCJ) will be much better delimited than the one of normal metaplasia, since the morphological transition between atypical and columnar cells is sharp. The caudal limit will mostly appear well defined, the atypical epithelium being thicker than the normal one. In the early phases, the outer limits of the normal transformation zone are still detectable when lesions originate at advanced stages of the metaplastic process. However, the caudal progression of the lesion will invariably cover the entire T-zone. As mentioned before, whether intra-epithelial lesions may extend onto the native epithelium is a controversial subject. As compared to arrested metaplasia, atypical tissue towers over the normal one, corresponding to the progressive mechanical replacement. Thus, stereoscopic vision by colposcope (better than any picture!) will clearly show the difference between the two processes [3].

Colposcopy thus allows a complete understanding of the normal transformation zone, as well as the origin and progression of cervical intra-epithelial lesions. However, colposcopic abnormalities are not specific and a biopsy is always necessary to establish if abnormal colposcopic features correspond to an intra-epithelial lesion or simply to metaplasia. This motivates the adoption of the recently proposed new colposcopical terminology, which distinguishes minor and major changes within the transformation zone, thus underlying that minor abnormalties do not necessarily correspond to a histologically defined lesion [16].

REFERENCES

1. Burghardt, E. (1984) *KolposKopie and spezielle Zervixpathologie*. GT Verlag, Stuttgart.
2. Coppleson, M., Pixley, E. and Reid, B. (1971) *Colposcopy*, C.C. Thomas, Springfield, MA.
3. Barrasso, R. and Guillemotonia, A. (1997) Cervix and vagina: diagnosis. In *Human Papillomavirus Infection: A Clinical Atlas* (eds G. Gross and R. Barrasso), Wiesbaden Ullstein Mosby, Berlin, pp. 145–274.
4. Coupez, F. (1983) Colposcopie des viroses du col. *Gynécologie*, **34**, 177–81.
5. Jordan, J.A. and Singer, A. (1976) *The Cervix*, W.B. Saunders, London.
6. Richart, R.M. (1968) Natural history of cervical intraepithelial neoplasia. *Clinical Obstetrics and Gynecology*, **10**, 748–84.
7. The Revised Bethesda System for Reporting Cervical/Vaginal Cytologic Diagnoses: Report of the 1991 Bethesda Workshop (1992) *Acta Cytologica*, **36**, 273–6.
8. Richart, R.M. (1990) Clinical commentary. A modified terminology for cervical intraepithelial neoplasia. *Obstetrics and Gynecology*, **75**, 131–3.
9. Wright, T.C., Kurman, R.J. and Ferenczy, A. (1994) Precancerous lesions of the cervix. In *Blausteins's Pathology of the Female Genital Tract*, 4th edn (ed. R. Kurman), Springer, New York, pp. 229–77.
10. Bergeron, C., Barrasso, R., Beaudenon, S., Flamant, P. Croissant, O. and Orth, G. (1992) Human papillomavirus associated with cervical intraepithelial neoplasia: great diversity and distinct distribution of types in low and high grade lesions. *American Journal of Surgical Pathology*, **16**, 641–9.
11. Lorincz, A.T., Reid, R., Jenson, A.B., *et al.* (1992) Human papillomavirus infection on the cervix: relative risk associations of 15 common anogenital types. *American Journal of Obstetrics and Gynecology*, **79**, 328–37.
12. Gardeil, F., Barry-Walsh, C., Prendiville, W., Clinch, J. and Turner, M.J. (1997) Persistent intraepithelial neoplasia after excision for cervical intraepithelial neoplasia grade III. *Obstetrics and Gynecology*, **89**, 419–22.
13. Burghardt, E. and Ostor, A.G. (1983) Site and origin of squamous cervical cancer: a histomorphologic study. *Obstetrics and Gynecology*, **62**, 117–27.
14. Johnson, L.D. (1969) The histopathological approach to early cervical neoplasia. *Obstetrics and Gynecology Surgical*, **24**, 735–67.
15. Barrasso, R. (1992) Colposcopic diagnosis of HPV cervical lesions. In *The Epidemiology of Cervical Cancer and Human Papillomavirus* (eds N. Munoz, F.X. Bosch, K.V. Shah and A. Meheus), IARC, Lyon, pp. 67–76.
16. Stafl, A. and Wilbanks, G.D. (1991) An international terminology of colposcopy: report of the Nomenclature Committee of the International Federation of Cervical Pathology and Colposcopy. *Obstetrics and Gynecology*, **77**, 313–14.

H. Fox

INTRODUCTION

In this chapter the term 'invasive carcinoma' will be taken as implying an invasive lesion which is at least International Federation of Gynaecology and Obstetrics (FIGO) Stage Ib. The term 'microinvasive carcinoma' will be considered as being synonymous with a FIGO Stage Ia invasive lesion. Only primary malignant tumours of epithelial origin will be discussed.

MICROINVASIVE CARCINOMA

MICROINVASIVE SQUAMOUS CARCINOMA

In conceptual terms, a microinvasive squamous cell carcinoma is easily defined as a neoplasm which, whilst no longer confined within the limits of the epithelium, is invading the stroma only to an extent which makes the possibility of nodal metastases extremely unlikely; a corollary of this is that a microinvasive carcinoma can be treated by conservative surgery. Putting this concept into practical terms has, however, proved difficult and over the years there have been many definitions of a microinvasive carcinoma, which have largely differed from each other in terms of the maximum permitted depth of stromal invasion; other controversial points have been the significance or otherwise of vascular space invasion and the importance of a confluent growth pattern. Implicit in all these definitions, however, is the belief that pathologists can accurately recognize the presence of invasion; in reality, this is far from being true and all surveys in which central stringent pathological review of collected cases from various centres has been performed have shown a high incidence of diagnostic error, this usually being because of overdiagnosis of a non-invasive lesion which is involving the crypts [1, 2].

During the last two decades the most widely used definition of a microinvasive squamous carcinoma has been that proposed by the Society of Gynecologic Oncologists (SGO) in 1977 [3], namely 'one in which neoplastic epithelium invades the stroma in one or more places to a depth of 3 mm or less below the base of the epithelium and in which lymphatic or blood vessel involvement is not demonstrated'. Particular points of note in this definition are the insistence upon depth of invasion as the prime diagnostic criterion, the somewhat vague description of the point from which measurements should be made, the absence of any comment on the pattern of growth and

Cancer and Pre-cancer of the Cervix
Edited by D.M. Luesley and R. Barrasso. Published in 1998 by Chapman & Hall, London. ISBN 0 412 56600 1.

the negation of a diagnosis of microinvasive carcinoma if vascular or lymphatic space involvement is detected.

Pathologists using this definition have encountered difficulties in knowing from where to measure depth of invasion, this being a particular problem when invasion occurs from epithelium within a crypt; in such cases should the measurement be from the base of the crypt epithelium or from the base of the surface epithelium? It has also, until recently, often been difficult to make the crucial distinction between true vascular space invasion and artefacts caused by tissue shrinkage around groups of tumour cells; this problem has, however, now been resolved by the introduction of specific immunocytochemical markers for endothelium, such as Factor 8 and Ulex europica.

The SGO definition has been criticized for its insistence upon depth of invasion as a defining feature [4, 5], it being pointed out that in the cervix increasing depth of invasion does not involve the breaching of any tissue boundaries, that vascular or lymphatic spaces lie very close to the basement membrane of the squamous epithelium and that the metastatic potential of a neoplasm is related principally to its volume. Arguments such as these led to an increasing belief that a microinvasive carcinoma should be regarded as a small carcinoma rather than as a tumour which is not deeply invasive and, in 1989, to a redefinition of Stage Ia by FIGO [6] and its subdivision into two groups:

- Stage Ia$_1$: Early stromal invasion, minimal microscopically evident stromal invasion.
- Stage Ia$_2$: Lesions detected microscopically can that be measured. The upper limit of the measurement should not show a depth of invasion of more than 5 mm from the base of the epithelium, either surface or glandular, from which it originates. A second dimension, the horizontal spread must not exceed 7 mm.

It will be noted that this definition, quite apart from splitting Stage Ia into two groups,

gives a clear statement as to the point from which depth of invasion should be measured, recognizes that two measurements, which give an indication of tumour volume, albeit possibly a rather inaccurate one, are required for the definition of Stage Ia$_2$ and does not allow vascular space invasion to invalidate the diagnosis (though the FIGO definition does insist that the presence or absence of vascular space invasion should be documented).

Stage Ia$_1$ lesions, as defined by FIGO in 1989, are usually easily identified and are seen as epithelial buds, or tongues, of neoplastic epithelium which often appear better differentiated than the epithelium from which they originate (Figure 3.1); there is usually a localized lymphocytic response to the invading cells and focal loosening, oedema and increased vascularity of the stroma: all these features should, and usually do, allow for a distinction between an invasive bud and tangentially cut cervical intra-epithelial neoplasia (CIN) involving a crypt. The depth of invasion of a Stage Ia$_1$ was not defined by FIGO but it has been generally accepted that invasion should not be to a depth of more than 1 mm. A Stage Ia$_2$ lesion is simply a small squamous cell carcinoma and the only diagnostic challenge it presents is that of accurate measurement.

It cannot be claimed that the 1989 FIGO definition solved all the problems of microinvasive carcinoma. Is the lack of emphasis on

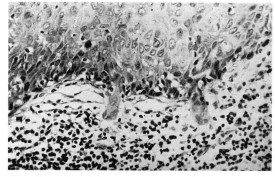

Figure 3.1 A Stage Ia$_1$ invasive carcinoma of the cervix. Two buds of epithelium are invading the stroma (H & E ×140).

vascular space invasion truly justified? Many have not been convinced that vascular space invasion can be safely disregarded [3, 7–13], but whilst it is true that the sensitivity of vascular space invasion as a risk factor has generally been found to be high its specificity is almost certainly too low for its presence to obviate a diagnosis of microinvasive carcinoma [14–17]: even the sensitivity of this finding may have been exaggerated, for in one recent very large study of microinvasive cervical carcinomas vascular space invasion was not associated with lymph node metastases or an increased incidence of recurrence [18].

Neither the SGO nor the FIGO definitions pay any attention to the pattern of growth of the microinvasive lesion, despite claims that a confluent growth pattern excludes a diagnosis of microinvasive carcinoma [10, 12]. In practice, a confluent growth pattern is difficult to define, the recognition of confluency is highly subjective and, as with vascular space invasion, the specificity of this feature is far too low for it to be of any real value [17].

The most contentious point, however, is the depth of invasion, this despite the inability of anybody to explain why depth of invasion should, in itself, be important. Many, particularly those who use the SGO definition, regard the FIGO definition as too liberal and maintain that whilst nodal metastases are exceedingly rare with tumours invading to a depth of less than 3 mm there is a significant incidence of both recurrence and node metastases in neoplasms invading to a depth of between 3 and 5 mm [8, 12, 14, 16, 19–21]. This latter contention is, however, meaningless unless the extent of horizontal spread is also documented, for tumours of similar depth of invasion may have widely varying horizontal diameters [20, 22]. When horizontal spread has been measured in these reports, this being the exception rather than the rule, it has, in cases with 3–5 mm depth of invasion associated with nodal metastases, almost invariably been such as to put the lesion outside of the defined FIGO category of Stage Ia$_2$.

It has to be accepted that a dilemma exists about the definition of microinvasive carcinoma. It is possible to produce a definition which is sufficiently strict that its clinical usage will result in a 100% survival [8]: such a definition will, however, lead to many women being subjected to unnecessary radical surgery with its resulting morbidity and, to some extent, mortality. Thus, in a series from the Norwegian Radium Hospital [23], 643 women were treated for microinvasive carcinoma of the cervix as defined by FIGO. Four patients died of cervical carcinoma (all Stage Ia$_2$), a mortality rate of 0.6%, and it is of note that in one of these fatal cases the depth of invasion was less than 3 mm. If the SGO definition of microinvasive carcinoma had been used three of these patients would have been treated more radically but this would have meant that a further 157 women would have been subjected to unnecessary radical surgery. It has to be added that if such a small tumour can kill a patient then it must be of an extremely aggressive nature and there is no proof that death would necessarily have been averted in the three fatal cases by more extensive surgery; certainly, patients have died of microinvasive carcinomas despite having been treated by radical hysterectomy [10, 11].

Recently, the situation has been made even more confused by a new FIGO definition of Stage Ia invasive carcinoma [24]. In this Stage Ia$_1$ is defined as a tumour measuring no more than 7 cm in diameter and invading to a depth of less than 3 mm, whilst Stage Ia$_2$ is a tumour of similar width which is invading to between 3 and 5 mm. This new definition has little to recommend it for there is an effective elimination of cases of early stromal invasion, these being subsumed into the group of patients with tumours invading to a depth of 3 mm. Hence Stage Ia$_1$ now includes a subset of very low risk cases and a subset of cases in which the risk of nodal metastases, whilst still low, is nevertheless greater and thus prognostic precision appears to have been blunted [25]. Time is required to assess how this new definition works out in practice but an initial reaction must be that it is a retrogressive step.

To some extent the basic concept of a microinvasive carcinoma is a mistaken one if it is assumed that the tumour has been removed at a particular stage along a progressive pathway of inexorable invasion: too many carcinomas are seen at this stage for this assumption to be true. An intra-epithelial neoplasm probably remains confined to the epithelium because a balance has been attained between host defence mechanisms and the inherent aggressiveness of the tumour cells.

It is almost certain that examples of early stromal invasion (Stage Ia$_1$) represent repetitive attempts at invasion by the neoplastic cells, that these elicit an immunological response and usually fail. It is also probable that a Stage Ia$_2$ neoplasm represents another stalemate in which the tumour has attained a restricted degree of invasion but is not progressing, a second 'latent' stage [26]. It might therefore be preferable to revert to the old term of 'microcarcinoma' to describe these small neoplasms that are only detectable histologically and to cease quibbling about minor differences in depth of invasion, minor differences which are subject, even with the aid of a measuring grid, to a degree of subjective interpretation and which can be subtly altered by differences in fixation technique. It could be argued that the re-introduction of the term microcarcinoma would mean that some very small Stage Ib tumours would be included in this category; this may well be true but very small Stage Ib carcinomas have the same prognosis as do Stage Ia$_2$ neoplasms, even when treated only by cone biopsy or simple hysterectomy [17]. The problem would then resolve itself into the recognition of those microcarcinomas, a very small minority, which are likely to pursue a highly malignant course. It is doubtful if this can be achieved by morphological techniques alone and the possibility that such tumours could be identified by study of their expression of proto-oncogenes or of their content of tumour suppressor genes should be explored.

MICROINVASIVE ADENOCARCINOMA

An entity of adenocarcinoma *in situ* with early stromal invasion, the equivalent of a stage Ia$_1$ squamous carcinoma, has been described [27, 28]; this is characterized by solid cellular buds, arising from the glandular epithelium, which project into the stroma, have a somewhat eosinophilic squamoid appearance and evoke a local inflammatory response.

An agreed definition of a true microinvasive adenocarcinoma of the cervix, comparable to a Stage Ia$_2$ squamous lesion, has not emerged [29, 30]; indeed, some doubt that a definition will ever be achieved [31] whilst others consider that the existence of such an entity is debateable [32, 33]. Some workers have used the term 'microinvasive' to indicate very superficial extension of the tumour into the endocervical stroma without measurement of a specific depth [34], others have defined the entity in terms of the SGO definition for squamous microinvasive carcinoma [35, 36], whilst yet others classify as microinvasive those neoplasms which are limited to the most superficial 2 mm of the endocervical stroma [37–39]; some have used a 5 mm depth of invasion from the mucosal surface as their defining criterion [40–42]. It has also been argued, quite convincingly, that the criteria for recognition of a microinvasive adenocarcinoma should be the same as those for a Stage Ia$_2$ squamous carcinoma, namely a maximum depth of 5 mm and a horizontal spread of up to 7 mm [26]; the major problem with this proposal is that it is often far from clear as to where depth of invasion should be measured from in small adenocarcinomas [43], for many of these arise from the glands rather than from the surface.

In our present state of knowledge it seems reasonable to accept the entity of adenocarcinoma *in situ* with early stromal invasion but to refrain from the use of the term 'microinvasive adenocarcinoma'. The concept and significance of small adenocarcinomas (microadenocarcinomas) requires further study; there is, at the moment, only minimal data about the behaviour of such neoplasms and these are conflicting [26, 44].

INVASIVE CARCINOMA

The various malignant epithelial neoplasms of the uterine cervix are listed in Table 3.1. The conventional view is that between 85 and 90% of cervical carcinomas are squamous in type and that adenocarcinomas make up most of the residue; this is, however, incorrect and it is now known that squamous carcinomas account for, at the most, 70% of cervical neoplasms and adenocarcinomas and adeno-squamous carcinomas, in roughly equal proportions, for about 25% [45]. This change of emphasis has resulted from inclusion of a mucin stain as part of the routine histopathological examination of cervical neoplasms. The use of such a stain, preferably PAS/Alcian blue,

Table 3.1 Histological classification of cervical carcinoma

Squamous carcinoma
  Well-differentiated
  Moderately differentiated
  Poorly differentiated
  Verrucous carcinoma
  Papillary squamous carcinoma
  Lymphoepithelioma-like carcinoma

Adenocarcinoma
  Endocervical
  Minimal deviation
  Villoglandular
  Endometrioid
  Papillary serous
  Clear cell
  Mesonephric
  Enteric

Mixed tumours
  Adenosquamous carcinoma
  Glassy cell carcinoma
  Adenoid cystic carcinoma
  Adenoid basal carcinoma

Miscellaneous
  Small cell carcinoma
  Transitional cell carcinoma

has revealed that many poorly differentiated adenocarcinomas and adenosquamous carcinomas escape detection on haematoxylin and eosin stained sections and masquerade as poorly differentiated squamous carcinomas [46–48].

MACROSCOPIC FEATURES

Carcinomas developing on the ectocervix are commonly exophytic and tend to grow in either a polypoidal or papillary fashion. Such tumours may achieve a considerable bulk but often only infiltrate the cervical stroma to a minor degree, extending mainly into the vagina. These exophytic neoplasms show less tendency to invade the paracervical tissues than do those growing in an endophytic fashion.

Carcinomas developing within the endocervical canal, or originating on the ectocervix but having an endophytic growth pattern, tend to expand the cervix to create a barrel-shaped deformity; these usually extensively infiltrate the cervical stroma and often invade paracervical tissues. Some of these neoplasms form multinodular masses but others show extensive ulceration.

It should be noted that neither the site of origin nor the pattern of growth of a cervical neoplasm offer any firm guide to its histological nature.

HISTOLOGICAL FEATURES

Squamous cell carcinoma

These are traditionally divided into large cell keratinizing, large cell non-keratinizing and small cell non-keratinizing types [49, 50]; it is, however, more useful to classify squamous carcinomas as well, moderately or poorly differentiated.

Well-differentiated squamous carcinomas (Figure 3.2) tend to form bands or discrete islands and are characterized by the presence of intercellular bridges and well-formed epithelial pearls and often, though far from invariably, show relatively little pleomorphism, cytological atypia or mitotic activity.

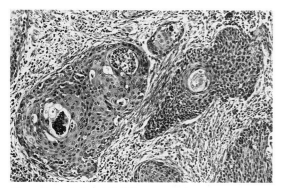

Figure 3.2 A well-differentiated squamous cell carcinoma of the cervix showing well-marked keratinization (H & E ×186).

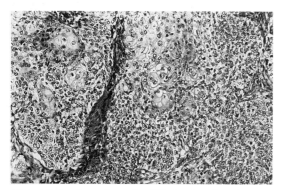

Figure 3.3 A moderately differentiated squamous cell carcinoma of the cervix showing focal keratinization (H & E ×186).

Moderately differentiated tumours (Figure 3.3) tend to grow in an infiltrative fashion or to form solid sheets: intercellular bridges are scanty and ill-formed, and whilst individually keratinized cells are present epithelial pearls are not seen. Pleomorphism, atypia and mitotic activity are more marked than in their better differentiated counterparts.

Poorly differentiated squamous cell carcinomas, which may be large or small celled, can only be recognized as such if some evidence of squamous differentiation, such as occasional keratinized cells, is found; in the absence of such evidence, and if mucin production is not demonstrable, the tumour can only be classed

as an undifferentiated carcinoma. It should be noted that most small-cell carcinomas of the cervix are not squamous in nature.

The appearances of a squamous carcinoma, of any degree of differentiation, may be altered by the presence of large multinucleated cells, foci of acantholysis or an infiltration by lymphocytes or eosinophils. Occasional squamous cell carcinomas are formed predominantly of spindle-shaped cells and these may be misdiagnosed as sarcomas; such tumours are recognized by their positive staining reactions for epithelial markers, such as cytokeratins and epithelial membrane antigen. Some squamous cell carcinomas show extensive hyalinization, a feature that can give rise to considerable diagnostic difficulty [51].

Verrucous carcinoma

These uncommon tumours [52–57] grow in an exophytic manner to form bulky, warty, fungating masses. Histologically they have a surprisingly bland appearance with papillary fronds formed of well-differentiated squamous epithelium showing little cytological atypia or mitotic activity: the base of the tumour is well circumscribed, broad and bulbous, commonly appearing to compress, rather than invade the stroma. Despite these reassuring histological appearances verrucous carcinomas are aggressive neoplasms which are locally invasive and commonly recur: they do not, however, metastasize. Very often a correct diagnosis is not achieved until multiple biopsies have revealed the contrast between the banal histological appearance of the neoplasm and its relentless clinical course.

Papillary squamous carcinoma

This has been described [58] as a variant of a squamous cell carcinoma that presents as a polypoidal lesion. A superficial biopsy shows papillae with connective tissue cores and a covering epithelium showing intra-epithelial neoplasia; there may be invasion of the stroma of a papillary frond but usually it is only when a deeper biopsy is taken is an invasive squamous cell carcinoma seen. It is debatable whether this is a true entity; it appears highly possible that it is nothing more than CIN in a papillary ectopy with an associated invasive carcinoma.

Lymphoepithelioma-like carcinoma

These tumours are composed of uniform round cells with clear or eosinophilic cytoplasm and vesicular nuclei with prominent nucleoli; the cells are arranged singly or in small poorly defined islands within a dense chronic inflammatory cell infiltrate [59–64]. The tumour cells stain positively for cytokeratins and most of the inflammatory cells are positive for T-cell markers. The clinical course of these neoplasms has not yet been fully defined although there is some tentative evidence that they may have an unusually favourable prognosis [59].

Adenocarcinoma

Endocervical-type adenocarcinoma

Tumours of recognizably endocervical-type account for about 70% of cervical adenocarcinomas [65]. In well-differentiated neoplasms (Figure 3.4) there are clearly defined glandular and branching cleft-like structures that are lined by a single or stratified layer of tall columnar cells with basal nuclei; intracellular and intraglandular mucus is usually present and sometimes there is excess mucus secretion with spillage into the stroma ('colloid' adenocarcinoma). In less well-differentiated tumours the degree of atypia increases and mucin production diminishes; some show loss of glandular differentiation, grow in a solid fashion and are only recognizable as adenocarcinomas by their content of intracytoplasmic mucin (Figure 3.5).

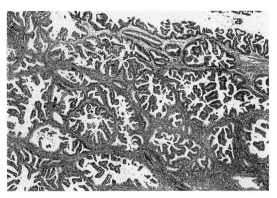

Figure 3.4 A well-differentiated endocervical-type adenocarcinoma of the cervix (H & E ×37).

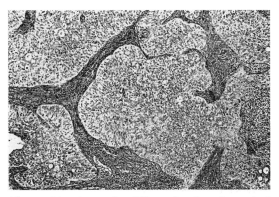

Figure 3.5 A poorly differentiated endocervical adenocarcinoma with a solid pattern resembling that of a squamous carcinoma; glands are not present but most of the cells contain intracytoplasmic mucin (H & E ×97).

Endocervical adenocarcinomas are associated, more often than would be the case by pure chance, with mucinous tumours of the ovary [66–68].

Minimal deviation adenocarcinoma

This neoplasm (also known as an adenoma malignum) is considered by some to be simply an extremely well-differentiated endocervical adenocarcinoma [65], but others regard it as a distinct entity [51], a view bolstered by the description of endometrioid-type minimal deviation adenocarcinomas [70–71]. In these neoplasms (Figure 3.6), which account for about 10% of cervical adenocarcinomas [72], there is an increased complexity of the glandular crypts which are often sharply angulated and extend unusually deeply into the stroma [69, 73–76]. The glands are lined most commonly by an endocervical-type epithelium which shows little in the way of cytological atypia and appears deceptively benign; less commonly they are lined by an epithelium of indifferent or endometrioid type. Histological recognition of a minimal deviation adenocarcinoma can be extremely difficult, but extensive sampling will usually reveal minor foci of atypia. In doubtful cases demonstration of cytoplasmic immunoreactivity for carcinoembryonic antigen [75] and human milk factor globulin [77] may be of value. There is a clear association of minimal deviation adenocarcinomas with the Peutz–Jeghers syndrome [78–80].

The prognosis for minimal deviation adenocarcinomas is in some dispute: it is widely considered that they have an unduly poor prognosis [70, 76, 79] but some have maintained that this is simply because of late diagnosis and inadequate treatment [74].

Villoglandular adenocarcinoma

These tumours [82, 83] occur in relatively young women and present as friable papillary or polypoid masses. Histologically they consist

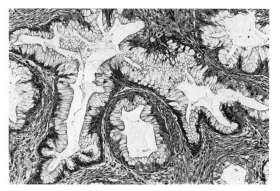

Figure 3.6 A minimal deviation adenocarcinoma of the cervix (H & E ×186).

of exophytic elongated fibrovascular papillae which are covered most commonly by endocervical-type epithelium; less frequently the epithelium is endometrioid or enteric in nature. Characteristically the epithelium shows very little in the way of atypia or mitotic activity. There is usually an invasive component at the base of the neoplasm. The prognosis for these tumours is uniformly good.

Papillary serous adenocarcinoma

These rare neoplasms [84–86] are identical to a papillary serous adenocarcinoma of the ovary. They have irregular fine fibrous papillae covered by a serous-type epithelium which shows marked cytological atypia and considerable mitotic activity: psammoma bodies may be present, although these are not specific for this type of adenocarcinoma. The natural history of these tumours has not yet been defined.

Endometrioid adenocarcinoma

Tumours of this type are histologically identical to endometrioid adenocarcinomas of the endometrium and usually have an acinar pattern; intraglandular squamous metaplasia is quite common. Some, but by no means all, endometrioid adenocarcinomas may arise in foci of cervical endometriosis [87].

Clear cell adenocarcinoma

These are histologically identical to clear cell adenocarcinomas of the ovary, endometrium and vagina [88–90]. They have a complex mixed solid, tubular, papillary and microcystic histological pattern, the constituent cells having large vesicular nuclei, prominent nucleoli and clear, granular or eosinophilic cytoplasm. The tubules are often lined by cells containing 'hob-nail' nuclei.

A proportion, but certainly not all [90, 91], of these neoplasms have occurred in young women exposed prenatally to diethylstilboestrol [90, 92], the tumour probably having developed from foci of tubo-endometrial cervical adenosis.

Mesonephric adenocarcinoma

These are rare tumours that arise deep in the lateral wall of the cervix, in the area where mesonephric duct remnants are usually found [93–99]: they often develop against a background of florid mesonephric hyperplasia. Histologically the appearances are variable but a tubuloglandular pattern is characteristic; the glands and tubules tend to be small and round and their lumens frequently contain eosinophilic mucin-negative material. In occasional cases the histological picture is complicated by areas of endometrioid differentiation [100]. These tumours appear to have a poor prognosis although too few authentic cases have been recorded for their natural history to be fully known.

Enteric adenocarcinomas

Focal enteric differentiation is seen quite frequently in endocervical-type adenocarcinomas [101, 102], but occasional cervical adenocarcinomas are formed entirely of enteric-type epithelium and histologically closely resemble intestinal tumours [103–105]. Such neoplasms contain goblet cells and secrete intestinal-type mucins: argyrophil and Paneth cells may also be present.

Enteric adenocarcinomas probably derive from foci of gastrointestinal metaplasia in endocervical epithelium.

Transitional cell carcinoma

A small number of transitional cell carcinomas, identical in all respects to those more commonly encountered in the bladder, have recently been described [106]; it is presumed that these arise in foci of transitional cell metaplasia. Too few examples have been recorded in the cervix for their prognosis to be determined but they are certainly malignant.

Mixed tumours

Neoplasms of mixed type represent biphasic differentiation of tumours derived from pluripotential undifferentiated cervical cells ('reserve cells'). A variety of entities are included under this general heading but it is probable that all represent variations on a common theme.

Adenosquamous carcinoma

An adenosquamous carcinoma is a neoplasm which either shows both squamous and glandular differentiation (Figure 3.7) or one in which a tumour morphologically appearing to be a squamous carcinoma also shows mucin production: the previously held difference between an adenosquamous carcinoma, i.e. one containing morpholologically identifiable squamous and glandular structures, and a mucoepidermoid carcinoma, i.e. an apparently pure squamous carcinoma without obvious glands but with demonstrable mucin production no longer seems valid [47, 48], although it has to be admitted that not everybody shares this view [107, 108].

When glands are recognizable in a mixed tumour of this type they are almost invariably of endocervical type.

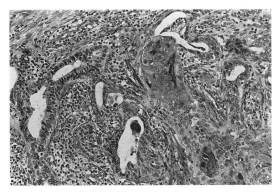

Figure 3.7 Well-differentiated adenosquamous carcinoma of the cervix. Well-formed glands are present together with well-differentiated keratinizing squamous carcinoma (H & E ×186).

Glassy cell carcinoma

This term is applied to uncommon aggressive tumours that tend to occur in relatively young women and are formed, either partially or wholly, of large polygonal cells with very distinct limiting margins, abundant finely granular cytoplasm, large vesicular nuclei and prominent nucleoli [109–118]. Some glassy cell tumours show focal keratinization or tentative attempts at glandular formation and not uncommonly there is a transition to a well-differentiated or moderately differentiated adenosquamous carcinoma. There is widespread agreement that glassy cell carcinomas are poorly differentiated adenosquamous carcinomas and whether they should be regarded as a separate entity or as a 'manifestation of taxonomic overenthusiasm' [119] is a moot point.

Adenoid cystic carcinoma

These rare neoplasms [120–128] usually occur in elderly women but a few cases have been described in patients under the age of 40 [127]; they are highly aggressive tumours which are associated with a poor prognosis. They are formed of small, uniform, or only mildly pleomorphic, basaloid cells with scanty cytoplasm and irregular, darkly staining nuclei. These cells are arranged most characteristically in nests, cords or islands which have a cribriform pattern, although sheets or trabeculae may also be present. Eosinophilic hyaline material, composed of basal lamina, encloses the cellular islands and forms the hyaline cylinders which fill the spaces within the cell nests. Some adenoid cystic carcinomas lack this cribriform appearance and are formed of solid groups of cells [128].

Cervical adenoid cystic carcinomas, despite some morphological resemblance, differ in many respects from the adenoid cystic carcinoma which is more commonly found in the salivary glands, are frequently associated with cervical intra-epithelial neoplasia and are often intimately admixed with a squamous cell carcinoma or an adenocarcinoma. This has led some to deny that the adenoid cystic carcinoma is a true entity, arguing that it simply represents a non-specific pattern which may be adopted, in part or in whole, by a squamous or glandular carcinoma. Proponents of this view therefore regard these neoplasms as 'cervical carcinomas with an adenoid cystic pattern' [129] or recommend the term 'adenoid cystic' carcinoma [65]. An alternative view is that adenoid cystic carcinomas are neoplasms derived from pluripotential (reserve) cells of the cervix and that an adenoid cystic pattern represents a poorly differentiated form of a mixed tumour.

Adenoid basal carcinoma

These neoplasms [126, 130–133], which probably fall into the 'mixed' category, tend to occur in post-menopausal women and, although capable of causing cervical ulceration, are often an incidental, unsuspected, finding in the cervix of a woman being treated for CIN or an invasive squamous cell carcinoma. The tumour is formed of small round or oval nests of cells that resemble those of a basal cell carcinoma of the skin with palisading of the cells at the periphery of the nests. The cells are small and uniform with dark oval nuclei and little cytoplasm. In some nests small acini may be present; these are lined by a single layer of cubocolumnar cells with clear cytoplasm and may contain mucin. In other nests there may be focal squamous differentiation. There is usually a striking lack of any stromal response to the tumour.

Adenoid basal cell carcinomas have, with only extremely rare exceptions, an excellent prognosis.

Small cell tumours

These are a heterogenous group of neoplasms. A small minority are true small cell squamous cell carcinomas but most are neuroendocrine neoplasms which are often called 'small cell undifferentiated carcinomas' [134]; this seems an inappropriate terminology for neoplasms that quite clearly show neuroendocrine differentiation and in this account they will be classed as neuroendocrine tumours.

Neuroendocrine tumours

These neoplasms [134–142] constitute about 2% of cases of malignant cervical disease, occur most commonly in the fifth decade and are usually clinically and macroscopically indistinguishable from a squamous cell carcinoma. Despite the wide range of hormones that can be detected immunohistochemically in these neoplasms clinical features of an endocrine disturbance are rare, only occasional patients having had Cushing's disease [143–147], a carcinoid syndrome [148, 149] or inappropriate anti-diuretic hormone secretion [150].

Histologically, these are densely cellular neoplasms which bear a close resemblance to either an 'oat cell' carcinoma of the bronchus or a poorly differentiated carcinoid tumour; they commonly show an admixture of diffuse, insular and trabecular patterns whilst occasional rosettes and small acini may be present. The tumour cells are small and darkly staining with a high nucleo-cytoplasmic ratio and indistinct limiting membranes. Most of these tumours contain argyrophil cells and neuroendocrine granules are seen on electron microscopy [135, 151]: immunohistochemistry reveals their ability to synthesize a variety of peptide hormones [139, 151–156].

The cervical epithelium contains a population of neuroendocrine cells [157, 158] and it is possible that neuroendocrine neoplasms are derived from such cells. Cervical neuroendocrine tumours may, however, be admixed with an adenocarcinoma or show focal squamous differentiation, findings which suggest that they are derived from the same pluripotential cells as other cervical neoplasms but differentiate along a predominantly neuroendocrine pathway.

Cervical neuroendocrine tumours are highly malignant and are associated with an extremely poor prognosis [137, 139, 142].

DEFINITION OF PROGNOSTIC FACTORS IN CERVICAL CARCINOMA

Delineation of prognostic factors for patients with invasive cervical carcinoma is probably only of real value in those with Stage Ib or IIa disease. In such cases it is generally held that the two most important factors governing the prognosis are the extent of the local disease and the presence or otherwise of nodal metastases.

The extent of the local disease is usually defined by clinical examination, i.e. clinical staging, although it is recognized that this can be, and indeed often is, highly inaccurate. Stage offers, however, only a very rough prognostic guide for the individual patient as within any single clinical stage there will be women who suffer markedly contrasting fortunes [159].

To a very significant extent the prognosis depends upon whether nodal metastases are present or not and the pathologist, when attempting to define histopathological tumour features indicative of a poor prognosis, is usually really trying to identify those neoplasms most likely to give rise to nodal metastases.

A major drawback to the traditional definition of prognostic factors is, however, that these do not identify those patients with node-negative disease who have a poor prognosis: although women without nodal metastases have an overall prognosis which is better than that for patients with nodal spread, it is nevertheless true that 50% of women with early stage disease whose tumours recur (half of whom will eventually die) will have been node-negative at the time of initial treatment [160]. Attempts have recently been made to define prognostic variables, in classical clinicopathological terms, for node-negative patients who relapse and die [161], but nevertheless it is for node-negative patients that newer techniques of prognostic assessment may prove particularly useful.

HISTOPATHOLOGICAL FINDINGS OF PROGNOSTIC SIGNIFICANCE

Histological type

There are some uncommon types of cervical carcinoma, such as small cell neuroendocrine tumours, which clearly have an unusually poor prognosis. Debate still continues, however, as to whether there are significant prognostic differences between the three major types of cervical neoplasm, namely, squamous cell carcinoma, adenocarcinoma and adenosquamous carcinoma.

It has been maintained that, stage for stage, the survival rate for women with a cervical adenocarcinoma is appreciably poorer than is that for patients with squamous neoplasms [162–167], but in some studies no difference in prognosis between these two tumour types could be demonstrated [168–173], whilst in one large series women with cervical adenocarcinomas had a better survival rate than did those with squamous neoplasms [174]. Prognostic comparisons between adenocarcinomas and squamous carcinomas have to be viewed, however, with some scepticism, for in many of these reported studies the diagnosis of adenocarcinoma had been made on purely morphological grounds without recourse to mucin stains; the recognized adenocarcinomas were therefore only those which were well-differentiated whilst poorly differentiated adenocarcinomas were probably subsumed within the diagnostic category of poorly differentiated squamous cell carcinoma. Furthermore, it is probably too simplistic to regard even well-differentiated adenocarcinomas as a single entity; in prognostic terms for example endometrioid adenocarcinomas of the cervix have a much better prognosis than do other types of cervical adenocarcinoma [175].

Adenosquamous carcinomas of the cervix, both those of conventional type and those distinguished from squamous carcinomas only by the presence of mucin, are more likely to metastasize to lymph nodes than are pure squamous cell tumours or adenocarcinomas [46–48] and in some studies have been associated with an unduly poor survival rate [170, 176–178], although this has not been everyone's experience [46, 169, 172, 173, 179].

It would seem reasonable, when reviewing these conflicting results, to conclude that within the common forms of cervical neoplasia the histological type is probably of little or no prognostic significance.

Tumour size

Tumour size is an important prognostic factor, and it is clear that the larger the neoplasm the greater is the chance of nodal metastasis and recurrence [172, 180–183]. Features such as the proportion of cervical stroma occupied by neoplastic tissue, the depth of invasion [184] and the width of the tumour-free rim of cervical stroma [185] are largely a reflection of tumour volume and are probably not of independent prognostic significance.

Tumour differentiation

Tumour differentiation, or grade, has appeared to be of no prognostic significance in some studies of cervical squamous cell carcinoma [48, 160, 172, 186] but has emerged as a separate prognostic indicator in others [173, 187–189]. This difference of opinion may well be a reflection of the inherent subjectivity and inconsistency of tumour grading. Nevertheless for cervical adenocarcinomas there has been a greater degree of consensus that poorly differentiated tumours have a gloomier prognosis than do their well-differentiated counterparts [163, 169, 170, 190–193] although this has, admittedly, not been confirmed in every study [194, 195].

Lymphatic/vascular space invasion

There has been widespread agreement that tumour invasion of vascular or lymphatic spaces in the cervix correlates highly with the presence of nodal metastases and with a poor survival rate [48, 160, 172, 175, 186, 196–205], there being only occasional dissidents from this view [206].

Nodal metastases

Patients with lymph node metastases have, as a group, a significantly worse prognosis than those without nodal spread, although, as already remarked, the absence of metastatic disease does not necessarily guarantee a good outlook. The adverse effect of nodal metastases is certainly true for squamous cell carcinomas and is even more striking for adenocarcinomas [168, 172, 179, 190, 207, 208], but the prognostic importance of nodal spread appears to be much less for patients with adenosquamous carcinomas [172].

There is conflicting evidence as to the prognostic importance of the number of nodal metastases some finding that the prognosis worsens as the number of nodal metastases increases [180, 209–211] and others noting that the absolute number of metastases is of little or no importance [190, 212, 213]. There has been a similar lack of agreement about the prognostic significance of bilateral nodal involvement [212, 214, 215], the size of the nodal metastases [208, 210, 216] and the involvement of nodal groups at different levels within the pelvis [180, 208, 212, 213].

These disagreements suggest the probability that it is the mere presence of metastatic disease which is the predominant adverse prognostic factor and that the size and extent of such disease is of much less relevance.

NEWER TECHNIQUES OF ASSESSMENT OF PROGNOSTIC FACTORS

A wide range of new techniques has become available to the pathologist within the last decade, which allow for a finer estimation of individual tumour malignancy than do conventional histopathological methods. Most of these are still being evaluated and few, if any,

are currently in routine usage: eventually some will, however, undoubtedly become an integral component of standard prognostic assessment.

Flow cytometric DNA analysis

A high proportion of invasive cervical carcinomas show DNA aneuploidy [217–219]. It has been claimed that neoplasms with a DNA index of above 1.5 are associated with a better prognosis than are those with a lower index [220, 221], but attempts to show that aneuploid tumours have a worse prognosis than diploid or near-diploid neoplasms have generally either proved wholly unsuccessful or have yielded equivocal, and usually not statistically significant, results [218, 219, 222–229].

Expression of oncogenes and growth factors

Overexpression of the *c-myc* oncogene occurs in a significant proportion of cervical carcinomas [230, 231] and it has been claimed that an increased level of *c-myc* transcripts is strongly indicative of a poor prognosis in early stage lesions, irrespective of other prognostic factors [232–234]. By contrast, overexpression of the *c-myc* oncogene does not appear to be of prognostic significance in advanced disease [235].

The *ras* proto-oncogene is also overexpressed in many cervical carcinomas [236, 237] and prognostic import for such overexpression has been both claimed [238] and denied [235].

Immunohistochemical studies of the proto-oncogene c-erbB-2 and of epidermal growth factor receptor have suggested that cervical carcinomas in which these can be demonstrated are associated with an adverse prognosis, overexpression of both these proteins being independent of other prognostic factors and being particularly of value in identifying those node-negative tumours associated with a poor prognosis [239–243].

Tumour suppressor genes

Mutation of the tumour suppressor gene *p53* occurs in many forms of neoplasia and tumours in which such a mutation has occurred generally appear to be associated with a poor prognosis. It has not been shown, however, that either *p53* mutation or HPV-mediated inhibition of wild *p53* (events which are both detectable by imunocytochemical means) is of any prognostic significance in cervical carcinoma [244, 245].

Oestrogen and progesterone receptors

These receptors can be detected immuno-cytochemically. Although cervical neoplasms are commonly receptor-positive there is no significant difference in survival between patients with receptor-positive and receptor-negative neoplasms [246–248].

Cell proliferation markers

Proliferating cell nuclear antigen (PCNA) is a marker for proliferating cells and can be detected with the monoclonal antibody PC10; studies of this substance in cervical carcinomas have, however, yielded conflicting results, one finding the PCNA index to be of considerable prognostic import [249] and another being unable to show that this index is of any prognostic value [250].

The antibody Ki67 also identifies a protein found only in replicating cells: no relationship has been shown between the proportion of cervical carcinoma cells staining positively for this substance and prognosis [251, 252].

Conflicting views have been expressed about the prognostic value of estimating the number of cells in the S phase on flow cytometry (i.e. those cells actively synthesizing DNA prior to mitosis). In one study the S phase fraction proved to be of no prognostic value [228] whilst in another it emerged as a highly significant prognostic variable [225].

ACKNOWLEDGEMENTS

Figures 3.2, 3.4–3.7 are reproduced from *Gynecologic Oncology*, 2nd edn (ed. M. Coppleson), Churchill Livingstone, Edinburgh (1992) by kind permission of the editor and publishers.

REFERENCES

1. Sedlis, A., Sall, S., Tsukada, Y., *et al.* (1979) Microinvasive carcinoma of the uterine cervix: a clinical–pathologic study. *American Journal of Obstetrics and Gynecology*, **133**, 64–74.
2. Ebeling, K., Bilek, K., Johannsmeyer, D., *et al.* (1989) Microinvasive stage Ia cancer of the uterine cervix – results of a multicenter clinic based analysis. *Geburtshilfe und Frauenheilkunde*, **49**, 776–81.
3. Seski, J.C., Abell, M.R. and Morley, G.W. (1977) Microinvasive squamous cell carcinoma of the cervix: definition, histological analysis, late results of treatment. *Obstetrics and Gynecology*, **50**, 410–14.
4. Burghardt, E. and Holzer, E. (1977) Diagnosis and treatment of microinvasive carcinoma of the cervix uteri. *Obstetrics and Gynecology*, **49**, 641–53.
5. Burghardt, E. (1984) Microinvasive carcinoma in gynaecological pathology. *Clinics in Obstetrics and Gynaecology*, **11**, 239–57.
6. Shepherd, J.H. (1989) Revised FIGO staging for gynaecological cancer. *British Journal of Obstetrics and Gynaecology*, **96**, 889–92.
7. Mussey, E., Soule, E.H. and Welch, J.S. (1969) Microinvasive carcinoma of the cervix: late results of operative treatment in 91 cases. *American Journal of Obstetrics and Gynecology*, **104**, 738–44.
8. Averette, H.E., Nelson, J.H. Jr, Ng, A.B.P., Hoskins, W.J., Boyce, J.G. and Ford, J.H. Jr (1976) Diagnosis and management of microinvasive (stage Ia) carcinoma of the uterine cervix. *Cancer*, **38**, 414–25.
9. Sedlis, A., Sall, S., Tsukada, K., *et al.* (1979) Microinvasive carcinoma of the cervix uteri: a clinicopathologic study. *American Journal of Obstetrics and Gynecology*, **133**, 63–74.
10. Yajima, A. and Noda, K. (1979) The results of treatment of microinvasive carcinoma (stage Ia) of the uterine cervix by means of simple and extended hysterectomy. *American Journal of Obstetrics and Gynecology*, **135**, 685–8.

11. van Nagell, J.R. Jr, Greenwell, N., Powell, D.F., Donaldson, E.S., Hanson, M.B. and Gay, E.C. (1983) Microinvasive carcinoma of the cervix. *American Journal of Obstetrics and Gynecology*, **145**, 981–91.

12. Tsukomoto, N., Kaku, T., Matsukama, K., *et al.* (1989) The problem of stage Ia (FIGO, 1985) carcinoma of the uterine cervix. *Gynecologic Oncology*, **34**, 1–6.

13. Johnson, N., Lilford, R.J., Jones, S.E., *et al.* (1992) Using decision analysis to calculate the optimum treatment for microinvasive cervical cancer. *British Journal of Cancer*, **65**, 717–22.

14. Roche, W.D. and Norris, H.J. (1975) Microinvasive carcinoma of the cervix: the significance of lymphatic invasion and confluent pattern of stromal growth. *Cancer*, **36**, 180–6.

15. Creasman, W.T., Fetter, B.F., Clarke-Paerson, D.L., Kaufmann, L. and Parker, R.T. (1985) Management of stage Ia carcinoma of the cervix. *American Journal of Obstetrics and Gynecology*, **153**, 164–72.

16. Simon, N., Gore, H., Shingleton, H.M., Soong, S.J., Orr, J.W. and Hatch, K.D. (1986) Study of superficially invasive carcinoma of the cervix. *Obstetrics and Gynecology*, **68**, 19–24.

17. Burghardt, E., Girardi, F., Lahousen, M., Pickel, H. and Tamussino, K. (1991) Microinvasive carcinoma of the uterine cervix (FIGO stage Ia). *Cancer*, **67**, 987–95.

18. Ostor, A.G. and Rome, R.M. (1994) Microinvasive squamous cell carcinoma of the cervix: a clinico-pathologic study of 200 cases with long-term follow-up. *International Journal of Gynecological Cancer*, **4**, 257–64.

19. Greer, B.E., Figge, D.C., Tamimi, H.K., Cain, J.M. and Lee, R.B. (1990) Stage Ia2 squamous carcinoma of the cervix: difficult diagnosis and therapeutic dilemma. *American Journal of Obstetrics and Gynecology*, **162**, 1406–11.

20. Maiman, M.A., Fructer, R.G., DiMaio, T.M. and Boyce, J.G. (1988) Superficially invasive carcinoma of the cervix. *Obstetrics and Gynecology*, **72**, 399–403.

21. Sevin, B.U., Nadji, M., Avarette, H.E., Hilsenbeck, S., Smith, D. and Lampe, B. (1992) Microinvasive carcinoma of the cervix. *Cancer*, **70**, 2121–8.

22. Larsson, G., Alm, P., Gullberg, P. and Grundsell, H. (1983) Prognostic factors in early invasive carcinoma of the uterine cervix. *American Journal of Obstetrics and Gynecology*, **146**, 145–53.

23. Kolstad, P. (1989) Follow-up study of 232 patients with stage Ia1 and 411 patients with stage Ia2 squamous cell carcinoma of the cervix (microinvasive carcinoma). *Gynecologic Oncology*, **33**, 265–72.

24. Shepherd, J.H. (1995) Staging announcement. FIGO staging of gynecologic cancers: cervical and vulva. *International Journal of Gynecological Cancer*, **5**, 319.

25. Burghardt, E., Ostor, A.G. and Fox, H. (1997) A critique of the 1994 FIGO definition of Stage Ia cervical cancer. *Gynecologic Oncology* **65**, 1–5.

26. Burghardt, E. (1992) Pathology of early invasive squamous and glandular carcinoma of cervix (FIGO stage Ia). In *Gynecologic Oncology*, 2nd edn (eds M. Coppleson, J.M. Monaghan, C.P. Morrow and M.H.N. Tattersall). Churchill Livingstone, Edinburgh, pp. 609–29.

27. Krumins, I., Young, Q., Pacey, F., Bousefield, L. and Mulhearn, L. (1977) The cytologic diagnosis of adenocarcinoma *in situ* of the cervix uteri. *Acta Cytologica*, **21**, 320–9.

28. Rollason, T.P., Cullimore, J. and Bradgate, M.G. (1989) A suggested columnar cell morphological eqivalent of squamous carcinoma *in situ* with early stromal invasion. *International Journal of Gynecological Paholology*, **8**, 230–6.

29. Jaworski, R.C. (1990) Endocervical glandular dysplasia, adenocarcinoma *in situ* and early invasive (microinvasive) adenocarcinoma of the uterine cervix. *Seminars in Diagnostic Pathology*, **7**, 190–204.

30. Young, R.H., Clement, P.B. and Scully, R.E. (1993) Premalignant and malignant glandular lesions of the uterine cervix. In *Tumors and Tumor-like lesions of the Uterine Corpus and Cervix* (eds P.B. Clement and R.H. Young). Churchill Livingstone, New York, pp. 85–136.

31. Yeh, I.-T., LiVolsi, V.A. and Noumoff, J.S. (1991) Endocervical carcinoma. *Pathology Research and Practice*, **187**, 129–33.

32. Buscema, J. and Woodruff, J.D. (1984) Significance of neoplastic atypicalities in endocervical epithelium. *Gynecologic Oncology*, **17**, 356–62.

33. Brand, E., Berek, J.S. and Hacker, N.F. (1988) Controversies in the management of cervical adenocarcinoma. *Obstetrics and Gynecology*, **71**, 261–9.

34. Tase, T., Okagaki, T., Clark, B.A., Twiggs, L.B., Ostrow, R.S. and Paras, A.J. (1989) Human papillomavirus DNA in adenocarcinoma *in situ*, microinvasive adenocarcinoma of the uterine cervix, and coexisting cervical squamous intraepithelial neoplasia. *International Journal of Gynecological Pathology*, **8**, 8–17.

35. Zuna, R.E. (1984) Association of condylomas with intraepithelial and microinvasive cervical neoplasia: histopathology of conization and

hysterectomy specimens. *International Journal of Gynecological Pathology*, **2**, 364–72.

36. Ayer, B., Pacey, F. and Greenberg, M. (1988) The cytologic diagnosis of adenocarcinoma *in situ* of the cervix uteri and related lesions. II. Microinvasive adenocarcinoma. *Acta Cytologica*, **32**, 318–24.

37. Betsill, W.I. Jr. and Clark, A.H. (1986) Early endocervical glandular neoplaia. I. Histomorphology and cytomorphology. *Acta Cytologica*, **30**, 115–26.

38. Clark, A.H. and Betsill, W.I. Jr (1986) Early endocervical glandular neoplasia. II. Morphometric analysis of the cells. *Acta Cytologica*, **30**, 127–34.

39. Matsukama, K., Tsukamato, N.M., Kaku, T. and Tsunehisa, K. (1989) Early adenocarcinoma of the uterine cervix: its histologic and immunopathologic study. *Gynecologic Oncology*, **35**, 38–43.

40. Quizilbash, A.H. (1975) *In situ* and microinvasive adenocarcinoma of the uterine cervix: a clinical, cytologic and histologic study of 14 cases. *American Journal of Clinical Pathology*, **64**, 155–70.

41. Noda, K., Kimura, K., Ikeda, M. and Teshima, K. (1983) Studies on the histogenesis of cervical adenocarcinoma. *International Journal of Gynecological Pathology*, **1**, 336–46.

42. Teshima, S., Shimosato, Y., Kishi, K., Kasamatsu, T., Ohmi, K. and Uei, Y. (1985) Early stage adenocarcinoma of the uterine cervix: histopathologic analysis with consideration of histogenesis. *Cancer*, **56**, 167–72.

43. Fu, Y.S., Berek, J.S. and Hilborne, L.H. (1987) Diagnostic problems of *in situ* and invasive adenocarcinomas of the uterine cervix. *Applied Pathology*, **5**, 47–56.

44. Coppleson, M. (1992) Early invasive squamous and adenocarcinoma of cervix (FIGO Stage Ia); clinical features and management. In *Gynecologic Oncology*, 2nd edn (eds M. Coppleson, J.M. Monaghan, C.P. Morrow and M.H.N. Tattersall) Churchill Livingstone, Edinburgh, pp. 632–48.

45. Buckley, C.H. and Fox, H. (1989) Carcinoma of the cervix. In *Recent Advances in Histopathology 14* (eds P.P. Anthony and R.N.M. MacSween). Churchill Livingstone, Edinburgh, pp. 63–78.

46. Benda, J.A., Platz, C.E., Buchsbaum, H. and Lifschitz, F. (1985) Mucin production in defining mixed carcinoma of the uterine cervix: a clinicopathologic study. *International Journal of Gynecological Pathology*, **4**, 314–27.

47. Ireland, D., Cole, S., Kelly, P. and Monaghan, J.M. (1987) Mucin production in cervical intraepithelial neoplasia and in stage Ib carcinoma of the cervix with pelvic lymph node metastases. *British Journal of Obstetrics and Gynaecology*, **94**, 467–72.

48. Buckley, C.H., Beards, C.S. and Fox, H. (1988) Pathological prognostic indicators in cervical cancer with particular reference to patients under the age of 40 years. *British Journal of Obstetrics and Gynecology*, **95**, 47–56.

49. Wentz, W.B. and Reagan, J.W. (1959) Survival in cervical cancer with respect to cell type. *Cancer*, **12**, 384–8.

50. Reagan, J.W. and Ng, A.B.P. (1973) The cellular manifestations of uterine carcinogenesis. In *The Uterus* (eds N.J. Norris, A.T. Hertig and M.R. Abell), Williams & Wilkins, Baltimore, MD, pp. 320–47.

51. Lawrence, W.D. (1991) Advances in the pathology of the uterine cervix. *Human Pathology*, **27**, 792–806.

52. Demian, S.D.E., Bushkin, F.L. and Echevarria, R.A. (1973) Perineural invasion and anaplastic transformation of verrucous carcinoma. *Cancer*, **32**, 395–401.

53. Isaacs, J.H. (1976) Verrucous carcinoma of the female genital tract. *Gynecologic Oncology*, **4**, 259–69.

54. Tiltman, A.J. and Atad, J. (1982) Verrucous carcinoma of the cervix with endometrial involvement. *International Journal of Gynecological Pathology*, **1**, 221–30.

55. Benedet, J.L. and Clement, P.B. (1980) Verrucous carcinoma of the cervix and endometrium. *Diagnostic Gynecology and Obstetrics*, **2**, 197–203.

56. Raheja, A., Katz, D.A. and Dermer, M.S. (1983) Verrucous carcinoma of the endocervix. *Obstetrics and Gynecology*, **62**, 535–8.

57. Degefu, S., O'Quinn, A.G., Lacey, C.G., Merkel, M. and Barnard, D.E. (1986) Verrucous carcinoma of the cervix: a report of two cases and literature review. *Gynecologic Oncology*, **25**, 37–47.

58. Randall, M.E., Kim, J.-A., Mills, S.E., Hahn, S.S. and Constable, W.C. (1986) Uncommon variants of cervical carcinoma treated with radical irradiation: a clinicopathologic study of 66 cases. *Cancer*, **57**, 816–22.

59. Hasumi, K., Sugano, H., Sakamoto, G., Masubachu, K. and Kubo, H. (1977) Circumscribed carcinoma of the uterine cervix with marked lymphocytic infiltration. *Cancer*, **39**, 2503–7.

60. Hafiz, M.A., Kragel, P.J. and Toker, C. (1985) Carcinoma of the uterine cervix resembling a lymphoepithelioma. *Obstetrics and Gynecology*, **66**, 829–31.

61. Mills, S.E., Austin, M.B. and Randall, M.E. (1985) Lymphoepithelioma-like carcinoma of the uterine cervix: a distinctive undifferentiated carcinoma with inflammatory stroma. *American Journal of Surgical Pathology*, **9**, 883–9.

62. Halpin, T.F., Hunter, R.E. and Cohen, M.B. (1989) Lymphoepithelioma of the uterine cervix. *Gynecologic Oncology*, **34**, 101–5.

63. Weinberg, E., Hoisington, S., Eastman, A.Y., Rice, D.K., Malfetano, J. and Ross, J.S. (1993) Uterine cervical lymphoepithelial-like carcinoma; absence of Epstein–Barr virus genomes. *American Journal of Clinical Pathology*, **99**, 195–9.

64. Walsh, C.B., Kay, E., Prendiville, W., Turner, M. and Leader, M. (1993) Lymphoepithelioma-like carcinoma of the uterine cervix with c-erbB2, p53 oncoprotein expression and DNA quantification. *Histopathology*, **23**, 592–3.

65. Young, R.H. and Scully, R.E. (1990) Invasive adenocarcinoma and related tumors of the uterine cervix. *Seminars in Diagnostic Pathology*, **7**, 205–27.

66. LiVolsi, V.A., Merino, M.J. and Schwartz, P.E. (1983) Coexistent endocervical adenocarcinoma and mucinous adenocarcinoma of the ovary: a clinicopathologic study of four cases. *International Journal of Gynecological Pathology*, **1**, 391–402.

67. Kaminski, P.F. and Norris, H.J. (1984) Coexistence of ovarian neoplasms and endocervical adenocarcinoma. *Obstetrics and Gynecology*, **64**, 553–6.

68. Young, R.H. and Scully, R.E. (1988) Mucinous tumors of the ovary associated with mucinous adenocarcinoma of the ovary: a clinicopathologic analysis of 16 cases. *International Journal of Gynecological Pathology*, **7**, 99–111.

69. Kaminski, P.F. and Norris, H.J. (1983) Minimal deviation carcinoma (adenoma malignum) of the cervix. *International Journal of Gynecological Pathology*, **2**, 141–53.

70. Rahilly, M.A., Williams, A.R.W. and Al-Nafussi, A. (1992) Minimal deviation endometrioid adenocarcinoma of cervix: a clinicopathological and immunohistochemical study of two cases. *Histopathology*, **20**, 351–4.

71. Young, R.H. and Scully, R.E. (1993) Minimal-deviation endometrioid adenocarcinoma of the uterine cervix: a report of five cases of a distinctive neoplasm that may be interpreted as benign. *American Journal of Surgical Pathology*, **17**, 660–5.

72. Hurt, W.G., Silverberg, S.G. and Frable, W.J. (1977) Adenocarcinoma of the cervix: histopathologic and clinical features. *American Journal of Obstetrics and Gynecology*, **120**, 304–15.

73. McKelvey, J.L. and Goodlin, R.R. (1963) Adenoma malignum of the cervix. *Cancer*, **16**, 549–57.

74. Silberberg, S.G. and Hurt, W.G. (1975) Minimal deviation adenocarcinoma ('adenoma malignum') of the cervix. *American Journal of Obstetrics and Gynecology*, **123**, 971–5.

75. Michael, H., Grawe, L. and Kraus, F.T. (1984) Minimal deviation endocervical adenocarcinoma: clinical and histologic features, immunohistochemical staining for carcinoembryonic antigen, and differentiation from confusing benign lesions. *International Journal of Gynecological Pathology*, **3**, 261–76.

76. Gilks, C.B., Young, R.H., Aguirre, P., DeLellis, R.A. and Scully, R.E. (1989) Adenoma malignum (minimal deviation adenocarcinoma) of the uterine cervix: a clinicopathological and immunohistochemical analysis of 26 cases. *American Journal of Surgical Pathology*, **13**, 717–29.

77. Bulmer, J.N., Griffin, N.R., Bates, C., Kingston, R.E. and Wells, M. (1990) Minimal deviation adenocarcinoma (adenoma malignum) of the endocervix: a histochemical and immunohistochemical study of two cases, *Gynecologic Oncology*, **36**, 139–46.

78. McGowan, L., Young, R.H. and Scully, R.E. (1980) Peutz–Jeghers syndrome with 'adenoma malignum' of the cervix: a report of two cases. *Gynecologic Oncology*, **10**, 125–33.

79. Kaku, T., Hachisuga, T., Toyoshima, S., *et al.* (1985) Extremely well-differentiated adenocarcinoma ('adenoma malignum') of the cervix in a patient with Peutz Jeghers syndrome. *International Journal of Gynecological Pathology*, **4**, 266–73.

80. Chen, K.T.K. (1986) Female genital tract tumors in Peutz–Jeghers syndrome. *Human Pathology*, **17**, 856–61.

81. Kaku, T. and Enjoji, M. (1983) Extremely well-differentiated adenocarcinoma ('adenoma malignum') of the cervix. *International Journal of Gynecological Pathology*, **2**, 28–41.

82. Young, R.H. and Scully, R.E. (1989) Villoglandular papillary adenocarcinoma of the uterine cervix: a clinicopathologic analysis of 13 cases. *Cancer*, **63**, 1773–9.

83. Jones, M.W., Silverberg, S.G. and Kurman, R.J. (1993) Well differentiated villoglandular adenocarcinoma of the uterine cervix: a clinicopathological study of 24 cases. *International Journal of Gynecological Pathology*, **12**, 1–7.

84. Gilks, C.B. and Clement, P.B. (1992) Papillary serous adenocarcinoma of the uterine cervix: a report of three cases. *Modern Pathology*, **5**, 426–31.

85. Shintaku, M. and Ueda, H. (1993) Serous papillary adenocarcinoma of the uterine cervix. *Histopathology*, **22**, 506–7.

86. Rose, P.G. and Reale, F.R. (1993) Serous papillary carcinoma of the cervix. *Gynecologic Oncology*, **50**, 361–4.

87. Chang, S.H. and Maddox, W.A. (1971) Adenocarcinoma arising within cervical endometriosis and invading the adjacent vagina. *American Journal of Obstetrics and Gynecology*, **110**, 1015–17.

88. Hasumi, K. and Ehrmann, R.L. (1978) Clear cell carcinoma of the uterine cervix with an *in situ* component. *Cancer*, **42**, 2435–8.

89. Herbst, A.L., Cole, P., Norusis, M.J., Welch, W.R. and Scully, R.E. (1979) Epidemiologic aspects and factors related to survival in 384 Registry cases of clear cell adenocarcinoma of the vagina and cervix. *American Journal of Obstetrics and Gynecology*, **135**, 876–86.

90. Kaminski, P.F. and Maier, R.C. (1983) Clear cell adenocarcinoma of the cervix unrelated to diethylstilbestrol exposure. *Obstetrics and Gynecology*, **62**, 720–6.

91. Anderson, M.C. and Robboy, S.J. (1996) Glandular lesions of the cervix. *Current Diagnostic Pathology*, **3**, 99–108.

92. Puri, S., Fenoglio, C.M., Richart, R.M. and Townsend, D.E. (1977) Clear cell carcinoma of cervix and vagina in progeny of women who received diethylstilbestrol: three cases with scanning and transmission electron microscopy. *American Journal of Obstetrics and Gynecology*, **128**, 550–5.

93. Hart, W.R. and Norris, H.J. (1972) Mesonephric adenocarcinomas of the cervix. *Cancer*, **29**, 106–13.

94. Buntine, D.W. (1979) Adenocarcinoma of the uterine cervix of probable Wolffian origin. *Pathology*, **11**, 713–18.

95. Valente, P.T. and Susin, M. (1987) Cervical adenocarcinoma arising in florid mesonephric hyperplasia: report of a case with immunohistochemical studies. *Gynecologic Oncology*, **27**, 58–68.

96. Lang, G. and Dallenbach-Hellweg, G. (1990) The histogenetic origin of cervical mesonephric hyperplasia and mesonephric adenocarcinoma of the uterine cervix studied with immunohistochemical methods. *International Journal of Gynecological Pathology*, **9**, 145–57.

97. Ferry, J.A. and Scully, R.E. (1990) Mesonephric remnants, hyperplasia, and neoplasia in the uterine cervix. *American Journal of Surgical Pathology*, **14**, 1100–11.

98. Stewart, C.J.R., Taggart, C.R., Brett, F. and Mutch, A.F. (1993) Mesonephric adenocarcinoma of the uterine cervix with focal endocrine differentiation. *International Journal of Gynecological Pathology*, **12**, 264–9.

99. Clement, P.B., Young, R.H., Keh, P., Ostor, A.G. and Scully, R.E. (1995) Malignant mesonephric neoplasms of the uterine cervix – a report of eight cases, including four with a malignant spindle cell component. *American Journal of Surgical Pathology*, **19**, 1158–71.

100. Nogales, F. (1995) Mesonephric (Wolffian) tumours of the female genital tract: is mesonephric histogenesis a mirage and trap? *Current Diagnostic Pathology*, **2**, 94–100.

101. Azzopardi, J.G. and Hou, L.T. (1965) Intestinal metaplasia with argentaffin cells in cervical adenocarcinoma. *Journal of Pathology and Bacteriology*, **90**, 686–90.

102. Savargaonkar, P.R., Hale, R.J., Pope, R., Fox, H. and Buckley, C.H. (1993) Enteric differentiation in cervical adenocarcinomas and its prognostic significance. *Histopathology*, **23**, 275–7.

103. Fox, H., Wells, M., Harris, M.M., McWilliam, L.J. and Anderson G.S. (1988) Enteric tumours of the lower female genital tract: a report of three cases. *Histopathology*, **12**, 167–76.

104. Alvaro, T. and Nogales, F. (1988) Villous adenoma and invasive adenocarcinoma of the cervix. *International Journal of Gynecological Pathology*, **7**, 96.

105. Lee, K.R. and Trainer, T.D. (1990) Adenocarcinoma of the uterine cervix of intestinal type containing numerous Paneth cells. *Archives of Pathology and Laboratory Medicine*, **114**, 731.

106. Albores-Saavedra, J. and Young, R.H. (1995) Transitional cell neoplasms (carcinomas and inverted papillomas) of the uterine cervix: a report of five cases. *American Journal of Surgical Pathology*, **19**, 1138–45.

107. Thelmo, W.M., Nicastri, A.D., Fruchter, R., Spring, H., DiMaio, T. and Boyce, J. (1990) Mucoepidermoid carcinoma of the uterine cervix stage Ib. *International Journal of Gynecological Pathology*, **9**, 316–24.

108. Colgan, J.T., Auger, M. and McLaughlin, J.R. (1993) Histopathologic classification of cervical carcinomas and recognition of mucin-secreting squamous carcinomas. *International Journal of Gynecological Pathology*, **12**, 64–9.

109. Littman, P., Clement, P.B., Henriksen, B., *et al.* (1976) Glassy cell carcinoma of the cervix. *Cancer*, **37**, 2238–46.

110. Paulsen, S.M., Hansen, K.C. and Nielsem, V.T. (1980) Glassy cell carcinoma of the cervix: case

report with light and electron microscopy study. *Ultrastructural Pathology*, **1**, 377–84.

111. Richard, L., Guralnick, M. and Ferenczy, A. (1981) Ultrastructure of glassy cell carcinoma. *Diagnostic Gynecology and Obstetrics*, **3**, 31–8.

112. Zaino, R.J., Nahhas, W.A. and Mortel, R. (1982) Glassy cell carcinoma of the uterine cervix: an ultrastructural study and review. *Archives of Pathology and Laboratory Medicine*, **106**, 250–4.

113. Ulbright, T.M. and Gersell, D.J. (1983) Glassy cell carcinoma of the uterine cervix: a light and electron microscopic study of five cases. *Cancer*, **51**, 2255–63.

114. Pak, H.Y., Yokota, S.B., Paladuga, R.R. and Agliozzo, C.M. (1983) Glassy cell carcinoma of the cervix. *Cancer*, **52**, 307–12.

115. Tamini, H.K., Ek, M., Hesla, J., *et al.* (1988) Glassy cell carcinoma of the cervix redefined. *Obstetrics and Gynecology*, **71**, 837–41.

116. Talerman, A., Alenghat, E. and Okagaki, T. (1991) Glassy cell carcinoma of the uterine cervix. *Acta Pathologica Microbiologica et Immunologica Scandinavica*, **23**, 119–25.

117. Lotocki, R.J., Krepart, G.V., Paraskevas, M., Vadas, G., Heywood, M. and Fung, F.K. (1992) Glassy cell carcinoma of the cervix: a bimodal treatment strategy. *Gynecologic Oncology*, **44**, 254–9.

118. Kenny, M.B., Unger, E.R., Chenggis, M.L. and Costa, M.J. (1992) *In situ* hybridization for human papillomavirus DNA in uterine adenosquamous carcinoma with glassy cell features (glassy cell carcinoma). *American Journal of Clinical Pathology*, **98**, 180–7.

119. Wells, M. and Brown, L.J.R. (1986) Glandular lesions of the uterine cervix: the present state of our knowledge. *Histopathology*, **10**, 777–92.

120. Miles, P.A. and Norris, H.J. (1971) Adenoid cystic carcinoma of the cervix: an analysis of 12 cases. *Obstetrics and Gynecology*, **38**, 103–10.

121. Fowler, W.C., Miles, P.A., Surwit, E.A., Edelman, D.A., Walton, L.A. and Photopulos, G.J. (1978) Adenoid cystic carcinoma of the cervix: report of 9 cases and a reappraisal. *Obstetrics and Gynecology*, **52**, 337–42.

122. Hoskins, W.J., Averette, H.E., Ng, A.P.B. and Yon, J.L. (1979) Adenoid cystic carcinoma of the cervix uteri: report of six cases and review of the literature. *Gynecologic Oncology*, **7**, 371–84.

123. Mazur, M.T. and Battifora, H.A. (1982) Adenoid cystic carcinoma of the uterine cervix: ultrastructure, immunofluorescence and criteria for diagnosis. *American Journal of Clinical Pathology*, **77**, 494–500.

124. Musa, A.G., Hughes, R.R. and Coleman, S.A. (1985) Adenoid cystic carcinoma of the cervix: a report of 17 cases. *Gynecologic Oncology*, **22**, 167–73.

125. Berchuck, A. and Mullin, T.J. (1985) Cervical adenoid cystic carcinoma associated with ascites. *Gynecologic Oncology*, **22**, 201–11.

126. Ferry, J.A. and Scully, R.E. (1988) 'Adenoid cystic' carcinoma and adenoid basal carcinoma of the uterine cervix: a study of 28 cases. *American Journal of Surgical Pathology*, **12**, 134–44.

127. King, L.A., Talledo, O.E., Gallup, D.G., Melhus, O. and Otken, L.B. (1989) Adenoid cystic carcinoma of the cervix in women under age 30. *Gynecologic Oncology*, **32**, 26–30.

128. Albores-Savedra, J., Manivel, C., Mora, A., Vuitch, F., Milchgrub, S. and Gould, E. (1992) The solid variant of adenoid cystic carcinoma of the cervix. *International Journal of Gynecological Pathology*, **11**, 2–10.

129. Shingleton, H.M., Lawrence, W.D. and Gore, H. (1877) Cervical carcinoma with adenoid cystic pattern: a light and electron microscopic study. *Cancer*, **40**, 1112–21.

130. Baggish, M.S. and Woodruff, J.D. (1971) Adenoid basal lesions of the cervix. *Obstetrics and Gynecology*, **37**, 807–19.

131. Daroca, P.J. and Dhurandhar, H.N. (1980) Basaloid carcinoma of the uterine cervix. *American Journal of Surgical Pathology*, **4**, 235–9.

132. van Dinh, T. and Woodruff, J.D. (1985) Adenoid cystic and adenoid basal carcinomas of the cervix. *Obstetrics and Gynecology*, **65**, 705–8.

133. Layton-Henry, J., Scurry, J., Planner, R., Allen, D., Sykes, P., Garlands, S. and Borg, A. (1996) Cervical adenoid basal carcinoma, five cases and literature review. *International Journal of Gynecological Cancer*, **6**, 193–9.

134. Clement, P.B. (1990) Miscellaneous primary tumors and metastatic tumors of the uterine cervix. *Seminars in Diagnostic Pathology*, **7**, 228–48.

135. Gersell, D.J., Mazoujian, G., Mutch, D.G. and Rudloff, M.A. (1988) Small-cell undifferentiated carcinoma of the cervix: a clinicopathologic, ultrastructural, and immunocytochemical study of 15 cases. *American Journal of Surgical Pathology*, **12**, 684–98.

136. Sheets, E.E., Berman, M.L., Hrountas, C.K., Liao, S.Y. and DiSaia, P.J. (1988) Surgically treated, early stage neuroendocrine small-cell cervical carcinoma. *Obstetrics and Gynecology*, **71**, 10–14.

137. Van Nagell, J.R. Jr, Powell, D.E. and Gallion, H.H. *et al.* (1988) Small cell carcinoma of the uterine cervix. *Cancer*, **62**, 1586–93.

138. Walker, A.N., Mills, S.E. and Taylor, P.T. (1988) Cervical neuroendocrine carcinoma: a clinical

and light microscopic study of 14 cases. *International Journal of Gynecological Pathology*, **7**, 64–74.

139. Silva, E.G., Gershenson, D., Sneige, N., Brock, W.A., Saul, P. and Copeland, L.J. (1989) Small cell carcinoma of the uterine cervix: pathology and prognostic factors. *Surgical Pathology*, **2**, 105–13.

140. Miller, B., Dockter, M., El Torky, M. and Photopulos, G. (1991) Small cell carcinoma of the cervix: clinical and flow-cytometric study. *Gynecologic Oncology*, **42**, 27–33.

141. Abeler, V.M., Holm, R., Nesland, J.M. and Kjorstadt, K.E. (1994) Small cell carcinoma of the cervix. *Cancer*, **73**, 672–77.

142. Perrin, L. and Ward, B. (1995) Small cell carcinoma of the cervix. *International Journal of Gynecological Cancer*, **5**, 200–3.

143. Berthelot, P., Benhanou, J.P. and Fauvert, R. (1961) Hypercorticisme et cancer de l'uterus. *Presse Medicale*, **69**, 189–90.

144. Jones, H.W. III, Plymate, S., Gluck, F.B., Miles, P.A. and Green, J.F. Jr (1976) Small cell nonkeratinizing carcinoma of the cervix associated with ACTH production. *Cancer*, **38**, 1629–35.

145. Matsuyama, M., Inoue, T., Ariyoshi, Y., *et al.* (1979) Argyrophil carcinoma of the uterine cervix with ectopic production of ACTH, beta-MSH, serotonin, histamine, and amylase. *Cancer*, **44**, 1813–23.

146. Lojek, M.A., Fer, M.F., Kasselberg, S.G., *et al.* (1980) Cushing's syndrome with small cell carcinoma of the uterine cervix. *American Journal of Medicine*, **69**, 140–4.

147. Iemura, K., Sonoda, T., Hayakawa, A., *et al.* (1991) Small cell carcinoma of the uterine cervix showing Cushing's syndrome caused by ectopic adrenocorticotrophin hormone production. *Japanese Journal of Clinical Oncology*, **21**, 293–8.

148. Driessens, J., Clay, A., Adenis, L. and Demaille, A. (1964) Tumeur cervico-uterine et syndrome biologique de carcinoidose. *Archives d'Anatomie Pathologique*, **12**, 200–3.

149. Stockdale, A.D., Leader, M., Phillips, R.H. and Henry, K. (1986) The carcinoid syndrome with multiple hormone secretion associated with a carcinoid tumour of the uterine cervix: case report. *British Journal of Obstetrics and Gynaecology*, **93**, 397–401.

150. Kothe, M.J.C., Prins, J.M., De Wit, R., Velden, K.V.D. and Schellekens, P.T.A. (1990) Small cell carcinoma of the cervix with inappropriate antidiuretic hormone secretion: case report. *British Journal of Obstetrics and Gynaecology*, **97**, 647–8.

151. Fuji, S., Konishi, I., Ferenczy, A., Imai, K., Okamura, H. and Mori, T. (1986) Small cell undifferentiated carcinoma of the uterine cervix: histology, ultrastructure and immunohistochemistry of two cases. *Ultrastructural Pathology*, **10**, 337–46.

152. Inoue, T., Yamaguchi, K., Suzuki, H., Abe, K. and Chihara, T. (1984) Production of immunoreactive polypeptide hormones in cervical carcinoma. *Cancer*, **53**, 1509–14.

153. Inoue, A., Ueda, G. and Nakajima, T. (1985) Immunohistochemical demonstration of neuron-specific enolase in gynecologic malignant tumors. *Cancer*, **55**, 1686–90.

154. Ulich, T.R., Liao, S., Layfield, L., Romansky, S., Cheng, L. and Lewis, K.J. (1986) Endocrine and tumor differentiation markers in poorly differentiated small-cell carcinoma of the cervix and vagina. *Archives of Pathology and Laboratory Medicine*, **110**, 1054–7.

155. Ueda, G., Shimizu, C., Shimizu, H., *et al.* (1989) An immunohistochemical study of small-cell and poorly differentiated carcinomas of the cervix using neuroendocrine markers. *Gynecologic Oncology*, **34**, 164–9.

156. Anbros, R.A., Park, J., Shah, K.V. and Kurman, R.J. (1991) Evaluation of histologic, morphometric, and immunohistochemical criteria in the differential diagnosis of small cell carcinomas of the cervix with particular reference to human papillomavirus types 16 and 18. *Modern Pathology*, **4**, 586–93.

157. Fox, H., Kazzaz, A. and Langley, F.A. (1964) Argyrophil and argentaffin cells in the female genital tract and in ovarian mucinous cysts. *Journal of Pathology and Bacteriology*, **88**, 479–88.

158. Fetissof, F., Serres, G., Arbeille, B., de Muret, A., Sam-Giao, M. and Lansac, J. (1991) Argyrophil cells and ectocervical epithelium. *International Journal of Gynecological Pathology*, **10**, 177–90.

159. van Bommel, P.F.J., Lindert, A.C.M., Kock, H.C.L.V., Leers, W.H. and Neijt, J.P. (1987) A review of prognostic factors in early-stage carcinoma of the cervix (FIGO IB and IIA) and implications for treatment strategy. *European Journal of Obstetrics, Gynecology and Reproductive Biology*, **26**, 69–84.

160. Smiley, L.M., Burke, T.W., Silva, E.G., Morris, M., Gershenson, D.M. and Wharton, J.T. (1991) Prognostic factors in stage IB squamous cervical cancer patients with low risk for recurrence. *Obstetrics and Gynecology*, **77**, 271–5.

161. Stockler, M., Russell, P., McGahan, S., Elliot, P.M., Dalrymple, C. and Tattersall, M. (1996) Prognosis and prognostic factors in node-

negative cervix cancer. *International Journal of Gynecological Cancer*, **6**, 477–82.

162. Berek, J.S., Castaldo, D.W., Hacker, N.F., Petrelli, E.S., Lagasse, L.D. and Moore, J.G. (1981) Adenocarcinoma of the uterine cervix. *Cancer*, **48**, 2734–41.

163. Milsom, I. and Friberg, L.G. (1983) Primary adenocarcinoma of the uterine cervix: a clinical study. *Cancer*, **52**, 942–7.

164. Kjorstad, K.E. and Bond, B. (1984) Stage Ib adenocarcinoma of the cervix; metastatic potential and patterns of dissemination. *American Journal of Obstetrics and Gynecology*, **150**, 297–9.

165. Moberg, P.J., Einhorn, N., Silfverswand, C. and Soderberg, G. (1986) Adenocarcinoma of the uterine cervix. *Cancer*, **57**, 407–10.

166. Kleine, W., Ran, K., Schwoerer, D. and Pfleiderer, A. (1989) Prognosis of adenocarcinoma of the cervix uteri: a comparative study. *Gynecologic Oncology*, **35**, 145–9.

167. Eifel, P.J., Burke, T.W., Morris, M. and Smith, T.L. (1995) Adenocarcinoma as an independent risk factor for disease recurrence in patients with stage IB cervical carcinoma. *Gynecologic Oncology*, **59**, 38–44.

168. Ireland, D., Hardiman, P. and Monaghan, J. (1985) Adenocarcinoma of the uterine cervix: a study of 73 cases. *Obstetrics and Gynecology*, **65**, 82–5.

169. Kilgore, L.C., Soon, S.-J., Gore, H., Shingleton, H.M., Hatch, K.D. and Partridge, E.E. (1988) Analysis of prognostic factors in adenocarcinoma of the cervix. *Gynecologic Oncology*, **31**, 137–48.

170. Davidson, S.E., Symonds, R.P., Lamont, D. and Watson, E.R. (1989) Does adenocarcinoma of the uterine cervix have a worse prognosis than squamous carcinoma treated by radiotherapy? *Gynecologic Oncology*, **33**, 23–6.

171. Vestrinen, E., Forss, M. and Nieminen, U. (1989) Increase in cervical adenocarcinoma: a report of 520 cases of cervical carcinoma including 112 tumors with glandular elements. *Gynecologic Oncology*, **33**, 49–53.

172. Hale, R.J., Wilcox, F.L., Buckley, C.H., Tindall, V.R., Ryder, W.D.J. and Logue, J.P. (1991) Prognostic factors in uterine cervical carcinoma: a clinicopathological analysis. *International Journal of Gynecological Cancer*, **1**, 19–23.

173. Macleod, A., Kitchener, H.C., Parkin, D.E., *et al.* (1994) Cervical carcinoma in the Grampian region (1980–1991): a population-based study of survival and cervical cytology history. *British Journal of Obstetrics and Gynaecology*, **101**, 797–803.

174. Sivanesaratnam, V., Sen, D.K., Kayalakshni, P.

and Ongs, G. (1993) Radical hysterectomy and pelvic lymphadenectomy for early invasive cancer of the cervix – 14 year experience. *International Journal of Gynecological Cancer*, **3**, 231–8.

175. Saigo, P.E., Cain. J.M., Kim, W.S., Gaynor, J.J., Johnson, K. and Lewis, J.L. Jr (1986) Prognostic factors in adenocarcinoma of the uterine cervix. *Cancer*, **57**, 1584–93.

176. Swan, D.S. and Roddick, J.W. (1973) A clinical–pathological correlation of cell type classification for cervical cancer. *American Journal of Obstetrics and Gynecology*, **116**, 666–70.

177. Gallup, D.G., Harper, R.H. and Stock, R.J. (1985) Poor prognosis in patients with adenosquamous cell carcinoma of the cervix. *Obstetrics and Gynecology*, **65**, 416–22.

178. Bethwaite, P., Yeong, M.L., Holloway, L., Robson, B., Duncan, G. and Lamb, D. (1992) The prognosis of adenosquamous carcinomas of the uterine cervix. *British Journal of Obstetrics and Gynaecology*, **99**, 745–50.

179. Helm, C.W., Jinney, W.K., Lawrence, W.D., *et al.* (1993) A matched study of surgically treated stage IB adenosquamous carcinoma and adenocarcinoma of the uterine cervix. *International Journal of Gynecological Cancer*, **3**, 245–9.

180. Burghardt, E., Pickel, H., Haas, J. and Lahousen, M. (1987) Prognostic factors and operative treatment of stages Ib to IIb cervical cancer. *American Journal of Obstetrics and Gynecology*, **156**, 988–96.

181. Alvarez, R.D., Potter, M.E., Soong, S.G., *et al.* (1991) Rationale for using pathologic tumor dimensions and nodal status to subclassify surgically treated stage IB cervical cancer patients. *Gynecologic Oncology*, **43**, 108–12.

182. Burghardt, E., Baltzer, J., Tulusan, H. and Haas, J. (1992) Results of surgical treatment of 1028 cervical cancers studied with volumetry. *Cancer*, **70**, 648–55.

183. Werner-Wasik, M., Schmid, C.H., Bornstein, L., Ball, H.G., Smith D.M. and Madoc-Jones, H. (1995) Prognostic factors for local and distant recurrence in stage I and II cervical carcinoma. *International Journal of Radiation Oncology, Biology and Physics*, **32**, 1309–17.

184. Inoue, T. (1984) Prognostic significance of the depth of invasion of cervical carcinoma relating to nodal metastases, parametrial extension, and cell types. *Cancer*, **54**, 3035–42.

185. Inoue, T., Casanova, H.A., Morita, K. and Chihara, T. (1986) The prognostic significance of the minimum thickness of ininvolved cervix in patients with cervical carcinoma stages IB, IIA and IIB. *Gynecologic Oncology*, **24**, 220–9.

186. Zaino, R.J., Ward, S., Delgado, G., *et al.* (1992) Histopathologic predictors of the behaviour of surgically treated IB squamous cell carcinoma of the cervix: a Gynecologic Oncology Group study. *Cancer*, **69**, 1750–8.
187. Prempree, T., Patanaphan, V., Sewchand, W. and Scott, R.M. (1983) The influence of patient's age and tumor grade on the prognosis of carcinoma of the cervix. *Cancer*, **51**, 1764–71.
188. Kapp, D.S., Fischer, D., Gutierrez, E., Kohorn, E.I. and Schwartz, P.E. (1983) Pretreatment prognostic factors in carcinoma of the uterine cervix: a multivariate analysis of the effect of age, stage, histology and blood counts on survival. *International Journal of Radiation Oncology Biology and Physics*, **9**, 445–55.
189. Hopkins, M.P. and Morley, G.W. (1991) Squamous cell cancer of the cervix: prognostic factors related to survival. *International Journal of Gynecological Cancer*, **1**, 173–7.
190. Hopkins, M.P., Sutton, P. and Roberts, J.A. (1987) Prognostic features and treatment of endocervical adenocarcinoma. *Gynecologic Oncology*, **27**, 69–75.
191. Hopkins, M.P., Schmidt, R.W., Roberts, J.A. and Morley, G.W. (1988) The prognosis and treatment of stage I adenocarcinoma of the cervix. *Obstetrics and Gynecology*, **72**, 789–95.
192. Goodman, H.M., Buttlar, C.A., Niloff, J.M., *et al.* (1989) Adenocarcinoma of the uterine cervix: prognostic factors and patterns of recurrence. *Gynecologic Oncology*, **33**, 241–7.
193. Raju, K.S., Kjorstad, K.E. and Abeler, V. (1991) Prognostic factors in the treatment of stage IB adenocarcinoma of the cervix. *International Journal of Gynecological Cancer*, **1**, 69–74.
194. Leminen, A., Paavonen, J., Forss, M., Wahlstrom, T. and Vesterinen, E. (1990) Adenocarcinoma of the uterine cervix. *Cancer*, **65**, 53–9.
195. Attanoos, R., Nahar, K., Bigrigg, A., Roberts, S., Newcombe, R.G. and Ismail, S.M. (1995) Primary adenocarcinoma of the cervix: a clinicopathologic study of prognostic variables in 55 cases. *International Journal of Gynecological Cancer*, **5**, 179–86.
196. Barber, H.R.K., Sommers, S.C., Rotterdam, H. and Kwan, T. (1978) Vascular invasion as a prognostic factor in stage Ib cancer of the cervix. *Obstetrics and Gynecology*, **52**, 343–8.
197. van Nagell, J.R., Donaldson, E.S., Wood, E.G. and Parker, J.C. (1978) The significance of vascular invasion and lymphocytic infiltration in invasive cervical cancer. *Cancer*, **41**, 228–34.
198. Baltzer, J., Lohe, K.J., Koepcke, W. and Zander, J. (1982) Histological criteria for the prognosis in patients with operated squamous cell carcinoma of the cervix. *Gynecologic Oncology*, **13**, 184–94.
199. Fuller, A., Elliot, N., Kosloff, C. and Lewis, J. (1982) Lymph node metastases from carcinoma of the cervix, stages Ib and IIa: implications for prognosis and treatment. *Gynecologic Oncology*, **13**, 165–74.
200. Noguchi, H., Shiozawa, K., Tsukamoto, T., Tsukahara, Y., Iwai, S. and Fukuta, T. (1983) The postoperative classification for uterine cancer and its clinical evaluation. *Gynecologic Oncology*, **16**, 219–31.
201. Boyce, J.G., Fruchter, R.G., Nicastri, A.D., *et al.* (1984) Vascular invasion in stage I carcinoma of the cervix. *Cancer*, **53**, 1175–80.
202. Crissman, J.D., Makuch, R. and Budhraja, M. (1985) Histopathologic grading of squamous cell carcinoma of the uterine cervix. *Cancer*, **55**, 1590–6.
203. Kenter, G.G., Ansink, A.C., Heinz, A.P.M., Delemarre, J., Aarsten, E.J. and Hart, A.A.M. (1988) Low stage invasive carcinoma of the uterine cervix stage I-IIA: morphological prognostic factors. *European Journal of Surgical Oncology*, **14**, 187–92.
204. Delgado, G., Bundy, B., Zaino, R. and Creasman, W.T. (1990) Prospective surgical–pathological study of disease-free interval in patients with stage IB squamous cell carcinoma of the cervix; a Gynecologic Oncology Group study. *Gynecologic Oncology*, **38**, 352–7.
205. Kamura, T., Tsukamoto, N., Tsuruchi, N., *et al.* (1993) Histopathologic prognostic factors in stage IIb cervical carcinoma treated with radical hysterectomy and pelvic-node dissection – an analysis with mathematical statistics. *International Journal of Gynecological Cancer*, **3**, 219–25.
206. Nahhas, W., Sharkey, F., Whitney, C., Husseinzadeh, N., Chung, C. and Mortel, R. (1983) The prognostic significance of vascular channel involvement and deep stromal penetration in early cervical carcinoma. *American Journal of Clinical Oncology*, **6**, 259–64.
207. Berek, J.S., Hacker, N.F., Fu, Y.S., Sokale, J.R., Leuchter, R.C. and Lagasse, L.D. (1985) Adenocarcinoma of the uterine cervix: histologic variables associated with lymph node metastasis and survival. *Obstetrics and Gynecology*, **65**, 46–51.
208. Hale, R.J., Buckley, C.H., Fox, H., Wilcox, F.L., Tindall, V.R. and Logue, J.P. (1991) The morphology and distribution of lymph node metastases in stage IB/IIA cervical carcinoma; relationship to prognosis. *International Journal of Gynecological Cancer*, **1**, 233–7.

209. Matsuyama, T., Inoue, I., Tsykamoto, N., *et al.* (1984) Stage IB, IIA and IIB cervix cancer: postsurgical staging and prognosis. *Cancer*, **54**, 3072–7.

210. Tinga, D.J., Timmer, P.R., Bouma, J. and Aalders, J.G. (1990) Prognostic significance of single versus multiple lymph node metastases in cervical carcinoma stage IB. *Gynecologic Oncology*, **39**, 175–80.

211. Inoue, T. and Morita, K. (1990) The prognostic significance of number of positive nodes in cervical carcinoma stages IB, IIA and IIB. *Cancer*, **65**, 1923–7.

212. Martinbeau, P.W., Kjorstad, K.E. and Iversen, T. (1982) Stage IB carcinoma of the cervix, the Norwegian Radium Hospital. II. Results when pelvic nodes are involved. *Obstetrics and Gynecology*, **60**, 215–18.

213. Terada, K.Y., Morley, G.W. and Roberts, J.A. (1988) Stage IB carcinoma of the cervix with lymph node metastases. *Gynecologic Oncology*, **31**, 389–95.

214. Webb, M.J. and Symmonds, R.E. (1979) Wertheim hysterectomy: a reappraisal. *Obstetrics and Gynecology*, **54**, 140–5.

215. Shingleton, H.M. and Orr, J.W. (1987) *Cancer of the Cervix: Diagnosis and Treatment*, Churchill Livingstone, Edinburgh.

216. Pilleron, J.P., Durand, J.C. and Hamelin, J.P. (1974) Prognostic value of node metastases in cancer of the uterine cervix. *American Journal of Obstetrics and Gynecology*, **119**, 458–62.

217. Strang, P., Stendahl, U., Frankendal, B. and Lindgren, A. (1986) Flow cytometric DNA patterns in cervical carcinoma. *Acta Radiologica*, **25**, 249–54.

218. Strang, P. (1989) Cytogenetic and cytometric analyses in squamous cell carcinoma of the uterine cervix. *International Journal of Gynecological Pathology*, **8**, 54–63.

219. van Dam, P.A., Watson, J.V., Lowe, D.G. and Shepherd, J.H. (1992) Flow cytometric DNA analysis in gynecological oncology. *International Journal of Gynecological Cancer*, **2**, 57–65.

220. Jakobsen, A. (1984) Prognostic impact of ploidy level in carcinoma of the cervix. *American Journal of Clinical Oncology*, **7**, 475–80.

221. Jakobsen, A., Bichel, P. and Vaeth, M. (1985) New prognostic factors in squamous cell carcinoma of cervix uteri. *American Journal of Clinical Oncology*, **8**, 39–45.

222. Davis, J.R., Aristizabal, S., Way, D.L., Weines, S.A., Hicks, M.J. and Hagaman, R.M. (1989) DNA ploidy, grade, and stage in prognosis of uterine cervical cancer. *Gynecologic Oncology*, **32**, 4–7.

223. Kenter, G.G., Cornelisse, C.J., Aartsen, E.J., *et al.* (1990) DNA ploidy level as prognostic factor in low stage carcinoma of the uterine cervix. *Gynecologic Oncology*, **39**, 181–5.

224. Ji, H.X., Syrjanen, S., Klemi, P., Chang, F., Tosi, P. and Syrjanen, K. (1991) Prognostic significance of human papillomavirus (HPV) type and nuclear DNA content in invasive cervical cancer. *International Journal of Gynecological Cancer*, **1**, 59–67.

225. Strang, P., Stendahl, U., Bergstrom, R., Frankendal, B. and Tribuait, R. (1991) Prognostic flow cytometric information in cervical squamous cell carcinoma: a multivariate analysis of 307 patients. *Gynecologic Oncology*, **43**, 3–8.

226. Zanetta, G.M., Katzmann, J.A., Keeney, G.L., Kinney, W.K., Cha, S.S. and Podratz, K.C. (1992) Flow-cytometric DNA analysis of stages IB and IIA cervical carcinoma. *Gynecologic Oncology*, **46**, 13–19.

227. Jarrell, M.A., Heintz, N., Howard, P., *et al.* (1992) Squamous cell carcinoma of the cervix: HPV 16 and DNA ploidy as predictors of survival. *Gynecologic Oncology*, **46**, 361–66.

228. Connor, J.P., Miller, D.S., Bauer, K.D., Murad, T.M., Rademaker, A.W. and Lurain, J.R. (1993) Flow cytometric evaluation of early invasive cervical cancer. *Obstetrics and Gynecology*, **81**, 367–71.

229. Pfisterer, J., Kommoss, F., Sauerbrei, W., *et al.* (1996) DNA flow cytometry in stage IB and II cervical carcinoma. *International Journal of Gynecological Cancer*, **6**, 54–60.

230. Ocadiz, R., Sauceda, R., Cruz, M., Graef, A.M. and Gariglio, P. (1987) High correlation between molecular alterations of the *c-myc* oncogene and carcinoma of the uterine cervix. *Cancer Research*, **47**, 4173–7.

231. Kohler, M., Janz, I., Wintzer, H.O., Wagner, E. and Bauknecht, T. (1989) The expression of EGF receptors, EFF-like factors and *c-myc* in ovarian and cervical carcinomas and their potential clinical significance. *Anticancer Research*, **9**, 1537–47.

232. Riou, G., Barrois, M., Le, M.G., George, M., Doussal, V.L. and Haie, C. (1987) C-myc proto-oncogene expression and prognosis in early carcinoma of the uterine cervix. *Lancet*, **i**, 761–3.

233. Sowani, A., Ong, G., Dische, S., *et al.* (1989) C-myc oncogene expression and clinical outcome in carcinoma of the cervix. *Molecular and Cell Probes*, **3**, 117–23.

234. Bourhis, J., Le, M.G., Barrois, M., *et al.* (1990) Prognostic value of c-myc-proto-oncogene overexpression in early invasive carcinoma of

the cervix. *Journal of Clinical Oncology*, **8**, 1789–96.

235. Symonds, R.P., Habeshaw, T., Paul, J., *et al.* (1992) No correlation between ras, c-myc and c-jun proto-oncogene expresssion and prognosis in advanced carcinoma of the cervix. *European Journal of Cancer*, **28A**, 1616–17.

236. Riou, G., Barrois, M., Sheng, Z.M., Duvillard, P.M. and Lhomme, C. (1988) Somatic deletions and mutations of *c-Ha-ras* gene in human cervical cancers. *Oncogene*, **3**, 329–33.

237. Wong, Y.F., Chung, T.K., Cheung, T.H., Lam, S.K., Xu, Y.G. and Chang A.M. (1995) Frequent ras gene mutations in squamous cell cervical cancer. *Cancer Letters*, **95**, 29–32.

238. Sagae, S., Kuzumaki, N., Hisada, T., Mugikura, Y., Kudo, R. and Hashimoto, M. (1989) *ras* Oncogene expression and prognosis of invasive squamous cell carcinoma of the uterine cervix. *Cancer*, **63**, 1577–82.

239. Hale, R.J., Buckley, C.H., Fox, H. and Williams, J. (1992) Prognostic value of c-erbB-2 expression in uterine cervical carcinoma. *Journal of Clinical Pathology*, **45**, 594–6.

240. Hale, R.J., Buckley, C.H., Gullick, W.J., *et al.* (1993) Prognostic value of epidermal growth factor receptor expresssion in uterine cervical carcinoma. *Journal of Clinical Pathology*, **46**, 149–53.

241. Stellwag, B., Scheidel, P., Pfeiffer, D. and Hepp, H. (1993) Der-EGF Rezeptor und die EGF-ahnliche Aktivitat als Prognosefaktoren beim Zervixcarzinom. *Geburtshilfe und Frauenheilkunde*, **53**, 177–81.

242. Kihana, T., Tsuda, H., Teshima, S., *et al.* (1994) Prognostic significance of the overexpression of c-erbB-2 protein in adenocarcinoma of the uterine cervix. *Cancer*, **73**, 148–53.

243. Oka, K., Nakano, T. and Arai, T. (1994) C-erbB-2 oncoprotein expression is associated with poor prognosis in squamous cell carcinoma of the cervix. *Cancer*, **73**, 664–671.

244. Oka, K., Nakano, T. and Arai, T. (1993) p53CM1 expression is not associated with prognosis in uterine cervical carcinoma. *Cancer*, **72**, 169–4.

245. Manek, S. and Wells, M. (1996) The significance of alterations in p53 expression in gynaecological neoplasms. *Current Opinion in Obstetrics and Gynecology*, **8**, 52–5.

246. Hunter, R.E., Longcope, C. and Keough, P. (1987) Steroid hormone receptors in carcinoma of the cervix. *Cancer*, **60**, 392–6.

247. Harding, M., McIntosh, J., Paul, J., *et al.* (1990) Oestrogen and progesterone receptors in carcinoma of the cervix. *Clinical Oncology*, **2**, 313–17.

248. Darne, J., Soutter, W.P., Ginsberg, R. and Sharp, F. (1990) Nuclear and 'cytoplasmic' estrogen and progesterone receptors in squamous cell carcinoma of the cervix. *Gynecologic Oncology*, **38**, 216–19.

249. Oka, K., Hoshi, T. and Arai, T. (1992) Prognostic significance of the PC10 index as a prospective assay for cervical cancer treated with radiation therapy alone. *Cancer*, **70**, 1545–50.

250. Al-Nafussi, A.I., Klys, H.S., Rebello. G., Kelly, C., Kerr, G. and Cowie, V. (1993) The assessment of proliferating cell nuclear antigen (PCNA) immunostaining in the uterine cervix and cervical squamous neoplasia. *International Journal of Gynecological Cancer*, **3**, 154–8.

251. Cole, D.J., Brown, D.C., Crossley, E., Alcock, C.J. and Gatter, K. (1992) Carcinoma of the cervix uteri: an assessment of the relationship of tumour proliferation to prognosis. *British Journal of Cancer*, **65**, 783–5.

252. Oka, K. and Arai, T. (1996) MIB1 growth fraction is not related to prognosis in cervical squamous cell carcinoma treated with radiotherapy. *International Journal of Gynecological Pathology*, **15**, 23–7.

EPIDEMIOLOGY OF CERVICAL DYSPLASIA AND NEOPLASIA

F.X. Bosch, N. Muñoz and X. Castellsagué

DESCRIPTIVE EPIDEMIOLOGY OF CERVICAL CANCER

Excluding non-melanocytic skin cancers, cancer of the cervix is the second most common cancer in women after cancer of the breast, and the commonest in developing countries. Each year, there are approximately 437,000 new cases of invasive cervical cancer diagnosed (about 12% of cancers in women) and in excess of 200,000 deaths from the disease. When men and women are considered together, cancer of the cervix is the fifth most common cancer worldwide, after cancers of the lung, stomach, breast and large bowel, accounting for an estimated 5.7% of all cancers [1, 2].

For each death from cancer of the cervix, it has been estimated that between 14 and 20 potential years of life before 70 years of age are lost. Assuming, therefore, an average of about 17 years of life lost per death, this gives a yearly worldwide estimate of more than 3.4 million woman-years of life before 70 years of age lost due to cancer of the cervix.

The highest incidence rates of cervical cancer, corresponding to relative frequencies between 20 and 30% of all cancers in women, occur in the developing areas of the world, particularly parts of Asia, South America and Africa. Intermediate rates are particularly evident in areas of eastern, northern and western Europe, while the lowest rates are seen in Australia and New Zealand, southern Europe, North America, and western Asia (the Middle East) [1].

Tables 4.1–4.3 show a selection of the age-adjusted incidence rates (AAIR) of cancer of the cervix as presented in the publication *Cancer Incidence in Five Continents*, vol. VI [3]. Table 4.1 includes registries in Europe and North America, Table 4.2 includes selected registries in South America and Africa and Table 4.3 shows a selection of registries in Asia and Oceania.

From Tables 4.1–4.3 it can be readily seen that the risk of cervical cancer varies within ranges of 10-fold between the low rates in Israel, Spain or Finland and the high rates in Colombia or Paraguay. The tables also show that within the USA, strong differences are observed across ethnic groups.

In most developed countries, mortality from cervical cancer has been decreasing steadily for at least the last half of the century. However, in the past decade, increases in incidence and mortality have been reported among young women in countries with reasonably developed screening programmes such as the UK, some areas in the USA, Australia and New Zealand. Some studies have suggested that the increase in incidence among young women is largely due to increases in the rarer adenocarcinomas of the cervix, a clinical entity that could more easily escape the standard screening tests [4–8].

Cancer and Pre-cancer of the Cervix
Edited by D.M. Luesley and R. Barrasso. Published in 1998 by Chapman & Hall, London. ISBN 0 412 56600 1.

Table 4.1   Cervical cancer incidence: selected registries in Europe and North America (1982–1988)

Europe *Registry/Country*	*Cervical cancer*		*North America* *Registry/Country*	*Cervical cancer*	
	AAIR	*CR (%)*		*AAIR*	*CR (%)*
Finland	4.4	0.5	US, White[g]	7.3	0.7
Spain[a]	6.7	0.7	US, Filipino[h]	8.8	0.9
Italy[b]	8.4	0.9	US, Chinese[h]	9.1	0.9
Switzerland[c]	8.6	0.9	Canada	10.2	1.0
France[d]	10.9	1.2	US, Black[g]	11.7	1.2
UK, England and Wales	11.9	1.2	US, Los Angeles[i]	17.4	2.2
Denmark	15.9	1.6	US, Los Angeles[j]	18.4	1.9
Poland[e]	18.1	1.9			
Germany[f]	21.8	2.2			

AAIR = age-adjusted incidence rates.
CR = cumulative rate to age 74.
Combined estimates from local registries as follows: [a]Basque country, Tarragona, Granada, Murcia, Navarra, Zaragoza; [b]Florence, Genoa, Latina, Varese, Parma, Ragusa, Romagna, Torino, Trieste; [c]Basel, Geneva, Neuchatel, St Gall, Vaud, Zurich; [d]Bas-Rhin, Calvados, Doubs, Isère, Somme, Tarn; [e]Cracow city, Lower Silesia, Nowy Sacz, Opole, Warsaw city, Warsaw rural; [f]GDR, Saarland; [g]SEER programme; [h]Los Angeles, Hawaii; [i]White, Spanish surname; [j]Korean.

Table 4.2   Cervical cancer registration in Latin America and Africa (1982–1988)

Latin America *Registry/Country*	*Cervical cancer*		*Africa* *Registry/Country*	*Cervical cancer*	
	AAIR	*CR (%)*		*AAIR*	*CR (%)*
Cuba	20.0	2.0	Algeria, Setif	10.3	
Costa Rica	26.1	2.7	Mali, Bamako	23.4	2.4
Brazil[a]	38.2	4.0			
Colombia, Cali	42.2	4.7			
Paraguay, Asunción	47.1	4.9			

AAIR = age adjusted incidence rates.
CR = cumulative rate to age 74.
[a]Combined estimates from cancer registries in Goiania and Porto Alegre.

Table 4.3   Cervical cancer registration in Asia (1983–1987) and Oceania (1986–1989)

Europe *Registry/Country*	*Cervical cancer*		*North America* *Registry/Country*	*Cervical cancer*	
	AAIR	*CR (%)*		*AAIR*	*CR (%)*
Israel[a]	4.0	0.4	Kyrgystan	15.6	1.8
China[b]	5.4	0.7	Singapore[g]	16.3	1.7
Kuwait[c]	7.3	0.8	Hong Kong	19.2	2.2
Australia[d]	10.8	1.1	Philippines[h]	23.3	2.6
Japan[e]	11.9	1.3	India[i]	27.7	3.0
New Zealand[f]	12.8	1.3	Thailand[j]	27.8	2.9

AAIR= Age adjusted incidence rates.
CR Cumulative rate to age 74.
Combined estimates from local registries as follows: [a]all Jews, non-Jews; [b]Qidog, Shanghai, Tianjin; [c]Kuwaitis, non-Kuwaitis; [d]Capital Territory, New South Wales, South Australia, Tasmania, Victoria, Western Australia; [e]Hiroshima, Myyagi, Nagasaki, Osaka, Saga, Yamagata [f]Maoris, Non-Maoris; [g]Chinese, Malay, Indian; [h]Manila, Rizal; [i]Ahmedabad, Bangalore, Bombay, Madras; [j]Chiang Mai, Khon Kaen.

Moreover, some data from Africa seem to indicate that in the absence of any screening, incidence and mortality due to cervical cancer remain steady or is, in fact, increasing [9, 10].

THE VARIATION OF CERVICAL CANCER INCIDENCE WITHIN DEVELOPED COUNTRIES

Figure 4.1 shows some examples of the range of AAIR of cervical cancer reported by different registries within developed countries. Given the difficulties in adjusting incidence rates by screening practices in most populations, intra-country comparisons may be the most reliable means of describing minor fluctuations in incidence (assuming equalitarian distribution of medical services and quality of the registries). In some countries like Switzerland, France or Italy 1.5–2-fold variation in incidence rates can be observed between registries. This suggests that in addition to the gross between-country variability, incidence rates can be further modulated by other factors. Some of them are known, such as variation by social class. However, some of the critical ones, such as the differences in the prevalence of human papillomavirus (HPV) DNA in defined sub-groups, the predominant sexual behaviour patterns or the efficacy of the local screening strategies, are still poorly described.

HPV TESTING: A CRITICAL ISSUE FOR EPIDEMIOLOGY

HPV testing was the critical factor in epidemiologic studies to establish the causal role between the infection and the disease [11–14]. Methods based on signal amplification such as the polymerase chain reaction (PCR) have been accepted as the more sensitive and specific for HPV detection and typing [15]. Other systems with or without amplification are in different stages of development and evaluation and it is expected that standardized methods will be made available and facilitate the transfer of technology to the clinic [16–18].

In addition to the quality of the test, the type

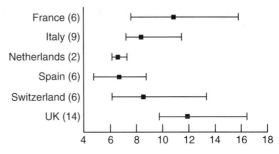

H■H Pooled estimate and range of incidence rates

Figure 4.1 Age-standardized incidence of cervical cancer within countries.

of specimen is of interest to ensure HPV detection. In one study conducted in Barcelona (Spain), Margall *et al.* (1993) compared the detection rates of HPV DNA in exfoliated cervical cells and biopsies in women with different cervical abnormalities (from chronic inflammatory lesions to cervical intra-epithelial neoplasia (CIN) III) [19]. The method used to detect HPV DNA was PCR with probes for HPV types 16 and 18. The prevalence of either viral type using biopsies was consistently higher than the one reported on cervical scrapes. These findings are consistent with a similar comparison conducted in 62 invasive cancer cases from the Colombia and Spain study. In this exercise, it was found that among HPV DNA-negative cases in the exfoliated specimen, 24% were HPV-positive using the biopsy material (Shah *et al.*, personal communication). Some ongoing studies should provide relevant information concerning the variation in the HPV detection rates by type of specimen among women with a normal cervix. This issue is relevant to epidemiological studies in which only cells are obtainable from controls and to screening programmes wishing to incorporate HPV testing as part of the testing protocols.

HPV AND CERVICAL NEOPLASIA

To date, the epidemiological evidence relating HPV infections to cervical cancer includes an impressive and consistent body of results from prevalence surveys, case control studies and some follow-up studies with pre-invasive neoplasia as the end point. We will confine this review to studies that included advanced hybridisation methods for HPV detection and typing.

HPV AND CIN LESIONS: CASE CONTROL STUDIES

Table 4.4 summarizes the results of selected case control studies investigating HPV and pre-invasive cervical lesions (CIN II–III), in which hybridization assays with amplification techniques were used. Some of the studies used PCR systems based on consensus primers in the L1 region of HPV 16/18 and included up to 25 HPV type specific probes as well as a generic probe [20–22]. Other studies used a PCR system based on a combination of a $GP5^+/6^+$ general primers and 31 type specific probes on crude cell suspensions [23, 24] and still others used tailor-made variations on the amplification methods [25]. During the period in which these studies were completed, the number of HPV specific probes and the strategies to collect, store and analyse specimens progressed considerably. Therefore, the variation in the prevalence observed across studies may be partially an artifact due to differences in the hybridization methodology employed. The odds ratios (OR) and corresponding proportions of attributable fractions (AF%) displayed in Table 4.4 are all statistically significant and very high. The AFs for HPV range between 60 and 90% and for HPV 16 within the range 40–80%. It is highly unlikely that these risk estimates could be explained by chance or any extreme form of bias. The consistency of results in different populations using different study designs and laboratory methods adds to the strength of the findings.

Some studies attempted to provide risk estimates by intensity of the HPV viral load. In the study in Colombia and Spain, viral load was quantified by combining the results of a dot-blot HPV assay (Virapap™) which is capable of detecting HPV DNA when present in high copy numbers with the results of the PCR (detection limits about 5–200 viral copies per specimen). In both countries there was a significant trend of increasing risk with increasing levels of HPV DNA. The ORs for women with both tests positive was 246 (35–1750) in Spain and 25 (8.8–72) in Colombia [20]. In the USA, Morrison *et al.* (1991) attempted to quantify the intensity of the HPV viral load by visual inspection of the signal and by measuring the size of the band [26]. The dose–response relationship was then assessed by stratifying the results of the Southern blot and the PCR assays into three categories (negative, weak and strong). The estimates of the ORs were again higher in direct relation to the viral load estimates.

HPV AND INVASIVE CERVICAL CANCER: PREVALENCE OF VIRAL MARKERS AND SELECTED CASE CONTROL STUDIES

Prevalence and geographic distribution of HPV types in cervical neoplasia

The International Biological Study on Cervical Cancer (IBSCC) reported on a survey that included over 1000 biopsy specimens of invasive cervical cancer from 22 countries. A PCR-based method including 26 type specific probes and a generic probe was used. This large study identified HPV DNA in 93% of the specimens. HPV 16 represented about 50% of the viral types identified and the sum of HPV types 16, 18, 31 and 45 comprised 80% of the viral types. Only one putative new type was identified, suggesting that most of the HPV types related to cervical cancer have been identified [27]. The results of this large survey are consistent with over 40 other reports on invasive cancer and 20 reports on pre-invasive disease. In these studies, HPV DNA prevalence ranges from 70 to 100% and the variability in the results are highly dependent on the methods used to detect HPV DNA (for a review see [15]).

Table 4.4 Case-control studies of CIN II–III lesions. HPV hybridization assays including amplification (PCR) methods

Reference	Cases (type, number)	Controls[a] (number)	HPV	HPV prevalence (%)		OR (95% CI)[b]	HPV[c] AF (%)
				Cases	Controls		
Bosch et al. [20] Spain (nine provinces)	CIN III, n=157	n=193	HPV 16	70.7 49.0	4.7 0.5	56.9 (24.8–130.6) 295.5 (44.8–1946.6)	72.4 59.6
Cali, Colombia	CIN III, n=125	n=181	HPV 16	63.2 32.8	10.5 3.3	15.5 (8.2–29.4) 27.1 (10.6–69.5)	60.3 46.3
Schiffman et al. [21] Portland, Oregon	CIN II–III, n= 50	n=433	HPV 16/18	90 62	17.7 2.9	42 (15.3–124.3) 180 (49–630)	87.9 83.8
Van den Brule et al. [23][d] Amsterdam	PAP III–IV=177	n=1762	HPV 16/18	76.3 47.5	6.1 2.2	49.7 (32.8–75.6) 39.9 (25.3–63.1)	74.8 46.1
Becker et al. [22] USA	CIN II–III, n=176	n=311	HPV 16	93.8 52.4	42.1 8.6	20.8 (10.8–40.2) 9.8 (5.4–18.3)	89.0 44.0
Kjaer et al. [24][d] Denmark	HGSIL, n=79	n=1000	HPV 16, 18, 31, 33	77.2 62.0	15.4 8.2	13.6 (9.4–19.7) 32.9 (17.3–59.8)	66.0 72.3
Olsen et al. [25] Norway	CIN II–III, n=103	n=234	HPV 16	91.0	15.0	72.8 (27.6–192) 182.4 (54–616)	92 92

[a] Largely recruited among participants in screening programmes, family planning clinics and gynaecology clinics.
[b] OR = odds ratio; CI = confidence interval.
[c] Attributable fraction for HPV.
[d] Some of the values in the table have been calculated by the authors of the review from the published data.

The IBSCC project identified some geographic variation in the HPV type distribution that deserve further consideration and confirmation. HPV 18 was the most common HPV type in Indonesia, where it comprised 50% of the viral types identified. HPV 18 was also relatively prevalent in specimens from other countries in south-east Asia. HPV 45 was prevalent in western Africa. HPV types 39 and 59 were found almost exclusively in Latin America, where HPV 31 and 33 were also relatively prevalent.

In most studies, cervical adenocarcinomas show HPV DNA prevalence similar to the one observed for squamous cell carcinomas and the majority of studies, including the IBSCC, found HPV 18 as the most common viral type.

Case control studies on invasive cervical cancer

Table 4.5 includes the strongest evidence to date generated by case control studies relating HPV to invasive cervical cancer. In at least three of the studies included in the table there is (i) an underlying and defined epidemiological design, (ii) a large sample size, (iii) a comprehensive questionnaire, (iv) a statistical analysis including multivariate methods, and (v) the HPV detection was done using highly sensitive and specific PCR systems [28, 29].

In Spain and Colombia, Muñoz *et al.* (1992) conducted two population based case control studies including 436 incident cases of squamous cell invasive cervical cancer and 387 population controls [28]. HPV detection was done using PCR methods based on the L1 region of HPV 16. Hybridization was performed sequentially with probes to HPV 6, 11, 16, 18, 31, 33 and 35 under high stringency conditions. Subsequently the filters were screened with a generic probe containing a mixture of amplimers of HPV 16 and 18, as described in Bauer *et al.* (1991) [30]. The ORs for HPV DNA, HPV DNA 16 and the corresponding AF are shown in Table 4.5. In this study over 65% of the cases could be attributed to HPV. ORs were

also calculated by other type specific HPVs as follows: HPV 31, 33 or 35 OR = 21.3 (6.1–75.6); HPV X OR = 79.6 (11.1–572.4). Further testing of the cytology specimens with additional HPV types (to a total of 25 HPV types) resulted in a global HPV prevalence of 76.3% in cervical cancer cases. The corresponding ORs were in Spain OR = 51.7 for HPV DNA and AF = 75%. In Colombia, the corresponding figures were HPV prevalence = 75%, OR = 14.9 and AF = 70%. If we add to these results the additional testing of the biopsies from the cases, the ORs increase to OR = 71.4 in Spain and OR = 16.7 in Colombia, with AFs of 80.6 and 72.5%, respectively (Shah, personal communication).

In Brazil, Eluf-Neto *et al.* (1994) conducted a hospital based case-control study including 199 histologically confirmed consecutive cervical cancer cases and 225 age frequency matched controls from a diversity of diagnoses [29]. A PCR system was used directly on crude cell suspensions by a combination of general primer-mediated and type specific PCR [31]. The OR and HPV attributable fractions calculated from this study were OR = 37.1 (19.6–70.4) for HPV, OR = 74.9 (32.5–173) for HPV 16 and AF = 86.0% for HPV DNA. This study clearly indicates that with more sensitive PCR methods (as suggested by the higher fraction of cases testing positive), the risk and the AF estimates greatly increases to levels above 85% (Tables 4.4 and 4.5). Such high estimates with imperfect measurement are statistically compatible with a prevalence of HPV in cervical cancer cases of 100%.

The study in Brazil is part of an International Agency for Research on Cancer (IARC) coordinated multicentric study completed also in Thailand, the Philippines, Paraguay, Mali and Morocco. The strength of the design is that the protocol and questionnaires are identical in all settings and that the specimens collected are being analysed using the same method in a unique research laboratory. The preliminary results of this project indicate that (i) the HPV DNA prevalence in cases ranged between 83 and 95%, (ii) the HPV DNA prevalence among

Table 4.5 Case-control studies of invasive cervical cancer. HPV hybridization assays including amplification (PCR) methods

Reference	Cases (type, number)	Controls[a] (number)	HPV	HPV prevalence (%)		OR (95% CI)[a]	HPV[b] AF (%)
				Cases	Controls		
Muñoz et al. [28]	Incident cases	Population-based					
Spain	250	238	HPV	69	4.6	46.2 (18.5–11.1)	67.5
			16	45.8	3.1	14.9 (5.0–49.5)	30.1
Colombia	186	149	HPV	72.4	13.3	15.6 (6.9–34.7)	66.0
			16	50.6	9.2	5.5 (2.4–12.9)	29.3
Eluf-Neto et al. [29] Brazil	199 Hospital-based	255 Hospital-based	HPV	84.0	17.0	37.1 (19.6–70.4)	86.0
			16	53.8	5.3	74.9 (32.5–173)	79.7
Peng et al. [64] China	101 Hospital-based	106 Clinic-based	16/33	34.7	1.4	32.9 (7.7–141.1)	31.0

[a]OR = odds ratio; CI = confidence interval.
[b]Attributable fraction for HPV.
Some of the values in the table have been calculated by the authors of the review from the published data.

controls correlates grossly with the incidence of cervical cancer in the country, (iii) the ORs estimates are consistently greater than 20 and statistically different from one, (iv) the HPV attributable fraction is consistently greater than 90%, and (v) the effects of any other risk factors are likely to be of marginal quantitative importance.

The magnitude of the ORs and AFs shown in Tables 4.4 and 4.5, and the consistency of the results in studies of invasive cancer and CIN III (see review by Moreno *et al.* 1995 [32]), suggests that the association is perhaps the strongest ever identified for any human cancer.

It has been suggested that HPV and cervical cancer may well be the first known example of a necessary cause of cancer applying universally [33].

HPV AND CERVICAL DYSPLASIA: FOLLOW-UP STUDIES

Case-control studies have the intrinsic uncertainty of establishing the sequence over time between HPV exposure and cervical neoplasia. Opportunistic infection, reactivation of latent infection or increased detectability of the neoplastic tissues cannot, strictly speaking, be ruled out as an alternative interpretation to the findings of case control studies. The case control designs are, in fact, HPV DNA surveys at the time of enrolment of subjects as either cases or controls.

To circumvent this difficulty, one approach derived from case-control studies has been to explore the HPV DNA prevalence in relation to the interval since last sexual intercourse. This procedure assumes that, among adults, the presence of HPV DNA in the cervix is largely acquired through sexual intercourse. In the studies in Colombia and Spain, the HPV DNA prevalence in cases was remarkably constant, independent of the duration of the intervals since last intercourse. Among controls, HPV positivity was also detected occasionally among women with long intervals (i.e. >10 years) since last intercourse [20, 28].

The second approach to resolve the time-sequence dilemma is to conduct active follow up of women characterized as to their HPV status for the occurrence of cervical neoplasia. In the past, numerous follow up studies were completed in women presenting to gynaecological clinics or to screening programmes. However, only recently have some studies that used hybridization methods and PCR to characterize HPV status been reported.

In one of the studies in the USA, the risk of progression from cytologically normal to biopsy confirmed CIN II–III was increased by over 10-fold among HPV DNA carriers as compared to HPV DNA negatives. Progression was largely attributable to HPV 16. In this study HPV DNA detection was done using dot-blot or Southern blot. HPV and cytology tests were repeated every 4 months for an average follow-up time of 25 months. The risk of progression was restricted to women who were persistent carriers of HPV DNA as assessed by repeated testing. The cumulative risk of progression among HPV positive women was 28% at 2 years compared to 3% among HPV-negative women [34].

In the Netherlands a cohort of 342 women with abnormal cytology (with Pap class 3B or lower, i.e. CIN III or lower) were followed-up every 3–4 months during an average follow-up period of 16 months [35]. During the follow-up visits the following examinations were performed: cytology, colposcopy without biopsy and HPV DNA testing for 27 HPV types using an accurate PCR technique. Nine (3.0%) of the 298 women with an original cytological diagnosis of Pap 3A (CIN I–II) progressed to CIN III (diagnosed by colposcopy and histology) and all of them were HPV DNA-positive for high-risk types at enrolment and during the follow-up.

In a second study from the USA, 206 women (173 with low-grade squamous intraepithelial lesions (SIL) and 33 with high-grade SIL) who participated in an intervention trial were followed every 2 months over a 6 month period. HPV DNA 16 was detected at study entry and at each follow-up examination by Southern blot. Multivariate methods were used to adjust for age, race, smoking, oral contraceptive (OC) use and plasma levels of micronutrients. HPV 16 was found to be related to progression to high-grade SIL with a RR of 1.19 (95% CI = 1.03–1.38) [36]. In an intensive follow up study, Ho *et al.* (1995) identified HPV type, HPV persistence and viral load as factors related to progression of HPV infection to neoplasia [37]. With some variation in the magnitude of the risk estimates and the time to progression, these results are consistent in all settings where follow-up studies are being conducted and suggest that persistent infection with high-risk HPV types precedes and predicts the development of CIN II–III. The main limitation of this type of study design is that in most settings follow-up is interrupted at stages CIN II–III for treatment. Therefore, the role of HPV in the progression to invasive cancer cannot be directly investigated. It is known that a certain proportion of CIN II–III lesions will regress spontaneously and these cases would limit the interpretation of most follow-up studies.

OTHER RISK FACTORS FOR CERVICAL DYSPLASIA AND CERVICAL CANCER

All the investigations above referenced identified HPV DNA as the single major etiologic agent for CIN III and cervical cancer. If allowances are made for sampling errors and laboratory limitations, one could speculate that most, if not all, cases of cervical cancer have been exposed to HPV and retain viral markers in the neoplastic tissue. This is in marked contrast with the prevalences observed in the corresponding controls, typically under 15%. In the presence of such powerful association, other exposures have little chance to be identified in classical case-control studies even if their association with cervical cancer is causal in nature. This is the case for OC use, smoking or herpes virus type 2 infections, all of which have shown in the early studies moderate associations with cervical cancer (i.e. ORs between 1 and 2).

Several of the classical risk factors for cervical cancer such as number of sexual partners, age at first sexual intercourse or even some of the reproductive variables are now seen as surrogates of HPV exposure. It can be predicted that as HPV detection methods progress, less independent effects would be observed for any of the surrogates.

Other exposures such as use of OC or smoking have been repeatedly reported as risk factors in pre-HPV studies and as potentially independent factors. These variables are difficult to evaluate because some collinearity exists with sexual practices (i.e. women who smoke or use OCs tend to have more partners). Similar difficulties apply when closely related variables have different biological implications. For example, age at first sexual intercourse and age at first birth are two variables closely associated in some settings. However, whereas first sexual intercourse can easily be interpreted as a surrogate of early HPV infection, age at first birth has in addition several implications in terms of hormonal balance. Because of these concurrent circumstances and the inherent difficulties in assessing with precision the exposure to some of the risk factors of interest (i.e. variables related to sexual behaviour) the impact of risk factors other than HPV in cervical cancer is unlikely to be fully revealed by case control studies.

Two approaches have been used in case-control studies to evaluate the role of other risk factors adjusting away the powerful effect of HPV. One is the use of multivariate techniques of adjustment of the risk estimates, the other is the restriction of the analyses to the subsets of cases and controls that are HPV-positive. We will briefly review the results of selected studies and risk factors.

SEXUAL BEHAVIOUR

An analysis stratified by HPV status in the Colombia and Spain study showed that among HPV-negative cases the risk factors identified were still related to sexual behaviour. The number of sexual partners, the key risk factor for cervical cancer, is no longer related to the disease among women who are HPV-positive and remains a strong risk factor among HPV-negatives. This finding suggests that number of partners is a surrogate measure (perhaps the best surrogate) of HPV infection. It also suggests that among women who are HPV-positive, further increasing the number of partners would not add to the risk of developing cervical cancer because the key exposure has already occurred. The study provided evidence on an increased risk linked to early age at first sexual intercourse and to early pregnancy, both surrogate measures of early HPV infection. This is consistent with the notion that early (adolescent) as opposed to adult HPV infections convey a higher probability of evolving to chronic HPV infection and to cervical cancer.

The conclusions are remarkably consistent across studies that reported both HPV-adjusted and HPV-stratified analysis relating number of partners and cervical neoplasia [21, 24, 29].

OTHER STDS

Antibodies to *Chlamydia trachomatis, Neisseria gonorrhoeae, Treponema pallidum, herpes simplex virus type-2* (HSV-2) and *Cytomegalovirus* (CMV) were measured in the study in Colombia and Spain. The only consistent association with cervical cancer was for *Chlamydia trachomatis* and CIN III. The association was of moderate intensity (ORs were 2.2 and 1.8 in Spain and Colombia, respectively) and showed some relationship with the antibody titre. Cases of invasive cancer in Spain showed higher antibody positivity rates than their controls to *Neisseria gonorrhoeae* (OR = 9.2) HSV-2 (OR 2.4) and *Chlamydia trachomatis* (OR = 7.6). No relationships to cervical cancer were found for seropositivity to HSV-2 or to CMV [38,39]. Among HPV-positive women,

none of the associations remained statistically significant. As described above, all and each of the STDs, including HPV, are more likely to occur among women with multiple partners, therefore it is not surprising that occasionally some are found associated to cervical cancer.

A particularly significant interaction has been repeatedly reported between HPV and the human immunodeficiency virus (HIV). Women (and men) who are exposed to HIV are more likely to be HPV positive and to develop HPV related neoplasia [40]. Further, the grade of neoplasia (cervical or anal) is related to the intensity of the immunosupression as measured by the level of $CD4^+$ counts (reviewed by IARC [15] and Melbye *et al.* [41]). The presence of cervical cancer has been recommended as an independent criteria to establish the diagnosis of AIDS in women with HIV infection.

ORAL CONTRACEPTIVES AND PARITY

In the Colombia/Spain study and also in the study in Brazil, the only significant differences among HPV-positive cases and controls were the use and the duration of use of OCs. In the three countries, the analysis by years of use of OCs suggests a positive trend, although the conclusion is severely limited by the small number of control women that were HPV DNA positives. Parity shows inconsistent results, with contrasting findings in Colombia (no association with invasive cancer and moderate with CIN III) and Brazil (strong association with invasive cancer), both being countries at high risk for cervical cancer and of high parity rates in Latin America. An association of parity (with cervical cancer was also found in Spain with invasive cancer but not with CIN III). Among HPV positive women there was no association with parity in the Spain and Colombia studies while a significant trend with increasing parity was observed in Brazil. These results suggest that hormonal factors may play a role as a cofactor in the acquisition of the HPV chronic carrier state and/or in the progression from HPV to HPV-related neoplasia [29, 42, 43].

TOBACCO SMOKING AND CERVICAL CANCER

Cigarette smoking showed inconsistent and weak associations in the studies in Spain, Colombia and Brazil, not supporting an independent effect of smoking. However, another PCR-based study reported a two-fold increased risk of SIL for women who were current smokers and HPV-positive [24]. Cigarette smoke contains powerful non-organ-specific carcinogens and smoke metabolites have been detected in the cervical mucosa [44]. Therefore the hypothesis of a carcinogenic effect in the cervix cannot be conclusively ruled out although the effect, if any, should be of minor quantitative importance.

If we consider the small fraction of cervical cancers in which HPV DNA cannot be detected (between 5 and 10%) as truly HPV-negative cases and having established that only a minority of HPV infections will evolve to invasive cancer one should conclude that other determinants of the evolution of HPV infections are in force. In addition to viral determinants such as viral type, viral load or perhaps variant types, some environmental factors such as hormonal treatments are being intensively studied. However, host factors such as genetic determinants (HLA or MHC haplotypes), genetic or induced immuno-supression, or endogenous hormones have been poorly addressed so far by epidemiological studies.

THE ROLE OF MEN IN THE HPV EPIDEMIOLOGICAL CHAIN

In 1982, a model was proposed whereby the high rates of cervical cancer in Latin America could be explained by a large number of sexual partners among males, including frequent contacts with prostitutes, coupled with monogamy or few sexual partners among women [45]. Case control studies assessing the contribution of male's sexual behaviour and genital HPV DNA to the risk of developing cervical neoplasia have yielded inconsistent results. The effect of the number of husbands' sexual partners was more apparent in countries at low risk of cervical cancer than in countries at high risk. The effect of the number of prostitutes was inconsistent and the early HPV studies in the penis were all negative (reviewed by Bosch *et al.* [46]).

The IARC studies in Spain and Colombia were the first to show a strong relationship between HPV DNA in the male penis/urethra and the risk of cervical cancer in their wives. The prevalence of HPV DNA in the penis was strongly related to sexual behaviour, and the number of sexual partners and of prostitutes reported by the husbands was higher in Colombia than in Spain [47]. In Spain, a country traditionally at low risk for cervical cancer, the presence of HPV DNA in the husband's penis conveyed a five- to seven-fold risk of cervical cancer to their wives. The risk was nine-fold for spouses of carriers of HPV 16. The risk of cervical cancer was strongly related to the number of extramarital sexual partners (OR = 11.0, 95% CI = 3.0–40.0, for 21 vs 1), and to the number of extramarital prostitutes as sexual partners (OR = 8.0, 95% CI = 2.9–22.2, for 10 vs none). The presence of antibodies to *Chlamydia trachomatis* and an early age at first sexual intercourse of the husband were both associated with a significant three-fold increased risk of cervical cancer in their wives [46].

In Colombia, a high risk country for cervical cancer, limited education and presence of antibodies to *Chlamydia trachomatis* were the only identified 'male' risk factors for cervical neoplasia. The prevalence of HPV DNA in the penis was 25.7% among husbands of case women and 18.9% among husbands of control women (OR = 1.2, 95% CI = 0.6–2.3). Neither the lifetime number of sexual partners (OR = 1.0, 95% CI = 0.4–2.6, for >50 partners vs 1–5) nor the lifetime number of prostitutes reported by the husbands (OR = 1.2, 95% CI = 0.7–2.0, for 21 prostitutes vs 1–5) were associated with the risk of cervical cancer in their wives [48].

In Spain, the study supports the role of men as vectors of the HPV types that are related to cervical cancer. Lifetime number of sexual partners, number of prostitutes as sexual partners and detection of HPV DNA in the penis at the time of the study are interpreted as surrogate markers of exposure to HPV during marriage. The results in Colombia are compatible with the hypothesis that in the high-risk population of Cali, exposure to HPV among young men is common and mediated by contacts with a high number of sexual partners and prostitutes. These widespread sexual practices limit the power of case-control studies to detect significant associations between men's sexual behaviour and cervical cancer risk. In this population, HPV DNA detection in the penis of adult men is a poor reflection of lifetime or of the etiologically relevant exposure to HPV. The role of *C. trachomatis* in cervical carcinogenesis deserves further investigation.

The results of the studies describing the role of men in the epidemiology of cervical cancer strongly confirm that the HPV types related to cervical cancer are a widespread sexually transmitted disease. Furthermore, they suggest that men can operate as HPV carriers in the epidemiological chain. At present there is no obvious recommendation concerning clinical management of male HPV carriers. Detection requires testing for HPV DNA in exfoliated cells from the penis, not an easy task both technically and socially. Colposcopic inspection using acetic acid painting has been recommended and minute HPV-related lesions are often unveiled among partners of women with CIN or HPV infections. Finally, there is at present no reliable treatments for HPV and it has not been shown that condoms would prevent HPV transmission. In spite of these difficulties, any comprehensive approach to HPV control should include research to further elucidate the male role in cervical carcinogenesis and to devise adequate intervention strategies.

HPV AS A PROGNOSTIC FACTOR IN CERVICAL CANCER

Several projects have discussed the value of HPV, presence/absence or HPV type, as an indicator of the prognosis in cervical cancer patients with controversial findings. The clinical follow up of the cases recruited in the case-control study in Colombia and Spain has recently been completed with the objectives of evaluating several HPV markers, including HPV, HPV type and antibodies to HPV 16-derived peptides, as prognostic factors for recurrence and survival in cervical cancer patients.

The study included an active follow-up of 471 incident cases of invasive cervical cancer recruited in 1985–1987 and followed to September 1992. Cases were tested for HPV DNA, including PCR methods, on cervical scrapes and tumour biopsies. Serum specimens were tested in ELISA format assays for antibodies against three peptides derived from overlapping regions of the E7 protein (E7/1 , 2, 3) and two peptides derived from selected regions of E2 and L2. Follow-up was achieved in 424 of the 471 cases recruited into the study (90%). Cox models were used to compare the survival probabilities adjusted by stage and treatment, the only independent factors found to be significantly related to prognosis. The 3- and 5-year survival rates were 72.2 and 66.9%, respectively. HPV DNA in cervical specimens was identified in 79.1% of the cases. Neither presence nor absence of HPV DNA, HPV type or different estimates of the HPV viral load were related to survival. However, HPV 18 and HPV type unspecified cases had (non-statistically significant) poorer survival than HPV 16 related cases. As compared to patients with no serologic response to any of the three E7-derived peptides, a poorer prognosis was recorded for patients with antibodies to E7. In a stratified analysis by stage, the prognostic value of E7 antibodies was significant within stages I and II but not in more advanced stages, suggesting that some of the clinically considered early stages

(I and II) may in fact be clinically silent more advanced stages [49]. The prognostic value linked to HPV 18 [50] or the presence of HPV 16 E7 antibodies have also been suggested by other authors [51, 52] as early markers of lymph node metastasis.

The IBSCC study found no relationship between HPV status and clinical stage, although HPV 18-related cases tended to be more undifferentiated than HPV 16-related cases [27]. An ongoing follow up of the cases recruited into this study should better describe in the near future the survival experience of the HPV 18-related cases.

IMPLICATIONS FOR CANCER PREVENTION: SCREENING AND VACCINATION

HPV appears as the central cause of cervical cancer and most of the disease worldwide seems to be explained by the viral infection. Hormonal factors seem to interact with HPV in the progression from self controlled infection to neoplasia.

Early age at infection and perhaps some other circumstances around the time at infection, such as concurrent pregnancy or OC use, seem to be relevant and deserve further investigation. It is unlikely that any additional factor plays a major independent role although minor effects cannot be ruled out for *Chlamydia trachomatis* and smoking.

USE OF HPV TESTING IN CERVICAL CANCER SCREENING PROGRAMMES

HPV testing fulfils some but not all of the requirements for a screening test. (i) HPV infection precedes morphological lesions for a prolonged period of time (most probably for years) during which the viral presence is detectable. (ii) HPV detection is a highly reliable test. New technology being developed should, in the near future, standardize PCR-based methods including quantitative estimates of the viral load. HPV detection may be transferred to clinical laboratories and it is expected that it will show low false-negative rates. (iii) Specimens for HPV DNA detection are obtained through non-invasive procedures and should be acceptable by the population at large as much as cytology based screening programmes are. Furthermore, the screening logistics are superimposable to the ones already in place in developed countries. (iv) The cost of HPV-based screening programmes should be reduced from current levels with standardization and commercialization of the assays.

Two arguments can be raised against the use of HPV tests in screening programmes. One, perhaps the strongest, is the lack of treatment once infection is detected. Increased surveillance of high-risk type HPV carriers using cytology seems the only option to date. Use of condoms has not been shown to prevent HPV transmission and to recommend reducing the number of sexual partners does not seem a feasible option. Secondly, in spite of the fact that cervical cancer is almost inevitably preceded by HPV infection, the natural history of the infection is not well understood. Some of the open questions that have a bearing in defining screening strategies can be summarized as follows: (i) it is established that most HPV infections will resolve spontaneously with time but it is unclear which factors intervene in establishing a chronic carrier state; (ii) The potential for progression of a LGSIL lesion (perhaps the earliest visible manifestation of the HPV infection) is highly variable and morphologicaily it is impossible to predict progression; (iii) the paradigm of cervical cancer progression from low-grade lesions to invasive cancer through intermediate discrete stages (CIN I–II–III–CX) has been challenged. New interpretations have been proposed suggesting that *de novo* high-grade lesions can appear within short time intervals as a consequence of unknown circumstances of the HPV infection (HPV type, viral load, immune status, etc.).

The earliest trials of HPV detection as screening tests have included HPV testing as adjuvant to cytology to decide on low grade lesions and on the diagnosis of abnormal cytology of uncertain significance (ASCUS). The results are encouraging and suggests that HPV testing may in fact be of use in predicting high grade lesions. Cuzick *et al.* (1992) investigated a group of women referred to colposcopy clinics because of mild to moderate dyskariosis [53]. The final diagnosis was established by colposcopy-directed biopsy and HPV 16 testing was done using a semi-quantitative PCR system. In this study of 55 women with mild to moderate dysplasia, 27 were normal on the biopsy, of whom 26 gave a negative or low intensity HPV 16 hybridization signal. In contrast 18 women had biopsy proven CIN III of whom 11 (61%) showed high levels of HPV 16 DNA at recruitment. These findings were extended and confirmed suggesting that the predictive value of HPV testing was similar to that of 'moderate dyskariosis'. The authors conclude that HPV testing may be of complementary support to cytology and in the triage of low level morphological lesions [54]. For further discision, see Chapter 6.

The Free University in Amsterdam has pioneered a large population-based screening scheme in which the risk of progression to cervical cancer is predicted by either an abnormal cytology or by a positivity to high risk HPV types as detected by a highly sensitive PCR system. If the hypothesis is confirmed, women at low risk (95% of the screening population) could considerably space their screening intervals, reducing the costs of unnecessary testing and concentrating diagnostic resources on the population at high risk [55]. This important study should clarify in the near future several issues concerning the value of HPV-based screening programmes, the hybridization test to be used, the most adequate screening intervals and the cost benefits of the new strategy. An updated review on the value of HPV testing (and other new technologies) in cervical cancer screening has been recently published [56].

HPV VACCINES

The majority of, if not all, cases of cervical cancer are related to HPV infections. Perhaps half of the cancers of the vulva, vagina and penis are also related to the same HPV types and there are indications that other cancers may be partially involved (i.e. the oral cavity, the oesophagus, the skin, and the urinary bladder) (see literature by IARC [15] and Franceschi *et al.*, [57]). Definite evidence for these cancer sites is still lacking, but given the impact of cervical cancer alone, it can be claimed that of the known human carcinogens, HPV may be the first in importance among women. HPV vaccines are thus of great potential interest.

Three types of HPV vaccines are being developed: prophylactic to prevent HPV infection and its associated lesion, therapeutic to induce regression of HPV-associated lesions and combined prophylactic and therapeutic to induce both effects.

Prophylactic HPV vaccines are based on structural proteins or late antigens in the L1 or L2 regions. The synthetic production of virus-like particles (VLPs) of papillomavirus (PV) has lead to major advances in the development of such vaccines. VLPs of PV are assembled by over-expression of the major capsid protein (L1) in various expression vectors. VLPs have the same surface topography as infectious viral particles and present the conformational epitopes required for generating high titres of neutralizing antibodies but do not contain the potentially oncogenic viral DNA. Vaccination experiments have shown the protective efficacy of VLPs in rabbits and dogs against challenge with cottontail rabbit papillomavirus [58] and canine oral papillomavirus [59] respectively. Thus, the VLPs are currently the immunogen of choice for a HPV vaccine to prevent genital HPV infection, and various VLPs vaccines are now under development. The efficacy and safety of these vaccines should be assessed in phase I and II trials before they can be used in the general population.

Therapeutic vaccines target the E6 and E7 oncoproteins of the HPV. Various recombinant HPV vaccines are being developed and some of them are already being tested in small-scale phase I trials.

In the UK, the vaccine is based on a recombinant vaccinia vector expressing mutated E6 and E7 from both HPV 16 and 18. The mutated oncoproteins have lost their oncogenic potential but retain their immunogenic properties. It has been used in seven patients with invasive cervical cancer also being treated with chemo- and radiotherapy. No side-effects of vaccination have been observed after 9 months [60].

In Australia, a vaccine is based on bacterial fusion proteins for HPV 16 E7 with algammulin adjuvant. The product has been administrated to five patients with stage 2b, 3 or 4 cervical cancer for whom no potentially curative therapy is available. Three of the five patients made antibodies to E7 and no adverse effects of vaccination were observed [61].

A similar trial has started in the Netherlands using peptides related to HPV 16 E7 that bind to MHC class I molecules for use in HLA-A0201 cervical cancer patients [62]. Although previous vaccination experiments have shown that this peptide protected mice against challenge with a syngeneic HPV 16-carrying tumour cell line, it remains to be seen if this approach is successful in humans. This scepticism is justified by the evidence indicating that most cervical cancers have defects either in MHC class I expression or processing.

Combined prophylactic and therapeutic vaccine are based on full-length HPV16 L2-E7 chimeric proteins that are incorporated into L1 VLPs [63]. Women with normal or abnormal cytology positive for high-risk HPVs could be the target population for this vaccine.

ACKNOWLEDGEMENTS

This review was partially supported by grants from the International Agency for Research on Cancer (IARC) and the Dirección General de Investigación, Ciencia y Tecnologia (DGICYT) of the Spanish Government (SAF96-0323).

REFERENCES

1. Parkin D.M., Pisani, P. and Ferlay, J. (1993) Estimates of the worldwide incidence of eighteen major cancers in 1985. *International Journal of Cancer*, **54**, 594–606.
2. Pisani, P., Parkin, D.M., and Ferlay, J. (1993) Estimates of the worldwide mortality from eighteen major cancers in 1985. Implications for prevention and projection of future burden. *International Journal of Cancer*, **55**, 891–903.
3. Parkin, D.M., Muir, C.S., Whelan, S.L., *et al.* (eds) (1992) *Cancer Incidence in Five Continents*, vol. VI, IARC Scientific Publications No. 120, Lyon.
4. Devesa, S.S., Young, J.I., Brinton, L.A., *et al.* (1989) Recent trends in cervix uteri. *Cancer*, **64**, 2184–90.
5. Zheng, T., Holford, T., Ma, Z., *et al.* (1996) The continuing increase in adenocarcinoma of the uterine cervix: a birth cohort phenomenon. *International Journal of Epidemiology*, **25**(2), 252–7.
6. Bergström, R., Sparen, P. and Adami, H.O. (1996) Trends in cancer of the cervix uteri in Sweden following cytologic screening.
7. Beral, V., Hermon, C., Muñoz, N., *et al.* (1994) Cervical cancer. *Cancer Surveys*, **19/20** (Trends in Cancer Incidence and Mortality), 265–285.
8. Vizcaino, A.P., Moreno, V., Bosch, F.X., *et al.* (1998) International trends in the incidence of cervical cancer. I. Adenocarcinoma and adenosquamous cell carcinomas. *International Journal of Cancer*, in press.
9. Parkin, D.M., Vizcaino, A.P., Skinner, M.E.G. (1994) Cancer patterns and risk factors in the African population of southwestern Zimbabwe, 1963–1977. *Cancer Epidemiology, Biomarkers & Prevention*, **3**, 537–47.
10. Wabinga, H.R., Parkin, D.M., Wabwire-Mangen, F. *et al.* (1993) Cancer in Kampala, Uganda, in 1989–91: changes in incidence in the era of AIDS. *International Journal of Cancer*, **54**, 26–36.
11. Muñoz, N., Bosch, F.X. and Kaldor, J.M. (1988) Does human papillomavirus cause cervical cancer? The state of the epidemiological evidence. *British Journal of Cancer*, **57**, 1–5.
12. Guerrero, E., Daniel, R.W., Bosch, F.X., *et al.* (1992) Comparison of virapap, Southern hybridization and polymerase chain reaction methods for human papillomavirus identification in an epidemiological investigation of cervical cancer. *Journal of Clinical Microbiology*, **30**, 2951–9.

13. Schiffman, M.H. and Schatzkin, A. (1994) Test reliability is critically important to molecular epidemiology: an example from studies of human papillomavirus infection and cervical neoplasia. *Cancer Research*, **54**, 1944s–7s.

14. Bosch, F.X., de Sanjosé, S. and Muñoz, N. (1994a) Test reliability is critically important to molecular epidemiology (Letter). *Cancer Research*, **54**, 6288–9.

15. IARC Monographs on the evaluation of carcinogenic risks to humans (1995) Vol. 64: *Human Papillomaviruses*. International Agency for Research on Cancer, Lyon.

16. Walboomers, J.M.M., Jacobs, M.V. Snijders, P.J.F., *et al.* (1996) A general primer GP5$^+$/6$^+$ mediated polymerase chain reaction enzyme immunoassay method for rapid detection of 14 high risk and 6 low risk human papillomavirus genotypes in cervical scrapes. Abstract presented at the *15th International Papillomavirus Workshop*, December, 1996, Brisbane, Australia.

17. Lorincz, A. (1992) Diagnosis of human papillomavirus infection by the new generation of molecular DNA assays. *Clinical Immunology News*, **12**, 123–8.

18. Gravitt, P., Wheeler, C., Schiffman, M., *et al.* (1996) Strip-based type detection of amplified human papillomavirus from genital samples using immobilized oligonucleotide probes. Abstract presented at the *15th International Papillomavirus Workshop*, December 1996, Brisbane, Australia.

19. Margall, N., Matias-Guiu, X., Chillon, M., *et al.* (1993) Detection of human papillomavirus 16 and 18 DNA in epithelial lesions of the lower genital tract by *in situ* hybridization and polymerase chain reaction: cervical scrapes are not substitutes for biopsies. *Journal of Clinical Microbiology*, **31**(4), 924–30.

20. Bosch, F.X., Muñoz, N., de Sanjosé, S., *et al.* (1993) Human papillomavirus and cervical intraepithelial neoplasia grade III/carcinoma *in situ*: a case-control study in Spain and Colombia. *Cancer Epidemiology, Biomarkers & Prevention*, **2**, 415–22.

21. Schiffman, M.H., Bauer, H.M., Hoover, R.N., *et al.* (1993) Epidemiologic evidence showing that human papillomavirus infection causes most cervical intraepithelial neoplasia. *Journal of the National Cancer Institute*, **85**(2), 958–64.

22. Becker, T.M., Wheeler, C.M., McGough, N.S., *et al.* (1994) Sexually transmitted diseases and other risk factors for cervical dysplasia among Southwestern Hispanic and non-Hispanic white women. *Journal of the American Medical Association*, **271**(15), 1181–8.

23. Van den Brule, A.J.C., Walboomers, J.M.M., Du Maine, M., *et al.* (1991) Difference in prevalence of human papillomavirus genotypes in cytomorphologically normal cervical smears is associated with a history of cervical intraepithelial neoplasia. *International Journal of Cancer*, **48**, 404–8.

24. Kjaer, S.K., Van den Brule, A.J.C., Bock, J.E., *et al.* (1996) Human Papillomavirus – the most significant risk determinant of cervical intraepithelial neoplasia. *International Journal of Cancer*, **65**, 601–6.

25. Olsen, A.O., Gjøen, K., Sauer, T., *et al.* (1995) Human papillomavirus and cervical intraepithelial neoplasia grade II–III: a population-based case-control study. *International Journal of Cancer*, **61**, 312–15.

26. Morrison, E.A.B., Ho, G.Y..F, Vermund, S.H., *et al.* (1991) Human papillomavirus infection and other risk factors for cervical neoplasia: a case-control study. *International Journal of Cancer*, **48**, 404–8.

27. Bosch, F.X., Manos, M.M., Muñoz, N., *et al.* (1995) Prevalence of human papillomavirus in cervical cancer: a worldwide perspective. *Journal of the National Cancer Institute*, **87**, 796–802.

28. Muñoz, N., Bosch, F.X, de Sanjosé, S., *et al.* (1992) The causal link between human papillomavirus and invasive cervical cancer: a population-based case-control study in Colombia and Spain. *International Journal of Cancer*, **52**, 743–9.

29. Eluf-Neto, J., Booth, M., Muñoz, N., *et al.* (1994) Human papillomavirus and invasive cervical cancer in Brazil. *British Journal of Cancer*, **69**, 114–9.

30. Bauer, H.M., Ting, Y., Greer, C.E., *et al.* (1991) Genital human papillomavirus infection in female university students as determined by a PCR-based method. *Journal of the American Medical Association*, **265**, 472–7.

31. Walboomers, J.M.M., Melkert, P.W.J., Van Den Brule, A.J.C., *et al.* (1992) The polymerase chain reaction for human papillomavirus screening in diagnostic cythopathology of the cervix, in *Diagnostic Molecular Pathology. A Practical Approach* (eds C.S. Herrington and J.O.D. McGee). IRL Press, Oxford, pp. 153–72.

32. Moreno, V., Muñoz, N., Bosch, F.X., *et al.* (1995) Risk factors for progression of Cervical Intraepithelial Neoplasm Grade III to Invasive Cervical Cancer. *Cancer Epidemiology, Biomarkers & Prevention*, **4**, 459–67.

33. Franco, E.L. (1995) Cancer causes revisited: human papillomavirus and cervical neoplasia (Editorial). *Journal of the National Cancer Institute*, **87**(11), 779–80.

34. Koutsky, L.A., Holmes, K. K., Critchlow, C.W., *et al.* (1992) A cohort study of the risk of cervical intraepithelial neoplasia grade 2 or 3 in relation to papillomavirus infection. *New England Journal of Medicine*, **327**(18), 1272–8.

35. Remmink, A.J., Walboomers, J.M., Helmerhorst, T.J.M., *et al.* (1995) The presence of persistent high-risk HPV genotypes in dysplastic cervical lesions is associated with progressive diasease: natural history up to 36 months. *International Journal of Cancer*, **61**, 306–11.

36. Liu, T., Soong, S.-J., Alvarez, R.D., *et al.* (1995) A longitudinal analysis of human papillomavirus 16 infection, nutritional status, and cervical dysplasia progression. *Cancer Epidemiology, Biomarkers and Prevention*, **4**, 373–80.

37. Ho, G.Y.F., Burk, R.D., Klein, S., *et al.* (1995) Perisitent genital human papillomavirus infection as a risk factor for persistent cervical dysplasia. *Journal of the National Cancer Institute*, **87**, 1365–71.

38. de Sanjosé, S., Muñoz, N., Bosch, F.X., *et al.* (1994) Sexually transmitted agents and cervical neoplasia in Colombia and Spain. *International Journal of Cancer*, **56**, 358–63.

39. Muñoz, N., Katho, I., Bosch, F.X., *et al.* (1995) Cervical cancer and herpes simplex virus type 2: case-controls studies in Spain and Colombia, with special reference to immunoglobulin-G sub-classes. *International Journal of Cancer*, **60**, 438–42.

40. de Sanjosé, S., Palacio, V., Tafur, L., *et al.* (1993) Prostitution, HIV, and cervical neoplasia: a survey in Spain and Colombia. *Cancer Epidemiology Biomarkers & Prevention*, **2**, 531–5.

41. Melbye, M., Palefsky, J., Gonzales, J., *et al.* (1990) Immune status as a determinant of human papillomavirus detection and its association with anal epithelial abnormalities. *International Journal of Cancer*, **46**, 203–6.

42. Muñoz, N., Bosch, F.X., de Sanjosé, S., *et al.* (1993) Risk factors for cervical intraepithelial neoplasia grade III/carcinoma *in situ* in Spain and Colombia. *Cancer Epidemiology, Biomarkers & Prevention*, **2**, 423–31.

43. Bosch, F.X., Muñoz, N., de Sanjosé, S., *et al.* (1992) Risk factors for cervical cancer in Colombia and Spain. *International Journal of Cancer*, **52**, 750–8.

44. Schiffman, M.H., Haley, N.J., Felton, J.S., *et al.* (1987) Biochemical epidemiology of cervical neoplasia: measuring cigarette smoke constituents in the cervix. *Cancer Research*, **47**, 3886–8.

45. Skegg, D.C., Corwin, P.A., Paul, C., *et al.* (1982) Importance of the male factor in cancer of the cervix. *Lancet*, **2**, 581–3.

46. Bosch, F.X., Castellsagué, X., Muñoz, N., *et al.* (1996) Male sexual behaviour and human papillomavirus DNA: key risk factors for cervical cancer in Spain. *Journal of the National Cancer Institute*, **88**, 1060–7.

47. Bosch, F.X., Muñoz, N., de Sanjosé, S., *et al.* (1994b) Importance of human papillomavirus endemicity in the incidence of cervical cancer: an extension of the hypothesis on sexual behaviour. *Cancer Epidemiology, Biomarkers & Prevention*, **3**, 375–9.

48. Muñoz, N., Castellsagué, X., Bosch, F.X., *et al.* (1996) Difficulty in elucidating the male role in cervical cancer in Colombia, a high-risk area for the disease. *Journal of the National Cancer Institute*, **88**, 1068–75.

49. Viladiu, P., Bosch, F.X., Castellsagué, X. *et al.* (1997) HPV DNA and antibodies to HPV 16 E2, L2 and E7 peptides as predictors of survival in patients with squamous cell cervical cancer. *Journal of Clinical Oncology*, **15**, 610–19.

50. Burger, R.A., Monk, B.J., Kurosaki, T., *et al.* (1996) Human papillomavirus type 18: association with poor prognosis in early stage cervical cancer. *Journal of the National Cancer Institute*, **88**, 1361–8.

51. Baay, M.F.C., Duk, J.M., Buerger, M.P.M., *et al.* (1995) Follow-up of antibody responses to human papillomavirus type 16 E7 in patients treated for cervical carcinoma. *Journal of Medical Virology*, **45**, 342–7.

52. Czeglédy, J., Iosif, C., Hansson, B.G., *et al.* (1995) Can a test for E6/E7 transcripts of human papillomavirus type 16 serve as a diagnostic tool for the detection of mecrometastasis in cervical cancer? *International Journal of Cancer (Pred. Oncol.)*, **64**, 211–15.

53. Cuzick, J., Terry, C., Ho, L., *et al.* (1992) Human papillomavirus type 16 DNA in cervical smears as predictor of high-grade cervical cancer. *Lancet*, **339**, 959–60.

54. Cuzick, J., Szarewski, A., Terry, G., *et al.* (1995) Human papillomavirus testing in primary cervical screening. *Lancet*, **345**, 1533–6.

55. Meijer, C.J.L.M., Van den Brule, A.J.C., Snijders, P.J.F., *et al.* (1992) Detection of human papillomavirus in cervical scrapes by the polymerase chain reaction in relation to cytology: possible implications for cervical cancer screening. In *The Epidemiology of Human Papilloma Virus and Cervical Cancer* (eds N. Muñoz, F.X. Bosch, K.V. Shah, *et al.*), IARC Scientific Publications No. 119. International Agency for Research on Cancer, Lyon, pp. 271–81.

56. Franco, E., Syrjanen, K., de Wolf, C., *et al.* (1996) New developments in cervical cancer screening and prevention (Meeting Report). *Cancer Epidemiology, Biomarkers & Prevention*, **5**, 853–6.

57. Franceschi, S., Muñoz, N., Bosch, F.X., *et al.* (1996) Human Papillomavirus and cancers of the upper aero-digestive tract: a review of epidemiological and experimental evidence. *Cancer Epidemiology, Biomarkers & Prevention*, **5**, 567–75.

58. Breitburd, F., Kirnbauer, R., Hubbert, N.L., *et al.* (1995) Immunization with viruslike particles form cottontail rabbit papillomavirus (CRPV) can protect against experimental CRPV infection. *Journal of Virology*, **69**, 3959–63.

59. Suzich, J.A., Ghim, S.-J., Palmer-Hill, F.J., *et al.*, (1995) Systemic immunization with papillomavirus L1 protein completely prevents the development of viral mucosal papillomas. *Proceedings of the National Academy of Science USA*, **92**, 11,553–7.

60. Borysiewicz, L.K., Fiander, A., Nimako, M. *et al.* (1996) A recombinant vaccinia virus encoding human papillomavirus types 16 and 18, E6 and E7 proteins as immunotherapy for cervical cancer. *Lancet*, **347**, 1523–7.

61. Frazer, I., Dunn, LA., Fernando, G.J.P., *et al.* (1994) Animal and human studies on immunotherapeuthic HPV vaccines. *Proceedings of IARC/Mérieux Joint Meeting on HPV Vaccines and their Potential Use in the Prevention and Treatment of Cervical Neoplasia*, 12–14 December 1994, Annecy, France. International Agency for Research on Cancer and Fondation Marcel Merieux, Lyon.

62. Kast, W.M., Brandt, R.M., Sidney, J., *et al.* (1994) Role of HLA-A motifs in identification of potential CTL epitopes in human papillomavirus type 16 E6 and E7 proteins. *Journal of Immunology*, **152**, 3904–12.

63. Schiller, J.T. and Roden, R.B.S. (1995) Papillomavirus-like particles. *Papillomavirus Report*, **6**, 121–8.

64. Peng, H., Liu, S., Mann, V., *et al.* (1991) Human papillomavirus types 16 and 33, Herpes simplex virus type 2 and other risk factors for Cervical Cancer in Sichuan, province, China. *International Journal of Cancer*, **47**, 711–16.

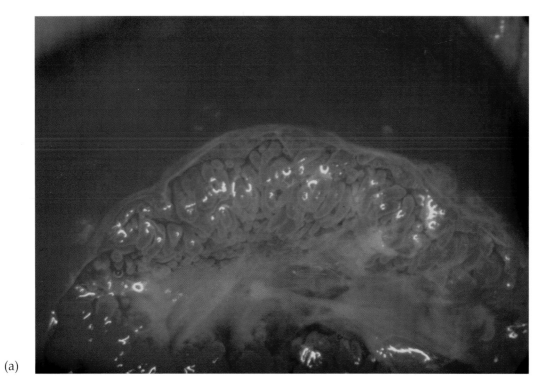

(a)

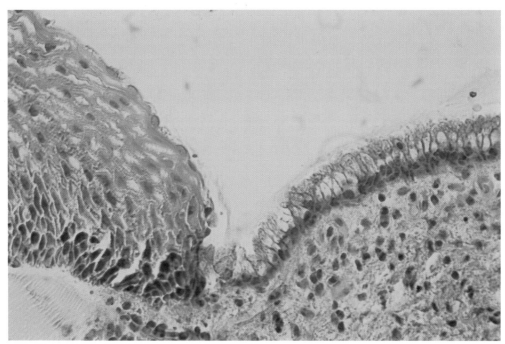

(b)

Plate 1   The normal squamo-columnar junction. (a) Colposcopy. (b) Histology.

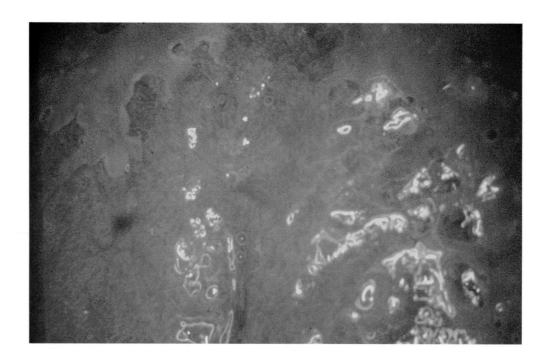

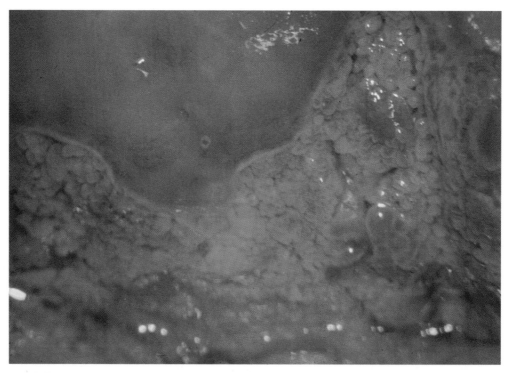

Plate 2 Colposcopy features of epithelialization.

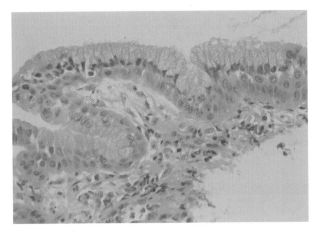

Plate 3 Reserve cells: subcolumnar reserve cells begin to proliferate beneath the endocervical epithelium.

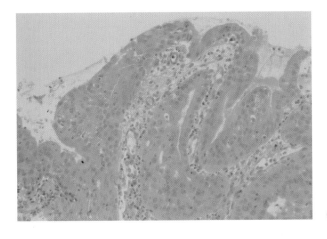

(a)

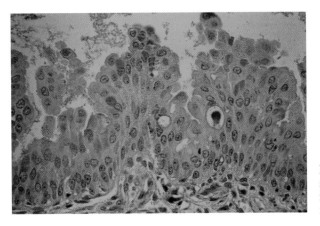

(b)

Plate 4 (a) Histology of immature metaplasia. (b) The endocervical epithelium above the immature squamous cells is stained by a mucin coloration (PAS).

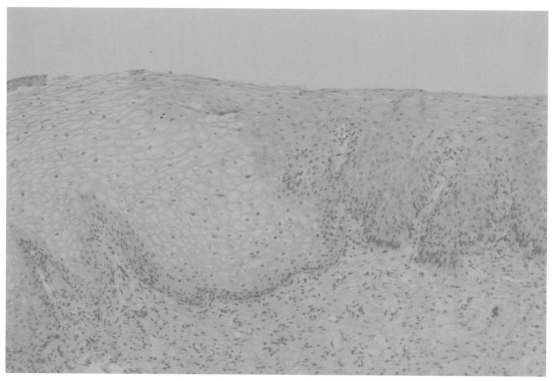

Plate 5 Histology of the squamo-squamous junction.

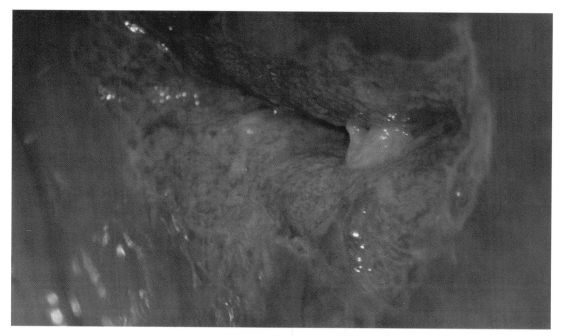

Plate 6 Colposcopic features of late metaplasia.

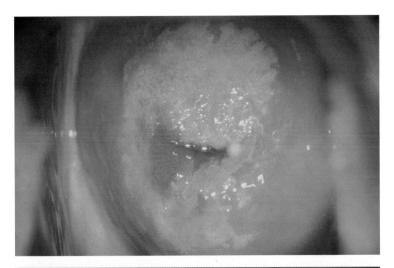

Plate 7 Colposcopic features of immature metaplasia: minor abnormal features within the transformation zone.

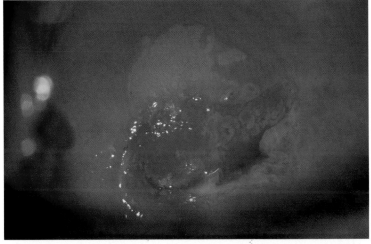

(a)

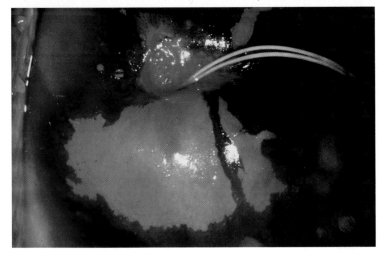

(b)

Plate 8 Colposcopic features of arrested metaplasia: (a) minor acetowhitening; (b) iodine-negative area with sharp and flat external borders.

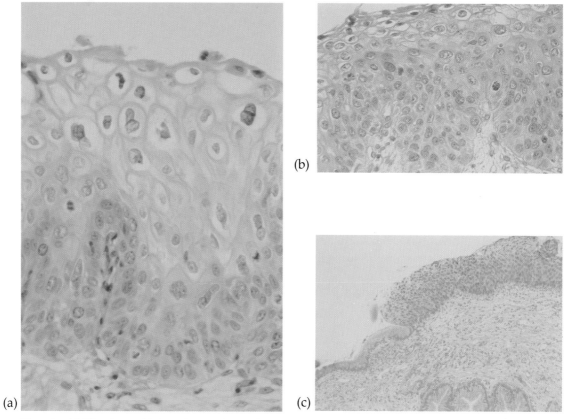

Plate 9   (a) Histology of CIN I. (b) Histology of CIN III. (c) CIN III originates at the squamo columnar junction of the transformation zone.

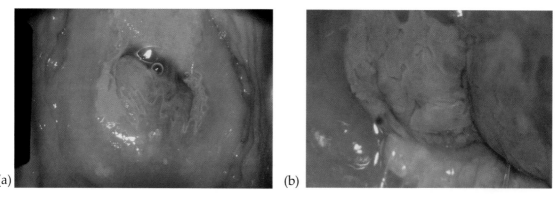

Plate 10   Colposcopic features of intra-epithelial neoplasia: (a) minor abnormal features within the transformation zone. Histology will show CIN I; (b) major abnormal features within the transformation zone. Histology will show CIN III.

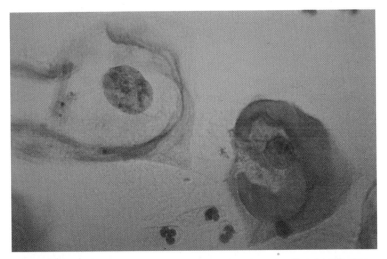

Plate 11 Koilocytosis with a large nucleus and a perinuclear halo.

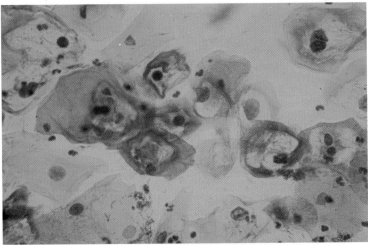

Plate 12 LGSIL with cytopathogenic effect reflecting an HPV infection.

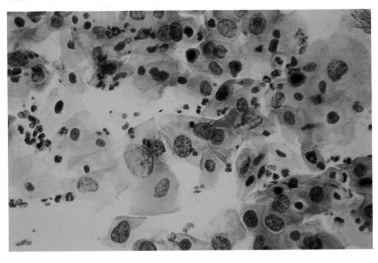

Plate 13 HGSIL (CIN II or moderate dysplasia) – abnormal basal cells are associated with koilocytosis.

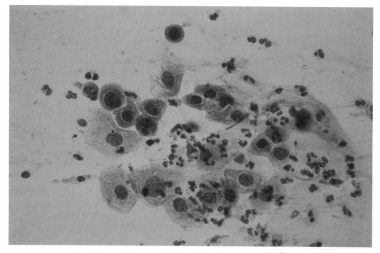

Plate 14 HGSIL (CIN III or severe dysplasia) – most of the basal cells are abnormal.

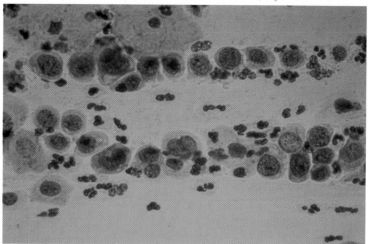

Plate 15 HGSIL (CIS) – all the basal cells are abnormal.

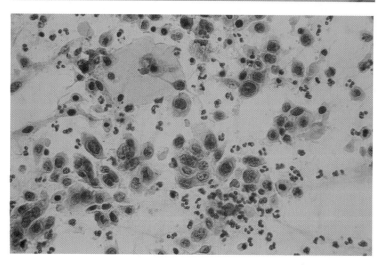

Plate 16 ASCUS – it is difficult to differentiate between basal metaplastic cells or abnormal basal cells.

Petr Skrabanek

INTRODUCTION

In 1851, Robert Hull marvelled that with the introduction of the vaginal speculum a veritable epidemic of uterine disease had appeared [1]. Screening for cervical cancer by means of exfoliative cytology was advocated by George Papanicolaou in the 1940s, and by the 1960s cervical cytological screening had become widespread, before any randomized controlled trials had been carried out. In 1976, a World Health Organization group of experts suggested that 'unless more hopeful evidence emerges from the available data in the near future, it may become ethical and necessary to conduct a randomized trial of cervical cancer screening' [2]. Twenty years later, we still have to rely on evidence which is conjectural, confusing and of poor quality. Critical analysis is frustrated by the lack of reliable data on the effectiveness of cervical screening and on the harm–benefit ratio. As pointed out in an editorial in the *British Medical Journal*, 'it is not a question of proving that screening has *no* value (this is always extremely difficult to show of any measure), but of deciding whether it has *sufficient* value to justify the risk and the efforts it entails' [3].

CRITERIA FOR SCREENING

In their monograph on the rational principles of screening, Wilson and Jungner stressed that screening tests should be validated before they are applied to populations, that the effect of screening should not be measured by surrogate outcomes but in terms of reduced morbidity and mortality, and that early detection of disease is unlikely to be cheaper than conventional curative medicine. They listed ten criteria which should be fulfilled before screening is used on a mass scale [4, 5].

In cervical screening, several criteria are not fulfilled, and the most problematic areas include, (1) the accuracy of the 'Pap' test; (2) uncertainties about the natural history of the disease; (3) disagreement about what constitutes the defined population at risk; (4) disagreement about the management of cervical abnormalities; and (5) the lack of reliable data for the assessment of cost-effectiveness.

THE TEST

In 1941, Papanicolaou suggested that exfoliated cells in the vaginal pool could be used for the early detection of uterine cancer. In 1947, Ayre devised a wooden spatula for obtaining cervical smears. As Leopold Koss emphasized, taking smears is not a simple procedure and their interpretation 'belongs to the most difficult areas of the microscopy' [6]. The smear must be rapidly fixed to avoid the formation of cellular artefacts, which mimic dyskaryosis, and increase the probability of a smear being reported as abnormal seven-fold [7]. A

Cancer and Pre-cancer of the Cervix
Edited by D.M. Luesley and R. Barrasso. Published in 1998 by Chapman & Hall, London. ISBN 0 412 56600 1.

negative smear is no guarantee that there is no serious abnormality. The smear is not a substitute for visual inspection of the cervix. Koss's earlier studies showed that the distinction between histologically diagnosed 'dysplasia' and 'carcinoma *in situ*' (CIS) is artificial, as 'there are no significant behavioural differences between these two groups of lesions' [6].

Experienced cytologists and histopathologists show marked intra-observer and inter-observer disagreements in interpretation of smears and biopsies [8–14].

The notion of the 'false-positive' result of a 'Pap' smear is meaningful only as far as it relates to the purpose of screening, that is, the detection of true premalignant abnormalities which would lead to invasive disease if left untreated. Thus all 'positive' smears which are in excess of the expected prevalence of the disease in the absence of screening, are false-positive. As this criterion applies to the majority of 'positive' smears, the positive predictive value of cervical cytology approaches zero.

False-negative cytology, as reported in the literature, refers to three kinds of failure: (a) an abnormal smear is reported as normal; (b) a smear is inadequate or the lesion does not shed abnormal cells; (c) invasive carcinoma arises after the last normal smear ('interval' cancer). The false-negative rate for lesions

with uncertain relationship to invasive carcinoma ('dysplasia', 'CIS', cervical intra-epithelial neoplasia (CIN)) ranges between 20 and 40% [15–16].

Berkeley *et al.* [17] described 10 patients, aged 20–39, who had 'normal' Pap tests within 10 months of the diagnosis of advanced cervical cancer. On review only five of the ten smears were truly negative; two smears showed 'dysplasia' and three smears 'malignant cells'. The literature abounds with similar reports, often in young women [18–29] (See Table 5.1).

In populations extensively screened, the percentage of women with invasive cancer and previously negative smears may be even higher. In British Columbia, '94% of invasive cancers in initially screened women in Cohort 1 [born 1914–18] and 82% in Cohort 2 [born 1929–33] were discovered within 5 years of a previously negative smear' [30].

NATURAL HISTORY

At the beginning of the century, cervical cancer was responsible for about 20% of all cancer deaths in women [31], and was known as *morbus miseriae* because it particularly afflicted the poor. The need for early diagnosis was acknowledged, but so were the dangers of overdiagnosis and overtreatment. In 1902 an editorialist wrote 'formidable looking yet in

Table 5.1 Invasive carcinoma in screened women with normal smears

Author(s)	Women with normal smears (%)	Years within diagnosis
Robertson *et al.* [18]	66	12
Bearman *et al.* [19]	37	3
Elismon and Chamberlain [20]	74	5
Paterson *et al.* [21]	32	3
Bain and Crocker [22]	25	3
Hall and Monaghan [23]	40	6
Walker *et al.* [24]	60	5
Featherston [25]	50	2
Adcock *et al.* [26]	54	3
Berkowitz *et al.* [27]	55	2
Clarke [28]	30–50	5
Rylander [29]	45	5

truth benign conditions may have caused more than one uterus to be condemned and executed as cancerous. Naturally no recurrences followed' [32].

CIS under various names was studied as early as the 1920s [33]. Later studies attempted to identify the precursor of CIS. Based on cytological follow-up, Richart and Barron [34] argued that there existed a continuum between 'dysplasia' (smear class 3), CIS and invasive carcinoma, and they proposed to use instead the term 'cervical intra-epithelial neoplasia' (CIN) with four grades of severity. This view has become generally accepted as the model of the natural history of cervical cancer on which the philosophy of screening has been based ever since.

Fundamental questions about the natural history, however, have not been clarified. Is the continuum model true? If true, what is the time scale of progressive changes? What proportion of lesions revert to normal if left untreated? Does treatment of precursors influence mortality from invasive disease? What is the role of human papillomaviruses in the genesis and progression of the disease?

The modal prevalence of dysplasia, CIS and invasive carcinoma compatible with the continuum model was reported by the Canadian Task Force, who postulated a 10-year interval between dysplasia and CIS, and a 35-year interval between CIS and invasive carcinoma [35]. Other studies however, failed to confirm this conjecture [36–38]. As screening leads to the diagnosis of dysplasia and CIS far in excess of the expected incidence of invasive carcinoma, many abnormal lesions must regress if left untreated. The higher the regression rate the higher risk of overtreatment. In his review of the natural history of cervical cancer, Miller stated that 'it is not desirable to assume that all abnormalities are part of a continuum … I do not find it helpful to use the concept of cervical intra-epithelial neoplasia' [39]. Leopold Koss, who in 1969 was certain that 'dysplasia was a stepping stone to cancer' [40], in 1986 rejected the continuum model: 'there is really

no good evidence that one morphological type of precancerous event can change or progress into another'. He further stated that 'invasive carcinoma may occur without any further transformation from [CIN I to III] or any of these [abnormalities] may return to normal. Why and how this happens we do not know today' [41]. More recently, Kirby *et al.* observed that 'our study does not support a pattern for the natural history of the disease in which there is steady progression from mild to moderate to severe dyskaryosis' [42].

This departure from the continuum model is the vindication of painstaking histopathological studies by Burghardt, who documented that 'dysplasia cannot be precursors of carcinoma *in situ* as they are differently located', and he showed that 'early invasive tumours arise not infrequently from dysplastic epithelium, even from its mild forms' [43]. The problem for the screener is the present inability to distinguish between dysplasia which is truly pre-cancerous and dysplasias which are unspecific proliferations [44].

It is not known what proportion of CIS are true precursors of invasive carcinoma. In the largest study of the natural history of CIS, in which over 1000 women were followed for up to 20 years [45], in 86% of women, usual treatment resulted in cure, while in the remaining 14%, the lesions continued to reappear or persist, regardless of the type of treatment used. As the authors noted, 'whether or not the lesion is completely excised does not appear to influence the possibility of invasion occurring subsequently', and 'regular clinical and cytology follow-up of the apparently successfully treated patients did not prevent invasive carcinoma development. In the majority of these women, the carcinoma arose either *de novo* or within a few months of the first new cytology abnormality, indicating that invasion had not progressed through the expected lengthy premalignant phase'. This New Zealand study was bitterly disputed by a specially set-up Committee of Inquiry in Auckland, without invalidating its findings [46].

Since in young women invasive carcinoma is rare and CIS is common, while in elderly women there is a deficit of CIS for the expected incidence of the invasive disease, the relationship between CIS and invasive disease is unclear. This discrepancy led Ashley to predict correctly in 1966 that 'the detection of all cases of *in situ* carcinoma is unlikely to result in a major reduction in the death rate from this condition' [47]. Coppleson and Brown reached a similar conclusion by noting that 'the fall in prevalence of dysplasia and carcinoma *in situ* before the major onset of invasive cancer must be fully appreciated and explained. The regression of carcinoma *in situ* and dysplasia to normal in large numbers must be accepted' [37]. More recently, Bergström *et al.* found no correlation between the detection rate (and treatment) of CIS in 24 Swedish counties and the subsequent reduction in invasive carcinoma 5 to 15 years later [48]. They calculated that the detection of an extra 100 cases of CIS per 100,000 less women would result in only one less case of invasive carcinoma 10 years later. This analysis was confined to younger women (aged 20–50 for CIS; 30–60 for invasive cancer) as older women were unlikely to have been screened.

POPULATION AT RISK

The US guidelines drawn up by the American Cancer Society, in association with the National Cancer Institute, the American Medical Association and the American College of Obstetricians and Gynecologists, recommend that 'all women who are, or have been, sexually active, or have reached the age of 18, have an annual Pap test and pelvic examination' [49]. Van Ballegooijen *et al.* suggested that annual screening between the age of 20 and 35 years and every 5 years between 35 and 60 years (as recommended by Canadian Task Force in 1982) would result in an unfavourable harm–benefit ratio as compared with more efficient policies; they calculated that 11,800 smears would have to be taken for one death prevented [50].

The British guidelines require local health authorities to 'invite all women aged 20 to 64 for screening within five years, with recall at least every five years' [51]. National screening programmes in other countries use different screening intervals and invite women in different age categories, e.g. the Netherlands, 35–54, every 3 years; Sweden, 30–49, every 4 years; Finland, 30–55, every 5 years; or Iceland, 29–69, every 2–3 years. The consensus of 25 experts was that women aged 30–60 should be screened at least every 3 years [52].

Yet the World Health Organization guidelines for cytological screening estimate the typical sequence from CIN I to invasion as lasting minimally 23 years and maximally 38 years [53]. 'If the interval from onset to incurability is 7 years, theoretically, smears every 7 years will suffice to prevent 100% of the deaths and no shorter intervals can improve on this' [54].

Van Wijngaarden and Duncan came up with a paradoxical proposal that screening should stop at the age of 50, since most CIN lesions are discovered before that age [55]; even though 20 out of 28 (71%) of their cases of invasive carcinoma occurred in women older than 50. Fletcher, on the other hand, argued that as 40% of all deaths from cervical cancer occur in women aged over 65, they should not be excluded from screening programmes [56]. Even though only about 5% of the 1500 annual deaths from cervical cancer occur in the UK. in women under the age of 35, some advocates of screening would like to start screening in early adolescence: 'although we are unlikely to find premalignant disease in the sexually active girls under 16, we think that cytological screening is useful to educate them about the need for regular cervical smears' [57].

WHAT IS THE EARLIEST PREMALIGNANT STAGE OF THE DISEASE?

Combining cytology and colposcopy, Giles *et al.* found that among 200 women from a predominantly middle-class general practice offered screening, there were 18 cases of CIN I or CIN II and four cases of CIN III [16]. The authors pointed out that they were 'unable, cytologically, colposcopically, or histologically to predict which lesions will progress to invasive disease'. Thirteen out of the 22 cases had had a normal smear within the past 2 years, and cytology alone would miss about half of the lesions. Griffiths, commenting on these findings, confirmed that about one-third of normal healthy women under the age of 35 had an abnormality (virus infection or CIN I) on colposcopy [58].

The screener is faced with a Hobson's choice: either to rely on cytology and follow-up minor abnormalities cytologically, with a risk of missing important abnormalities, or to refer all minor cytological abnormalities for colposcopy, which overwhelms the screening system and increases overtreatment. The problem with the Pap smear is that it is not a suitable screening test. Bigrigg *et al.* found that 24% of smears showing only mild dyskaryosis came from cervices with the histological diagnosis of CIN or invasion [59]. Philip found among 299 patients with mild dyskaryosis 54 cases (18%) with CIN on colposcopy [60]. Wilson *et al.* reported that 12 out of 96 patients (13%) with 'inflammatory' smears showed CIN abnormalities on colposcopy, including six CIN III and one adenocarcinoma [61].

Repeat cytology is not an answer to misleading Pap smears, as even mildly atypical smears, normal when repeated, may be obtained in the presence of CIN II or CIN III [62]. In an Australian programme, 123/256 (48%) of class II smears (without evidence of dysplasia) were subsequently rediagnosed as dysplasia (62 cases), CIS (53 cases) and eight cases of invasive carcinoma [63]. Jones *et al.* found that 58/236 (25%) class II smears ('atypical but not dysplastic') originated from CIN lesions (48 cases of CIN I, seven cases of CIN II, three cases of CIN III) [64]. Borderline smear cytology may be present in as many as 5% of all smears [65]. Hirschowitz *et al.* followed up 437 cases with such smears for up to 9 years: 98 (= 22%) developed a high-grade dyskaryosis,

including 69 cases of CIN II or III, and seven microinvasive carcinomas. An additional three invasive carcinomas occurred among 339 patients who showed no progression to high-grade dyskaryosis [65]. Interestingly; the progression was more likely to occur in younger women, particularly in those without HPV infection. Jenkins and Toy reported that 25% of women with borderline dyskaryosis had CIN III lesions [66]. An editorialist in the *British Medical Journal* warned that 'the presence of only mildly dyskaryotic cells in a smear should be treated with the utmost seriousness' [67].

The implications of this line of reasoning on referral policies for colposcopy, with gross overtreatment of benign lesions in young women, are staggering [68, 69]. A 1% change in the rate of referral for colposcopy will represent about 40,000 women annually [70]. Smith *et al.* estimated that between 33 and 500 women would have to be referred for an expensive and possibly hazardous procedure for everyone who is at risk of developing serious disease [70]. Coleman, criticizing an editorial in *The Lancet*, which recommended that all cases with mild dyskaryosis should be colposcoped, calculated that since only 4000 new cases of invasive cancer occur each year, but about 200,000 women are found to have low-grade dyskaryosis, 196,000 women will be referred for colposcopy and biopsy unnecessarily [71]. Woodman and Jordan concluded in their review of colposcopic services that there were no rational criteria for referral, treatment, or follow-up, resulting in dissatisfaction and chaos [72].

COST-EFFECTIVENESS ANALYSIS

The most detailed study of the cost-effectiveness of cervical screening was carried out by Eddy [73]. The usefulness of Eddy's analysis is limited by his unrealistic assumption, in the absence of randomized controlled trials, that the effectiveness of screening is 80–90%. In Eddy's model, cost per year of life expectancy gained, for screening every 4 years, was

$10,000. The marginal cost per year of life expectancy for screening every 3 years after four annual negative smears was $681,000. As cervical cancer is a rare cause of death, the increase in life expectancy for a woman screened every 1–3 years between the ages of 20 and 75 years was approximately 10 days.

A more realistic analysis was carried out by Charny *et al.* [74]. They assumed that the effectiveness of screening in Britain was 5%, i.e. about 100 deaths prevented annually by screening, and estimated the cost per life saved at £270,000–£285,000. The cost was based on an average cost of the smear test, without costing the follow-up of abnormal smears, iatrogenic morbidity and the consequences of unnecessary anxiety and fear.

An editorialist in *The Lancet*, however, commented that 'since cytological screening was introduced on a large scale around 1964, mortality has declined at 1% a year; but that seems to be the rate at which it had been falling for several decades previously. There was no obvious change' [75]. Conceding at most 2.5% effectiveness, the editorialist calculated that for every life saved by screening, 40,000 smears and 200 excision biopsies were carried out – 'a grievously poor cost–benefit ratio'. Even when the effectiveness is doubled, as proposed by Charny *et al.*, only one death will be prevented annually per 160,000 women aged 25 and over.

SCREENING PROGRAMMES

The failure of mass cervical screening in various countries has been attributed to lack of organization. However, even in countries with organized programmes evidence is lacking that the introduction of screening accelerated the decline in mortality which was recorded before the onset of screening. The popular explanation for this failure to accelerate the decline is the 'epidemic' hypothesis (see Figure 5.1). The hypothesis is falsified by the absence of an epidemic of cervical cancer in unscreened women.

In the USA, the same slope of logarithmic decline is observed in all ages, regardless of screening activity, and preceding the onset of widespread screening (see Figure 5.2).

In eastern European countries, cervical screening was organized centrally with frequent screening intervals and obligatory attendance. No evidence has emerged from these countries that screening was effective [76].

In Auckland, New Zealand, a mass screening programme was initiated in the early 1960s. Despite intense screening of young women, mortality in this group increased slightly, while in elderly women, who were unlikely to be screened, mortality declined [77–78].

In this chapter the discussion of mass screening programmes will be limited to British Columbia, the 'best' (Iceland) and the 'worst' (Norway) cases in Scandinavia, and to the UK experience.

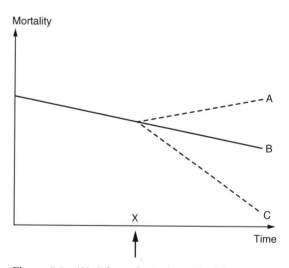

Figure 5.1 (A) A hypothetical trend without mass screening in the presence of an epidemic. (B) The actual trend (with mass screening and with an alleged epidemic). (C) A hypothetical trend (with mass screening in the absence of of an epidemic). (X) The onset of screening and of an alleged epidemic.

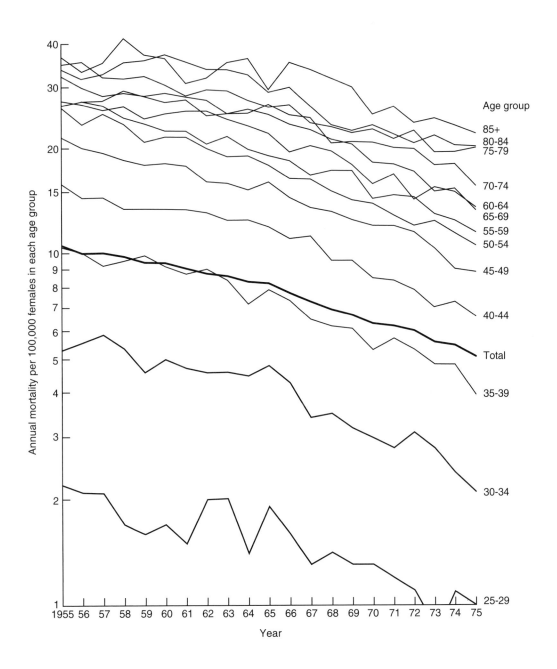

Figure 5.2 Mortality from cancer of uterine cervix in American women aged 25 and over, 1955–1975. (From *The Lancet*, 1989, **ii**, 629.)

CANADA

Mass screening started in British Columbia around 1960 and the initial reports were so enthusiastic that the case for cervical screening was generally accepted as proved. Yet objective evidence for the benefit of screening was lacking. No correlation was found between screening intensity in British Columbia and other Canadian provinces, and mortality from the disease [79]. Mortality was declining in all provinces at the same rate both before and after screening was introduced [80].

The latest update on the British Columbia programme has failed to provide new evidence that screening accelerated the decline in mortality observed before the onset of screening. Between 1955 and 1970, incidence was declining linearly (1.3 cases/100,000/year), followed by a new, slower linear trend between 1970 and 1985 (0.6 cases/100,000/ year). Mortality was also declining linearly between 1958 and 1985 at a rate of 0.4 cases per 100,000 women per year [81]. 'Despite the success of the programme in British Columbia an average of 65 new cases of invasive carcinoma of the cervix are found each year' [81]. In 1981, Miller acknowledged that there was a more limited potential for the application of screening ... than had been anticipated previously' [79]. In 1991, Miller wrote 'previous and continuous analyses have confirmed that the screening programmes for cancer of the cervix have not been as successful in Canada as in other countries with organized programmes' [82]. For example, in Prince Edward Island province, between 1981 and 1986, 35% (13/37) of women with invasive carcinoma had at least one cervical smear done within 3 years of the diagnosis [83].

NORWAY

It is claimed that Norway lagged behind other Scandinavian countries in reducing mortality from cervical cancer because they had a pro-

gramme operating in only one county [84]. What is not made clear is that this county (Østfold) was the first area in Scandinavia in which an attempt was made to evaluate the efficacy of cervical screening. The programme started in 1959 and all women aged 25–59 were invited for screening at 2–3 years intervals. After two rounds of screening, 23 cases of invasive cancer were found, of which 18 (= 78%) had had a 'normal' smear within 2 years of diagnosis and five were not screened; on review six of the 18 cases had 'definitely negative' smears [85]. The organizers found no effect of screening on the annual incidence of invasive cancer, despite the fact that the detection of CIS increased eight-fold and they asked 'why the elimination of hundreds of cases of dysplasia and CIS has failed so far to bring about a drastic reduction of invasive cases in the study population'. After four rounds, there was still to evidence for a reduction in mortality when compared to the trend in the whole country [86]. As in other countries, in Norway, 'the mortality was steadily decreasing since the early 1950s' (Per Kolstad, personal communication, 28 January 1985).

ICELAND

'In Iceland, where the nationwide programme has the widest target age range, the fall in mortality was greatest (80%)' [87]. This claim was disputed [87–88]. There was no change in the incidence of cervical cancer between 1955 and 1959 (68 cases) and 1980and 1984 (77 cases) [89]. Between 1975 and 1984 the mortality no longer fell, and was the same as it was in 1955–1959, that is, before the onset of screening [89]. Mortality rates provided by the Icelandic Cancer Registry for 1951–1955 and l976–1980 were 3.8 per million and 3.6 per million, respectively [90]. In the latest update on screening programmes in Scandinavian countries, mortality data in Iceland are omitted [91]. Also missing are data on incidence in the age groups 25–39 and 45–59. The incidence on invasive cancer in the age group 20–24 increased, in the age group 40–44 it remained static, and in the age group 60–64 it decreased [91], exactly the same pattern as observed in New Zealand [77, 78].

BRITAIN

According to a press release from the Imperial Cancer Research Fund in December 1992, there was a 15% decrease in deaths from cervical cancer between 1985 and 1991, which was the 'beginning of a downward trend' and evidence that 'screening does work'. As can be seen from Figure 5.3, this downward trend did not start with the onset of screening and no attempt has been made to disentangle the possible contribution of screening from spontaneous decline. The clinical director of BUPA Health Management stated that the UK cervical cancer programme was brought into disrepute as 'it did not seem to save lives in its first 20 years of operation', and he surmised that it

'may take another 20 years to show benefit' [92]. Murray concluded that in the West Midlands 'the screening programme has been a relative failure in terms of reducing mortality from cervical cancer' [93]. In the area administered by Bristol and District Health Authority, 60,000 women are tested annually, with an uptake of 81%. Each year about 3000 women are found to have abnormal smears for the first time and more than half are referred for colposcopy. Yet, there is no evidence in the region that between 1975 and 1992 mortality from cervical cancer decreased [94]. Comparing cervical cancer incidence and mortality with screening intensity in 16 health regions in Britain, Murphy *et al.* concluded that screening was unsuccessful [95]. There was no correlation between screening effort and the rate of detection of CIS, and no correlation between crude smear rates and outcome in young women in whom the smear rates were the highest.

The 1992 Report of the National Audit Office noted that there was a wide discrepancy in the percentage of abnormal smears reported by regional health authorities (see Figure 5.4). In the Mersey region, for example, at district level, abnormal smears ranged from zero to 18%: 'three [of the five] districts classified a wide range of smears as abnormal when they were not' [51]. The report notes that there is no nationally agreed protocol for investigating and treating various abnormalities. Waiting time for recalls after abnormal smears ranged from 2 to 14 weeks. The cost of running the programme was not known and the efficiency and effectiveness of the programme could not be monitored. It is questionable whether after 30 years of running a national programme, with some 60 million smears taken, and without clear evidence of benefit, there is much to be hoped for in the future.

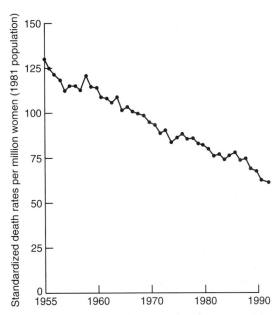

Figure 5.3 Mortality from cancer of uterine cervix in England and Wales, 1950–1992. The graph was kindly provided by Dr Peter Sasieni, Imperial Cancer Research Fund, London.

ADVERSE EFFECTS OF SCREENING

With 4–5 million smears taken annually in Britain, the potential for iatrogetic harm is

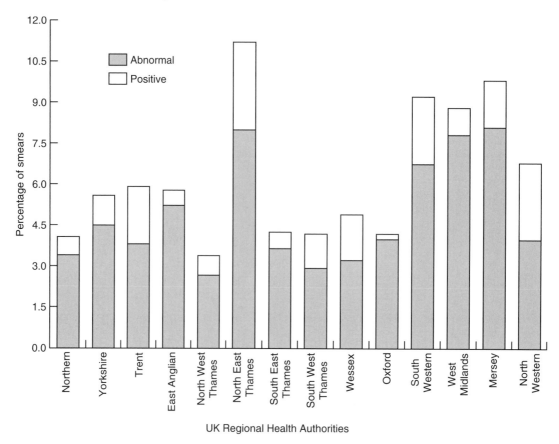

Figure 5.4   Percentages of smear test results reported as abnormal or positive by region, 1990–1991. (Source: Department of Health.)

enormous. Dr Elaine Farmery, previous director of the National Cervical Cancer Screening Network, was quoted in *The Guardian* (9 November 1993) as saying: 'to get it right 100% of the time is impossible, to get it right 95% of the time is acceptable'. Five per cent of 5 million smears, however, represents 250,000 wrongly reported smears annually! Blunders occur due to a faulty technique of smear-taking, due to false-negative and false-positive results, due to failure to follow-up abnormal smears, and due to overtreatment of benign lesions.

In Birmingham, incorrect smear technique led to a recall of over 1000 women (*The Times*, 11 September 1993). 'Panic over cervical cancer reached new heights last week with the revelation that 911 women in Liverpool had been

wrongly diagnosed', was a headline in *The Sunday Times* (27 September 1987). In Inverclyde Royal Hospital, Scotland, 18,000 smear tests had to be re-examined, because the laboratory gave false-negative results in 70% of cases (*The Times*, 10 September 1993). Such scandals seriously undermine the confidence of women in the national screening programme.

Serious psychological and psychosexual adverse effects of being told that one's test is 'positive' have been documented [96–98]. Cone biopsy may be followed by infection, or primary or secondary haemorrhage, at times severe enough to require transfusion, while other patients may develop symptomatic cervical stenosis and infertility [99].

Medico-legal matters arising from screening can be also considered among the adverse effects of screening. The husband of a patient who died of cervical cancer after two positive smears which were not followed up was awarded £87,000 damages against three doctors (*The Times*, 14 December 1991). Gynaecologists who will 'follow up' rather than immediately refer for colposcopy and treatment of patients with mild to moderate dyskaryosis could conceivably be sued for relying on unreliable cytological diagnosis. As one gynaecologist observed, 'the point has been reached where it is necessary to consider the medico-legal threat from virtually all patients … there is definitely no place for doing more clinical work than is reasonable' [100].

ETHICAL ASPECTS

'The advocates of screening, usually for impeccable motives, conclude that the pre-existing evidence plus commonsense – in the face of the ongoing toll of disability and untimely death – demand mass screening programme … In "keeping the faith", screening advocates may find themselves forced to accept or reject evidence not so much on the basis of its scientific merits as on the extent to which it supports or rejects the stand that screening is good' [101]. The exaggeration of benefits and the suppression of harms when inviting women to participate in a screening programme is unethical. As Cochrane and Holland warned, there is an ethical difference between offering screening to a healthy population, without strong evidence that the benefits outweigh the harms, and the ordinary medical practice of responding to patients' requests for help. In the latter case, the doctor is not responsible for defects in medical knowledge and the treatment he or she offers reflects the current practice, often without proved benefit.

When healthy people are asked to attend screening, however, it is imperative that a quantitative assessment of expected benefits and possible harms is available, so that the person can make a rational choice between accepting or rejecting the offer of screening [103–105]. Such information is currently not available for cervical screening. 'When we know more about the causes and natural history of cervical cancer, we may have a test which validly identifies those at special risk. This time has not yet come' [105].

REFERENCES

1. Hull, R. (1851) On the speculum vaginae. *Dublin Medical Press*, **26**, 216–17.
2. World Health Organization Study Group (1976) *Randomised Trials in Preventive Medicine and Health Service Research*, World Health Organization Regional Office for Europe, Copenhagen, p. 15.
3. Anonymous (1973) Uncertainties of cervical cytology (editorial). *British Medical Journal*, **4**, 501–2.
4. Wilson, J.M.G. and Jungner, G. (1968) *Principles and Practice of Screening for Disease*, World Health Organization, Geneva.
5. Skrabanek, P. (1990) Screening for disease – possibilities and problems. *West of England Medical Journal*, **106**, 37–45.
6. Koss, L.G. (1989) The Papanicolaou test for cervical cancer detection. A triumph and a tragedy. *Journal of the American Medical Association*, **261**, 737–43
7. Buntinx, F., Knottnerus, J.A., Crebolder, H.G.J.M. *et al.* (1992) Relation between quality of cervical smears and probability of abnormal results. *British Medical Journal*, **304**, 1224.
8. Seybolt, J.F. and Johnson, W.D. (1971) Cervical cytodiagnostic problems: a survey. *American Journal of Obstetrics and Gynecology*, **109**, 1089–103.
9. Bellina, J.H., Dunlop, W.P. and Riopelle, M.A. (1982) Reliability of histopathological diagnosis of cervical intraepithelial neoplasia. *Southern Medical Journal* , **75**, 6–8.
10. Langley, F.A. (1984) Consistency in the diagnosis of squamous carcinoma of the cervix and its precursors. In *Cancer of the Uterine Cervix: Biochemical and Clinical Aspects*, (eds D.C.H. McBrien and T.F. Slater). Academic Press, London, pp. 115–38.
11. Yobs, A.R., Plott, A.E. Hicklin, M.D. *et al.* (1987) Retrospective evaluation of gynecological cytodiagnosis. II. Interlaborotory reproducibility as shown in rescreening large consecutive samples of reported cases. *Acta Cytologico*, **31**, 900–10.

12. Ismail, S.M., Colclough, A.B., Dinnen, J.S., *et al.* (1989) Observer variation in histopathological diagnosis and grading of cervical intra-epithelial neoplasia. *British Medical Journal,* **298,** 707–10.

13. Jenkins, D. and Jarmulowicz, M.R. (1989) Grading of cervical intraepithelial neoplasia. *British Medical Journal,* **298,** 1030–1.

14. Robertson, A.J., Anderson, J.M., Beck, J.S., *et al.* (1989) Observer variability in histopathological reporting of cervical biopsy specimens. *Journal of Clinical Pathology,* **42,** 231–8.

15. Coppleson, L.W. and Brown, B. (1974) Estimation of the screening error rate from the observed detection rates in repeated cervical cytology. *American Journal of Obstetrics and Gynaecology,* **119,** 953–8.

16. Giles, J.A., Hudson, E., Crow, J., *et al.* (1988) Colposcopic assessment of the accuracy of cervical cytology screening. *British Medical Journal,* **296,** 1099–102.

17. Berkeley, A.S., LiVolsi, V.A. and Schwartz, P.E. (1980) Advanced squamous cell carcinoma of the cervix with recent normal Papanicolaou tests. *Lancet,* **ii,** 375–6.

18. Robertson, J.H. and Woodend, B. (1993) Negative cytology preceding cervical cancer: causes and prevention. *Journal of Clinical Pathology,* **46,** 700–2.

19. Bearman, D.M., MacMillan, J.P. and Creasman, W.T. (1987) Papanicolaou smear history of patients developing cervical cancer: on assessment of screening protocols. *Obstetrics and Gynecology,* **69,** 151–155.

20. Elisman, R. and Chamberlain, J. (1984) Improving the effectiveness of cervical cancer screening. *Journal of the Royal College of General Practitioners,* **34,** 537–41.

21. Paterson, M.E.L., Peel, K.R. and Joslin, C.A.F. (1984) Cervical smear histories of 500 women with invasive cervical cancer in Yorkshire. *British Medical Journal,* **289,** 896–8.

22. Bain, R.W. and Crocker, D.W. (1983) Rapid onset of cervical cancer it on upper socioeconomic group. *American Journal of Obstetrics and Gynecology,* **146,** 366–71.

23. Hall, S.W. and Monaghan, J.M. (1983) Invasive carcinoma of the cervix in young women. *Lancet,* **ii,** 731.

24. Walker, E.M., Hare, M.J. and Cooper, P. (1983) A retrospective review of cervical cytology in women developing invasive squamous cell carcinoma. *British Journal of Obstetrics and Gynaecology,* **90,** 1087–91.

25. Fetherston, W.C. (1983) False-negative cytology in invasive cancer of the cervix. *Clinical Obstetrics and Gynecology,* **26,** 929–37.

26. Adcock, L.L., Juliat, T.M., Okogaki, T., *et al.* (1982) Carcinoma of the uterine cervic FIGO stage I-B. *Gynecological Oncology,* **14,** 199–208.

27. Berkowitz, R.S., Ehrmann, R.L., Lavizzo-Mourey, R., *et al.,* (1979) Invasive cervical carcinoma in young women. *Gynecological Oncology,* **8,** 311–16.

28. Clarke, E.A. and Anderson, T.W. (1979) Does screening by 'Pap' smears help prevent cervical cancer? *Lancet,* **ii,** 1–4.

29. Rylander, E. (1976) Cervical cancer in women belonging to a cytologically screened population. *Acta Obstetrica Gynaecologica Scandinavica,* **55,** 361–6.

30. Boyes, D.A., Morrison, D., Knox, E.G. *et al.* (1982) A cohort study of cervical cancer screening in British Columbia. *Clinical and Investigative Medicine,* **5,** 1–29.

31. Bonney, V. (1909) The bearing of pathology on the prevention, diagnosis and surgical care of carcinoma of the cervix. *Practitioner,* **i,** 737–8.

32. Anonymous. (1902) The address in obstetrics (editorial). *British Medical Journal,* **2,** 321–7.

33. Younge, P.A., Hertig, A.T. and Armstrong, D. (1949) A study of 135 cases of carcinoma *in situ* of the cervix at the Free Hospital for Women. *American Journal of Obstetrics and Gynecology,* **58,** 867–95.

34. Richart, R.M. and Barron, B.A. (1969) A follow-up study of patients with cervical dysplasia. *American Journal of Obstetrics and Gynecology,* **105,** 386–93.

35. Canadian Task Force [Walton's Report] (1976) Cervical cancer screening programs. I. Epidemiology and natural history of carcinoma of the cervix. *Canadian Medical Association Journal,* **114,** 1003–12.

36. Jordan, S.W., Munsick, R.A. and Stone, R.S. (1969) Carcinoma of the cervix in American Indian women. *Cancer,* **23,** 1227–32.

37. Coppleson, L.W. and Brown, B. (1975) Observations on a model of the biology of carcinoma of the cervix: a poor fit between observation and theory. *American Journal of Obstetrics and Gynecology,* **122,** 127–36.

38. Ishiguro, T., Yoshido, Y, Tenzaki, T., *et al.* (1983) Evaluation of mass screening for cervical cancer in Shiga prefecture. *Acta Obstetrica Gynaecologica Japonica,* **35,** 655–60.

39. Miller, A.B. (1985) Screening for cancer of the cervix. In *Screening for Cancer* (ed. A.B. Miller). Academic Press, New York, pp. 296–302.

40. Koss, L.G. (1969) Concept of genesis and development of carcinoma of the cervix. *Obstetrical and Gynecological Survey,* **24,** 850–60.

41. Koss, L.G. (1986) Chairman's concluding remarks: sequence of events in carcinogenesis of the uterine cervix. In *Viral Etiology of Cancer* (eds R. Peto and H. zur Hausen). Cold Spring Harbour, New York, pp. 179–184.

42. Kirby, A.J., Spiegelhalter, D.J., Day, N.E., *et al.* (1992) Conservative treatment of mild/moderate cervical dyskoryosis: long-term outcome. *Lancet*, **339**, 828–31.

43. Burghardt, E. (1985) Natural history of CIN. In *Cancer of the Uterine Cervix* (eds H.G. Bender and L. Beck). Gustav Fischer Verlag, Stuttgart, pp. 91–6.

44. Burghardt, E. and Östör, A.G. (1983) Site and origin of squamous cervical cancer: a histomorphological study. *Obstetrics and Gynecology*, **62**, 117–27.

45. McIndoe, W.A., McLean, M.R., Jones, R.W., *et al.* (1984) The invasive potential of carcinoma *in situ* of the cervix. *Obstetrics and Gynecology*, **64**, 451–8.

46. Skrabanek, P. (1991) Cervical cancer study. *New Zealand Medical Journal*, **104**, 77–8.

47. Ashley, D.J.B. (1966) The biological status of carcinoma *in-situ* on the uterine cervix. *Journal of Obstetrics and Gynaecology of the British Commonwealth*, **73**, 372–81.

48. Bergström, R., Adosmi, H.O., Gustafsson, L., *et al.* (1993) Detection of preinvasive cancer of the cervix and the subsequent reduction in invasive cancer. *Journal of the National Cancer Institute*, **85**, 1050–7.

49. Anonymous. (1988) USA. Pap test guidelines. *International Cancer News*, No. 11 (July), 18.

50. Van Ballegooijen, M., Koopsmanschap, M.A., vat Oortmarssen, G.J. *et al.* (1990) Diagnostic and treatment procedures induced by cervical cancer screening. *European Journal of Cancer*, **26**, 941–5.

51. National Audit Office. Report of the Controller and Auditor General (1992) *Cervical and Breast Cancer Screening in England*. HMSO, London.

52. IARC Working Group on Evaluation of Cervical Cancer Screening Programmes (1986) Screening for squamous cervical cancer: duration of low risk after negative results of cervical cytology and its implication for screening policies. *British Medical Journal*, **293**, 659–64.

53. World Health Organization (1988) *Cytological Screening in the Control of Cervical Cancer: Technical Guidelines*, World Health Organization, Geneva.

54. Spriggs, A.I. and Husain, O.A.N. (1977) Cervical smears. *British Medical Journal*, **1**, 1516–18.

55. Van Wijngaarden, W.J. and Duncan, I.D. (1993) Rationale for stopping cervical screening in women over 50. *British Medical Journal*, **306**, 967–71.

56. Fletcher, A. (1990) Screening for cancer of the cervix in elderly women. *Lancet*, **335**, 97–9.

57. Robinson, A.J., Tullett, J. and Wilson, J.D. (1988). Cervical intra-epithelial neoplasia. *British Medical Journal*, **297**, 359–60.

58. Griffiths, M. (1990) Implications of inflammatory changes on cervical cytology. *British Medical Journal*, **300**, 1076–7.

59. Bigrigg, M.A., Codling, B.W., Pearson, P., *et al.* (1990) Colposcopic diagnosis and treatment of cervical dysplasia at a single clinic visit. *Lancet*, **336**, 229–31.

60. Philip, G. (1990) Follow-up of women with dyskaryotic cervical smears. *British Medical Journal*, **301**, 983.

61. Wilson, J.D., Robinson, A.J., Kinghorn, S.A., *et al.* (1990) Implications of inflammatory changes on cervical cytology. *British Medical Journal*, **300**, 638–40.

62. Walker, E.M., Dodgson, J. and Duncan, I.D. (1986) Does mild atypia on a cervical smear warrant further investigation? *Lancet*, **ii**, 672–3.

63. MacCormac, L., Lew, W., King, G., *et al.* (1988) Gynaecological cytology screening in South Australia: a 23-year experience. *Medical Journal of Australia*, **149**, 530–6.

64. Jones, D.E.D., Creasman, W.T., Dombroski, R.A., *et al.* (1987) Evaluation of the atypical Pap smears. *American Journal of Obstetrics and Gynecology*, **157**, 544–9.

65. Hirschowitz, L., Raffle, A.E., Mackenzie, E.F.D., *et al.* (1992) Long-term follow-up of women with borderline cervical smear test results: effect of age and viral infection on progression to high grade dyskaryosis. *British Medical Journal*, **304**, 1209–12.

66. Jenkins, D. and Tay, S.K. (1987) Management of mildly abnormal cervical smears. *Lancet*, **i**, 748–9.

67. Fox, H. (1987) Cervical smears: new terminology and new demands. *British Medical Journal*, **294**, 1307–8.

68. Kitchener, H.C., Burnett, R.A., Wilson, E.S.B., *et al.* (1987) Colposcopy in a family planning clinic: a future model? *British Medical Journal*, **294**, 1313–15.

69. Chomet, J. (1987) Screening for cervical cancer: a new scope for general practitioner? Results of the first year of colposcopy in general practice. *British Medical Journal*, **294**, 1326–8.

70. Smith, A., Elkind, A. and Eardley, A. (1989) Making cervical screening work. *British Medical Journal*, **298**, 1992–4.

71. Coleman, D.V. (1987) Colposcopy. *Lancet*, **i**, 749.

72. Woodman, C.B.J. and Jordan, J.A. (1989) Colposcopy service in the West Midlands region. *British Medical Journal*, **299**, 899–901.

73. Eddy, D.M. (1990) Screening for cervical cancer. *Annals of Internal Medicine*, **113**, 214–26.

74. Charny, M.C., Farrow, S.C. and Roberts, C.J. (1987) The cost of saving a life through cervical cytology screening: implications for health policy. *Health Policy*, **7**, 345–59.

75. Anonymous (1985) Cancer of the cervix: death by incompetence (editorial). *Lancet*, **ii**, 363–4.

76. Napalkov, N.P. and Eckhardt, S. (1982) *Cancer Control in the Countries of the Council of Mutual Economic Assistance*, Akadesmiai Kiado, Budapest.

77. Green, G.H. (1978) Cervical cancer and cytology screening in New Zealand. *British Journal of Obstetrics and Gynaecology*, **85**, 881–6.

78. Green, G.H. (1988) Cervical cancer in New Zealand: a failure of cytology? *Asia-Oceania Journal of Obstetrics and Gynaecology*, **7**, 303–13.

79. Miller, A.B., Visentin, T. and Howe, G.R. (1981). The effects of hysterectomies and screening on mortality from cancer of the uterus in Canada. *International Journal of Cancer*, **27**, 651–7.

80. Miller, A.B. (1973) Evaluation of screening for carcinoma of the cervix. *Modern Medicine of Canada*, **28**, 1067–9.

81. Andersson, G.H., Boyes, D.A., Benedet, J.L., *et al.* (1988) Organisation and results of the cervical cytology programme in British Columbia, 1955–1985. *British Medical Journal*, **296**, 975–8.

82. Miller, A.B., Knight, J. and Narod, S. (1991) The natural history of cancer of the cervix and the implication for screening policy. In *Cancer Screening* (eds A.B. Miller, J. Chamberlain, N.E. Day, *et al.*). Cambridge University Press, Cambridge, pp. 141–52.

83. Sweet, L., Tesch, M., Dryer, D., *et al.* (1991) A review of cervical cytology screening history of PEI women diagnosed with carcinoma of the cervix, 1981–1986. *Chronic Diseases in Canada*, **12**(1), 1–3.

84. Läärä, E., Day, N.E. and Hakama, M. (1987) Trends in mortality from cervical cancer in Nordic countries: association with organised screening programmes. *Lancet*, **i**, 1247–9.

85. Pedersen, E., Høeg, K. and Kolstad, P. (1971) Mass screening for cancer in Østfold county, Norway: An experiment. Second report of the Norwegian Cancer Society. *Acta Obstetrica Gynaecologica Scandinavica*, Suppl. **11**, 5–18.

86. Hougen A. (1980) Mass screening for cancer of the uterine cervix in the county of Østfold, Norway. In *Prevention and Detection of Cancer* (ed H.E. Nieburgs). M. Dekker, New York, pp. 1875–84.

87. Skrabanek, P. (1987) Cervical cancer screening. *Lancet*, **ii**, 510.

88. Skrabanek, P. (1987) Cervical cancer screening. *Lancet*, **ii**, 1432–3.

89. Sigurdsson, K., Geirsson, G., Tulinius, H., *et al.* (1985) Mass screening for cervical cancer in Iceland, 1964–1984. Presented at *XIth World Congress of Gynaecology and Obstetrics*, 15–20 September, West Berlin.

90. Tulinius, H., Geirsson, G., Sigurdsson, K., *et al.* (1984) Screening for cervix cancer in Iceland. In *Cancer of the Uterine Cervix* (eds D.C.H. McBrien and T.F. Slater), Academic Press, London, pp. 55–76.

91. Hakama, M., Magnus, K., Pettersson, F., *et al.* (1991) Effect of organised screening on the risk of cervical cancer in Nordic countries. In *Cancer Screening* (eds A.B. Miller, J. Chamberlain, N.E. Day, *et al.*). Cambridge University Press, Cambridge, pp. 153–62.

92. Warren, D.J. (1992) Under scrutiny. *British Medical Journal*, **305**, 126.

93. Murray, S.J. (1990) Evaluation of the cervical cancer screening in the West Midlands region. *Public Health Physician*, **1**(4), 17.

94. Morgan, K. (1993) *The Health of the Population*. Report of the Director of Public Health. Bristol and District Health Authority, Bristol, pp. 18–20.

95. Murphy, M.F.G., Campbell, M.J. and Goldblatt, P.O. (1987) Twenty years' screening for cancer of the uterine cervix in Great Britain, 1964–1984: further evidence for its ineffectiveness. *Journal of Epidemiology and Community Health*, **42**, 49–53.

96. Posner, T. and Vessey, M. (1988) *Prevention of Cervical Cancer: The Patient's View*, King's Fund Publishing Office, London.

97. Campion, M.J., Brown, J.R., McCance, D.J., *et al.* (1988) Psychosexual trauma of an abnormal cervical smear. *British Journal of Obstetrics and Gynaecology*, **95**, 175–81.

98. Lerman, C., Miller, S.M. and Scarborough, R. (1991) Adverse psychological consequences of positive cytological cervical screening. *American Journal of Obstetrics and Gynaecology*, **165**, 658–62.

99. Luesley, D.M., McCrum, A., Terry, P.B., *et al.* (1985) Complications of cone biopsy related to the dimensions of the cone and the influence of prior colposcopic assessment. *British Journal of Obstetrics and Gynaecology*, **92**, 158–64.
100. Hamlett, J.D. (1988) Personal view. *British Medical Journal*, **296**, 1256.
101. Sackett, D. and Holland, W.W. (1975) Controversy in the detection of disease. *Lancet*, **ii**, 357–8.
102. Cochrane, A.L. and Holland, W.W. (1971) Validation of screening procedures. *British Medical Bulletin*, **27**, 3–8.
103. Skrabanek, P (1988) Cervical cancer screening: the time of reappraisal. *Canadian Journal of Public Health*, **79**, 86–9.
104. Skrabanek, P. (1988) The physician's responsibility to the patient. *Lancet*, **i**, 1155–7.
105. McCormick, J.S. (1989) Cervical smears: a questionable practice. *Lancet*, **ii**, 207–9.

THE ROLE OF HUMAN PAPILLOMAVIRUS TESTING

Jack Cuzick

INTRODUCTION

The premise that cervical cancer is caused by a sexually transmitted agent is well documented and evidence for it can be traced back as far as the work of Rigoni-Stern in the first half of the nineteenth century. The responsible agent has been more difficult to track down and at different times claims have been made for syphilis [1], herpes simplex virus [2–4], smegma [5, 6] and semen [7] as being the responsible factors. In 1983 Durst and colleagues isolated human papillomavirus (HPV) type 16 in an invasive cervical cancer and proceeded to show that it was commonly found in cancers but not in the normal cervix [8]. To date, over 70 HPV types have been discovered and several types have been shown to be linked with cervical cancer. Lorincz *et al.* (1992) [9] have examined the malignant potential of several genital types (Table 6.1 and Figure 6.1) and have shown that types 16 and 18 are most related to invasive cancer. High-grade pre-invasive disease (cervical intra-epithelial neoplasia (CIN III)) is also linked to types 31, 33, 35, 45, 51, 56 and 58, whereas types 6, 11, 42, 43 and 44 tend to be associated with benign conditions and low-grade pre-invasive disease (CIN I). A resumé of the epidemiological evidence is presented here and expanded upon in Chapter 9.

A large international study of almost 1000 cancers [10] has found that over 90% are HPV-positive when adequate material is available. HPV 16 was the most common type and was found throughout the world. Low-risk types were very rare in cancers and only one HPV 6-positive cancer was found (Table 6.2).

Recent case-control studies have found that up to 70% of all cancers contain one of the high-risk types and that the relative risk for cervical cancer in HPV-infected women is among the highest of all known risk factors for any cancer (Table 6.3). For example, in the International Agency for Research on Cancer (IARC) study the odds ratio for polymerase chain reaction (PCR) detected HPV was 15.6 in Columbia, a high-risk area and 46.2 in Spain, a low-risk area. In both cases about 70% of cases had detectable levels of HPV in smears and the major differences were in the level in controls (13.3% for Columbia, 4.6% for Spain) About 70% of the HPV infections were type 16 in both countries, both for cases and controls. Types 18, 31, 33 and 35 each comprise between 5 and 10% of the HPV-positive women. Of the positives, 15% were other types in Spain compared to 9% in Columbia. Similar results have been found in many other studies (Figure 6.2 and [11]).

Case-control studies have also shown that there is a strong relationship between CIN and HPV, and in addition to the results

Cancer and Pre-cancer of the Cervix
Edited by D.M. Luesley and R. Barrasso. Published in 1998 by Chapman & Hall, London. ISBN 0 412 56600 1.

Table 6.1 Specific disease associations of 15 HPV types at varying grades in the clinicopathologic spectrum (from [9])

HPV type	Normal*	Atypia of undetermined significance	Low-grade SIL	High-grade SIL	Invasive cancer	Total
None	1465	206	115	33	16	1835
6/11	8	6	63	8	0	85
16	16	12	61	123	72	284
18	5	5	15	13	36	74
31	7	5	19	27	8	66
33	3	1	13	12	2	31
35	2	2	10	11	2	27
42	1	0	4	1	0	6
43	2	2	4	0	0	8
44	3	5	5	2	0	15
45	2	0	4	1	3	10
51	6	1	9	5	1	22
52	8	1	7	6	1	23
56	1	3	8	3	2	17
58	4	1	5	1	2	13
Unclassified	33	20	35	15	8	111
Total	1566	270	377	261	153	2627

SIL, squamous intra-epithelial lesions.
[a]In six of the eight databases, subjects had cytology plus cervicography or colposcopy.
[b]Reprobing for HPV 58 was restricted to 194 of 283 specimens; hence, its frequency in each category was underestimated by about 31%.

Table 6.2 HPV positivity and type in a large international study (from [10])

HPV type		Number positive (%)
HPV 16 and related	HPV 16	465 (49.9)
	HPV 31	49 (5.3)
	HPV 33	26 (2.8)
	HPV 35	16 (1.7)
	HPV 52	25 (2.7)
	HPV 58	19 (2.0)
HPV 18 and related	HPV 18	128 (13.7)
	HPV 39	14 (1.5)
	HPV 45	78 (8.4)
	HPV 59	15 (1.6)
	HPV 68	11 (1.2)
Other	HPV 6	1 (0.1)
	HPV 11	1 (0.1)
	HPV 56	16 (1.7)
Miscellaneous		26 (2.8)
Undetermined		12 (1.3)
Negative		66 (7.1)
Total		932

Table 6.3 Selected case-control studies of HPV status and invasive cervical cancer

Study	No. of cases (% positive)	No. of controls (% positive)	OR (95% CI)	Method and type
Hong Kong [66]	30 (37)	17 (6)	9.3 (1.0–84.1)	Southern HPV 16
Uganda [67]	34 (50)	23 (4)	22.0 (5.1–104.3)	Southern HPV 16/18
Latin America [68]	721 (47)	1225 (18)	4.0 (3.3–5.0)	FISH HPV 16/18
Pakistan [69]	80 (69)	30 (10)	19.8 (5.8–66.8)	ISH HPV 16/18
Japan [69]	82 (68)	26 (19)	9.0 (3.2–25.7)	ISH HPV 16/18
China [70]	101 (35)	146 (1)	32.9 (7.7–141.1)	PCR HPV 16, 33
Columbia [71]	87 (72)	98 (13)	15.6 (6.9–34.7)	PCR L1 – consensus
Spain [71]	142 (69)	130 (5)	46.2 (18.5–115.1)	PCR L1 – consensus

OR = odds ratio; CI = confidence interval.

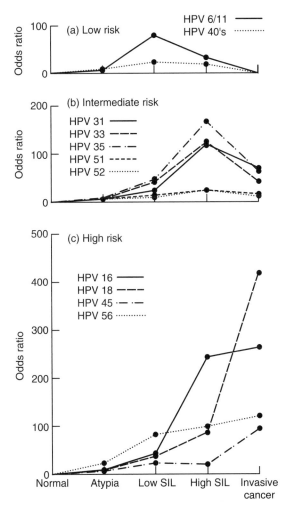

Figure 6.1 Odds ratios for different grades of cervical neoplasia for (A) low risk, (B) intermediate risk, and (C) high-risk HPV (from [9]).

Lorincz *et al.* (1992) cited in Table 6.1 [9], several other authors have observed a relation between the high-risk types (especially HPV 16 and 18) and high-grade CIN [11–14].

In addition to these studies relating certain HPV types to high-grade disease, there are also studies linking HPV to progression of disease. Campion *et al.* (1986) [15] observed 100 women at 4-monthly intervals with mild dysplasia whose lesions were judged to be CIN I by colposcopy, and found that after a follow-up period ranging from 10 to 30 months, 22% of HPV 16-positive women had progressed to CIN III compared to only 9% of HPV 6-positive women and 3% of women negative for both types (Figure 6.3) This result has been confirmed by Schneider *et al.* (1987) [16] and Kataja *et al.* (1990) [17]. Koutsky *et al.* (1992) have taken this one step further and

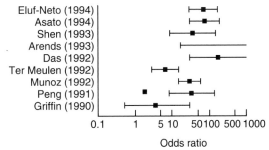

Figure 6.2 Odds ratios and 95% confidence intervals for associations found in case-control studies using PCR methods between HPV 16 (or its nearest surrogate) and invasive cervical cancers (from [11]).

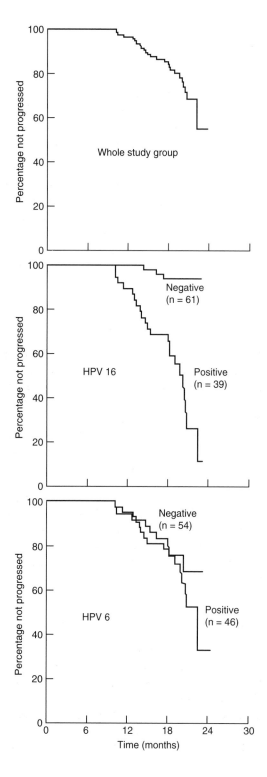

followed up women attending an sexually transmitted disease clinic (STD) clinic who were HPV-positive but cytologically negative [18]. They also found a strong relationship of HPV with the development of clinically detectable high-grade disease. This study is discussed in detail below.

Viral persistence has been found to be a key factor in the development of HPV-related neoplasia [19–22]. A working scheme is shown in Figure 6.4. Most HPV infections are transient and harmless, but occasionally the virus escapes immune surveillance and a persistent infection ensues. The factors related to persistence are not well understood but have been linked to HLA type [23], smoking [24], other venereal infections [25] and sequence variants of the HPV virus [22]. Once the virus becomes persistent, there is a much greater chance it can disrupt the cellular regulation functions, either by integration into a suitable location or by direct actions of the E6 and E7 proteins. Once cellular control mechanisms have been damaged, invasive uncontrolled proliferation can take place, leading to high-grade CIN and cancer. Other cocarcinogens are probably required, but to date none have been clearly identified, although smoking has been suggested by a multitude of studies.

Additionally laboratory studies have shown that HPV has transforming activity *in vitro* [26, 27] and that HPV 16 and HPV 18 can immortalize human foreskin keratinocytes, but this is not obtained for types 6 and 11 ([8, 28; Schlegel, quoted in [29]). A clear carcinogenic activity of related viruses in animals such as the bovine papilloma virus and the shope virus in cottontail rabbits also supports an oncogenic role for human papillomaviruses.

Taken together these observations provide strong evidence that high-risk types of HPV are the causative agent for the great majority

Figure 6.3 Percentage of patients showing progression from CIN I to CIN III as a function of follow-up time for HPV 16 and HPV 6 (from [15]).

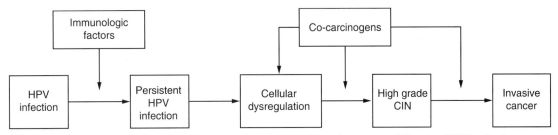

Figure 6.4 Conceptual schematic diagram of evolution of cervical cancer following HPV infection.

of cervical cancers. This conclusion has been affirmed by a large body of experts [11]. However, there has been some contradictory evidence as well, notably a high positivity rate in normal controls in some studies and a low positivity in invasive cancers in other studies [11]. Many of these inconsistencies can be accounted for by weaknesses in the methodology used to detect HPV, and before proceeding it is useful to briefly comment on some of the different assays that have been applied. These are more fully described in Chapter 9.

METHODOLOGY FOR HPV DNA DETECTION

SOUTHERN BLOT HYBRIDIZATION

This is the gold standard when sufficient tissue is available. Purified DNA is digested with restriction endonucleases and fragments are separated by electrophoresis on gel. Fragments are transferred on to nitrocellulose or nylon membranes and are detected by hybridization with radiolabelled or biotinylated probes. The method can be made extremely specific by using stringent hybridization conditions. Alternatively, by using low stringency, new types can be identified which have partial homology with known types. By using multi-cut restriction enzymes and looking at digest patterns, it can be used to discriminate between episomal and integrated HPV DNA and to look for subtypes and certain mutations. However, it has limited sensitivity and 10^5–10^6 copies of the virus are needed for reli-

able detection. This can be a problem when looking in cervical smears where the total cell count is of a similar magnitude, and many of these cells will be normal and uninfected. The method is also labour-intensive and not suited to large-scale clinical or epidemiological applications.

DOT-BLOT HYBRIDIZATION

This is a simplified method where extracted DNA is applied directly to the membrane and probed under stringent conditions to minimize problems with background. Because information in the banding patterns is no longer available, problems with background are much greater and it is difficult to reliably detect weak positives leading to a sensitivity somewhat less than for Southern blots. However, because of its speed and simplicity the test has been used extensively in epidemiological studies and a kit using an RNA probe (Virapap) is commercially available. This kit appears to have sensitivity at least equal to that for Southern blot.

FILTER *IN SITU* HYBRIDIZATION (FISH)

This is a further simplification of the dot-blot technique in which cells are directly deposited on the membrane and denatured in place. This eliminates the need for DNA extraction and purification. However, because the entire cellular material is still present, the test is more likely to produce false-positives. False-negatives also are more likely because of greater interference with the hybridization process. Because it is one of the easiest assay methods for HPV it has been widely used, especially in epidemiologic studies. However, its unreliability has helped to cause much of the confusion in HPV research and it is no longer recommended.

IN SITU HYBRIDIZATION (ISH)

This method can be used to test fresh or fixed histological material on specially treated glass slides. It has the advantage of allowing direct visualization of infected cells, but it is insensitive. Crum *et al.* (1984) [30] were able to find HPV in low-grade CIN, but not in high-grade CIN or cancer, but their method had a sensitivity limit of 500 copies per cell. Schneider *et al.* (1985) [31] used a sensitive single stranded ^{3}H-labelled RNA probe and were able to achieve a sensitivity of 20–50 copies per cell. With this method they were able to localize HPV 16 DNA in over 70% of biopsies with neoplasia, including high-grade CIN and invasive cancer, but their methods are very labour-intensive and exceedingly slow.

POLYMERASE CHAIN REACTION

This method allows selective exponential amplification of specific DNA sequences and is extremely sensitive. There are several possible approaches to amplification and they have been reviewed recently by Gravitt and Manos [32] and are also discussed in Chapter 9. Because of its sensitivity, only a small proportion of the cells from a cervical smear are necessary. However, this sensitivity has created new problems in interpretations. Contamination is a serious problem, which requires careful laboratory procedures, but more fundamentally the technique is capable of detecting very low level infections of doubtful clinical significance. To combat this, Terry *et al.* (1993) [33] have developed a semiquantitative PCR assay and Cuzick *et al.* (1992) [34] have applied this to show the clinical relevance of separating 'high level' from 'low level' infections.

A major difference between different PCR-based assays is the use of the consensus primer system versus type specific primers. Consensus systems allow the simultaneous amplification of several HPV types, which can then be distinguished by type-specific probing. Two systems have been widely used: one uses a degenerate pair of primers in the L1 region and amplifies fragments of length about 450 bp [35]. The other is also located in the L1 region and amplifies a smaller region of about 140–150 bp with a single non-degenerate pair of primers [36]. This system has a larger spectrum of detectable types, but both systems identify types 6, 11, 16, 18, 31, 33, 45 and 51. Consensus primers in the E1 region have also been used [37, 38].

For each system the amplification process is not equally efficient for the important common types and can be considerably less efficient than for type specific probes. A more restricted set of types can also be amplified simultaneously by mixing compatible sets of type specific primers. The different methods also can respond differently when more than one type is present or when inhibitors are present. Inhibitors in the collection medium can also affect the assay. To date, no comprehensive comparison of different PCR amplification and detection systems has been undertaken and this is urgently needed to evaluate different systems and develop standards for routine work.

HYBRID CAPTURE

This is a new DNA–RNA hybridization system that can quantitate the level of HPV by using a chemiluminescent assay for an antibody which recognizes the DNA–RNA hybridization product. Its sensitivity appears to be similar to the best dot-blot systems, but the ability to quantitate the product allows a much clearer separation of high-level and low-level infections. Sensitivity is much less than for PCR methods and it is not yet clear whether its sensitivity will be adequate for routine use with cervical smears.

COMPARISON OF DIFFERENT METHODS

There have been several unpublished studies comparing different combinations of Southern blot, dot-blot, filter *in situ* and PCR methods of HPV detection. Some of these have been reviewed in Schiffman [39]. The problem is most acute when looking at exfoliated cells because of the small amount of DNA available. The unsuitability of FISH in this context in one of our studies (Hollingworth, 1991, unpublished thesis) is shown in Table 6.4, where a kappa value of 0.41 was obtained for detecting HPV16 when compared to Southern blot. Note that both false-positives and false-negatives are apparent. The adequacy of dot-blots or Southern blots compared to PCR depends crucially on the amount of material collected. Schiffman *et al.* [40] have reported similar sensitivity of Southern blot and PCR when a cervicovaginal lavage was used for sample collection and Wheeler *et al.* (reported in [39]) have found excellent correlation between Southern blot and a dot-blot method (Virapap). However, when only aliquots were available (one-quarter of a scrape) Guerrero *et al.* [41] found a much higher detection rate for PCR (Table 6.5). Melchers *et al.* (1989) and others also have reported a much higher sensitivity for PCR using type-specific primers when the test was performed on aliquots of scrapes [42].

MANAGEMENT OF WOMEN WITH BORDERLINE OR MILDLY DYSKARYOTIC SMEARS

In the UK approximately 5% of all smears show borderline or mildly dyskaryotic changes. About 10–25% of these women will have high-grade underlying CIN lesions, a similar proportion will have low-grade lesions, and the remainder will have no detectable CIN at colposcopy. The

appropriate management of these women presents difficult problems. The Bethesda classification recognizes atypical squamous cells of undetermined significance (ASCUS), which poses similar difficulties. Current British guidelines [43] recommend cytological follow-up at 6–12 month intervals unless progression or persistence (two mild or three borderline smears) occurs. Return to routine 3–5 yearly screening is recommended after two consecutive normal smears at least 6 months apart. This approach leads to a large number of extra smears at short intervals, which are both costly and produce anxiety in women, who must live with a potentially cancerous lesion.

Table 6.4 Concordance of Southern blot hybridization and filter *in situ* hybridization for the detection of HPV 16 in cervical smears. kappa = 0.41 (from Hollingworth, 1991 unpublished thesis)

		Southern	
		+	–
FISH	+	20	11
	–	20	88

Table 6.5 Distribution of invasive carcinoma cases and controls by their HPV status by Virapap (VP), Southern blot (SB) and polymerase chain reaction (PCR) (from [41])

Result by PCR[a]	SB[b]	VP	No.	(%)
+	+	+	58	(11)
–	–	–	273	(54)
+	–	–	99	(19)
+	+	–	15	(3)
–	+	+	4	(1)
–	+	–	14	(3)
–	–	+	13	(3)
Total			510	(100)

[a]PCR-positive includes those specimens which were positive by type specific probes and/or generic probe.
[b]SB-positive includes those specimens which were positive under low-stringency and/or high-stringency conditions.

There are also an increasing number of reports of invasive cancer occurring in women who had a minor abnormality many years previously followed by a number of (presumably false) negative smears [44, 45]. Several authors have reported a high false-negative rate for cytology [46–49] and others have questioned the policy of following up women with previous abnormal smears by cytology alone [50–52]. The alternative approach of colposcoping all these women is even more expensive and results in overtreatment and the unnecessary anxiety of a hospital visit in the majority of cases.

Testing for HPV DNA on material taken at the time of the index smear may offer the possibility of better management for these women. Nuovo *et al.* [53] and others have shown that testing for HPV DNA in biopsies from women with abnormal smears can help to determine which women have CIN and Lungu *et al.* [12] showed that CIN II–III had a different type distribution than CIN I. Several studies [18, 39, 54, 55] have shown that HPV DNA can be reliably assayed from material obtained from a cervical smear, swab or lavage, and have suggested that this could be a useful adjunct to cytology in patient management. However, the problem of false-positives has remained a concern.

One approach is to require that the HPV infection is persistent and is detectable on two or more visits at least 6 months apart. This has many of the disadvantages of cytological follow-up in that it is slow and creates anxiety during the follow-up period. An alternative is to evaluate the extent or activity of the infection by estimating the amount of virus present. Viral load appears to be a good surrogate for persistence, especially for HPV 16 [56].

Terry *et al.* [33] developed a semiquantitative PCR assay and suggested that high levels of HPV DNA were an important discriminant for separating high-grade disease from minor lesions. Cuzick *et al.* [34, 57] applied this in a screening context and were able to show directly that HPV testing on material from a cervical smear could help predict the severity of the underlying lesion. They performed type specific PCR for types 16, 18, 31, 33, 35 in 133 women attending for colposcopy because of an abnormal smear. The results are summarized in Tables 6.6 and 6.7.

In Table 6.6 the relationship between referring cytology and histologic diagnosis is shown. Overall, just over half of the patients had CIN II–III. Severe dyskaryosis was highly specific for this, and was found in 29 of 38 instances (76%), but lacks sensitivity. However, the outcome was much more variable for mild dyskaryosis, which on a population basis is several times more common, and only one-third of these patients had high-grade disease.

Table 6.6   Relationship between referring cytology and histological diagnosis (from [57])

Histological diagnosis	Cytological diagnosis (%)			
	Severe	Moderate	Mild	Total
CIN III	26[a] (68)	24 (45)	10 (32)	60
CIN II	3 (8)	7 (13)	2 (6)	12
CIN I	2 (5)	6 (11)	3 (10)	11
HPV I	2 (5)	7 (13)	0 (0)	9
Normal	5 (13)	9 (17)	16 (52)	30[b]
Total	38 (100)	53 (100)	31 (100)	122

[a]Includes two cases of invasive cancer.
[b]Includes 12 women with no visible lesion on colposcopy who were not biopsied.

Table 6.7   Number of patients (and percentage) with high and low levels of HPV 16 by histological diagnosis (from [57])

Histology	No. of patients	HPV 16	
		Low	High
CIN III	61 (46)	5 (31)	36 (86)
CIN II	12 (9)	2 (13)	3 (7)
CIN I	13 (10)	3 (19)	0
HPV I	11 (8)	2 (13)	1 (2)
Normal	36 (27)	4 (25)	2 (5)
Total	133 (100)	16 (100)	42 (100)

The results for HPV testing in this population are shown in Table 6.7. High levels of HPV 16 were found in 42 cases and in 39 of these CIN II–III was found on histology giving a positive predictive value of 93% in this context. In contrast only seven of 16 patients with low-level HPV 16 infection had high-grade disease, which was similar to the group as a whole. However, only 59% of the CIN III lesions (and 53% of all high-grade lesions) were positive for HPV 16, so it is clearly not an adequate management tool by itself.

The detection rates for high levels of all the types measured are shown in Table 6.8. HPV types 31 and 33, the next most common in this series, and their positive predictive value for high-grade disease were of the order of 70%. Altogether 51 of the CIN III lesions (84%) and 58 of the CIN II–III lesions (79%) were positive at a high level for at least one of the five types. Of interest is the low prevalence of HPV 35 in general and the relatively low positive predictive value of HPV 18, compared to HPV 16.

Larger studies focusing exclusively on mild and borderline dyskaryosis are needed before firm conclusions can be drawn about the use of HPV testing in this context. In particular, the relative value of testing for different types and role of consensus tests in this context require further study. However, this study does supply evidence that women with high levels of HPV type 16 in the presence of any degree of dyskaryosis are very likely to have a high-grade lesion, and they should be referred for colposcopy without delay. Further support for this has also been reported by Bavin *et al.* [58]. The study also indicates that HPV testing is unlikely to replace cytology and that a diagnosis of severe or moderate dyskaryosis is grounds for referral independent of any HPV results.

HPV TESTING AS PART OF PRIMARY SCREENING

The ability of semi-quantitative tests for high levels of HPV DNA to discriminate between high- and low-grade disease in women with low-grade cytological abnormalities raises the possibility that this approach may also be of value as part of primary screening. However, this is a more difficult environment and because of the much lower disease prevalence the need to control the false-positive rate becomes much more severe. Ideally, the addition of this test to conventional cytology would improve the detection rate of high-grade CIN, reduce invasive cancer rates and also reduce the referral rate for colposcopy by safely allowing women who are negative for HPV but have borderline or mild dyskaryosis to be followed by routine cytology. A very large trial involving at least 250,000 women will be necessary to establish this conclusively. Factors which are likely to influence the cost–benefit analysis are age, number of types tested for, and level of HPV DNA detected. However, smaller studies can provide information on detection rates and length of protection.

Studies conducted thus far in this area have been encouraging. Reid *et al.* [59] screened 1012 women aged 18–35 by conventional cytology, cervicography and Southern blot hybridization for high-risk HPV types (16, 18, 31, 33, 35, 45, 51, 52, 56) and low-risk HPV types (6, 11, 42, 43, 44). About two-thirds of the population were taken from STD clinics and had a high risk of HPV infection and cervical disease. Altogether 23 women with high-grade lesions were found by one or more of these modalities.

Cytology was positive (mild atypia or worse) in 88 cases and within this group 12 CIN II–III lesions were found (apparent sensitivity 52%, positive predictive value 14%) whilst HPV tests for high-risk types were positive in 55 women and 13 cases of CIN II–III were found in this group (apparent sensitivity 56%, positive predictive value 24%). Both tests together detected 19 (83%) high-grade CIN lesions. The best performance was obtained when HPV testing and cervicography were both used to decide on referral for women with minor cytological abnormalities, but no

Table 6.8 Number of patients with high and low levels of specific HPV types by histological diagnosis (from [57])

		HPV positivity and level									
		HPV 16		HPV 18		HPV 31		HPV 33		HPV 35	
Histology	No. of patients	Low	High	Low	High	Low	High	Low	High	Low	High
CIN III	61	5 (1)	36 (10)	8 (6)	4 (2)	5 (3)	14 (5)	1 (0)	7 (5)	0	2 (0)
CIN II	12	2 (1)	3 (2)	1 (0)	2 (1)	0	0	0	4 (1)	0	0
CIN I	13	3 (1)	0	3 (2)	0	1 (0)	3 (0)	2 (0)	1 (0)	1 (0)	0
HPV 1	11	2 (0)	1 (1)	0	0	0	2 (1)	0	2 (2)	0	0
Normal	36	4 (1)	2[a] (1)	3 (1)	3 (1)	0	1 (0)	1 (0)	2 (0)	0	2 (2)
Total	133	16 (4)	42 (14)	15 (9)	9 (4)	6 (3)	20 (6)	4 (0)	16 (8)	1 (0)	4 (2)

Numbers in parentheses indicate the number of patients with multiple infections, where the other type(s) were 'high level'. Multiple infections are included for each positive type and thus are represented at least twice in the table.
[a] One patient was pregnant.

Table 6.9 Positive predictive value for CIN II–III of semiquantitative HPV testing and cytology among 1985 evaluable women (from [49])

HPV testing		Cytology	
Type	True positive/test high-level positive (%)	Grade	True-positive/ test positive (%)
HPV 16	36/66 (55)	Severe	17/18 (94)
HPV 18	5/24 (33)	Moderate	10/23 (43)
HPV 31	20/46 (43)	Mild	10/29 (34)
HPV 33	7/22 (32)	Borderline	8/58 (14)
HPV 16 or 31	55/111 (50)	Moderate or severe	27/41 (66)
HPV 16, 18, 31 or 33	61/146 (42)	Any dyskaryosis	45/128 (35)

data were given for the combined use of HPV and cytology without cervicography. In particular, the value of HPV positivity in the presence of negative cytology was not assessed.

In another study, Koutsky *et al.* [18] reported on the predictive value of HPV testing for the development of CIN II–III in 241 women with negative cytology. Again, this was a high-risk STD clinic population so the performance cannot be directly extrapolated to the general population. The women were followed by colposcopy every 4 months, and initial and repeat smears were assayed for HPV by a method which detected types 6, 11, 16, 18, 31 and some others (the method changed in the middle of the study). HPV positivity was treated as a time-dependent covariate, so analysis was based on the time from HPV positivity to detection of high-grade CIN. Altogether 110 women were positive for HPV at some stage and 28% of these women developed CIN II–III within 2 years, compared to only 3% of women who remained negative for HPV (Figure 6.5). Women with types 16 or 18 were most predictive of high-grade disease and over 40% of women infected with these types

developed CIN II–III within 2 years of infection. Altogether 78% of all high-grade disease was attributed to HPV and 52% to types 16 and 18. Types 6 and 11, in the absence of types 16 or 18, accounted for 13% of cases and types 31, 33 and 35 for 9%. Altogether 45% of HPV infections could not be attributed to these types but these untyped infections were associated with only 5% of the high-grade CIN lesions, suggesting that most of them are not very oncogenic.

A Dutch group has also proposed using HPV testing as part of primary screening. To date, their studies have been observational and have focused on the correlation between HPV positivity and cytological abnormality, both at the time of testing and on subsequent screens. Meijer *et al.* [13] have shown a clear relationship between HPV positivity and severity of disease (Figure 6.6) and have shown that severe dyskaryosis is associated with persistent HPV infections, whereas minor abnormalities more often are associated with fluctuating or transient infections. They have suggested that for women aged over 35 years primary screening for HPV (using a con-

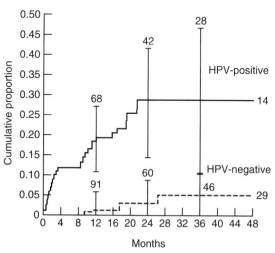

Figure 6.5 Cumulative proportion of women with negative cytology at enrolment who developed CIN II–III according to HPV status at entry (from [18]).

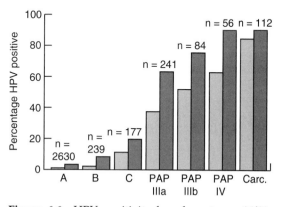

Figure 6.6 HPV positivity based on types 16/18 and a wide spectrum consensus system for three cytologically negative groups with different risks of disease (A – asymptomatic, B – patients without history of cervical disease, C – patients with history of cervical disease), and patients with different grades of cytological abnormalities (from [13]).

sensus primer system) every 5 years may be adequate, and only women positive on this test would need cytological testing. Referral for colposcopy would depend on the type of HPV present and the cytological result. Such a programme assumes very few women with high-grade lesions will be negative for HPV as detected by their consensus system, and further studies are needed to confirm this.

Cuzick *et al.* [49] evaluated HPV testing for types 16, 18, 31, and 33 on material taken at the time of a cervical smear in 1985 evaluable women having routine screening. Women with any degree of dyskaryosis or high levels of one of these HPV types were referred for colposcopy. Most of the women in this study were aged 20–45. Of the CIN lesions of grade II–III detected 44% (33/81) had negative cytology and were found only by HPV testing. A further 22% of the CIN II–III lesions were positive for HPV but showed only borderline or mild cytological changes. The positive predictive value of HPV testing was 42%, which was similar to that for moderate dyskaryosis. HPV types 16 and 31 were more sensitive and specific for CIN II–III than were types 18 or 33. However, 25% of the CIN II–III lesions were not detected by these four HPV tests (Table 6.9). HPV positivity decreased with age, but CIN II–III did so in a similar manner, so that the positive predictive value was not affected by age in this study (Figure 6.7).

These results suggest that HPV testing could usefully augment but not replace conventional cytology. It is possible that a longer interval between screens could be used if HPV testing was introduced. A much larger randomized trial is needed to assess the impact of these improved CIN II–III detection rates on the subsequent incidence of invasive cancer.

Most of the women in this study were under the age of 40. We are currently undertaking a study of routine smears testing by GPs in 3000 women over the age of 35, to determine the value of this test in an older population.

HPV IN FOLLOW-UP AFTER CIN

No studies have been reported to date on the value of HPV testing in following-up women after treatment for CIN. The persistence of HPV positivity could prove a valuable and inexpensive way of assessing treatment failure and a study in this area is urgently needed.

HPV IN INVASIVE CANCER

When sensitive wide spectrum tests are used, approximately 90% of all invasive cervix cancers are positive for some type of HPV. However, we know very little about the role of HPV in the subsequent behaviour of invasive cancers. The HPV E6 protein of HPV 16 and 18 is known to bind to p53 leading to its inactivation. In cancers where HPV involvement cannot be found, a genetic defect in p53 has been found [60] and this appears to have prognostic significance. At least three groups [61–63] have now reported that patients with HPV-negative tumours have a worse prognosis than those which are positive. Undoubtedly there is much more to be learned in this area and even simple comparisons of the prognosis of tumours related to different types have not

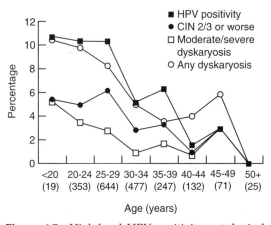

Figure 6.7 High-level HPV positivity, cytological abnormalities, and histological diagnosis of CIN II–III by age. Numbers in parentheses under each group indicate the number of women studied (from [49]).

as yet been reported. The taxonomy of sequence variants for HPV 16 has been fairly extensively mapped out [64] and a few reports have suggested a role for certain sequence variants in pathogenesis [22, 65–67]. However, much work is still required to make sense of this.

In summary, the case for introducing HPV testing into clinical practice is strongest for women with mild or borderline cytologic abnormalities. Here, HPV testing may refine the decision about referral for colposcopy and thus reduce unnecessary expense and anxiety associated with prolonged intensive cytologic follow-up. A role for HPV testing in primary screening also looks very promising and potentially cost-effective, although the details of optimal use still need to be worked out. This may be a viable approach in the developing world, since there may be a reduced requirement for highly trained personnel. HPV testing may also be useful in follow-up of treated CIN patients and prognosis of cancer patients, but little has been done to evaluate this.

REFERENCES

1. Levin, M.L., Kress, L.C. and Goldstein, H. (1942) Syphilis and cancer. *N.Y. State Journal of Medicine*, **42**, 1737–45.
2. Adam, E., Rawls, W.E. and Melnick, J.L. (1974) The association of herpes virus type 2 infection and cervical cancer. *Preventative Medicine*, **3**, 122–41.
3. Graham, S., Rawls, W., Swanson, M., *et al.* (1982) Sex partners and herpes simplex virus type 2 in the epidemiology of cancer of the cervix. *American Journal of Epidemiology*, **115**, 729–35.
4. Rawls, W.E., Lavery, C., Marrett, L.D., *et al.* (1986) Comparison of risk factors for cervical cancer in different populations. *International Journal of Cancer*, **37**, 537–46.
5. Wynder, E.L., Cornfield, J., Schroff, P.D., *et al.* (1954) A study of environmental factors in carcinoma of the cervix. *American Journal of Obstetrics and Gynecology*, **68**, 1016–47.
6. Pratt-Thomas, H.R., Heins, H.C., Latham, E., *et al.* (1956) The carcinogenic effect of human smegma: an experimental study. *Cancer*, **9**, 671–80.
7. Reid, B.L. (1965) Cancer of the cervix uteri: review of causal factors with an hypothesis for its origins. *Medical Journal of Australia*, **1**, 375–83.
8. Durst, M., Dzarlieva-Petrusevska, R.T., Boukamp, P., *et al.* (1987) Papillomavirus sequences integrate near cellular oncogenes in some cervical carcinoma. *Journal of the National Academy of Science*, **84**, 1070–4.
9. Lorincz, A.T., Reid, R., Jenson, A.B., *et al.* (1992) Human papillomavirus infection of the cervix: relative risk associations of 15 common anogenital types. *Obstetrics and Gynecology*, **79**, 328–37.
10. Bosch, F.X., Manos, M.M., Muñoz, N., *et al.* (1995) Prevalence of human papillomavirus in cervical cancer: a worldwide perspective. *Journal of the National Cancer Institute*, **87**, 796–802.
11. International Agency for Research on Cancer (1995) *IARC Monographs on the Evaluation of Carcinogenic Risks to Humans*, vol. 64, *Human Papillomaviruses*. IARC Scientific Publications, Lyon.
12. Lungu, O., Sun, X.W., Felix, J., *et al.* (1992) Relationship of human papillomavirus type to grade of cervical intraepithelial neoplasia. *Journal of the American Medical Association*, **267**, 2493–6.
13. Meijer, C.J., van den Brule, A.J., Snijders, P.J., *et al.* (1992) Detection of human papillomavirus in cervical scrapes by the polymerase chain reaction in relation to cytology: possible implications for cervical cancer screening. In *The Epidemiology of Cervical Cancer and Human Papillomavirus*, vol. 19 (eds N. Muñoz, F.X. Bosch, K.V. Shah and A. Meheus). IARC Scientific Publications, Lyon, pp. 271–81.
14. Schiffman, M.H., Bauer, H.M., Hoover, R.N., *et al.* (1993) Epidemiologic evidence showing that human papillomavirus infection causes most cervical intraepithelial neoplasia. *Journal of the National Cancer Institute*, **85**, 958–64.
15. Campion, M.J., McCance, D.J., Cuzick, J., *et al.* (1986) Progressive potential of mild cervical atypia: prospective cytological, colposcopic, and virological study. *Lancet*, **ii**, 237–40.
16. Schneider, A., Meinhardt, G., de Villiers, E.M., *et al.* (1987) Sensitivity of the cytologic diagnosis of cervical condyloma in comparison with HPV-DNA hybridization studies. *Diagnostic Cytopathology*, **3**, 250–5.
17. Kataja, V., Syrjänen, K., Syrjänen, S., *et al.* (1990) Prospective follow-up of genital HPV infections: survival analysis of the HPV typing data. *European Journal of Epidemiology*, **6**, 9–14.
18. Koutsky, L.A., Holmes, K.K., Critchlow, C.W., *et al.* (1992) A cohort study of the risk of cervical

intraepithelial neoplasia grade 2 or 3 in relation to papillomavirus infection. *New England Journal of Medicine*, **327**, 1272–8.

19. Hildesheim, A., Schiffman, M.H., Gravitt, P.E., *et al.* (1994) Persistence of type-specific human papillomavirus infection among cytologically normal women. *Journal of Infectious Diseases*, **169**, 235–40.

20. Ho, G.Y.F., Burk, R.D., Klein, S., *et al.* (1995) Persistent genital human papillomavirus infection as a risk for persistent cervical dysplasia. *Journal of the National Cancer Institute*, **87**, 1365–71.

21. Remmink, A.J., Walboomers, J.M.M., Helmerhorst, T.J.M., *et al.* (1995) The persistence of persistent high-risk HPV genotypes in dysplastic cervical lesions is associated with progressive disease: natural history up to 36 months. *International Journal of Cancer*, **61**, 306–11.

22. Londesborough, L., Ho, L., Terry, G., *et al.* (1996) Human papillomavirus (HPV) genotype as a predictor of persistence and development of high grade lesions in women with minor cervical abnormalities. *International Journal of Cancer*, **69**, 364–8.

23. Odunsi, K., Terry, G., Ho, L., *et al.* (1995) Association between HLA DQB1*03 and cervical intra-epithelial neoplasia. *Molecular Medicine*, **1**, 161–71.

24. Barton, S.E., Jenkins, D., Cuzick, J., *et al.* (1988) Effect of cigarette smoking on cervical epithelial immunity: a mechanism for neoplastic change? *Lancet*, **ii**, 652–4.

25 Lacey, C.J.N. (1992) Assessment of exposure to sexually transmitted agents other than human papillomavirus. In *The Epidemiology of Cervical Cancer and Human Papillomavirus*, vol. 19 (eds N. Muñoz, F.X. Bosch, K.V. Shah and A. Meheus). IARC Scientific Publications, Lyon, pp. 93–105.

26. Watts, S.L., Phelps, W.C., Ostrow, R.S., *et al.* (1984) Cellular transformation by human papillomavirus DNA *in vitro*. *Science*, **225**, 634–6.

27. McCance, D.J., Kopan, R., Fuchs, E., *et al.* (1988) Human papillomavirus type 16 alters human epithelial cell differentiation *in vitro*. *Proceedings of the National Academy of Sciences, USA*, **85**, 7169–73.

28. Pirisi, L., Yasumoto, S., Feller, M., *et al.* (1987) Transformation of human fibroblasts and keratinocytes with human papillomavirus type 16 DNA. *Journal of Virology*, **61**, 1061–6.

29. Shah, K.V. and Gissmann, L. (1989) Experimental evidence on oncogenicity of papillomaviruses. In *Human Papillomavirus and Cervical Cancer*, vol. 94 (eds N. Muñoz, F.X.

Bosch and O.M. Jensen). IARC Scientific Publications, Lyon, pp. 105–11.

30. Crum, C.P., Ikenberg, H., Richard, R.M. and Gissman, L. (1984) Human papillomavirus type 16 and early cervical neoplasia. *New England Journal of Medicine*, **310**, 880–3.

31. Schneider, A., Kraus, H., Schuhmann, R., *et al.* (1985) Papillomavirus infection of the lower genital tract: detection of viral DNA in gynecological swabs. *International Journal of Cancer*, **35**, 443–8.

32. Gravitt, P.E. and Manos, M.M. (1993) Polymerase chain reaction-based methods for the detection of human papillomavirus DNA. In *The Epidemiology of Human Papillomavirus and Cervical Cancer*, vol. 119 (eds N. Muñoz, F.X. Bosch, K.V. Shah and A. Meheus). IARC Scientific Publications, Lyon, pp. 121–33.

33. Terry, G., Ho, L., Jenkins, D., *et al.* (1993) Definition of human papillomavirus type 16 DNA levels in low and high grade cervical lesions by a simple polymerase chain reaction technique. *Archives of Virology*, **128**, 123–33.

34. Cuzick, J., Terry, G., Ho, L., *et al.* (1992) Human papillomavirus type 16 DNA in cervical smears as predictor of high-grade cervical intraepithelial neoplasia. *Lancet*, **339**, 959–60.

35. Manos, M.M., Wright, D.K., Lewis, A.J., *et al.* (1989) The use of polymerase chain reaction amplification for the detection of genital human papillomaviruses. In *Molecular Diagnostics of Human Cancer, Cancer Cells*, 7 (eds M. Furth and M. Greaves). Cold Spring Harbor Press, NY, pp. 209–14.

36. Snijders, P.J.F., van den Brule, A.J.C., Schrijnemakers, H.F.J., *et al.* (1990) The use of general primers in the polymerase chain reaction permits the detection of a broad spectrum of human papilloma genotypes. *Journal of General Virology*, **71**, 173–81.

37. Gregoire, L., Arella, M., Campione-Piccardo, J., *et al.* (1989) Amplification of human papillomavirus DNA sequences by using conserved primers. *Journal of Clinical Microbiology*, **27**, 2660–5.

38. van den Brule, A.J.C., Snijders, P.J.F., Gordijn, R.L.J., *et al.* (1990) General primer-mediated polymerase chain reaction permits the detection of sequenced and still unsequenced human papillomavirus genotypes in cervical scrapes and carcinomas. *International Journal of Cancer*, **45**, 644–9.

39. Schiffman, M.H. (1992) Validation of hybridization assays: correlation of filter *in situ*, dot-blot and PCR with Southern blot. IARC Scientific Publications No. 119, 169–79.

40. Schiffman, M.H., Bauer, H.M., Lorincz, A.T., et al. (1991) Comparison of Southern blot hybridization and polymerase chain reaction methods for the detection of human papillomavirus DNA. *Journal of Clinical Microbiology*, **29**, 573–7.

41. Guerrero, E., Daniel, R.W., Bosch, F.X., et al. (1992) Comparison of ViraPap, Southern hybridization, and polymerase chain reaction methods for human papillomavirus identification in an epidemiological investigation of cervical cancer. *Journal of Clinical Microbiology*, **30**, 2951–9.

42. Melchers, W., van den Brule, A., Walboomers, J., de Bruin, M., et al. (1989) Increased detection rate of human papillomavirus in cervical scrapes by the polymerase chain reaction as compared to modified FISH and Southern-blot analysis. *Journal of Medical Virology*, **27**, 329–35.

43. Duncan, I.D. (1993) *Guidelines for Clinical Practice and Programme Management*, National Coordinating Network, NHS Cervical Screening Programme, Oxford.

44. Stanbridge, C.M., Suleman, B.A., Persad, R.V., et al. (1992) A cervical smear review in women developing cervical carcinoma with particular reference to age, false negative cytology and histologic type of the carcinoma. *International Journal of Gynecologic Cancer*, **2**, 92–100.

45. Soutter, W.P. and Fletcher, A. (1994) Invasive cancer of the cervix in women with mild dyskaryosis followed up cytologically. *British Medical Journal*, **308**, 1421–3.

46. Giles, J.A., Hudson, E., Crow, J., et al. (1988) Colposcopic assessment of the accuracy of cervical cytology screening. *British Medical Journal*, **296**, 1099–102.

47. Tawa, K., Forsythe, A., Cove, J.K., et al. (1988) A comparison of the Papanicolaou smear and the cervigram: sensitivity, specificity, and cost analysis. *Obstetrics and Gynecology*, **71**, 229–35.

48. Szarewski, A., Cuzick, J., Edwards, R., et al. (1991) The use of cervicography in a primary screening service. *British Journal of Obstetrics and Gynaecology*, **98**, 313–17.

49. Cuzick, J., Szarewski, A., Terry, G., et al. (1995) Human papillomavirus testing in primary cervical screening. *Lancet*, **345**, 1533–6.

50. Soutter, W.P., Wisdom, S., Brough, A.K., et al. (1986) Should patients with mild atypia in a cervical smear be referred for colposcopy? *British Journal of Obstetrics and Gynaecology*, **93**, 70–4.

51. Jones, D.E.D., Creasman, W.T., Dombroski, R.A., et al. (1987) Evaluation of the atypical Pap smear. *American Journal of Obstetrics and Gynecology*, **157**, 544–9.

52. Flannelly, G., Anderson, D., Kitchener, H.C., et al. (1994) Management of women with mild and moderate cervical dyskaryosis. *British Medical Journal*, **308**, 1399.

53. Nuovo, G.J., Blanco, J.S., Leipzig, S., et al. (1990) Human papillomavirus detection in cervical lesions nondiagnostic for cervical intraepithelial neoplasia: correlation with Papanicolaou smear, colposcopy, and occurrence of cervical intraepithelial neoplasia. *Obstetrics and Gynecology*, **75**, 1006–11.

54. Bauer, H.M., Ting, Y., Greer, C.E., et al. (1991) Genital human papillomavirus infection in female university students as determined by a PCR-based method. *Journal of the American Medical Association*, **265**, 472–7.

55. van den Brule, A.J.C., Walboomers, J.M.M., du Maine, M., et al. (1991) Difference in prevalence of human papillomavirus genotypes in cytomorphologically normal cervical smears is associated with a history of cervical intraepithelial neoplasia. *International Journal of Cancer*, **48**, 404–8.

56. Cuzick, J., (1997) Viral load as a surrogate for persistence in cervical human papillomavirus infection. In *New Developments in Cervical Cancer Screening and Prevention* (eds. E. Franco and J. Monsonego). Blackwell Science, Oxford, 373–8.

57. Cuzick, J., Terry, G., Ho, L., et al. (1994) Type-specific human papillomavirus DNA in abnormal smears as a predictor of high-grade cervical intraepithelial neoplasia. *British Journal of Cancer*, **69**, 167–71.

58. Bavin, P.J., Giles, J.A., Deery, A., Crow, J., Griffiths, P.D., Emery, V.C. and Walker, P.G. (1993b) Use of semi-quantitative PCR for human papillomavirus DNA type 16 to identify women with high grade cervical disease in a population presenting with a mildly dyskaryotic smear report. *British Journal of Cancer*, **67**, 602–5.

59. Reid, R., Greenberg, M.D., Lorincz, A., et al. (1991) Should cervical cytologic testing be augmented by cervicography or human papillomavirus deoxyribonucleic acid detection? *American Journal of Obstetrics and Gynecology*, **164**, 1461–71.

60. Crook, T., Wrede, D., Tidy, J.A., et al. (1992) Clonal p53 mutation in primary cervical cancer: association with human-papillomavirus-negative tumours. *Lancet*, **339**, 1070–3.

61. Riou, G., Favre, M., Jeannel, D., et al. (1990) Association between poor prognosis in early-stage invasive cervical carcinomas and non-detection of HPV DNA. *Lancet*, **335**, 1171–4.

62. Higgins, G.D., Davy, M., Roder, D., *et al.* (1991) Increased age and mortality associated with cervical carcinomas negative for human papillomavirus RNA. *Lancet*, **338**, 910–13.

63. Ikenberg, H., Sauerbrei, W., Schottmuller, U., Spitz, C. and Pfleiderer, A. (1994) Human papillomavirus DNA in cervical carcinoma: correlation with clinical data and influence on prognosis. *International Journal of Cancer*, **59**, 322–6.

64. Yamada, T., Wheeler, C.M., Halpern, A.L., *et al.* (1995) Human papillomavirus type 16 variant lineages in United States populations characterised by nucleotide sequence analysis of the E6, L2 and L1 coding segments. *Journal of Virology*, **69**(12), 7743–53.

65. Bavin, P.J., Walker, P.G. and Emery, V.C. (1993a) Sequence microheterogeneity in the long control region of clinical isolates of human papillomavirus type 16. *Journal of Medical Virology*, **39**, 267–72.

66. Hecht, J.L., Kadish, A.S., Jiang, G., *et al.* (1995) Genetic characterization of the human papillomavirus (HPV) 18 E2 gene in clinical specimens suggests the presence of a subtype with decreased oncogenic potential. *International Journal of Cancer*, **60**, 369–76.

67. Ellis, J.R.M., Keating, P.J., Baird, J., *et al.* (1995) The association of an HPV 16 oncogene variant with HLA-B7 has implications for vaccine design in cervical cancer. *Nature Medicine*; **1**(5), 464–70.

68. Donnan, S.P.B., Wong, F.W.S., Ho, S.C., *et al.* (1989) Reproductive and sexual risk factors and human papilloma virus infection in cervical cancer among Hong Kong Chinese. *International Journal of Epidemiology*, **18**, 32–6.

69. Schmauz, R., Okong, P., de Villiers, E.M., *et al.* (1989) Multiple infections in cases of cervical cancer from a high-incidence area in tropical Africa. *International Journal of Cancer*, **43**, 805–9.

70. Reeves, W.C., Brinton, L.A., Garcia, M., *et al.* (1989) Human papillomavirus infection and cervical cancer in Latin America. *New England Journal of Medicine*, **320**, 1437–41.

71. Anwar, K., Inuzuka, M., Shiraishi, T., *et al.* (1991) Detection of HPV DNA in neoplastic and non-neoplastic cervical specimens from Pakistan and Japan by non-isotopic *in situ* hybridization. *International Journal of Cancer*, **47**, 675–80.

72. Peng, H., Liu, S., Mann, V., *et al.* (1991) Human papillomavirus types 16 and 33, herpes simplex virus type 2 and other risk factors for cervical cancer in Sichuan Province, China. *International Journal of Cancer*, **47**, 711–16.

73. Muñoz, N., Bosch, F.X., de Sanjose, S., *et al.* (1992) The casual link between human papillomavirus and invasive cervical cancer: a population-based case-control study in Colombia and Spain. *International Journal of Cancer*, **52**, 743–9.

Per-Anders Mårdh and Marius Domeika

THE ROLE OF GENITAL SECRETION COMPONENTS IN MICROBIAL DEFENCE

Mucus, among other components, consists of glycoproteins, which in the female genital tract are produced by epithelial cells of the endocervix and the endometrium. Hormonal changes during the menstrual cycle cause dramatic alterations in the viscosity of the mucus [1] and of the protective effect of the cervical 'plug' against ascending infections.

The cervical mucus creates a proper environment for phagocytic cells and cell-mediated immune actions. It also provides a vehicle for persistence of bacteria in the ecosystem of the genital tract. Mucus may also activate the bactericidal capacity by some redox-active agents and bactericidal proteins [2]. It also provides a means for the body to shed organisms, aside engaging immune mechanisms (see below).

POLYMORPHONUCLEAR LEUCOCYTES (PMN) AND MACROPHAGES

Both neutrophils and monocytes are involved in a variety of microbial defence mechanisms in the lower female genital tract, i.e. in chemotaxis as well as in binding and digestion of offending microorganisms.

In the majority of lower genital tract infections an increased number of PMN leucocytes can be found – which have the capacity to phagocytize causative agents. Thus interaction between *Neisseria gonorrhoeae*, *Chlamydia trachomatis*, *Treponema pallidum*, *Trichomonas vaginalis* and group B streptococci and phagocytes has been demonstrated [3].

A significant role for PMN in the clearance of indigenous bacteria in the lower genital tract is, however, questionable, as neutropenia is not associated with vaginitis and cervicitis.

Some strains of sexually transmitted disease (STD) agents are less susceptible to phagocytosis, e.g. gonococci not expressing OMP protein II [4], which is believed to facilitate opsonin-independent ingestion of bacteria [5].

The PMN that can phagocytize gonococci (a phenomenon which can be observed in stained smears and which has and still forms an important basis for the diagnosis of gonorrhoea) [6] possesses lectin-like receptors that recognize carbohydrate structures present in the cell wall of the gonococcus [7].

Gonococci that possess pili are less readily phagocytized than such bacteria lacking these structures. *In vitro*, phagocytosis of gonococci by PMN may be increased by gonococcal antipilus antibodies. In experimental studies in humans, using non-pilated gonococci, the infection rate was only 40% of that when using a piliated strain to challenge volunteers. Human monocytes are capable of phagocytizing non-pilated gonococci [8].

An inflammatory process causes an abundance of L^+-lactate secretion by phagocytes.

Cancer and Pre-cancer of the Cervix
Edited by D.M. Luesley and R. Barrasso. Published in 1998 by Chapman & Hall, London. ISBN 0 412 56600 1.

Gonococci use this substrate to consume oxygen more aggressively than the phagocytes [6]. Under these circumstances the phagocytes attacking gonococci cannot generate superoxide as shown *in vitro* studies. As the environment becomes anaerobic, gonococci use nitrite secreted by phagocytes as part of their microbicidal capacity, e.g. as a terminal electron acceptor [9]. Although oxygen-independent microbicidal proteins kill gonococci *in vitro*, survival of the organisms *in vivo* suggests that this defence is not very effective.

Macrophages represent a target for infection with *C. trachomatis* [10]. Replication of the organism can be prevented by using a critical substrate, i.e. tryptophan, which is mediated through the action of the enzyme indoleamine 2,3-dioxygenase. The enzyme is activated by interferon. *C. trachomatis* can be taken up by macrophages, where it may reproduce [11]. However the role of cytokines, produced by macrophages, and thereby a possible role of these cells in the regulation of the intracellular reproduction of chlamydiae remains to be established.

Macrophages (which occur in great numbers in the uterine mucosal lining) may be involved in eliminating microbial antibodies and immune complexes from the genital tract.

It is notable that there is a paucity of macrophages in the decidualized endometrium of listeria-infected women, which is in contrast to other sites affected by this bacterium. This may contribute to the tissue tropism of *Listeria monocytogenes* in pregnant women.

T-LYMPHOCYTES

Information on T-lymphocytes has been derived primarily from *in situ* staining of tissue with specific monoclonal antibodies. The endocervix is populated principally with the cytotoxic/suppressor type of T-lymphocytes (CD4/CD8 ratio=1/2). Few cells (<2%) show a phenotypic marker for activation (IL-2R) [12, 13].

The majority of T-lymphocytes in the cervical epithelium belongs to the T8 suppressor/cytotoxic subset, but there are also T-cells belonging to the T4 helper/inducer subset. The ratio has differed considerably in various studies. The T-cells are found exclusively in the basal layers of the squamous epithelium. However, they may also occur in high number in the transformation zone of the ectocervix.

Approximately 95% of circulating T-lymphocytes have antigen receptors on their surface that contain α- and β-chains. The rest of the cells contain such receptors composed of one γ- and one δ-chain. Unlike $\alpha\beta$ T-cells, most $\gamma\delta$ T-cells lack $CD4^+$ or $CD8^+$ surface molecules. $\alpha\beta$ T-cells and $\gamma\delta$ T-cells are $CD3^+$-positive. The function of the $\gamma\delta$ cells remains to be established, but there is some evidence that these cells may provide a first line of defence in mucosal infections. Their function may be to recognize a small number of specific alterations in epithelial cell surfaces rather than a wide repertoire of bacterial, viral, fungal, and protozoan antigens [14].

The T-cells of human peripheral blood contain two mononuclear cell types of opposing function. Dendritic cells act as antigen-presenting cells, while monocytes have a suppressive effect on antigen activation of T-cells. Patients with cervical carcinoma have an increased number of suppressive monocytes compared with healthy controls [15].

T-lymphocytes play an important role in the immune response to neoplasia. Normally about 60% of T-lymphocytes in the peripheral blood are $T4^+$ lymphocytes, while 20–30% are $T8^+$ ones. The T4/T8 ratio is usually 2/1. The average T4/T8 lymphocyte ratio in the peripheral blood of patients with leucokeratosis, cervical intra-epithelial neoplasia (CIN) and in those with cancer is 1.2–1.6. Healthy persons have an average T4/T8 ratio of 2.3 [16].

A significant T-cell depression has been observed in patients with severe dysplasia, cervical carcinoma *in situ*, as well as in micro- and invasive carcinoma. A low T-cell level was found in micro- and invasive carcinoma, which revealed an inverted OK T4$^+$/OK T8$^+$ ratio. These results support the hypothesis that cell-mediated immunity is of importance in premalignancy and in the early development of cervical cancer [17].

Patients with cervical cancer, as compared with controls, exhibit a decrease in the T- and B-lymphocyte count. An increase of lymphocyte reactivity to PHA is also seen. An increase in the PHA-induced lymphocyte ability of *in vitro* blast transformation occurs in patients with large tumours of the cervix or with metastases to the lymph nodes of the pelvis. In patients with small Stage I tumour, the mean indices of blast transformation do not differ from those observed for healthy controls. On the other hand, in patients with a large Stage I tumour and metastases to the lymph nodes of the pelvis, these indices exceed the average by 88% of the controls [18].

T-cells predominate over B-cells in infiltrates surrounding cervical cancers. The intensity of infiltration of T-cells is not correlated to the grade and prognosis of cervical cancer. The ratio of Leu 2a+ to Leu 3a+ cells tends to change with the spread of the cancer. Leu 2a+ cells are relatively predominant in early cases and cases with a good prognosis. There are very few B-cells close to cancer nests, and their presence is correlated with the grade and prognosis of cervical cancer [19].

In lymph nodes without tumour involvement in uterine cervical cancer patients, there are a greater number of lymphocytes of each subset in Stage Ib without nodal metastasis than in lymph nodes in Stage Ia and Ib and in Stage II with nodal metastasis, suggesting that the regional lymph nodes are immunocompetent in Stage Ib cancer without nodal metastasis. In lymph nodes with tumour involvement, there are great numbers of lymphocytes of each subset, especially of Leu 7+ cells, suggesting that strong immune reactions occur in these nodes. On the other hand, there are very few lymphocytes in the lymph nodes of diffused tumours. These findings suggest that regional lymph nodes in cervical cancer play an important role in the antitumour immune response [20].

HSV and also *C. trachomatis* have been discussed in the aetiology of cervical carcinoma. Patients with CIN and invasive cancer show approximately the same immune response to chlamydial and herpes simplex virus (HSV) antigens. About 58% of such patients showed a proliferative T-cell response to *C. trachomatis*, as was the case in 87% of those with HSV. Chlamydial antibodies were detected in 65% of the patients studied, while 81% were HSV-seropositive. A correlation between antibody titre and T-cell response was lacking. It is concluded that patients with CIN and invasive cervical cancer have an intact cellular immune response to both chlamydial and HSV antigens [21].

Neopterin is produced by human macrophages upon stimulation with interferon-γ and is therefore a measure for activated T-cells. In patients with cervical carcinoma and ovarian cancer, neopterin levels before therapy may be predictive of the clinical course; patients with pretherapeutically high neopterin levels are more likely to die earlier. Primary events in the immune surveillance are intact, but effect mechanisms might be damaged. Persistent activation of cell-mediated immunity appears to be unfavourable in cancer patients [22].

Spermatozoa can induce an increased concentration of T-lymphocytes, macrophages, and of polymorphonuclear leucocytes in the cervix [23]. These lymphoid cells migrate into the vagina after coitus.

LANGERHANS CELLS

The cellular constituents of the cervical mucosa make it unique, compared with other mucosal surfaces. The multiple layers of the epithelium do not contain bridging M-cells, but dendritic or so-called Langerhans cells, which are DR$^+$ and are thought to participate in antigen presentation [13, 24]. Furthermore, in the cervical mucosa 30% of Ig-producing cells produce IgG and 60% IgA [25]. This is lower than the 90% of IgA and 10% of IgG-producing B-cells found in the lamina propria of the gut.

Langerhans cells do occur in the lamina propria of the cervical squamous epithelium and stain positive for HLA-DR antigen and OK T6 [26]. Their concentration in the cervical epithelium is higher than that in the vagina. Langerhans cells express cell surface receptors for the Fc region of the IgG molecule and for complement factor C3. Class II major histocompability antigens (HLA-DR) are also present on the surface of Langerhans cells. This last property allows the cells to present antigens to T-lymphocytes and thereby initiate a specific immune response [27].

When a possible correlation between the amount of Langerhans cells and presence of human papillomavirus HPV 16 and HPV 18 was looked for, it was established that the presence of HPV reduced the number of Langerhans cells. Controversially, the absence of HPV was associated with an increased number of Langerhans cells in CIN. These findings suggest that the oncogenic potential of HPV 16 and HPV 18 in particular may be mediated by a specific effect on the efferent limb of the immune response [28].

In patients with various stages of CIN or in those who had koilocytosis and HPV infection of the uterine cervix, Langerhans cells were reduced in number and had a changed morphology. HLA-DR expression by Langerhans cells is significantly increased in koilocytic lesions and in CIN of grades I and II. HLA-DQ expression is also significantly increased in all grades of CIN, being most pronounced in CIN I. Columnar epithelium expresses major histocompatibility complex (MHC) class II antigens. These findings support the view that there is a localized disturbance of the immune function in both neoplastic cervical epithelium and in papillomavirus-infected epithelium [29].

The mean density of S100+ Langerhans cells in cervical carcinoma cases is significantly higher than in cervicitis cases. S100+

Langerhans cells seem to play a major role in immunosurveillance [30].

A correlation between infiltration of immunogenic cells in tumour tissues and prognosis of radiation therapy has been established. In patients with cervical cancer treated with radiation therapy alone, including patients with Stage III squamous carcinomas and adenocarcinomas of all stages, the 5-year-survival rate for patients with infiltration of Langerhans cells was significantly better than that of those without such detectable cells (78% versus 60%). The 5-year survival rate of patients with T-cell infiltration was significantly better than that of patients without such cell infiltration (83% versus 61%). Similar trends were observed in patients with adenocarcinoma.

The 5-year survival rates for patients with Langerhans cell infiltration and those without such cell infiltration were 49% and 25%, respectively. The survival rates for patients with and without T-cell infiltration were 50% and 33%, respectively. Failure of radiation therapy demonstrated that the favourable prognosis by Langerhans cell infiltration was attributable mainly to improvement of local control mechanisms, while that of T-cell infiltration was not. The immune response of individual cancer patients may be remarkably different at the first step of antigen recognition by Langerhans cells. The Langerhans cells may induce a T-cell-mediated antitumour response and improve local response in radiation therapy [31].

NATURAL KILLER (NK) CELLS

NK activity is low in patients with invasive cervical carcinoma (Stages I and II). However, differences in lymphocyte populations in patients with pre- and invasive carcinoma and controls are, however, not found [32]. NK cells are identified in the cervix in many cases of cervical cancer, but they are scattered and not correlated with grade and prognosis of cervical cancer [19].

Postoperative immunotherapy enhances NK

activity in patients with cervical carcinoma. In such cases NK activity is lower in patients with Stages I and II of cervical carcinoma both before and after such therapy. The number of cells carrying the marker of a non-major histocompatibility complex restrict cytotoxic lymphocytes in the peripheral blood and NK activity increases significantly after immunotherapy [15].

NK lymphocytes are present in most HPV-associated CIN lesions. HPV-positive cervical cancer cells and HPV-immortalized human cervical epithelial cells, which possess properties similar to cervical dysplasia, are, however, resistant to NK cells. They are sensitive to lymphokine-activated killer lymphocyte lysis. Sensitivity can be enhanced by treatment of cervical cells with leucoregulin, a cytokine secreted by lymphocytes. Treatment with leucoregulin and chemotherapeutic drugs, e.g. cisplatin, further enhances sensitivity of HPV-infected cells to lymphokine-activated killer lymphocyte lysis. In contrast, interferon-γ treatment of cervical cells can result in decreased sensitivity to lymphokine-activated killer lysis, illustrating the balance cytokines can exert in the immunologic control of cervical cancer [33].

Both early ($\leq$20 weeks) and late ($>$30 weeks) passage HPV-16-immortalized cells are resistant to NK lymphocyte cytotoxicity, but sensitive to lymphokine-activated killer lymphocyte cytotoxicity of a lymphocyte/cervical cell ratio ranging from 1/1 to 50/1. Treatment of early-passage HPV-16 DNA-immortalized cells with the NK lymphocytotoxicity-sensitizing lymphokine leucoregulin for 1 hour induced a modest sensitivity to NK cells. It up-regulated the lymphokine-activated killer sensitivity two- to three-fold. At later passages, leucoregulin up-regulation of the sensitivity to NK cells is lost, but remains so to lymphokine-activated killer lymphocytotoxicity. Similarly, the cervical carcinoma cell line, QGU, is also resistant to NK toxicity but sensitive to lymphokineactivated killer lymphocytotoxicity. Leucoregulin exposure did not confer sensitivity to

the NK-resistant QGU tumour cells, but 1.5–2-fold increased their sensitivity to lymphokine-activated killer lymphocytotoxicity [34].

The reduction of NK cell activity in cancer patients is not an irreversible phenomenon, since it can be restored by lipid administration. This is important in relation to immunotherapy by lymphokine-activated NK cells [35].

The NK cell activity in women with malignant tumours is lower than in healthy persons and in women with benign tumours, suggesting that there is a defective NK activity in patients with gynaecological malignancies. Such activity in patients with ovarian malignant tumours is much lower than in those with cervical cancer. Leucocyte interferon increases the NK cell activity to nearly normal levels in cases of gynaecological malignancies. Interferon has been used as immunotherapy for gynaecological malignancies [36].

Autologous serum factors can influence NK cell activity by depression. The reduction increases with the tumour load. Patients with poorly differentiated tumours have lower levels of NK cell activity than those with well differentiated tumours. An increase in circulating immune complexes is also evident. Poorly differentiated tumours show the highest levels of immune complexes [37].

A significant reduction in NK and antibody dependent cell-mediated cytotoxicity is observed in patients with disseminated cervical carcinoma as compared to localized cancers. In the majority of patients, who receive radiotherapy, both NK and antibody-dependent cell-mediated cytotoxicity activity recover after therapy. Furthermore, interferon-α can be demonstrated to augment NK activity in healthy donors as well as in cancer patients [38].

NK cells may play a role in HPV infection in patients with CIN. The frequency and pattern of distribution of NK cells is similar in all grades of CIN. The number of NK cells present is usually small, but the degree of infiltration by Leu-11 positive cells is pronounced in HPV infections and in patients with CIN 1 [39].

IMMUNOGLOBULINS

In humans and subhuman primates as well as in other mammalian species, the level of immunoglobulins in secretions of the different compartments of the female reproductive tract is low as compared to that of serum. The levels show typical cyclic changes under the influence of oestrogen and progesterone [40]. Cervical and vaginal secretions contain immunoglobulins of several Ig classes. The Fallopian tubes also contain Ig-producing cells which, as in the vagina and the cervix, are located subepithelially. A great number of immunoglobulin-producing cells are seen in the Fallopian tubes of cases of acute and chronic salpingitis [41].

Antibody levels in cervical mucus appear to be especially low during the pre-ovulatory period and, according to observations in primates, also in tubal fluid.

There are indications that the secretory immune system may be operational mainly in the cervical compartment of the female genital tract [40].

Many studies have been published concerning IgA produced in the lower female genital tract, while few publications have dealt with IgM and IgG production at the same site.

IgG makes up about one-third of the concentration of Ig in the cervico-vaginal secretion [24]. Subepithelial cervical immunocytes capable of synthesizing IgG increase in number when progesterone is in excess, as in the luteal phase of the menstrual cycle. Their number declines during the follicular phase. Of the immunoglobulin-producing cells in the endo- and ectocervix and in the vagina, 15, 11 and 7% produce IgG, respectively [42].

IgM makes up about 10% of the concentration of Ig in the cervicovaginal secretion, while the rest consists of sIgA and IgA. No obvious differences in this respect are found between the endo- and ectocervical secretion on one hand and the vaginal secretion on the other [24].

The majority of IgA found in the vagina is derived from the uterus and transported through the cervical canal to the vagina or is produced in the endocervical mucosa. In all mucosal secretions, including those of the cervix, the concentration of sIgA exceeds that of IgG [24]. In the cervico-vaginal secretion, the concentration of sIgA is more than double that of monomeric IgA [43].

Secretory IgA_1 is the dominating antiprotein antibody, although when the antigen is a lipopolysaccharide or a lipoteichoic acid, it is IgA_2. Analysis of cervical B-lymphocytes has relied primarily upon immunochemical studies of lumenal immunoglobulins, where potential contributions by Fallopian tube and uterine sources as well as serum transudation may cloud the interpretation. Within these limitations, about 60% of immunoglobulin is believed to be dimeric IgA containing J-chain, with the predominating IgA_2 subclass [24]. A wide range (0.5–1.4 µg/ml) of concentrations of IgA in cervical secretion have been reported [25]. This may to some extent reflect variations within the hormonal levels during the menstrual cycle. Nevertheless, the concentration of IgA in cervical secretion is low compared with that in serum and intestinal secretion (1.4–4.0 mg/ml) [44].

In serum, IgA circulates as a monomeric structure. By the addition of J-chain (formed by IgA committed plasma cells found adjacent to the mucosa), the formation of a IgA dimer can take place. This occurs when IgA traverses the mucosal epithelium of the lower genital tract. The addition of the J-chain or the so-called secretory component increases the molecular weight of sIgA to 395,000, as compared with 160,000 for monomeric serum IgA.

IgA antibodies are found in the cervico-vaginal secretion of 89% of patients with CIN, in 33% of patients with koilocytosis and condylomas (but without CIN), and in 25% of women with normal Pap-smear and colposcopy. The proportion of women with an IgA-containing cervical secretion was significantly higher in the CIN patients than in the control group [45]. However, in contrast to IgG, IgA levels decrease with age [43].

In patients with benign and malign lesions of the cervix, the oral cavity and of the breast, IgE was found to be elevated only in those with benign lesions. Serum IgA, IgD and IgE levels were elevated in all of these three types of cancer and the concentration increased with clinical stage. In carcinoma of the cervix, IgG levels are often also found to be elevated [46].

Immunoglobulins A and D return to normal after clinical cure, whereas the IgE level remains slightly elevated. IgD and IgE remain high in patients who have residual cancer [46].

The IgG, IgA and IgM levels are significantly higher in the cervical mucus of users of copper IUD as compared to combined oral contraceptive users and women not using any contraception [47].

CYTOKINES

Of the interleukins (IL), IL-5 selectively enhances IgA synthesis and this effect is synergized by IL-2. IL-5 also enhances IL-2 receptor expression on IgA-producing cells. IL-6 contributes to clonal expression of cells synthesizing IgA.

The levels of interferon-γ in genital secretion is age-related, being higher in older than younger women. Interferon-γ was found in higher levels (6.71 ± 2.8) in endocervical secretion of 47 women infected with *C. trachomatis* as compared to 52 currently non-infected women (1.4 ± 0.4) ($p \leq 0.002$) [48]. The levels of interferon-γ in plasma did not correlate with the presence of chlamydia organisms as detected by culture studies, or by the presence of signs of genital inflammation.

Interferon can induce intracellular production of nitrogen oxide, which interferes with intracellular reproduction of *C. trachomatis* [49].

COMPLEMENT

By complement fixation and IgG antibody binding, microbes can be eliminated from the cervix.

C3b fragments are recognized by gonococcal receptors on PMN cells. Antigonococcal IgM and IgG serum antibodies can activate complement.

INFLUENCE OF SEX HORMONES ON IMMUNE COMPETENCE

Concentration of sIgA in cervical secretion is not constant, but decreases to almost undetectable levels at the midpoint of the menstrual cycle. Levels of IgA are greater particularly in the progestation stage, suggesting that immunoglobulin synthesis or secretion is influenced by circulating hormones [50].

In an animal model, in which oestrogen was injected daily for 3 days in previously overectomized adult females, IgA and IgG levels rose at least 10-fold in the uterine secretion, while it decreased at least four-fold in the vaginal secretion. Similar trends were noted with antibodies to sheep erythrocytes. The antibody levels were measured 13 days after a second immunization with sheep erythrocytes, in animals in which estradiol had been given daily over 3 days prior to assay [51]. The authors suggested that uterine IgA is not derived from serum but from local synthesis, and that estradiol stimulates IgA-lymphocyte migration from mesenteric lymph nodes to the reproductive tract.

Sex hormones affect the function of T- and B-lymphocytes. Also, NK cells and macrophages are under the influence of such hormones. However, how this influence occurs is not known.

Information concerning the effect of hormones on the immune system in the cervical tissue has been generated from studies in patients receiving hormones as medication. The IgA concentration in cervical secretion was low during oestrogen administration and rose after administration of progesterone [50]. Increase of IgA plasma cells occurred after administration of oestrogen or progesterone [52]. These results suggest that a specific receptor, probably on the endothelium of post-capillary venules in the cervical mucosa, are susceptible to hormonal modulation, and can bind a complementary structure on the surface of the IgA-bearing B-lymphocytes.

In one study, the number of cervical IgA plasma cells was equally low (20/mm^2) in both early and late pregnancy. Patients in late gestation (40 weeks) had twofold or more cervical epithelial cells containing secretory component (J-chain) and IgA, as compared with those in early pregnancy [52].

The B-lymphocyte population of the cervix increases, irrespective of the hormone status, i.e. an increased number of IgM- and IgA-producing B-lymphocytes are found in cervical biopsy specimens of patients with genital infections with, for example *N. gonorrhoeae*, *T. vaginalis* and *Candida albicans* [53].

SEX HORMONES

The effect on the uptake of intracellular parasites in host cells, for example of *C. trachomatis*, has been shown both *in vitro* studies [54] and in animal models [55]. They also influence the spread of chlamydiae from the cervix to the Fallopian tubes [56].

Few lymphocytes are found in the genital tract of healthy women at any stage of the menstrual cycle. The few lymphocytes that may occur are found in the subepithelial layers [26]. Other immunocompetent cells, like granulocytes and macrophages, occur in large numbers during menses, but are few during the proliferate phase [57].

Ig RESPONSE TO VARIOUS INFECTIOUS AGENTS

Locally produced IgG antibodies to a variety of infectious agents have been demonstrated, for example to bacteria, chlamydiae, viruses, fungi and parasites [41].

Secretory IgA has a role of activating the complement system by the alternative pathway and thereby killing microorganisms by lysis. The role of IgA and sIgA is to facilitate agglutination of microorganisms and thereby inhibit their attachment to the mucosal lining. IgA deficiency, which compared to many other immunodeficiences is relatively common (1:600 persons). However, it does seem to result in an increased susceptibility to genital tract infections.

IgA antibodies to *N. gonorrhoeae* are produced by plasma cells localized in the lamina propria of the endocervix. IgA antibodies can inhibit attachment of *N. gonorrhoeae* to the mucosa of the cervix, although this effect is limited. Gonococcal OMP antibodies can, as already mentioned, interfere with the attachment of gonococci [58–60]. IgG$_1$ and IgG$_3$ antigonococcal LPS antibodies may lyse gonococci, but the protection of such antibodies is limited. Pili antigen protein II, a 23–33 kDa antigen, and presumably also LPS, are predominant gonococcal antigens which can react with gonococcal serum IgG antibodies. IgA reacts also with gonococcal protein II and 46–48 kDa proteins.

Antibodies to *C. albicans* occurring in cervico-vaginal secretion are mainly of the IgA class [61]. They, however, do not prevent *Candida* vulvovaginitis in susceptible individuals [62]. Antibody to purified macrophages incubated with *C. albicans* and lymphocytes can inhibit recognition by lymphocytes [63] and processing of *Candida* antigens by macrophages.

The cervical secretion of women with chlamydial cervicitis contains greater amounts of albumin, IgG, IgA and sIgA than in those with vulvovaginal candidiasis. A comparison of the IgG/IgA ratios in serum and cervical secretion indicates a local secretion of specific IgA antibodies, supported by the demonstration of sIgA [64].

The rate of sexual transmission of HIV appears to correlate to the level of lymphoid cells present during coitus in genital secretions [65]. However, comparing HIV-seropositive and -negative women, all of whom had leucorrhoea from presumed infection(s). CD4$^+$ T-lymphocytes were identified in combined vaginal and exocervical secretion from all subjects studied [66].

HPV antibody- (in 50% of cases) and T-cell responses (in 25% of cases) seen in patients with cervical intra-epithelial neoplasia correspond with a history of past or present skin warts. Although HPV antibodies are detected in about 50% of women who thought they had never had warts, only 8% with no known history of warts had positive stimulation indices in lymphocyte proliferation assays [67].

Cervical secretions in 62% of patients with condylomas and 29% of controls had IgA antibodies to the HPV E2 synthetic peptide. Fifty-nine per cent of patients with condylomas and 18% of controls had sIgA antibodies to the E7 peptide [3].

The association between the presence of certain types of HPV DNA and cervical carcinoma is well documented. However, less is known about the immune response to HPV infections and its relationship to cervical cancer. A higher prevalence of antibodies to the HPV-16 peptide E7 among women with cervical cancer compared with controls has been reported, but the possible reactivity to other antigens has not been systematically examined [68]. The HPV 16 E7 peptide seems to be a marker for invasive cervical cancer [69].

In cervical carcinoma, there is a significant increase in the total concentrations of IgG and IgA in Stage I, and of IgG and IgM in Stage II and of all three immunoglobulins in Stage III as compared to women with chronic cervicitis, but without signs of cancer. The occurrence of circulating immune complexes and T-cell depression is also stage-related [70].

Of cervical cancer patients, 63.0% were positive for antibodies to E4 and/or to E7 of HPV 16 and/or to E7 of HPV 18. In contrast, only 6.5% of non-genital cancer patients and 9.8% of healthy individuals were antibody-positive for HPV 16 to E4 or to E7, while antibodies to the homologous proteins of HPV 18 could not be detected. Prevalence rates of antibodies to the HPV 16/18 late proteins occurred in 54.3% of women with cervical carcinoma, in 41.9% in those with HPV non-genital cancer-related

types and in 43.9% in healthy individuals. Antibodies to HPV 6b late gene products ranged between 6.5 and 12.2% in the different patient groups studied. Antibodies to late proteins may indicate that, regardless of clinical stage, HPV infections are widespread in the female population. The striking difference between the prevalence rates of antibodies to early proteins of HPV 16 and HPV 18 among cervical cancer patients and controls supports the involvement of these HPV types in the carcinogenesis of the cervix [71].

The human cervical mucosa is a rich source of antibodies to HSV, particularly in patients with clinical lesions [72].

Among a group of women with first-episode genital herpes (positive for HSV by culture), sIgA antibody to HSV-2 was detected in 48%. The mean titre of sIgA antibody to HSV-2 peaked between days 9 and 16 days after onset of the disease, whereas the peak occurred at days 3–8 among women with recurrent genital HSV. HSV-2 was not isolated from the cervix from any samples taken when the titres of sIgA antibody to HSV-2 were greater than or equal to 1:2, compared with samples taken from patients with titres less than or equal to 1:2 [73].

In cervical secretions of women attending an STD clinic, 33% had sIgA HSV-2 antibodies. Twenty-five per cent of them also reacted with a type-specific HSV antigen, but only 10% had HSV-2 IgG antibodies detectable in serum. Secretory IgA against HSV was found in significantly more women with cervicitis than in controls. Local antibody production and immunity were triggered by previous exposure to HSV. Local mucosal immunity to HSV-2 can be detected in women who do not have a specific humoral antibody response to the virus [64].

Serum IgG levels to HSV are significantly elevated in women with mild and severe cervical dysplasia/carcinoma *in situ*. Serum IgA antibody titres to HSV-1 are significantly higher in women with cervical squamous cell carcinoma and other genital tumours than in controls. Significantly higher levels of serum IgA to HSV-2 are detectable in the two first groups [74].

In the series of patients with cervical carcinoma, 96% were HSV-positive as compared to 87% of controls. Antibodies to HSV-2 were found in the sera by immunoblotting tests from 24% of the patients and 17% of the controls. Patients and controls were typed for HLA-A, B, C and D/DR antigens, but no significant association was found [75].

Patients with CIN had significantly increased neutralizing antibody activity to HSV-2 in cervical mucus than in controls. While the patients studied differed from the controls with respect to the number of lifetime sexual partners, there was significant higher neutralizing antibody activity in patients when controlled for number of sexual partners. Also, socio-economic grouping shows such a correlation. The results lead the investigators to believe that there is a putative association between HSV-2 infection and pre-invasive and invasive carcinoma of the uterine cervix [76].

Under experimental conditions the T-cell response both to HSV-1 and purified protein derivative (PPD) is about the same in the two groups, but after removal of monocytes a significantly higher T-cell response is seen among the patients with advanced stages of carcinoma (International Federation of Gynaecology and Obstetrics (FIGO) Stages IIb–IVa) as compared with patients in early stages (FIGO Stages Ia–IIa) and healthy controls. These observations indicate that patients with cervical carcinoma have an increased number of T-cells reactive with HSV, and normal dendritic cell function. The relative suppression expressed on a per monocyte basis is the same in all groups, which indicates that the suppression in the cancer patients is caused by an increased number of monocytes, and not by changes in their activation state [15].

ANTISPERM ANTIBODIES

A significant association between sperm-bound IgA antibodies (less than 3.0 cm/2 h) and mucus sperm penetration is found. There is also such an association between poor sperm penetration and the presence of antibodies on the sperm tail mainpiece. In contrast, IgG antibodies do not show any association with poor penetration [77]. Sperm antibodies may impair the sperm migration through the cervical secretions [40].

Approximately 70% of the cervical mucus IgA antisperm antibodies are of the IgA_1 subclass. Cervical mucus IgG is primarily of the IgG_4 subclass [78].

Antisperm antibodies of the IgA_1 isotype in cervical secretion are susceptible to bacterial IgA_1 proteases [79]. Fab fragments of these antibodies, generated by proteolytic cleavage do not, however, interfere with sperm mobility due to their monovalency and consequent inability to cross-link antigens.

REFERENCES

1. Corbeil, L.B., Wunderlich, A.C., Lyons, J.M. and Braude, A.I. (1984) Specific cross-protective antigonococcal immunity in the murine genital tract. *Canadian Journal of Microbiology*, **30**, 482–7.
2. Dillner, L., Bekassy, Z., Johnsson, N., Moreno-Lopez, J. and Blomberg, J. (1989) Detection of IgA antibodies against human papilloma virus in cervical secretions from patients with cervical intraepithelial neoplasia. *International Journal of Cancer*, **43**, 36–40.
3. Dillner, L., Fredriksson, A., Persson, E., Forslund, O., Hansson, B.G. and Dillner, J. (1993) Antibodies against papilloma virus antigens in cervical secretions from condyloma patients. *Journal of Clinical Microbiology*, **31**, 192–7.
4. Kilian, M., Mastecky, J. and Russel, M.W. (1988) Defense mechanisms involving Fc-dependent functions of immunoglobulin A and their subversion by bacterial immunoglobulin A proteases. *Microbiology Reviews*, **52**, 296–303.
5. Cohen, M.S., Weber, R.D. and Mårdh, P.-A. (1990) Genitourinary mucosal defenses, in *Sexually Transmitted Diseases*, 2nd edn (eds K.K.

Holmes, P.-A. Mårdh, P.F. Sparling and P.J. Wiesner), McGraw-Hill, New York.

6. Dutertre, Y., Haglund, O., Fröman, G., Bioteau, O. and Mårdh, P.-A. (1993) Cycloheximide reduces nitric oxide production and increases infectivity of *Chlamydia trachomatis* in McCoy cells. *Proceedings of the Tenth International Meeting on STD Research*, University of Helsinki, 239.

7. Möller, B., Thorsen, P. and Mårdh, P.-A. (1991) Experimental inoculation of *Chlamydia trachomatis* in animal models, in *Chlamydial Infections of the Genital and Respiratory Tracts and Allied Conditions* (eds P.-A. Mårdh and P. Saikku), Uppsala University Centre for STD Research, Uppsala.

8. Milsom, I., Nilsson, L.A., Brandberg, A., Ekelund, P., Mellström, D. and Eriksson, O. (1991) Immunoglobulin A (IgA) levels in post-menopausal women: influence of oestriol therapy. *Maturitas*, **13**, 129–35.

9. Edvards, J.N. and Morris, H.B. (1985) Langerhan's cells and lymphocyte subsets in the female genital tract. *British Journal of Obstetrics and Gynaecology*, **92**, 9774–8.

10. Fowler, J.E. and Mariano, M. (1984) Longitudinal studies of prostatic fluid immunoglobulin in men with bacterial prostatitis. *Journal of Urology*, **131**, 363–9.

11. Rein, M.F. and Müller, M. (1990) *Trichomonas vaginalis* and trichomoniasis, in *Sexually Transmitted Diseases* (eds K.K. Holmes, P.-A. Mårdh, F.R. Sparling, P. Wiesner, J. Cates, S.M. Lemon and W.E. Stamm), McGraw-Hill, New York.

12. Becker, J., Behem, J., Loning, T.H., Reichart, P. and Geerlings, H. (1985) Quantitative analysis of immunocompetent cells in human normal oral and uterine cervical mucosa, oral papillomavirus and leukoplakias. *Archives in Oral Biology*, **30**, 257–64.

13. Roncalli, M., Sideri, M., Gie, P. and Servida, E. (1988) Immunophenotypic analysis of the transformation zone of human cervix. *Laboratory Investigations*, **58**, 141–9.

14. Bozner, P., Gombosova, A., Valent, M., Demes, P. and Alderete, J.F. (1992) Proteinases of *Trichomonas vaginalis*: antibody response in patients with urogenital trichomoniasis. *Parasitology*, **105**, 89–91.

15. Bjercke, S., Onsrud, M. and Gaudernack, G. (1986) Suppressive effect of monocytes *in vitro* in patients with carcinoma of the uterine cervix. *Acta Obstetrica et Gynecologica Scandinavia*, **65**, 619–24.

16. Kesic, V., Sulovic, V., Bujko, M. and Dotlic, R. (1990) T lymphocytes in non-malignant, pre-malignant and malignant changes of the cervix. *European Journal of Gynaecological Oncology*, **11**, 191–4.

17. Castello, G., Esposito, G., Stellato, G., Dalla Mora, L., Abate, G. and Germano, A. (1986) Immunological abnormalities in patients with cervical carcinoma. *Gynecologic Oncology*, **25**, 61–4.

18. Kietlinska, Z. (1984) T and B lymphocyte counts and blast transformation in patients with Stage I cervical cancer. *Gynecologic Oncology*, **18**, 247–56.

19. Okamoto, Y., Sano, T., Okamura, S., Ueki, M. and Sugimoto, O. (1987) [An immunohistochemical study with monoclonal antibodies on lymphocytes infiltrating in cervical cancer]. *Nippon Sanka Fujinka Gakkai Zasshi*, **39**, 925–32.

20. Kinugasa, M., Akahori, T., Mochizuki, M. and Hasegawa, K. (1991) [Distribution of lymphocyte subsets in regional lymph nodes in uterine cervical cancer and its immunological significance]. *Nippon Sanka Fujinka Gakkai Zasshi*, **43**, 383–90.

21. Qvigstad, E., Onsrud, M., Degre, M. and Skaug, K. (1985) Cell-mediated and humoral immune responses to chlamydial and herpesvirus antigens in patients with cervical carcinoma. *Gynecologic Oncology*, **20**, 184–9.

22. Fuchs, D.N., Fuith, L.C., Hausen, A., Hetzel, H., Reibnegger, G.J., Werner, E.R. and Wachter, H. (1988) Preactivated T cells in cancer patients with poor prognosis. *Cancer Detection and Prevention*, **12**, 97–103.

23. Berman, B.H. (1964) Seasonal allergic vulvovaginitis caused by pollen. *Annals of Allergy*, **22**, 594–7.

24. Kutteh, W.H., Hatch, K.D., Blackwell, R.E. and Mestecky, J. (1988) Secretory immune system of the female reproductive tract. I. Immunoglobulin and secretory component-containing cells. *Obstetrics and Gynecology*, **71**, 56–60.

25. Ogra, P.L., Yamanaka, T. and Losonsky, G.A. (1981) Local immunologic defenses in the genital tract. *Reproductive immunology*, 381–94.

26. Belec, L., Georges, A.J., Steenman, G. and Martin, P.M. (1989) Antibodies to human immunodeficiency virus in vaginal secretions of heterosexual women. *Journal of Infectious Diseases*, **160**, 385–91.

27. Hargreave, T.B., James, K., Kelly, R., Skibinski, G. and Szymaniec, S. (1993) Immunosuppressive factors in the male reproductive tract, in *Local Immunity in Reproductive Tract Tissues* (eds P.D. Griffin and P.M. Johnson), Oxford University Press, Oxford.

28. Hawthorn, R.J., Murdoch, J.B., MacLean, A.B. and MacKie, R.M. (1988) Langerhans cells and subtypes of human papillomavirus in cervical intraepithelial neoplasia. *British Medical Journal*, **297**, 643–6.

29. Hughes, R.G., Norval, M. and Howie, S.E. (1988) Expression of major histocompatibility class II antigens by Langerhans' cells in cervical intraepithelial neoplasia. *Journal of Clinical Pathology*, **41**, 253–9.

30. Mann, V.M., de Lao, S.L., Brenes, M., Brinton, L.A., Rawls, J.A., Green, M., Reeves, W.C. and Rawls, W.E. (1990) Occurrence of IgA and IgG antibodies to select peptides representing human papillomavirus type 16 among cervical cancer cases and controls. *Cancer Research*, **50**, 7815–19.

31. Nakano, T., Oka, K., Takahashi, T., Morita, S. and Arai, T. (1992) Roles of Langerhans cells and T-lymphocytes infiltrating cancer tissues in patients treated by radiation therapy for cervical cancer. *Cancer*, **70**, 2839–44.

32. Chou, C.Y., Hsieh, C.Y., Hsieh, K.H. and Chen, C.A. (1989) Decreased natural killer cell activity in patients with invasive cervical carcinoma. *Taiwan I Hsueh Hui Tsa Chih*, **88**, 128–31.

33. Evans, C.H., Flugelman, A.A. and DiPaolo, J.A. (1993) Cytokine modulation of immune defenses in cervical cancer. *Oncology*, **50**, 245–51.

34. Furbert-Harris, P.M., Evans, C.H., Woodworth, C.D. and DiPaolo, J.A. (1989) Loss of leukoregulin up-regulation of natural killer but not lymphokine-activated killer lymphocytotoxicity in human papillomavirus 16 DNA-immortalized cervical epithelial cells. *Journal of the National Cancer Institute*, **81**, 1080–5.

35. Garzetti, G.G., Cignitti, M., Marchegiani, F., Provinciali, M., Zaia, A.M., Fabris, N., Pieri, C. and Romanini, C. (1990) Improvement of decreased natural killer cell activity in cervical cancer patients by active lipids (AL 721). *European Journal of Gynaecology and Oncology*, **11**, 123–5.

36. Ma, D., Gu, M.J. and Liu, B.Q. (1990) A preliminary study on natural killer activity in patients with gynecologic malignancies. *Journal of the Tongji Medical University*, **10**, 159–63.

37. Pillai, M.R., Balaram, P., Abraham, T., Padmanabhan, T.K. and Nair, M.K. (1988) Natural cytotoxicity and serum blocking in malignant cervical neoplasia. *American Journal of Reproductive Immunology and Microbiology*, **16**, 159–62.

38. Satam, M.N., Suraiya, J.N. and Nadkarni, J.J. (1986) Natural killer and antibody-dependent cellular cytotoxicity in cervical carcinoma patients. *Cancer Immunology and Immunotherapy*, **23**, 56–9.

39. Tay, S.K., Jenkins, D. and Singer, A. (1987) Natural killer cells in cervical intraepithelial neoplasia and human papillomavirus infection. *British Journal of Obstetrics and Gynaecology*, **94**, 901–6.

40. Schumacher, G.F. (1988) Immunology of spermatozoa and cervical mucus. *Human Reproduction*, **3**, 289–300.

41. Roche, J.K. and Crum, C.P. (1991) Local immunity and the uterine cervix: implications for cancer associated viruses. *Cancer Immunology and Immunotherapy*, **33**, 203–9.

42. Hagman, M. and Danielsson, D. (1989) Increased adherence to vaginal epithelial cells and phagocytic killing of gonococci and urogenital meningococci associated with heat modifable proteins. *APMIS*, **97**, 839–44.

43. Waldman, R.H., Cruz, J.M. and Rowe, D.S. (1971) Immunoglobulin levels and antibody to *Candida albicans* in human servicovaginal secretions. *Clinical Experimental Immunology*, **10**, 427–34.

44. Eisen, H.N. (1990) Immunoubulins and immunoglobulin genes, in *General Immunology* (ed. H.N. Eisen), Lippincott, Philadelphia, PA.

45. Dillner, L., Bekassy, Z., Jonsson, N., Moreno-Lopez, J. and Blomberg, J. (1989) Detection of IgA antibodies against human papillomavirus in cervical secretions from patients with cervical intraepithelial neoplasia. *International Journal of Cancer*, **43**, 36–40.

46. Vijayakumar, T., Ankathil, R., Remani, P., Sasidharan, V.K., Vijayan, K.K. and Vasudevan, D.M. (1986) Serum immunoglobulins in patients with carcinoma of the oral cavity, uterine cervix and breast. *Cancer Immunology and Immunotherapy*, **22**, 76–9.

47. Eissa, M.K., Sparks, R.A. and Newton, J.R. (1985) Immunoglobulin levels in the serum and cervical mucus of tailed copper IUD users. *Contraception*, **32**, 87–95.

48. Kalo-Klein, A. and Witkin, S.S. (1991) Regulation of the immune response to *Candida aibicans* by monocytes and progesterone. *American Journal of Obstetrics and Gynecology*, **164**, 1351–4.

49. Witkin, S.S., Jeremias, J. and Ledger, W.J. (1989) Vaginal eosinophils and IgE antibodies to *Candida albicans* in women with recurrent vaginitis. *Journal of Medical and Vetinary Mycology*, **27**, 57–8.

50. Schumacher, G., Kimm, M.H. and Hosseinian A.H. (1977) Immunoglobulins, proteinase inhibitors, albumin and lysozyme in human

cervical mucus. *American Journal of Obstetrics and Gynecology*, **129**, 624–35.

51. Wira, C.R. and Sandoe, C.P. (1986) Origin of IgA and IgG antibodies in the female reproductive tract: regulation of the genital response by estradiol, in *Recent Advances in Mucosal Immunology: Part 1. Cellular Interactions* (ed. M. J.), Plenum Press, New York.

52. Murdoch, A.J.M., Buckley, C.H. and Fox, H. (1982) Hormonal control of the secretory immune system of the human cervix. *Journal of Reproductive Immunology*, **4**, 23–30.

53. Chipperfield, E.J. and Evans, B.A. (1972) The influence of local infection on immunoglobulin formation in the human endocervix. *Clinical Experimental Immunology*, **11**, 219–23.

54. Witkin, S.S. (1991) Immunologic factors influencing susceptibility to recurrent candidal vaginitis. *Clinical Obstetrics and Gynecology*, **34**, 662–8.

55. Witkin, S.S., Kalo-KIein, A., Galland, L., Teic, M. and Ledger, W.J. (1991) Effect of *Candida albicans* plus histamine on prostaglandin E$_2$ production by peripheral blood mononuclear cells from healthy women and women with recurrent candidal vaginitis. *Journal of Infectious Diseases*, **164**, 396.

56. Witkin, S.S. (1993) Immunology of the vagina. *Clinical Obstetrics and Gynecology*, **36**, 122–8.

57. Kinane, D.F., Weir, D.M., Blackwell, C.C. and Winstanley, F.P. (1984) Binding of *Neisseria gonorrhoeae* by lectin like receptors on human phagocytes. *Journal of Clinical and Laboratory Immunology*, **13**, 107–10.

58. O'Hagen, D.T., Rafferty, D., McKeating, J.A. and Illum, L. (1992) Vaginal immunization of rats with a synthetic peptide from human immunodeficiency virus envelope glycoprotein. *Journal of General Virology*, **73**, 2141–5.

59. O'Reilly, R.J., Lee, L. and Welch, B.G. (1976) Secretory IgA response to *Neisseria gonorrhoeae* in the genital secretions in infected females. *Journal of Infectious Diseases*, **133**, 113–25.

60. Miller, C.J., Kang, D.W., Marthas, M., *et al.* (1992) Genital secretory immune response to chronic simian immunodeficiency virus (SIV) infection: a comparison between intravenously and genitally inoculated Rhesus macaques. *Clinical Experimental Immunology*, **88**, 520–4.

61. Heisterberg, L., Branebjerg, P.E., Scheibel, J. and Hoj, L. (1977) The role of vaginal secretory immunoglobulin A, *Gardnerella vaginalis*, anaerobes and *Chiamydia trachomatis* in postabortal pelvic inflammatory disease. *Acta Obstetrica et Gynecologica Scandinavica*, **66**, 99–102.

62. Kalo-KIein, A. and Witkin, S.S. (1989) *Candida albicans*: cellular immune system interactions during different stages of menstrual cycle. *American Journal of Obstetrics and Gynecology*, **161**, 1132–6.

63. Rebello, R., Green, F.H. and Fox, H. (1975) A study of the secretory immune system of the female genital tract. *British Journal of Obstetrics and Gynaecology*, **82**, 812–16.

64. Persson, E., Eneroth, P. and Jeansson, S. (1988) Secretory IgA against herpes simplex virus in cervical secretions. *Genitourinary Medicine*, **64**, 373–7.

65. Koenig, S. and Fauci, A.S. (1990) Immunology of HIV infection, in *Sexually Transmitted Diseases* (eds K.K. Holmes, P.-A. Mårdh, P.F. Sparling, P.J. Wiesner, W. Cates, S.M. Lemin and W.E. Stamm), McGraw-Hill, New York.

66. Mathews, H.M., Moss, D.M. and Callaway, C.S. (1987) Human serologic response to subcellular antigens of *Trichomonas vaginalis*. *Journal of Parasitology*, **73**, 601–10.

67. Cubie, H.A. and Norval, M. (1988) Humoral and cellular immunity to papillomavirus in patients with cervical dysplasia. *Journal of Medical Virology*, **24**, 85–95.

68. Mandelson, M.T., Jenison, S.A., Sherman, K.J., Valentine, J.M., McKnight, B., Daling, J.R. and Galloway, D.A. (1992) The association of human papillomavirus antibodies with cervical cancer risk. *Cancer Epidemiology, Biomarkers & Prevention*, **1**, 281–6.

69. Meng, X. (1992) [Role of Langerhans' cells against cervical human papilloma virus infection and development of cervical carcinoma]. *Chung Hua I Hsueh Tsa Chih*, **72**, 155–7.

70. Agarwal, J., Gupta, S.C., Singh, P.A., Bisht, D. and Keswani, N.K.(1992) A study of humoral factors in carcinoma cervix. *Indian Journal of Pathology and Microbiology*, **35**, 5–10.

71. Kochel, H.G., Monazahian, M., Sievert, K., Hohne, M., Thomssen, C., Teichmann, A., Arendt, P. and Thomssen, R. (1991) Occurrence of antibodies to L1, L2, E4 and E7 gene products of human papillomavirus types 6b, 16 and 18 among cervical cancer patients and controls. *International Journal of Cancer*, **48**, 682–8.

72. Sharma, B.K., Gupta, M.M., Saha, K. and Luthra, U.K. (1987) Specificity of secretory immunoglobulins in cervical mucus of women with uterine cervical dysplasia. *Indian Journal of Medical Research*, **85**, 72–6.

73. Merriman, H., Woods, S., Winter, C., Fahnlander, A. and Corey, L. (1984) Secretory IgA antibody in cervicovaginal secretions from women with genital infection due to herpes simplex virus. *Journal of Infectious Diseases*, **149**, 505–10.

74. Dale, G.E., Coleman, R.M., Best, J.M., Benetato, B.B., Drew, N.C., Chinn, S., Papacosta, A.O. and Nahmias, A.J. (1988) Class-specific herpes simplex virus antibodies in sera and cervical secretions from patients with cervical neoplasia: a multi-group comparison. *Epidemiology and Infection*, **100**, 445–65.

75. Vass-Sorensen, M., Abeler, V., Berle, E., Pedersen, B., Davy, M., Thorsby, E. and Norrild, B. (1984) Prevalence of antibodies to herpes simplex virus and frequency of HLA antigens in patients with preinvasive and invasive cervical cancer. *Gynecological Oncology*, **18**, 349–58.

76. Murphy, J.F., Murphy, D.F., Barker, S., Mylotte, M.L., Coughlan, B.M. and Skinner, G.R. (1985) Neutralising antibody against type 1 and type 2 herpes simplex virus in cervical mucus of women with cervical intra-epithelial neoplasia.

Medical Microbiology and Immunology, **174**, 73–80.

77. Clarke, G.N. (1988) Immunoglobulin class and regional specificity of antispermatozoal autoantibodies blocking cervical mucus penetration by human spermatozoa. *American Journal of Reproductive Immunology and Microbiology*, **16**, 135–8.

78. Haas, G., Jr and D'Cruz, O.J. (1989) A radiolabeled antiglobulin assay to identify human cervical mucus immunoglobulin (Ig) A and G antisperm antibodies. *Fertility and Sterility*, **52**, 474–85.

79. Mårdh, P.-A. (1981) Some constituents of body fluids influencing the capability of *Chlamydia trachomatis* to multiply in McCoy cell cultures, in *International Proceedings of 4th Meeting of the International Society for STD Research*, Heidelberg.

LYMPHOKINES AND CYTOKINES IN LOWER GENITAL TRACT INTRA-EPITHELIAL NEOPLASIA

8

Slawomir Majewski, Jacek Malejczyk and Stefania Jablonska

INTRODUCTION

Infections with oncogenic and non-oncogenic genital human papillomaviruses (HPVs) of the lower genital tract (LGT) are substantially influenced by the host immune system, including cellular and humoral responses [1]. The immune surveillance mechanisms determine both susceptibility to HPV infection and regression of HPV-associated lesions [1–4].

In the multistage process of HPV oncogenesis with a long latency period between the primary infection and the malignant proliferation, both viral and host factors are of prime importance [5, 6]. It is well established, by epidemiologic studies, that the majority of HPV-induced genital lesions may regress spontaneously [6–8]. This is due to intracellular mechanisms suppressing viral transcription and function of viral oncoproteins [6], to cytokines acting in autocrine and paracrine ways, and to local and systemic immune mechanisms closely linked with cytokine interplay.

Both local and systemic complex immunosurveillance would explain the latency of genital HPV infection. Infection with potentially oncogenic HPV16 was disclosed in asymptomatic men and in a high proportion of CIN I and II, which often regress spontaneously in spite of association with potentially oncogenic HPVs [9, 10]. The latent HPV infections, up to 30–40% of asymptomatic sexually active college students [11], in asymptomatic women [12, 13] and in newborns [14], with no pathological changes, provides further evidence of the existing and operative defence mechanisms.

The best example of the role of host factors in the lower genital tract HPV infections presents as bowenoid papulosis, which in spite of association with HPV16 and histologic atypia usually has a benign course in young persons, with spontaneous regression. Our study on the role of TNFα, other cytokines and biologic response modifiers has been performed on the tumour cell lines developed from vulvar bowenoid papulosis (Skv cells). The difference in cytokine release (especially of TNFα) and response to their action between tumorigenic and non-tumorigenic sublines harbouring HPV16 provided a model to study the role of escape from local immunosurveillance in tumour progression. The interplay of cytokines and chemokines with cellular and humoral immune reactions is responsible for regression, persistence or progression of HPV-associated lesions.

This chapter addresses both the production of cytokines by HPV-infected keratinocytes

Cancer and Pre-cancer of the Cervix
Edited by D.M. Luesley and R. Barrasso. Published in 1998 by Chapman & Hall, London. ISBN 0 412 56600 1.

and mononuclear cells, and their anti-HPV and antitumour activity, stimulation of the immune system as an important defence mechanism, and inhibition of inflammatory processes by cytokines generated by Th2-lymphocytes. The assessment of biological activities of cytokines and chemokines may not only shed more light on HPV-associated oncogenesis and mechanisms of regression of HPV-induced lesions, but may also provide a base for novel therapeutic modalities. Some new possibilities of antitumour treatment with the use of cytokines and biological response modifiers although experimental, have already been introduced into the clinical practice.

MUCOSA-ASSOCIATED LYMPHOID TISSUE (MALT)

MALT consists of several populations of lymphoid cells [1], including multiple Langerhans cells (LC). These cells express class II MHC molecules and actively participate in antigen presentation [15, 16]. In addition, keratinocytes of the cervical epithelium, focally expressing class II MHC antigens, are also believed to be able to present antigens. The endocervix contains both T- and B-cells and intra-epithelial macrophages. T-lymphocytes consist of both CD4 and CD8 cells (ratio 1:2), some of them express the CD25 molecule, i.e. the marker of activation [16]. Studies on the function of cervical mucosal B-cells have revealed that the main immunoglobulin class generated locally is IgA (60%) as compared to IgG (30%) [1]. Immunoglobulins A contain J-chain, and the main subclass is A_2, which is characteristic of all mucosa [15].

Several studies suggest that HPV infection of the cervix evokes a local immune dysfunction as manifested by decrease of intra-epithelial antigen-presenting cells [17], and T-lymphocytes [18]. Thus HPV infection may interfere with the local immunosurveillance mechanisms including both induction phase (antigen presentation) and effector phase (generation of cytotoxic T-cells and antibody-producing B-cells). The exact mechanism of this interference is unknown, but recent data suggest that the local action of immunomodulatory cytokines may be of main importance [19].

THE ROLE OF CYTOKINES IN THE CONTROL OF HPV INFECTION AND DEVELOPMENT OF HPV-ASSOCIATED TUMOURS

Cytokines are products of various lymphoid and non-lymphoid cells and constitute an important network of soluble mediators which facilitate cell–cell interactions within MALT and skin-associated lymphoid tissue (SALT) [20, 21]. Some cytokines, e.g. IL-1, TNFα, etc., exert pleiotropic effects *in vivo* and are referred to us as 'primary cytokines'. They have the capability to induce synthesis of other mediators, i.e. 'secondary cytokines'. The latter are also known as chemokines, i.e. mediators responsible for lymphoid cell activation and chemotaxis [22]. Cytokines act through specific receptors, which may be variably expressed on HPV-infected cells and shed into an intracellular space and to the microcirculation. These soluble forms of cytokine receptors might markedly affect the immunoregulatory activity of cytokines. Cytokines not only mediate immune/inflammatory reactions but may also regulate cell proliferation and differentiation, and programmed cell death (apoptosis).

In the normal mucosa and skin, the cells of stratified epithelia and epidermal keratinocytes are the most important source of cytokines [22–24]. In addition to the cells of stratified epithelia, an important source of various cytokines in HPV-associated lesions are infiltrating leucocytes (monocytes/macrophages, lymphocytes, granulocytes) and other cells, including: melanocytes, dermal dendritic cells, endothelial cells, mastocytes and fibroblasts. The main cytokines with a potential role in the immunosurveillance against HPV infection and HPV-associated tumours are listed in Table 8.1 (for review see [21]).

Table 8.1 The main cytokines produced by antigen-activated or tumour promoter-activated cells of MALT and/or SALT involved in the control of HPV infection and growth of HPV-associated tumours

Cytokines	Cell source	Main biologic activities
IL-1α,β	KC, HPV-KC, Mon, Mac, EC	Down-regulation of HPV mRNA, activation of T- and B-lymphocytes and NK cells, induction of acute-phase proteins
IL-2	T-cells, NK cells	Growth factor for T- and B-cells, stimulation of anti-HPV cytotoxic lymphocytes
IL-3	T-cells, KC	Growth factor for mast cells and haemopoietic cells
IL-4	Th2-lymphocytes	Enhancement of proliferation of B- and T-cells, stimulation of Ig production (especially IgE)
IL-5	T-cells	Growth factor for eosinophils, enhancement of IgA production by B-cells
IL-6	KC, HPV-KC, Mon, FB, T-cells	Enhancement of IgG secretion, stimulation of anti-HPV cytotoxic cells and NK cells, induction of acute-phase proteins
IL-7	KC, BM	T- and B-cell growth factor
Chemokines IL-8	KC, HPV-KC, Mon, Mac	Enhancement of granulocyte and T-cell chemotaxis
MCP-1	KC, HPV-KC, EC	Stimulation of T-cell and monocyte chemotaxis
RANTES	KC, HPV-KC, FB	Stimulation of T-cell and monocyte chemotaxis
IL-9	T-cells	Stimulation of Th clones
IL-10	Th2-cells, KC	Inhibition of Th1, and IL-12 and IFNγ production, suppression of TNFα and proinflammatory cytokines, stimulation of B-cell growth and differentiation
IL-11	KC, BM	Stimulation of T- and B-cell proliferation, induction of acute-phase proteins
IL-12	T-cells, KC	Stimulation of Th1-cells and inhibition of Th2-cells, enhancement of T-cytotoxic cells and NK cells, induction of IFNγ production
IL-13	Th2-cells, KC	Induction of Ig synthesis (including IgE), inhibition of IL-1, IL-6, TNFα, IL-2, IFNγ production
TNFα	Mon, Mac, KC, HPV-KC	Down-regulation of HPV mRNA, stimulation of MCP-1, stimulation of effector phase of contact dermatitis, modulation of adhesion molecule expression
TNFβ	Th1-cells	Immunostimulatory effects (similar to TNFα)
Leucoregulin	T-cells, Mon	Down-regulation of HPV mRNA expression, enhancement of target cell sensitivity to cytotoxic cells
IFNα	Mononuclears	Down-regulation of HPV mRNA expression, inhibition of cell proliferation, stimulation of T-cells, cytotoxic cells
IFNβ	FB, EC, KC	Down-regulation of HPV mRNA expression, inhibition of cell proliferation, stimulation of NK cells

[*Table continued overleaf*]

Table 8.1   *Continued*

Cytokines	Cell source	Main biologic activities
IFNγ	Th1-cells, NK	Down-regulation of HPV mRNA expression, stimulation of NK cells and T-cytotoxic cells, stimulation of adhesion molecule expression
GM-CSF	T-cells, KC, EC, HPV-KC, Mon	Stimulation of proliferation of leucocytes, maintaining Langerhans cell viability
TGFβ	KC, HPV-KC, mononuclears	Down-regulation of HPV mRNA, inhibition of cell proliferation, inhibition of IL-1, stimulation of wound healing
EGF family: EGF, TGFα, Amphiregulin	Various normal and tumour cells KC, HPV-KC	Activation of EGF receptor, down-regulation HPV mRNA, stimulation of cell growth
αMSH	Pituitary gland, Mononuclears, KC, Mel	Stimulation of melanogenesis, inhibition of IL-1, IL-2, IFNγ, stimulation of IL-10, inhibition of contact and delayed-type hypersensitivity
VEGF	KC, HPV-KC, various tumours	Increase in vascular permeability induction of angiogenesis
bFGF	KC, tumour cells, mononuclears	Stimulation of tumour angiogenesis and wound healing stimulation of proliferation and migration of fibroblasts and endothelial cells

KC, keratinocyte; HPV-KC, HPV-harbouring keratinocyte; Mon, monocyte; Mac, macrophage; EC, endothelial cell; FB, fibroblast; BM, bone marrow cell.

PRODUCTION OF CYTOKINES BY HPV-HARBOURING CELLS

Since epithelial cells are the target for HPV infection, cytokines released by these cells might play a crucial role in the control of the infection and progression of HPV-induced tumours (Table 8.2). HPV-infected or HPV-transformed keratinocytes were found to constitutively express some cytokines, including IL-1α,β, IL-6, IL-8, TNFα, GM-CSF, TGFβ and MCP-1 [25–29]. Some of these cytokines (IL-1, IL-6, IL-8, TNFα) were found to be produced by cells of normal exo- and endocervical epithelia *in vivo*, whereas cervical carcinoma cell lines were shown to secrete lowered amounts of these cytokines [27]. TGFβ1 expression was reported to be decreased in cervical neoplastic epithelium [29], which would suggest that an early event in the malignant transformation of cervical epithelium may involve the loss of this growth factor

[29]. In contrast, the pattern and intensity of TGFβ2 expression, also reported to be increased in CIN lesions, did not correlate with the CIN grade [30]. Thus the local expression of some cytokines in cervical lesions is only of partial prognostic value.

MECHANISMS OF ANTI-HPV AND ANTI-HPV TUMOUR ACTIVITY OF CYTOKINES

Cytokines derived from HPV-harbouring cells and from other cell types constitute a complex network of local mediators exerting various autocrine and paracrine effects responsible for the control of the progression of HPV infection (Table 8.3).

Cytokines inhibiting HPV gene expression

The main cytokines capable of inhibiting HPV E6 and E7 gene expression are: TNFα, IL-1, TGFβ, IFNs, leucoregulin and EGF. Kyo *et al.*

Table 8.2 Cytokines produced by keratinocytes harbouring high-risk genital HPVs

Cytokine	References
IL-1α, IL-1β	Woodworth and Simpson [27]
IL-6	Malejczyk *et al.* [25]
IL-8	Woodworth and Simpson [27]
TNFα	Malejczyk *et al.* (1992) [26]; Woodworth *et al.* [36, 54]
MCP-1	Rösl *et al.* [28]
TGFβ	Woodworth and Simpson [27]; Braun *et al.* [128]; Commerci *et al.* [29]; Ho [42]
GM-CSF	Woodworth and Simpson (1993) [27]

Table 8.3 Possible cytokine-mediated mechanisms of surveillance against HPVs and HPV-associated tumours

Effects	Cytokines
Inhibition of E6 and/or E7 gene expression	TNFα, IL-1, TGFβ, IFNs, leucoregulin, EGF
Inhibition of cell proliferation and/or enhancement of apoptosis	TNFα, IFNs, TGFβ, leucoregulin
Stimulation of inflammatory infiltrate formation	MCP-1, RANTES, IL-8, IL-1, IL-6
Stimulation of effector cytotoxic cells	IL-1, IL-2, IL-12, TNFα, IFNs, leucoregulin and others
Inhibition of angiogenic capability and metastatic potential of HPV-associated tumours	IFNs, IL-12, TGFβ

[31], using *in vitro* CAT assay, found that, in contrast to IL-6 and IFNγ, TNFα and IL-1 repressed HPV16 early gene expression at the transcriptional level via the non-coding region (NCR) of the HPV genome. Using Northern blot analysis, the authors showed that TNFα and IL-1 inhibited HPV16 E6/E7 mRNA levels in an HPV16 immortalized keratinocyte cell line. Moreover, they found that TNFα- and IL-1-responsive elements in HPV16 NCR lie within the cell-specific enhancer endowed with several binding sites for nuclear factors (AP-1, NF-1, NF-IL-6) involved in regulation of HPV early gene expression. NF-IL-6 is activated by various cytokines, and this transcription factor inhibited mRNA coding for E6 and E7 oncoproteins of high-risk HPVs [31]. It is also possible that down-regulation of HPV mRNA expression is due to cytokine-induced changes in AP-1 composition leading to formation of repressor Jun/Fra-1 heterodimers that compete with native Tun/Fos complex for binding the enhancer sequences of HPV NRC [32].

TGFβ is another cytokine that rapidly and reversibly inhibits E6 and E7 expression of both HPV16 and 18 [33–35]. This suppression occurs at the level of transcription and requires *de novo* protein synthesis.

Interferons regulate transcription of a variety of genes involved in HPV infection. It was shown that IFNγ reduced expression of E6/E7 mRNA of HPV16, 18 and 33 in several human cervical cell lines [36]. Reduction in E6/E7 mRNA expression was accompanied by inhibition of cell proliferation simultaneously with an increase in epidermal transglutaminase activity, i.e. the marker of squamous differentiation. Similar effects on the expression of mRNA coding for E6 and E7 of HPV16, 18 and 33 were obtained by treatment of keratinocyte lines with leucoregulin, a 32 kDa acidic glycoprotein with the ability to inhibit cell transformation [36].

There is a controversy on the effect of IFNα on E6 and E7 gene transcription. In several human cervical epithelial cell lines immortalized by HPV16, 18 or 33, IFNα was not found to influence specific mRNAs [36]. However, immunofluorescence studies of HPV16 E6 and E7 proteins with the use of specific anti-E6 and anti-E7 monoclonal antibodies showed significant inhibition of E7 protein expression in the cells treated with IFNa [37]. The inhibition of E7 protein levels was further supported by Western blot analysis.

Recently IFNβ was proved to inhibit the expression of E6 and E7 mRNA of HPV16-transformed keratinocytes [38]. IFNβ but not IFNα was also shown to induce a marked cytopathic effect in HPV16-transformed cells.

Finally, it was reported that EGF elicits negative regulation of HPV16 E6 and E7 at the mRNA level in HPV16-immortalized human keratinocyte cell line [39, 40].

Thus inhibition of HPV oncogene expression by cytokines could be one of the mechanisms of clinical efficacy of some of them in the treatment of HPV-associated tumours.

Cytokines modulating proliferation of HPV-harbouring cells

Since some cytokines, e.g. TNFα and TGFβ, were reported to be expressed by HPV-harbouring tumour cell lines *in vitro* [26, 33–35] and *in vivo* [29, 41, 42], one might presume that these cytokines could exert an autocrine inhibitory effect on the growth of HPV-infected or HPV-transformed keratinocytes. In addition to TNFα and TGFβ, some other cytokines (mainly IFNs) produced by non-HPV-harbouring cells could also be of importance for the control of proliferation of HPV-infected keratinocytes.

TNFα has been found to inhibit proliferation of non-tumorigenic HPV16-harbouring

keratinocytes [26] and HPV16 and HPV18-immortalized keratinocyte cell lines [43]. The mechanism of proliferation inhibition by TNFα may be related to down-regulation of E6/E7 expression or to the inhibition of expression of some cellular proto-oncogenes, e.g. *c-myc* [44]. Therefore, decreased production of TNFα by HPV-harbouring keratinocytes could contribute to a progressive growth of HPV-associated tumours [27]. Another possibility of HPV-tumour progression could be loss of sensitivity of HPV-harbouring cells to autocrine or paracrine growth inhibitory effects of this cytokine.

Recently we found that tumorigenic progression of HPV16-harbouring Skv keratinocytes was associated with resistance to the inhibitory effects of endogenous and exogenous TNFα on cell growth and *c-myc* expression [45]. This resistance of HPV16-harbouring cells to autocrine growth limitation could be due to decreased levels of TNFα receptor expression (TNFR) or to an increased TNFR shedding. Both phenomena were found to correlate with the tumorigenic potential of Skv keratinocytes [45, 46]. The soluble form of TNFR may act as an endogenous inhibitor of TNFα activity, thus enabling the HPV-transformed cell to escape from TNFα-mediated surveillance [46]. In our recent studies we determined serum levels of TNFR type I in patients with cervical squamous cell carcinomas (SCC) of various stages, according to the International Federation of Gynaecology and Obstetrics (FIGO) classification (Teshima *et al.*, in preparation). We found that serum concentrations of TNFR type I in patients with cervical SCC of Stage 2b, 3b and 4 were significantly higher as compared to those in healthy individuals and in patients with other cervical or ovarian malignancies, irrespective of the stage of the disease. These results indicate that serum levels of TNFR type I correlate with the stage of progression of cervical SCC and depend on the origin and pathologic pattern of the tumours. Thus determination of serum levels of this receptor may be of value for early detection, follow-up and prognosis of cervical SCC. However, the significance of this phenomenon for the progression of the tumours requires further studies.

TGFβ is another important factor capable of inhibiting proliferation of HPV-harbouring cells [34, 35]. Similarly to TNFα, a loss of susceptibility to TGFβ could enhance growth and malignant phenotype of HPV-associated lesions. It was found that more malignant HPV-harbouring cell lines are resistant to TGFβ-mediated inhibition of cell proliferation without any effect on the expression of HPV and *c-myc* mRNA [33–35]. Thus the mechanism of escape from TGFβ-mediated surveillance seems to be very complex and, most likely, does not involve down-regulation of TGFβ receptors [34].

Interferons constitute a heterogenous group of biologic response modifiers which, in addition to direct anti-HPV effect, display a pronounced antiproliferative activity. These compounds are of special interest because of their application in the treatment of HPV-induced benign and malignant anogenital tumours. The antiproliferative effects of IFNs could be linked with their ability to decrease HPV gene expression (see above). Another mechanism of proliferation inhibition by IFNs may be related to their ability to induce the synthesis of other inhibitory cytokines in the treated cells, e.g. TGFβ [47]. However, some IFNs, e.g. IFNγ, had lower abilities to decrease proliferation of HPV-infected cells, whereas IFNα markedly decreased proliferation of HPV16-harbouring cells [37, 48]. These variations in the response of HPV-harbouring cells to IFNs could be due to differences in the stage of cell transformation or to different induction of growth inhibitory mediators in the cells [36].

Of practical importance is that the anti-HPV and antiproliferative action of IFNs can be markedly enhanced by derivatives of vitamin A (retinoids) and vitamin D3 (calcitriol) [49–51]. Such combinations proved to be effective in the treatment of advanced squamous cell carcinoma of the cervix and the skin [52, 53].

It is to be stressed that some cytokines, depending on the cell culture system used, can either inhibit or stimulate proliferation. It was reported that IL-1 and TNFα also enhance proliferation of various HPV-harbouring cell lines [54]. This effect is probably due to induction of expression and release of TGFα and amphiregulin (AR), both members of the EGF family, which, in turn, exert an autocrine stimulatory effect. A similar mechanism of stimulation of proliferation of cervical epithelial cells was described for IL-6 [55]. These observations suggest that growth factors of the EGF family may enhance progression of HPV-induced lesions. Moreover, EGF itself was shown to affect terminal differentiation of HPV-harbouring cells [40]. An increased expression of EGF receptors is a frequent feature, both *in vitro* and *in vivo*, of cells harbouring high-risk genital HPV types [56–60]. In spite of the growth-stimulating activity, EGF was reported to down-regulate transcription of the HPV E6/E7 oncoprotein mRNA [40].

The effects of cytokines on apoptosis in HPV-harbouring cells

The antiproliferative effects of some cytokines *in vitro* could be related to their capability to induce cytotoxicity or to affect apoptosis in HPV-infected cells. Also 'in vivo' two main mechanisms of cell loss in growing tumours are necrosis and apoptosis [61, 62]. The experimental data showed that regulation of apoptosis involves expression of several oncogenes or tumour-suppressor genes, such as *c-myc* [63], p53 [64], bcl-2 [65] and others [66, 67].

Using rabbit uterine epithelial cultures Rotello *et al.* [68] have shown that TGFβ1 inhibited cell proliferation with a concomitant increase in cells undergoing apoptosis. The role of TGFβ1 in apoptosis was confirmed in various experimental systems [66]. Recently it was found that TNFα decreases the expression of different cell-cycle regulatory proteins, such as cyclin A, cyclin B, etc., in normal keratinocytes and in non-tumorigenic, TNFα-sensitive HPV16-harbouring cells, leading to growth arrest in G0–G1 phase [69]. However, it is not known whether this effect of TNFα could contribute to the enhancement of apoptosis. Hagari *et al.* [70] found that in regressing Shope papillomas the expression of TNFα gene was associated with apoptotic cell death.

Two recent reports on apoptosis in cervical neoplasia provided somewhat conflicting results. Shoji *et al.* [71] have shown a significant positive correlation between apoptotic index and histologic malignant grading in CIN and invasive tumours; however, the proliferation index was similar in the cervical lesions, irrespective of histologic malignancy. Isacson *et al.* [72] have found that both apoptosis and proliferation were increased in the higher grade lesions, but there was no correlation with HPV type.

The role of chemokines

One of the important effects of cytokines produced by HPV-harbouring keratinocytes is activation and attraction of leucocytes at the site of HPV infection, thus leading to generation of immune effector mechanisms responsible for eradication of the infected and/or transformed cells. The primary cytokines (e.g. IL-1, TNFα) are not chemotactic *in vitro* and therefore direct leucocyte migration *in vivo* has to depend on generation and/or release of chemotactic factors, i.e. chemokines [22, 73, 74].

At least 17 different chemokines have been characterized to date and they are categorized into three types: CXC, CC, and C (which stands for the structural property of the motif containing cysteines in the amino acid terminus) [22]. The members of the CXC subfamily (e.g. IL-8, IFNγ-inducible protein: IP-10) are chemotactic for neutrophils, and partially for T-lymphocytes. The CC type includes MCP-1, RANTES, MIP and other chemokines that are responsible for attraction of monocytes, T-cells and their subsets.

Rösl *et al.* [28] showed that TNFα is capable of inducing expression and release of MCP-1 in non-tumorigenic HPV18-harbouring HeLa/fibroblast hybrids. In contrast, HeLa cells and tumorigenic HeLa/fibroblast segregants did not express MCP-1 after stimulation with TNFα. Although the JE gene (encoding MCP-1), including its regulatory region, was found to be intact in tumorigenic cells, no JE (MCP-1) mRNA was detected. Thus, due to an unknown mechanism, JE gene expression cannot be upregulated. *In vivo* experiments showed that inoculation of JE (MCP-1)-negative HeLa cells into nude mice caused rapidly growing tumours without macrophage infiltrations [75]. Transfection of these cells with the JE gene led to a strong macrophage infiltration at the site of injection of the tumour cells. These results suggest that the tumorigenicity of HPV-harbouring cells may depend, at least in a part, on the inability of HPV-infected cells to release chemokines, resulting in lack of recruitment of leucocytes responsible for anti-HPV local immunosurveillance. Our recent studies also showed that HPV16-harbouring vulvar Skv keratinocyte lines express lowered amounts of MCP-1 and RANTES, as compared to non-tumorigenic HaCat keratinocytes. Interestingly, Skv cells spontaneously produced IL-8, and their conditioned media displayed a marked granulocyte-chemotactic activity (unpublished).

Stimulation of anti-HPV effector cells by cytokines

Cytokines can affect various stages of anti-HPV immune reactions, which could serve as potential targets for new treatment modalities.

Cytokines stimulating expression of MHC and adhesion molecules

When attracted by chemokines, various leucocyte subpopulations become activated locally by proinflammatory/immunostimulatory cytokines released either by HPV-harbouring cells or lymphoid cells. Cytokines modulate both presentation of HPV-associated antigens and differentiation of effector T-cytotoxic lymphocytes or natural cytotoxic cells.

Antigen presentation and activation of T-cells requires preserved function of antigen-presenting Langerhans cells and an optimal expression of class I and class II MHC molecules. It was shown that decrease of Langerhans cells in cervical HPV infection is associated with enhanced replication of the virus [76]. The mechanism of this depletion is unknown but it could be due to a decreased production of some chemotactic factors, e.g. MCP-1 [28].

Various cytokines (e.g. TNFα, IFNs) are also stimulators of MHC antigen expression. In a high proportion of cervical carcinomas a decreased or absent expression of MHC class I antigens was repeatedly shown [77–79]. Loss of MHC class I expression in cervical lesions was found to be accompanied by a lowered number of CD8-positive T-cells [80], and was related with poor prognosis [81, 82]. Down-regulation of class I MHC antigen expression does not seem to be due to HPV oncoprotein expression [78] but is probably related to the loss of TAP-1 proteins [82] involved in antigenic peptide transportation in the endoplasmic reticulum. It is tempting to speculate that decreased MHC class I expression in some cervical tumours could also be related to the unresponsiveness to stimulatory cytokines, e.g. TNFα was found to upregulate expression of these molecules on HPV16-harbouring cells [48]. Expression of cytokine-inducible MHC class II was found to be upregulated in some lesions, including cervical carcinomas [78, 79] and associated with the presence of leucocyte infiltrates [76, 83, 84].

Cytokines may also stimulate expression of some adhesion molecules, e.g. ICAM-1 involved in various immune reactions [85]. Although ICAM-1 is not constitutively expressed on normal keratinocytes and cervical epithelium, its expression may be induced by TNFα, IFNs and IL-1 in cervical cell organotypic raft culture and in various HPV16-harbouring cervical and vulvar cell lines, including Skv, CaSki and SiHa [50,

86]. The mechanism of spontaneous expression of ICAM-1 on HPV-harbouring cells is unknown but our recent studies suggest that this phenomenon could be due to autocrine stimulation by TNFα (unpublished).

ICAM-1 expression was also found in various HPV-associated lesions, especially at sites of numerous infiltrating lymphoid cells, which probably serve as a source of proinflammatory cytokines [86, 87]. The upregulation of ICAM-1 expression by cytokines may facilitate T-cell binding to target keratinocytes [88, 89], and is necessary for T-cytotoxic cell activity [90] and the killing of HPV16-harbouring cells by NK cells (own data – unpublished).

The role of cytokines in differentiation of T-helper lymphocytes

Recent studies on the mechanism of T-helper (Th) cell differentiation disclosed two populations that differ substantially in the profile of cytokine production [91, 92]. The Th1 subpopulation produces IL-2, IFNγ and TNFβ, and stimulates cellular effector mechanisms, e.g. cytotoxic T-lymphocytes and NK cells, whereas the Th2 subpopulation generates IL-4, IL-5, IL-6, IL-10 and IL-13, and stimulates humoral responses.

Most importantly, some Th2-derived cytokines (IL-10, IL-4, IL-13) inhibit the function of Th1-lymphocytes, leading to a decrease in cell-mediated immune responses [91]. Thus enhanced activity of Th2 cells may be of some benefit for generation of HPV-neutralizing antibodies, however, it could have adverse effects on Th1-dependent cell-mediated immunity against HPV-infected or HPV-transformed cells. Studies on mononuclear cells isolated from CIN lesions revealed a frequent production of IL-10 by infiltrating T-cells [93]. The local production of immunosuppressive IL-10 might explain the apparent inability of large numbers of infiltrating T-lymphocytes to clear HPV infections in high-grade CIN.

IL-10 has been shown to inhibit production of TNFα, IFNγ, IL-1, IL-6, IL-8 and GM-CSF by lymphocytes and monocytes [94, 95]. Another important immunosuppressive factor, although not yet studied in HPV infection, is αMSH, which antagonizes the effects of pro-inflammatory cytokines such as IL-1, IL-6, IFNγ, and selectively induces production of Th2-associated IL-10 [96–98]. Recently Grabbe *et al.* [99] found that αMSH suppresses elicitation of contact hypersensitivity and induces hapten-specific immunotolerance. This would suggest that its presence or absence in the epithelia may determine the final outcome of immune response to some antigens and probably to epitheliotropic pathogens, including HPVs.

The discovery of the immunoregulatory mechanisms involved in Th differentiation makes possible the development of new modalities of adjuvant treatment of HPV infection and HPV-associated lesions. Th1 cytokines (IL-12 and IFNγ) were found to inhibit IL-10 production [100], and recently IL-12 was shown to induce a switch from a type 2 to a type 1 T-helper cell response [101]. Nabros *et al.* also showed that treatment with IL-12 resulted not only in a switch from Th2 to Th1 response but cured established *Leishmania major* infection in mice.

Inhibition of angiogenic capability and metastatic potential of HPV-associated tumours by cytokines

Various cytokines and growth factors produced by HPV-harbouring or HPV-transformed cells and by tumour infiltrating lymphoid cells could affect angiogenic capability and metastasis formation.

The role of angiogenesis in the invasiveness and metastases of solid tumours, including cervical carcinoma, is well established by clinical and experimental studies [102–104]. Angiogenesis, i.e. neovascularization of tumour tissue, was found to be associated with high-grade cervical dysplasia and with invasive SCC of the cervix [105–108]. And abnormal vasculature was suggested to be helpful in colposcopic diagnosis of such lesions [109, 110].

Tumour angiogenesis factors in HPV-associated tumours

One of the main tumour angiogenesis factors *in vivo* is VEGF (vascular endothelial growth factor) [111]. Neutralization of VEGF by monoclonal antibodies or blocking of its receptor leads to a decrease in experimental angiogenesis and inhibition of tumour growth *in vivo* [112]. In cervical neoplasia the highest expression of VEGF mRNA was detected in a majority of cases of invasive carcinoma and in nearly half of the cases of high-grade intra-epithelial lesions, in the areas of the most pronounced new blood vessel formation. This provides supportive evidence for the role of VEGF in the angiogenesis in these neoplasia [107].

Another tumour angiogenesis factor is bFGF, produced by a variety of neoplastic cells, and considered as a marker of malignant transformation [113]. It was found that progression of dermal fibromas to fibrosarcomas in transgenic mice containing the bovine papillomavirus genome was accompanied by the ability to secrete bFGF [114].

Tumour angiogenesis could also be stimulated by a variety of other cytokines originating from tumour-infiltrating lymphoid cells (IL-1, IL-6, Il-8, EGF, TNFα, etc.) (for review see [119]).

The angiogenesis inhibitory factors

A great interest in the inhibitors of tumour angiogenesis is due to their potential use for the treatment of advanced metastatic disease and for prevention of metastasis formation [115–119]. Some of the angiogenesis inhibitors are already undergoing clinical trials.

In our studies we found marked anti-angiogenic properties of retinoids [120, 121] and 1,25-dihydroxyvitamin D3 (VD3) [49, 122]. The inhibition of angiogenesis induced by HPV16-harbouring Skv cells in an experimental system in immunosuppressed mice was shown not to be due to the decreased cell proliferation, since some cell lines that were insensitive to the anti-angiogenic activity of retinoids were sensitive to the antiproliferative effect of these compounds [50]. We have also found that retinoids and VD3 markedly enhanced the anti-angiogenic properties of INFα [50, 122]. These synergistic anti-angiogenic and antiproliferative effects may explain the favourable results of combination therapy with retinoids and IFNα in cervical carcinomas and advanced squamous cell carcinomas of the skin [52, 53].

Another cytokine with pronounced antineoplastic and antimetastatic activities is IL-12. Until recently, these effects were attributed to the ability of this cytokine to stimulate T-cytotoxic cells and NK cells [123–125]. However, antitumour effects of IL-12 were also found in mice with a severe combined immunodeficiency (SCID mice) and in animals deprived of CD8-positive cells or NK cells [125]. These results suggested that antitumour and antimetastatic action of IL-12 could be due to 'non-immunological' mechanisms. One such mechanism seems to be the capability of IL-12 to decrease tumour-induced angiogenesis, as shown in our recent studies [115]. We found that recombinant IL-12 strongly inhibited angiogenesis induced in immunosuppressed mice by HPV16-harbouring cells, and this effect could be abolished by antibodies neutralizing IFNγ. These results suggest that IFNγ or IFNγ-inducible chemokines (e.g. IP-10) [126] are important mediators of the anti-angiogenic activity of IL-12.

CONCLUSIONS

Recent studies provide evidence for an important role of cytokines in the control of HPV infection and HPV-associated tumours of the genital tract. The effects of cytokines are pleiotropic and involve several immune and non-immune steps in the progression of the lesions. Poor prognostic value in the genital tract neoplasia have following cytokine-related parameters:

- increased cell proliferation and apoptosis
- decreased *in situ* expression of TGFβ1

- increased levels of soluble TNFαRI in circulation
- decreased production of MCP-1 chemokine by HPV-harbouring keratinocytes
- decreased or absent MHC class I molecules on tumour cells
- decreased or absent T-cell infiltrates
- predominance of Th2 response
- increased tumour-induced angiogenesis.

Some cytokines are capable of exerting complex direct or indirect antiviral, antiproliferative, immunostimulatory and anti-invasive effects and could find application as adjuvant therapy in genital tract tumours. The most promising seems to be a combined therapy with the use of cytokines, e.g. IL-12, IFNs, and other biologic response modifiers, e.g. retinoids, vitamin D₃ [127], and/or some non-toxic cytostatics.

REFERENCES

1. Roche, J.K. and Crum, C.P. (1991) Local immunity and the uterine cervix: implications for cancer-associated viruses. *Cancer Immunolology and Immunotherapy*, **33**, 203–9.
2. Vardy, D.A., Baadsgaard, O., Hansen, E.R., *et al.* (1990) The cellular immune response to human papillomavirus infection. *International Journal of Dermatology*, **29**, 603–10.
3. Frazer, I. and Tindle, R. (1992) Cell-mediated immunity to papillomaviruses. *Papillomavirus Report*, **3**, 53.
4. Wu, R., Coleman, N., Higgins, G., *et al.* (1994) Lymphocyte-mediated cytotoxicity to HPV16 infected cervical keratinocytes. In *Immunology of Human Papillomaviruses* (ed. M.A. Stanley). Plenum Press, New York, pp. 255–9.
5. zur Hausen, H. (1991) Human papillomaviruses in the pathogenesis of anogenital cancer. *Virology*, **184**, 9–13.
6. Koutsky, L.A., Holmes, K.K., Critchlow, C.W., *et al.* (1992) A cohort study of the risk of cervical intraepithelial neoplasia grade 2 or 3 in relation to papillomavirus infection. *New England Journal of Medicine*, **327**, 1272.
7. Schiffman, M.H. (1992) Recent progress in defining the epidemiology of human papillomavirus infection and cervical neoplasia. *Journal of the National Cancer Institute*, **84**, 394–7.

8. Schiffman, M.H. (1994) Epidemiology of cervical human papillomaviruses. In *Human Pathogenic Papillomaviruses* (ed. H. zur Hausen). Springer, Heidelberg, pp. 55–82.

9. Barrasso, R., de Brux, J., Croissant O., *et al.* (1987) High prevelence of papillomavirus-associated penile interepithelial neoplasia in sexual partners of women with cervical intraepithelal neoplasia. *New England Journal of Medicine*, **317**, 916.

10. Aynaud, O., Ionesco, M. and Barrasso, R. (1994) Penile intraepithelial neoplasia: specific clinical features correlate to histological and virological findings. *Cancer*, **74**, 1762–7.

11. Bauer, H.M., Ting, Y., Greer, O.E., *et al.* (1991) Genital human papillomavirus infection in female university students as determined by a PCR-based method. *Journal of the American Medical Association*, **265**, 472–4.

12. Tidy, J.A., Parry, G.C.N., Ward, P., *et al.* (1989) High rate of human papillomavirus type 16 infection in cytologically normal cervices. *Lancet*, **1**, 434.

13. Young, I.S., Bevan, I.S., Johnson, M.A., *et al.* (1989) The polymerase chain reaction: a new epidemiological tool for investigating cervical human papillomavirus infection. *British Medical Journal (Clinical Research)*, **298**, 14–18.

14. Cason, J., Kaye, J.N., Jewers, R.J., *et al.* (1995) Perinatal infection and persistence of human papillomavirus types 16 and 18 in infants. *Journal of Medical Virology*, **47**, 209–18.

15. Kutteh, W.H., Hatch, K.D., Blackwell, R.E., *et al.* (1988) Secretory immune system of the female reproductive tract: I. Immunoglobulin and secretory component-containing cells. *Obstetrics and Gynaecology*, **71**, 56–60.

16. Roncalli, M., Sideri, M., Paolo, G., *et al.* (1988) Immunophenotypic analysis of the transformation zone of human cervix. *Laboratory Investigation*, **58**, 141–9.

17. McArdle, J.P. and Muller, H.K. (1986) Quantitative assessment of Langerhans cells in human cervical intraepithelial neoplasia and wart virus infection. *American Journal Obstetrics and Gynaecology*, **154**, 509–15.

18. Tay, S.K., Jenkins, D., Maddox, P., *et al.* (1987) Lymphocyte phenotypes in cervical intra-epithelial neoplasia and human papillomavirus infection. *British Journal Obstetrics and Gynaecology*, **94**, 16–21.

19. Majewski, S., Malejczyk, J. and Jablonska, S. (1996) The role of cytokines and other factors in HPV infection and HPV-associated tumors. *Papillomavirus Report*, **7**, 143–54.

20. Streilein, J.W. (1993) Sunlight and skin-associated lymphoid tissue (SALT): if UVB is the trigger and TNFa is its mediator, what is the message? *Journal of Investigative Dermatology*, **100**, 47S–52S.

21. Grabbe, S. and Luger, T. (1995) The skin as an immunologcal organ as well as a target for immune responses. In *Multi-systemic Auto-immune Diseases: An Integrated Approach. Dermatological and Internal Aspects* (eds L. Kater and H.B. de la Faille). Elsevier, Amsterdam, pp. 17–41.

22. Schröder, J.-M. (1995) Cytokine networks in the skin. *Journal of Investigative Dermatology*, **105**, 20S–4S.

23. Luger, T.A. and Schwarz, T. (1990) Epidermal cell-derived cytokines. In *Skin Immune System (SIS)* (ed. J.D. Boss). CRC Press, Boca Raton, FL, pp. 257–91.

24. Stadnyk, A.W. (1994) Cytokine production by epithelial cells. *FASEB Journal*, **8**, 1041–7.

25. Malejczyk, J., Malejczyk, M., Urbanski, A., *et al.* (1991) Constitutive release of IL-6 by human papillomavirus type 16 (HPV16)-harboring keratinocytes: a mechanism augmenting the NK-cell-mediated lysis of HPV-bearing neoplastic cells. *Cellular Immunology*, **136**, 155–64.

26. Malejczyk, J., Malejczyk, M., Köck, A., *et al.* (1992) Autocrine growth limitation of human papillomavirus type 16-harboring keratinocytes by constitutively released tumor necrosis factor-a. *Journal of Immunology*, **149**, 2702–8.

27. Woodworth, C.D. and Simpson, S. (1993) Comparative lymphokine secretion by cultured normal human cervical keratinocytes, papillomavirus, and carcinoma cell lines. *American Journal of Pathology*, **142**, 1544–55.

28. Rösl, F., Lengert, M., Albrecht, J., *et al.* (1994) Differential regulation of the JE gene encoding the monocyte chemoattractant protein (MCP-1) in cervical carcinoma cells and derived hybrids. *Journal of Virology*, **68**, 2142–50.

29. Commerci, J.T., Runowicz, C.D., Fianders, K.C., *et al.* (1996) Altered expression of transforming growth factor-beta1 in cervical neoplasia as an biomarker in carcinogenesis of the uterine cervix. *Cancer*, **77**, 1107–14.

30. Tervahauta, A., Syrjanen S., Yliskoski M., *et al.* (1994) Expression of transforming growth factor-b1 and -b2 in human papillomavirus-associated lesions of the uterine cervix. *Gynaecologic Oncology*, **54**, 349–56.

31. Kyo, S., Inoue, M., Nishio, Y., *et al.* (1993) NF-IL6 represses early gene expression of human papillomavirus type 16 through binding to the noncoding region. *Journal of Virology*, **67**, 1058–66.

32. Rösl, F., Das, B.C., Lengert, M., *et al.* (1997) Antioxidant-induced changes in AP-1 composition result in a selective suppression of human papillomavirus transcription. *Journal of Virology,* **71**, 362–70.

33. Woodworth, C.D., Notario, V. and DiPaolo, J.A. (1990) Transforming growth factors beta 1 and 2 transcriptionally regulate human papillomavirus (HPV) type 16 early gene expression in HPV-immortalized human genital epithelial cells. *Journal of Virology,* **64**, 4767–75.

34. Braun, L., Dürst, M., Mikumo, R., *et al.* (1990) Differential response of nontumorigenic and tumorigenic human papillomavirus type 16-positive epithelial cells to transforming growth factor β_1. *Cancer Research,* **50**, 7324–32.

35. Braun, L., Dürst, M. and Mikumo, R. (1992) Regulation of growth and gene expression in human papillomavirus-transformed keratinocytes by transforming growth factor-b: implications for the control of papillomavirus infection. *Molecular Carcinogeneis,* **6**, 100–11.

36. Woodworth, C.D., Lichti, U., Simpson, S., *et al.* (1992) Leucoregulin and gamma-interferon inhibit human papillomavirus type 16 gene transcription in human papillomavirus-immortalized human cervical cells. *Cancer Research,* **52**, 456–63.

37. Khan, M.A., Tolleson, W.H., Gangemi, J.D., *et al.* (1993) Inhibition of growth, transformation, and expression of human papillomavirus type 16 E7 in human keratinocytes by alpha interferons. *Journal of Virology,* **67**, 3396–403.

38. De Marco, F. and Marcante, M.L. (1995) Cellular and molecular analyses of interferon β cytopatic effect on HPV-16 *in vitro* transformed human keratinocytes (HPK-IA). *Journal of Biological Regulators and Homeostatic Agents,* **9**, 24–30.

39. Yasumoto, S., Taniguchi, A. and Sohma, K. (1991) Epidermal growth factor (EGF) elicits down-regulation of human papillomavirus type 16 (HPV-16) E6/E7 mRNA at the transcriptional level in an EGF-stimulated human keratinocyte cell line: functional role of EGF-responsive silencer in the HPV-16 long control region. *Journal of Virology,* **65**, 2000–9.

40. Vambutas, A., Di Lorenzo, T.P. and Steinberg, B.M. (1993) Laryngeal papilloma cells have high levels of epidermal growth factor receptor and respond to epidermal growth factor by a decrease in epithelial differentiation. *Cancer Research,* **53**, 910–14.

41. Majewski, S., Hunzelmann, N., Nischt, R., *et al.* (1991) TGFb1 and TNFa expression in the epidermis of patients with epidermodysplasia verruciformis. *Journal of Investigative Dermatology,* **97**, 862–7.

42. Ho, L., Terry, G., Mansell, B., *et al.* (1994) Detection of DNA and E7 transcripts of human papillomavirus type 16, 18, 31 and 33, TGFb and GM-CSF transcripts in cervical cancers and precancers. *Archives of Virology,* **139**, 79–85.

43. Villa, L.L., Vieira, K.B.L., Pei, X.-F., *et al.* (1992) Differential effect of tumor necrosis factor on proliferation of primary human keratinocytes and cell lines containing human papillomavirus types 16 and 18. *Molecular Carcinogenesis,* **6**, 5–9.

44. Yarden, A. and Kimchi, A. (1986) Tumor necrosis factor reduces *c-myc* expression and cooperates with interferon-gamma in HeLa cells. *Science,* **234**, 1419–21.

45. Malejczyk, J., Malejczyk, M., Majewski, S., *et al.* (1994) Increased tumorigenicity of human papillomavirus type 16-harboring keratinocytes is associated with resistance to endogenous tumor necrosis factor-α-mediated growth limitation. *International Journal of Cancer,* **56**, 593–8.

46. Malejczyk, J., Malejczyk, M., Breitburd, F., *et al.* (1996) Progressive growth of human papillomavirus type 16-transformed keratinocytes is associated with an increased release of soluble tumour necrosis factor (TNF) receptor. *British Journal of Cancer,* **74**, 234–9.

47. Arany, I., Rady, P. and Tyring, S.K. (1994) Interferon treatment enhances the expression of underphosphorylated (biologically-active) retinoblastoma protein in human papilloma virus-infected cells through the inhibitory TGFβ1/IFNβ cytokine pathway. *Antiviral Research,* **23**, 131–41.

48. Majewski, S., Breitburd, F., Orth, G., *et al.* (1994) Regulation of MHC class I, class II and ICAM-1 expression by cytokines and retinoids in HPV-harboring keratinocyte lines. In *Immunology of Human Papillomaviruses* (ed. M.A. Stanley). Plenum Press, New York, pp. 207–10.

49. Majewski, S., Marczak, M., Szmurlo, A., *et al.* (1993) Inhibition of tumor cell-induced angiogenesis by retinoids, 1,25-dihydroxyvitamin D3 and their combination. *Cancer Letters,* **75**, 35–9.

50. Majewski, S., Szmurlo, A., Marczak, M., *et al.* (1994) Synergistic effect of retinoids and interferon-α on tumor-induced angiogenesis: antiangiogenic effect on HPV-harboring tumor cell lines. *International Journal of Cancer,* **57**, 81–5.

51. Bollag, W., Majewski, S. and Jablonska, S. (1994) Biological interactions of retinoids with cyto-

kines and vitamin D analogs as a basis for cancer combination chemotherapy. In *Retinoids: from Basic Science to Clinical Applications* (eds M.A. Livera and G. Vidaldi), Birkhauser Verlag. Basel, pp. 267–80.

52. Lippman, S.M., Kavanagh, J.J., Paredes-Espinoza, M., *et al.* (1992) 13-*cis* Retinoic acid plus interferon alfa-2a: highly active systemic therapy for squamous carcinoma of the cervix. *Journal of the National Cancer Institute*, **84**, 241–5.

53. Lippman, S.M., Parkinson, D.R., Itri, L.M., *et al.* (1992) 13-*cis* Retinoic acid and interferon alfa-2a: effective combination therapy for advanced squamous cell carcinoma of the skin. *Journal of the National Cancer Institute*, **84**, 235–41.

54. Woodworth, C.D., McMullin, E., Iglesias, M., *et al.* (1995) Interleukin 1a and tumor necrosis factor-α stimulate autocrine amphiregulin expression and proliferation of human papillomavirus-immortalized and carcinoma-derived cervical epithelial cells. *Proceedings of the National Academy of Sciences of the USA*, **92**, 2840–4.

55. Iglesias, M., Plowman, G.D. and Woodworth, C.D. (1995) Interleukin-6 and interleukin-6 soluble receptor regulate proliferation of normal, human papillomavirus-immortalized, and carcinoma-derived cervical cells *in vitro*. *American Journal of Pathology*, **146**, 944–52.

56. Gullick, W.J., Marsden, J.J., Whittle, N., *et al.* (1986) Expression of epidermal growth factor receptors on human cervical, ovarian, and vulvar carcinomas. *Cancer Research*, **46**, 285–92.

57. Bauknecht, T., Kohler, M., Janz, I., *et al.* (1989) The occurrence of epidermal growth factor receptors and the characterization of EGF-like factors in human ovarian, endometrial, cervical and breast cancer. *Journal of Cancer Research and Clinical Oncology*, **115**, 193–9.

58. Goppinger, A., Wittmaack, F.M., Wintzer, H.O., *et al.* (1989) Localization of human epidermal growth factor receptor in cervical ontraepithelial neoplasias. *Journal of Cancer Research and Clinical Oncology*, **115**, 259–63.

59. McGlennen, R.C., Ostrow, R.S., Carson, L.E., *et al.* (1991) Expression of cytokine receptors and markers of differentiation in human papillomavirus-infected cervical tissues. *American Journal Obstetrics and Gynaecology*, **165**, 696–705.

60. Zyzak, L.L., MacDonald, L.M., Batova, A., *et al.* (1994) Increased levels and constitutive tyrosine phosphorylation of the epidermal growth factor receptor contribute to autonomous growth of human papillomavirus type 16

immortalized human keratinocytes. *Cell Growth and Differentiation*, **5**, 537–47.

61. Vaux, D.L. (1993) Toward an understanding of the molecular mechanisms of physiological cell death. *Proceedings of the National Academy of Sciences of the USA*, **90**, 786–9.

62. Wyllie, A.H. (1993) Apoptosis (The 1992 Frank Rose memorial lecture). *British Journal of Cancer*, **67**, 205–8.

63. Evans, C.H., Flugelman, A.A. and DiPaolo, J.A. (1993) Cytokine modulation of immune defenses in cervical cancer. *Oncology*, **50**, 245–51.

64. Clarks, A.R., Purdic, C.A., Harrison, D.J., *et al.* (1993) Thymocyte apoptosis induced by p53-dependent and independent pathways. *Nature*, **362**, 849–52.

65. Saegusa, M., Takano, Y., Hashimura M., *et al.* (1995) The possible role of bcl-2 expression in the progression of tumors of the uterine cervix. *Cancer*, **76**, 2297–303.

66. Polakowska, R.H. and Haake, A.R. (1994) Apoptosis: the skin from a new perspective. *Cell Death and Differentiation*, **1**, 19–31.

67. Arends, M.J., McGregor, A.H. and Wyllie A.H. (1994) Apoptosis is inversely related to necrosis and determines net growth in tumors bearing constitutively expressed myc, ras, and HPV oncogenes. *American Journal of Pathology*, **144**, 1045–57.

68. Rotello, R.J., Lieberman, R.C., Purchio, A.F., *et al.* (1991) Coordinated regulation of apoptosis and cell proliferation by transforming growth factor b1 in cultured uterine cells. *Proceedings of the National of Academy Sciences of the USA*, **88**, 3412–15.

69. Vieiro, K.B., Goldstein, D.J. and Vill, L.L. (1996) Tumor necrosis factor alpha interferes with the cell cycle of normal and papillomavirus-immortalized human kenetinocytes. *Cancer Research*, **56**, 2452–7.

70. Hagari, Y., Budgeon, L.R., Pickel, M.D., *et al.* (1995) Association of tumor necrosis factor-α gene expression and apoptotic cell death with regression of Shope papillomas. *Journal of Investigative Dermatology*, **104**, 526–9.

71. Shoji, Y., Saegusa, M. and Takano, Y. (1996) Correlation of apoptosis with tumour cell differentiation, progression, and HPV infection in cervical carcinoma. *Journal of Clinical Pathology*, **49**, 134–8.

72. Isacson, Ch., Kessis, T.D., Hedrick, L., *et al.* (1996) Both cell proliferation and apoptosis increase with lesion grade in cervical neoplasia but do not correlate with human papillomavirus type. *Cancer Research*, **56**, 669–74.

73. Schröder, J.M., Sticherling, M., Smid, P., *et al.*

(1994) Interleukin 8 and structurally related cytokines. In *Epidermal Growth Factors and Cytokines* (eds T.A. Luger and T. Schwarz). Marcel Dekker, New York, pp. 89–112.

74. Baggiolini, M., Dewald, B. and Moser, B. (1993) Interleukin-8 and related chemotactic cytokines-CXC and CC chemokines. *Advances in Immunology*, **55**, 97–107.

75. Kleine, K., König, G., Kreutzer, J., *et al.* (1995) Expression of the JE (MCP-1) gene encoding the monocyte chemoattractant protein-1 results in partial tumor suppression of HeLa-cells in nude mice. *Molecular Carcinogenesis*, **14**, 179–89.

76. Lehtinen, M., Rantala, I., Toivonen, A., *et al.* (1993) Depletion of Langerhans cells in cervical HPV infection is associated with replication of the virus. *APMIS*, **101**, 833–7.

77. Connor, M.E. and Stern, P.L. (1990) Loss of MHC class-I expression in cervical carcinomas. *International Journal of Cancer*, **46**, 1029–34.

78. Cromme, F.V., Meijer, C.J.L.M., Snijders, P.J.F., *et al.* (1993) Analysis of MHC class I and II expression in relation to presence of HPV genotypes in premalignant and malignant cervical lesions. *British Journal of Cancer*, **68**, 1372–80.

79. Glew, S.S., Conner, M.E., Snijders, P.J., *et al.* (1993) HLA expression in pre-invasive cervical neoplasia in relation to human papilloma virus infection. *European Journal of Cancer*, **29A**, 1963.

80. Hildres, C.G., Houbiers, J.G., van Ravenswaay-Claasen, H.H., *et al.* (1993) Association between HLA-expression and infiltration of immune cells in cervical carcinoma. *Laboratory Investigation*, **69**, 651–9.

81. Torres, L.M., Cabrera, T., Concha, A., *et al.* (1993) HLA class I expression and HPV-16 sequences in premalignant and malignant lesions of the cervix. *Tissue Antigens*, **41**, 65–71.

82. Cromme, F.V., van Bommel, P.F.J., Walboomers, J.M.M., *et al.* (1994) Differences in MHC and TAP-1 expression in cervical cancer lymph node metastases as compared to the primary tumours. *British Journal of Cancers*, **69**, 1176–81.

83. Coleman, H., Birley, H.D., Renton, A.M., *et al.* (1994) Immunological events in repressing genital warts. *American Journal of Clinical Pathology*, **102**, 768–73.

84. Jahmus, I., Durst, M., Reid, R., *et al.* (1993) Major histocompatibility complex and human papillomavirus type 16 E7 expression in high-grade vulvar lesions. *Human Pathology*, **24**, 519.

85. Springer, T.A. (1990) Adhesion receptors of the immune system. *Nature*, **346**, 425–34.

86. Coleman, N., Greenfield, I.M., Hare, J., *et al.* (1993) Characterization and functional analysis of the expression of intercellular adhesion molecule-1 in human papillomavirus-related disease of cervical keratinocytes. *American Journal of Pathology*, **143**, 355–67.

87. Morelli, A.E., Belardi, G., DiPaola, G., *et al.* (1994) Cellular subsets and epithelial ICAM-1 and HLA-DR expression in human papillomavirus infection of the vulva. *Acta Dermato Venereological*, **74**, 45–9.

88. Dustin, M.L., Singer, K.H., Tuck, D.T., *et al.* (1988) Adhesion of T lymphoblasts to epidermal keratinocytes is regulated by interferon γ and is mediated by intercellular adhesion molecule 1 (ICAM-1). *Journal of Experimental Medicine*, **167**, 1323.

89. Caughman, S.W., Lian-Jie, L. and Degitz, K. (1990) Characterization and functional analysis of interferon-γ-induced intercellular adhesion molecule-1 expression in human keratinocytes and A-431 cells. *Journal of Investigative Dermatology*, **94**, 22S–5S.

90. Symington, F.W. and Santos, E.B. (1991) Lysis of human keratinocytes by allogeneic HLA class I-specific cytotoxic T cells: keratinocyte ICAM-1 (CD54) and T cell LFA-1 (CD11a/CD18) mediate enhanced lysis of IFN-γ-treated keratinocytes. *Journal of Immunology*, **146**, 2169–75.

91. Seder, R.A. and Paul, W.E. (1994) Acquisition of lymphokine-producing phenotype by $CD4^+$ T cells. *Annual Review in Immunology*, **12**, 635.

92. Reiner, S.L. and Seder, R.A. (1995) T helper cell differentiation in immune response. *Current Opinion in Immunology*, **7**, 360–6.

93. Crowley-Nowick, P.A., Bell, M.C., Bull, R., *et al.* (1993) Cytokine expression by T cells isolated from cervical intraepithelial neoplasia lesions. In *12th International Papillomavirus Conference*, 26 September–1 October, 1993, Johns Hopkins Medical Institution, Baltimore, MD, p. 7.

94. de Waal Malefyt, R., Abrams, J., Bennett, C., *et al.* (1991) IL-10 inhibits cytokine synthesis by human monocytes: an autoregulatory role of IL-10 produced by monocytes. *Journal of Experimental Medicine*, **174**, 1209–15.

95. Hsu, D.H., Moore, K.W. and Spits, H. (1992) Differential effects of interleukin-4 and -10 on interleukin-2-induced interferon-γ synthesis and lymphokine-activated killer activity. *International of Immunology*, **4**, 563.

96. Cannon, J.G., Tatro, J.B., Reichlin, S., *et al.* (1986) α-Melanocyte-stimulating hormone inhibits immunostimulatory and inflamatory action of interleukin 1. *Journal of Immunology,* **137**, 2232–8.

97. Hiltz, M.E., Catania, E.A. and Lipton, J.M. (1992) α-MSH peptides inhibit acute inflammation induced in mice by rIL-1β, rIL-6, rTNFα and endogenous pyrogen but not that caused by LTB4, PAF and rIL-8. *Cytokine,* **4**, 320–8.

98. Bhradwaj, R.S., Arange, Y., Becher, E., *et al.* (1994) Alpha melanocyte stimulating hormone differentially regulates IL-10 by human peripheral blood mononuclear cells. *Journal of Investigative Dermatology,* **102**, 586–90.

99. Grabbe, S., Bhardwaj, R.S., Mahnke, K., *et al.* (1996) α-Monocyte-stimulating hormone induces hapten-specific tolerance in mice. *Journal of Immunology,* **156**, 473–8.

100. Chomarat, P., Rissoan, M.C., Banchereau, J., *et al.* (1993) Interferon γ inhibits interleukin 10 production by monocytes. *Journal of Experimental Medicine,* **177**, 523–7.

101. Nabros, G.S., Afonso, L.C.C., Farreli, J.P., *et al.* (1995) Switch from a type 2 to a type 1 T helper cell response and cure of established *Leishmania major* infection in mice is induced by combined therapy with interleukin 12 and pentostan. *Proceedings of the National Academy of Sciences of the USA,* **92**, 3142–6.

102. Folkman, J. (1992) What is the evidence that tumors are angiogenesis dependent? *Journal of the National Cancer Institute,* **82**, 4–6.

103. Folkman, J. (1995) Angiogenesis in cancer, vascular, rheumatoid and other disease. *Nature Medicine,* **1**, 27–31.

104. Craft, P.S. and Harris, A.L. (1994) Clinical prognostic significance of tumor angiogenesis. *Annals of Oncology,* **5**, 305–11.

105. Stafl, A. and Matringly, R.F. (1975) Angiogenesis of cervical neoplasia. *American Journal of Obstetrics and Gynaecology,* **121**, 845–52.

106. Chomette, G., Auriol, M., Tranbaloc, P., *et al.* (1989) Intraepithelial neoplasm of the uterine cervix and angiogenesis: morphologic study. *Archives D Anatomie et de Cytologie Pathologiques,* **37**, 73–9.

107. Smith-McCune, K.K. and Weidner, N. (1994) Demonstration and characterization of the angiogenic properties of cervical dysplasia. *Cancer Research,* **54**, 800–4.

108. Guidi, A.J., Abu-Jawdeh, G., Berse, B., *et al.* (1995) Vascular permeability factor (vascular endothelial growth factor) expression and angiogenesis in cervical neoplasia. *Journal of the National Cancer Institute,* **87**, 1237–45.

109. Sillman, F., Boyce, J. and Fruchter, R. (1981) The significance of atypical vessels and neovascularization in cervical neoplasia. *American Journal Obstetrics and Gynaecology,* **139**, 154–9.

110. Burke, L., Antonioli, D. and Ducarman, B. (1991) *Colposcopy Textbook and Atlas,* Appleton & Lange, East Norwalk, CT.

111. Klagsburn, M. and Soker, S. (1993) VEGF/VPF: the angiogenesis factor found? *Current Biology,* **3**, 699–702.

112. Kim, K.J., Li, B., Weiner, J., *et al.* (1993) Inhibition of vascular endothelial growth factor-induced angiogenesis suppresses tumour growth *in vivo. Nature,* **363**, 841–4.

113. Klagsburn, M. and Folkman, J. (1990) Angiogenesis. *Handbook of Experimental Pharmacology,* **95**, 549–86.

114. Kandel, J., Bossy-Wetzel, E., Radvanyi, F., *et al.* (1991) Neovascularization is associated with a switch to the export of bFGF in the multistep development of fibrosarcoma. *Cell,* **66**, 1095–104.

115. Majewski, S., Marczak, M., Szmurlo, A., *et al.* (1996) Interleukin-12 inhibits angiogenesis induced by human tumor cell lines *in vivo. Journal of Investigative Dermatology,* **106**, 1114–18.

116. Baillie, C.T., Winslet, M.C. and Bradley, N.T. (1995) Tumor vasculature – a potential therapeutic target. *British Journal of Cancer,* **72**, 257–67.

117. Denekamp, J. (1993) Review article: Angiogenesis, neovascular proliferation and vascular pathophysiology as targets for cancer therapy. *British Journal of Radiology,* **66**, 181–96.

118. Karp, J.E. and Broder, S. (1995) Molecular foundations of cancer: new targets for intervention. *Nature Medicine,* **1**, 309–20.

119. O'Reilly, M.S., Holmgren, L., Shing, Y., *et al.* (1994) Angiostation a novel angiogenesis inhibitor that mediastes the suppression of metastases by a Lewis lung carcinoma. *Cell,* **69**, 315–28.

120. Majewski, S., Polakowski, I., Marczak, M., *et al.* (1986) The effects of retinoids on lymphocyte- and transformed cell line-induced angiogenesis. *Clinical and Experimental Dermatology,* **2**, 317–18.

121. Majewski, S., Marczak, M. and Jablonska, S. (1989) Effects of retinoids on angiogenesis and on the proliferation of endothelial cells *in vitro.* In *Pharmacology of Retinoids in the Skin* (eds U. Reichert and B. Shroot). Karger, Basel, pp. 94–5.

122. Majewski, S., Marczak, M., Szmurlo, A., *et al.* (1995) Retinoids combined with interferon alpha or 1,25-dihydroxyvitamin D3 synergistically inhibit angiogenesis induced by non-HPV-harboring tumor cell lines. *Cancer Letters*, **89**, 117–24.

123. Brunda, M.J., Luistro, L., Warrier, R.R., *et al.* (1993) Antitumor and antimetastatic activity of interleukin 12 against murine tumors. *Journal of Experimental Medicine*, **178**, 1223–30.

124. Brunda, M.J. and Gately, M.K. (1994) Antitumor activity of interleukin-12. *Clinical Immunology and Immunopathology*, **71**, 253–5.

125. Banks, R.E., Patel, P.M. and Selby, P.J. (1995) Interleukin-12: a new clinical player in cytokine therapy. *British Journal of Cancer*, **71**, 655–9.

126. Sgadari, C., Angiolillo, A.L. and Tosato, G. (1996) Inhibition of angiogenesis by interleukin-12 is mediated by the interferon-inducible protein 10. *Blood*, **87**, 3877–82.

127. Majewski, S., Skopinska, M., Marczak, M., *et al.* (1996) Vitamin D3 is a potent inhibitor of tumor cell-induced angiogenesis. *Journal of Investigative Dermatology Symposium Proceedings*, **1**, 97–101.

VIRAL INFECTIONS AND CERVICAL NEOPLASIA

9

Penelope I. Blomfield and Suzanne Garland

INTRODUCTION

Both epidemiological and experimental data support the hypothesis that viruses are involved in carcinogenesis. The basis for proposing viruses as oncogenic agents emerged with the realization that they have the potential for altering cellular phenotype. In 1911, Rous found that agents from tumour cells of chicken sarcomas were not only capable of inducing tumours but also imprinted the phenotypic properties of the original tumour on the recipient transformed cell [1]. It is now recognized that the transmissible agent in these experiments was a Rous sarcoma virus. In the 1950s and 1960s the rapid expansion of viral research led to the realization that other viral groups also have transforming properties. These included papillomaviruses, polyomaviruses, hepadnaviruses, adenoviruses, herpes viruses and pox viruses. During the 1970s and 1980s, with the advent of new molecular biological techniques, it became clear that viruses play important roles in the development and behaviour of a number of major malignancies in both animals and humans, e.g. HTLV1 (adult T-cell leukaemia), hepadnavirus (hepatocellular carcinoma) and Epstein–Barr virus (nasopharyngeal carcinoma, Burkitt's lymphoma).

With respect to cervical cancer, investigators have long recognized the importance of sexual behaviour in the aetiology of the disease. Early studies tended to describe sexual behaviour in terms of marital status and reproductive history. Subsequent studies specifically identify the total number of lifetime sexual partners and age at first intercourse as two main risk factors for the development of cervical neoplasia, irrespective of marital status (reviewed by Muñoz and Bosch, 1989 [2]). It has also become clear that the sexual background of male partners can influence the risk of disease development [3]. The relationship between different aspects of sexual behaviour and the risk of cervical neoplasia led to the hypothesis that a sexually transmitted agent was, at least in part, responsible for disease development. Over the years *Trichomonas vaginalis*, *Neisseria gonorrhoea*, the genital mycoplasmas and *Chlamydia trachomatis* have all been proposed as possible causative agents in cervical cancer [4]. Typical of these studies was the observation of an increase in syphilitic infection in patients with cervical cancer as compared to control populations [5, 6]. For most of the proposed agents no clear epidemiological link has been established and the *in vitro* data to substantiate such suggestions data are lacking.

Viral research in cervical cancer has in the past focused on the epidemiological characteristics of viral infection, the isolation of viral genes or gene products from neoplastic cervical

Cancer and Pre-cancer of the Cervix
Edited by D.M. Luesley and R. Barrasso. Published in 1998 by Chapman & Hall, London. ISBN 0 412 56600 1.

tissue and viral transforming activity *in vitro*. During the 1950s to the 1970s, herpes simplex virus was considered a likely causative agent. However, whilst herpes simplex type 2 (HSV2) is sexually transmissible, remains latent in the body and can subsequently be isolated intermittently from cervical secretions, attempts to consistently find HSV2 DNA from cervical neoplastic biopsies were unsuccessful. The strongest viral contender during the last decade has been specific genotypes of the human papillomavirus (HPV), as first proposed by zur Hausen in 1977 [7].

A review of the current evidence supporting the hypothesis that specific HPV types may be implicated in the development of cervical neoplasia follows.

MOLECULAR BIOLOGY OF HPV AND *IN VITRO* EVIDENCE SUPPORTING A ROLE FOR SPECIFIC VIRAL TYPES IN CERVICAL CARCINOGENESIS

Major advances have been made by molecular biologists in the understanding of the molecular and cellular mechanisms by which certain HPV types may be involved in cervical carcinogenesis.

BACKGROUND TO PAPILLOMAVIRUS BIOLOGY

Papillomaviruses are classified as a genus of the Papovaviridae family along with the genus Polyomavirus [8]. There are, however, fundamental differences in the biology and genomic organization of the two genera and they may well be classified in the future as two distinct families of viruses. Papillomaviruses are well known for inducing benign epidermal proliferation in the skin and mucosa of animals and humans (*papilla* from the Latin meaning nipple or pustule, *oma* from the Greek meaning tumour). Whilst they infect many vertebrates, the papillomaviruses are highly species-specific and have cellular tropism for squamous epithelial cells, where

they replicate in the nucleus. Papillomaviruses measure 55 nm in diameter, are non-enveloped and consist of an icosahedral protein capsule compiled of 72 capsomers, surrounding a double stranded circular DNA. The DNA molecule consists in the majority of cases of 7900 base pairs and exists in two predominant forms, either as superhelical twists or a relaxed open circle form. One of the characteristics of papillomaviruses is that all the viral genes are located on one strand of DNA, i.e. only one strand serves as a template for transcription. Figure 9.1 is an electronmicrograph of human papillomaviruses.

Classification of papillomaviruses

Each virus is first named after its natural host and then classified according to sequence homology.

Genomic organization

The genomic organization of papillomaviruses sequenced thus far show remarkable similarity. The complete nucleotide sequences have been determined for many of the papillomaviruses, including HPV types 1, 5, 6, 8, 11, 16, 18, 31 and 33. There are approximately 8–10 designated putative protein coding sequences (open reading frames) and all are of comparable size and in similar positions for

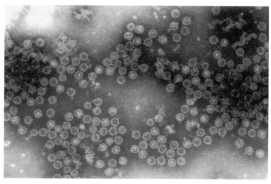

Figure 9.1   Electron micrograph of human papillomaviruses, reproduced with permission from C. Laverty, Australia.

each papillomavirus type. The individual frames are classified as 'early' (E)$_{1-8}$ or 'late' (L)$_{1-2}$ by analogy with other DNA viruses, where the genes are transcribed according to a specific time schedule in the course of a permissive infection (see Figure 9.2). The early genes are expressed shortly after infection and prior to DNA replication and many of their functions have now been defined. There are at least two proteins encoded by the E1 open reading frames (ORF) and both have subsequently been shown to be involved in DNA replication and plasmid maintenance. The E2 ORF encodes for two proteins, one which acts as an activator and the other as a repressor of viral mRNA synthesis. The E4 protein, unlike the typical early viral proteins, is produced in large amounts within HPV-induced lesions [9] and is involved in viral particle maturation [10]. The E5 ORF is lacking in some HPVs, but when present has a transforming function. Key transforming activity is usually attributed to the early genes E6 and E7. The late genes (L1 and L2) encode for the two structural proteins of viral particles and are activated during the final stages of the viral cycle. The overall ratio of L1:L2 is 4:1 and L1 acts as the group antigen whilst L2 is thought to be type specific. These are only expressed in permissively infected cells. Upstream of the early genes and spanning the junction of the terminal start and end of the circular genome is a non-coding region referred to as the upstream regulatory region (URR), non-coding region (NCR) or long control region (LCR). This region contains the most important early promoter, and many of the important enhancers and binding sites, for trans-acting factors which regulate transcription of the early genes [11].

HUMAN PAPILLOMAVIRUS GENOTYPES

Since the first study of HPV DNA in 1965 [12], it has been well recognized that human papillomaviruses are a large and diverse group of viruses with over 70 genotypes, with certain groups showing tropism for different anatomical sites, e.g. HPV 1–4 with skin, HPV 6, 11, 16, 18, 31, 33 and 35 with the lower genital tract epithelium. Due to the lack of a serological classification system, the definition of an HPV type has been based upon the degree of DNA sequence homology to other known papillomaviruses. The present definition of a new type requires cloning of the entire genome and a new genotype is based on the nucleotide sequence of the E6, E7 and L1 showing less than 90% nucleotide sequence homology to well-established HPV types [13]. Recently, sequencing and subsequent phylogenetic analysis has allowed organization of the HPVs into specific groups encompassing known tissue tropism and oncogenic potential [14]. Furthermore, this has allowed subgrouping of genital HPV types into those of 'low' malignant potential,

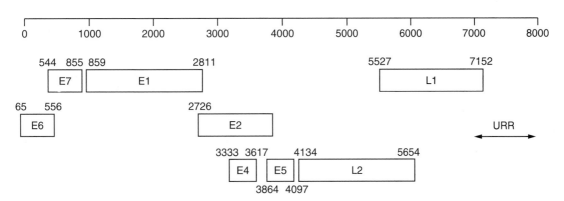

Figure 9.2 Genomic organization of HPV 16.

usually responsible for condylomata, i.e. HPV 6, 11, 13, 42, and those of 'high' malignant potential linked to cervical intra-epithelial neoplasia (CIN) and cancers, i.e. HPV 16, 18, 31, 33, 35, 39, 51 and 58.

ONCOGENIC POTENTIAL OF HUMAN PAPILLOMAVIRUSES: *IN VITRO* TRANSFORMATION STUDIES

Rodent cells were initially used to study the transforming properties (i.e. production of a tumorigenic phenotype) of human papillomaviruses. For the oncogenic HPV types 16 and 18, the E6 and E7 ORFs were identified as the transforming genes [15, 16]. In transformation assays using primary rodent cells, the HPV 16 ORFs E6 and E7 act individually and in combination with the EJ-*ras* oncogene [17]. For human keratinocytes and fibroblasts, however, both E6 and E7 in co-operation with an oncogene are required for transformation. Similar findings were observed with other high-risk HPV types 18, 31 and 33 but not with HPV 6 or 11 [18].

HPV 16 and 18 have also been shown to cause loss of senescence in human foreskin keratinocytes *in vitro* and the establishment of immortalization (i.e. normal cells are converted to cells of unrestricted life span) [19]. This finding is of particular importance since papillomaviruses are strictly epitheliotrophic and are dependent upon the environment provided by terminally differentiating keratinocytes for completion of their life cycle. If the HPV 16 immortalized human keratinocytes are allowed to stratify into a multilayer epithelium in a collagen raft system, they display the histopathological features of CIN [20]. HPV 16 and 18 have also now been shown to immortalize primary human cervical epithelial cells [21]. Cells transfected with cloned viral DNAs from HPV types 16 and 18 acquired indefinite life-spans, distinct morphological alterations, and anchorage-dependent growth, and contain transcriptionally active viral genomes. This was in contrast to cells transfected with HPV 6, an HPV type of low malignant poten-

tial. So far unlimited growth in culture has only been achieved with DNA of high risk HPV types 16 or 18, and not with the low risk types 6 or 11 [22]. Although immortalization of human keratinocytes by HPV 16 and 18 is now reproducibly demonstrated, it appears that massive tumour growth in nude mice only occurs after the cells have been kept in culture for a number of passages prior to inoculation of the animals or with the co-operation of an activated *ras* oncogene [23].

The E7 ORF has been shown to be responsible for these immortalizing effects in rodent cells. However, in human cells when E6 plus E7 ORF are used for transfection, the efficiency of immortalization is enhanced [24, 25]. Both oncoproteins E6 and E7 of high-risk HPV types have been shown *in vitro* to bind the cell-regulating proteins p53 and the retinoblastoma gene product (p105-RB), respectively, providing a plausible mechanism for the altered cellular behaviour [26]. The retinoblastoma and p53 proteins act as tumour suppressor genes negatively controlling entry or progress through the cell cycle, and hence are important in maintaining normal cell growth. Binding of the viral oncoproteins inhibits the functions of the cell proteins and relieves the cells of their 'normal' brakes on cell division. In the case of the oncoproteins of the high-risk HPV types 16 and 18, this interaction has been shown to be much stronger than that of low-risk HPV types.

VARYING STATUS OF THE PAPILLOMAVIRUS GENOME

HPV DNA remains stable and is transcriptionally active after many decades in cell lines derived from cervical cancer biopsies (e.g. HeLa, Caski), despite many hundreds of passages. In these cell lines, HPV DNA is always integrated at non-specific sites of the host chromosomes and the same copy number of HPV DNA is retained over the years. In contrast, *in vivo* integration generally occurs by the interruption of the E1 and E2 ORFs with consequent disruption of the

E2 transcriptional regulatory circuit, but leaving the E6 and E7 ORFs intact. In HPV-associated cervical cancers, investigators have confirmed that HPV types 16, 18 and 31 may be found integrated into the host cellular DNA, and in some, this integration is in the vicinity of cellular oncogenes. In pre-malignant lesions, however, viral DNA persists as episomal although integration has been rarely reported [27]. In women infected with high-risk types and who have no evidence of cervical neoplasia, integration is not observed.

EARLY OBSERVATIONS OF A RELATIONSHIP BETWEEN CERVICAL NEOPLASIA AND HPV INFECTION

The cytomorphological characteristics of koilocytes were described in 1949 by Ayre [28] and linked to HPV infection by Meisels and Fortin [29]. In 1977, zur Hausen first suggested that HPV may be the sexually transmitted aetiological agent in cervical neoplasia and several authors subsequently reported the frequent association of the cytopathological features of HPV with cervical disease [7]. Over the following 10 years a large number of prevalence studies, based upon DNA hybridization techniques, were reported and by the mid-1980s a pattern had emerged, linking a high prevalence of different HPV types and cervical disease. It became evident that cervical infection with HPV types 16 or 18 was seen more frequently in women with high-grade cervical abnormalities compared to women with normal cervices or low-grade lesions. In contrast, infection with HPV types 6 or 11 was noted to be uncommon in invasive disease, but frequently observed in low-grade CIN lesions. With the recognition that a percentage of cytologically normal women in fact harbour HPV infection, these early studies were widely criticized for poor design, lack of control groups, inadequately described controls, and the lack of standardized HPV assays [30]. The need for prospective case-control and cohort studies was thus established.

Epidemiologists subsequently also investigated the risk factors for HPV infection; the rationale being that genital HPV infection should share similar risk factors as cervical neoplasia, if the virus was implicated. These investigations proved more difficult than first anticipated, largely because of the problems in identifying an appropriate standardized method for assessing type-specific viral infection that is applicable to large scale epidemiological investigations.

DETECTION SYSTEMS FOR HPV INFECTION

VIRAL DETECTION METHODS

Due to the stringent conditions papillomaviruses require for replication they cannot be cultured using traditional laboratory methods. Moreover, methods for viral identification based on conventional serological assays used with other viral infections have not been achieved. Earlier techniques for identifying HPV infection included light microscopy, electron microscopy and immunohistochemistry. However, these methods are relatively insensitive and have a degree of subjectivity and poor reproducibility. Electron microscopy and immunohistochemistry, which detects the group-specific capsid antigen only, rely on the presence of a large number of intact virions and are unable to differentiate the HPV genotypes. With the advent of molecular biological techniques such as nucleic acid hybridization, HPV DNA has been detected by dot-blot, Southern blot and filter *in situ* hybridization methods. Southern blot hybridization, although considered by some to be the gold standard for HPV detection, is a labour-intensive and complicated technique requiring larger amounts of DNA than cervical scrapes can reliably render (10 μg) [31]. Under ideal conditions the sensitivity of the assay is in the order of 1 copy of a genome in 10–100 cells. Whilst the technique allows typing and also gives information regarding the physical state of the virus (whether it is episomal or integrated), interlaboratory variation has been demonstrated with its use [32].

Filter *in situ* hybridization (FISH) was the first assay specifically designed for large population studies. This technique involves the filtration of exfoliated cells directly onto a filter rather than transferring DNA from a gel, denaturation of the DNA, followed by hybridization with labelled probes under stringent conditions. Although the method was initially well received and widely used, it was subsequently heavily criticized because of its proneness to false-positive and negative results. Initial validation investigations confirmed good intralaboratory reproducibility [33] and favourable comparison to Southern blot [34]. However, more recent comparisons have shown a much poorer correlation than first recognized [35, 36]. The technique is also insensitive; the level of detection being in the order of 50 genome copies per cell.

In situ hybridization has the advantage of siting infection at the cellular level within a tissue section. It is, however, relatively insensitive, requiring around 20 viral copies per infected cell.

The dot-blot technique involves the direct transfer of extracted DNA onto a membrane and subsequent hybridization with labelled DNA/RNA probes. It does not require the lengthy gel electrophoresis and transfer steps of Southern blot hybridization and large numbers of samples can be examined simultaneously. Dot-blot in the form of Virapap/Viratype (RNA probes) has been popularized in the USA and is commercially available. Dot-blot was initially also criticized for being prone to false-positives, however, recent studies comparing the Virapap/Viratype system to Southern blot appear to produce extremely favourable comparisons [37]. Despite this, the system has not been widely used outside the USA, largely because of its cost.

The polymerase chain reaction

Over the last 10 years the polymerase chain reaction (PCR) has emerged as an exquisitely sensitive technique for identifying the pres-

ence of different types of HPV DNA. The PCR is based upon the recognition and amplification of a specific DNA target sequence using the thermostable *Taq* polymerase enzyme. A set of nucleotide primers complementary to the target region is designed and repeated cycles of denaturation, annealing and elongation (usually 40) results in up to 10^{10} copies of the flanked region. It was originally described as a method for the diagnosis of inherited genetic diseases such as sickle cell anaemia [38], but was soon found to be a useful tool in many areas of biomedical research, as well as infectious diseases. Here, it is especially useful for the identification of non-culturable, slow growing, poor yielding pathogens.

At the outset of development of PCR widely varying prevalence rates of cervical HPV infection were reported in cytologically normal women (5–80%), and it soon became evident that PCR was extremely vulnerable to false-positive results due to contamination [39]. The complete process, from sample collection in the clinical setting, to completion of the laboratory technique, is susceptible to contamination by minute amounts of recombinant plasmids, PCR product carryover and other clinical samples that may give rise to false-positive results. Many precautions are therefore necessary to prevent contamination at each stage in the procedure and have recently been reviewed [40]. They include: careful collection of the specimen from the patient and transfer to the laboratory in individual packaging; physical separation of the cloning and PCR preparation procedures; separation of the pre- and post-amplification procedures and instruments; the inclusion of appropriate positive and negative controls. The amplification process can be entirely automated and detection of the amplified product can be visualized using ethidium bromide staining of an agarose gel. Confirmation can be by Southern blotting with oligonucleotide probes or restriction endonuclease fragment analysis [41].

Several investigators have attempted to prevent contamination of PCR reactions with cloned HPV plasmids by using primers that flank the HPV cloning site, so called anti-contamination primers [42]. Alternating nested PCR can also be used to enhance sensitivity and specificity. Here there is an initial round of amplification with one set of primers, then further amplification of that primer product with another set of primers designed internal to the first pair. Finally, a commercially available technique that avoids product carryover has been described, which involves the incorporation of deoxyuridine triphosphate (dUTP) and the enzyme uracil-N-glycosylase.

It has now become clear that even with the above precautions, major influences upon prevalence rates reported not only include the population source and age group concerned, but also the differing sampling techniques and PCR primers used [43]. There is still an obvious need for collaborative studies by different laboratories to standardize techniques and for the adoption of a standard set of primers that can be used with confidence in epidemiological studies.

Compared to other hybridization techniques the PCR is extremely sensitive, having the theoretical potential to detect a single copy of a viral genome in 10^5 cells. It has been used for the detection of HPV from fresh cervical biopsy material, cervical scrapes and lavages as well as archival and paraffin-embedded material and will detect latent, subclinical and clinically evident infection. With the increased sensitivity of PCR detection of HPV, cervical infection rates are higher than in previous population studies using Southern blot and mixed infections have also been noted at a greater rate.

Techniques using PCR continue to be refined and adapted. Application of PCR to *in situ* hybridization has been described and can be applied to genome or mRNA detection. It has been stated to be approximately 200 times more sensitive than standard *in situ* hybridization. Also the PCR has been adapted to give a semiquantitative assessment of HPV DNA [44].

Hybrid capture

Persistence of HPV, as well as viral load, now seems to be important in the development of cervical neoplasia. Hybrid capture is a relatively new technique, involving a semiquantitative detection and typing system of viral DNA. It also has the advantages of not using radioactive probes, employing RNA probes and a chemiluminescent method of detecting RNA–DNA hybrids that are captured onto a solid phase. Hybrid capture is commercially available. As compared to Southern blot it is a sensitive and specific test, provides same-day results and assesses the amount of HPV DNA present in a sample.

Most recent large studies have used PCR alone or in combination with either dot blot or Southern blot techniques. Initial validation studies using hybrid capture have been promising and it has been proposed as a test suitable to triage women for colposcopy when a cervical smear reports atypical squamous cells of undetermined significance (ASCUS) [45] or low-grade cytological abnormality [46].

Serology

Recently, serological assays such as Western blot or enzyme-linked immunosorbent assays that use antigen targets such as bacterial fusion proteins, synthetic peptides and virions have been developed. An association between antibodies to epitopes on HPV 16 E6 and especially on E7 and cervical cancer has been reported [47]. A similar type-specific reactivity has been proposed with the HPV 18 E7 peptide in cases of HPV 18-associated invasive cervical cancer. More recent results have suggested that serum antibody to HPV 16 virus-like particles free of DNA (VLPs) is a relatively sensitive indicator of persisting cervical HPV 16 infection [48]. Using VLP-ELISA assays, development of antibodies in incidental HPV infection appears to occur slowly (median of 12 months), well after clinical signs of HPV infection have abated. However, serology is still in its infancy as varying results are being reported depending on the type of antigen targeted and its role in diagnosis is yet to be defined.

EPIDEMIOLOGICAL CONSIDERATIONS

Despite difficulties in establishing a reproducible, sensitive and efficient detection method for cervical and genital HPV infection, the emergence of PCR and the refining of traditional hybridization techniques has allowed a number of large prevalence surveys and prospective case-control studies to be performed. The results of these studies go a long way to confirming the involvement of high-risk HPV types in cervical neogenesis.

PREVALENCE SURVEYS OF HPV DNA IN CERVICAL NEOPLASIA

The most impressive study investigating the prevalence of HPV DNA in cervical cancer is the International Biology Study on Cervical Cancer [49]. This study used PCR-based methods to investigate the prevalence of HPV DNA in cervical cancers from 22 participating countries. They confirmed that HPV DNA was identified in 92.9% of over 1000 cancer biopsy specimens and that HPV 16 was present in 51.5%. HPV types 16, 18, 31 and 45 accounted for 79.8% of the viral types identified. An apparent geographic variation was reported, with HPV 18 as the most common HPV type in Indonesia (comprising 50% of the viral types identified). It was also more commonly found in other south-east Asian countries. HPV types 39 and 59 were only found in Latin America and HPV 45 in western Africa. As with other prevalence studies that have investigated the role of HPV in adenocarcinomas of the cervix, this large study found HPV 18 to be the most common HPV type in these tumours. Such data strongly support an aetiological role for HPV in cervical carcinogenesis and HPV-negative cervical neoplasia is rare (<10%). If one accepts the definition that the cancer-associated group

of genital HPV types is defined as those found with appreciable prevalence alone in invasive cervical cancers, then the current list includes types 16, 18, 26, 31, 33, 35 39, 51, 52, 55, 56, 58, 59, 64 and 68 [50].

CASE CONTROL STUDIES

Studies utilizing multivariate statistical analysis and including known risk factors for cervical neoplasia, as well as detection of cervical HPV DNA, have now been published. Initial studies were criticized because of small numbers, inadequate control of potential confounding variables, and use of less-sensitive HPV hybridization methods [51–54]. However, despite these inadequacies, all such studies showed a large and significant association between cervical HPV infection with high-risk genotypes and cervical cancer risk. A study utilizing FISH for the detection of HPV infection showed the weakest association and this method has now been criticized on the grounds of relatively poor sensitivity and specificity [52]. The use of FISH tends towards a lower HPV prevalence amongst cases and a higher HPV prevalence amongst controls, artificially reducing the risk estimates reported [55]. Recent large well-designed case-control studies using PCR systems for HPV detection and multivariate analysis have confirmed these initial findings and are summarized in Table 9.1. All studies confirmed an independent and strong association of oncogenic HPVs with cervical cancer with large odds ratios.

The relationship between HPV infection and CIN has also been explored in case-control studies. All recent studies utilizing PCR-based hybridization techniques have demonstrated a strong association of HPV positivity, especially for high-risk types, with both low- and high-grade intra-epithelial neoplasia [56–58]. A number of these studies are summarized in Table 9.2.

The strength of association of high-risk HPV types with cervical cancer and CIN observed in the majority of these case-control

Table 9.1 Selected case-control studies of invasive cervical cancer and human papillomavirus (HPV) infection that have used multivariate analysis and included polymerase chain reaction (PCR) based methodology for HPV detection

Reference source and country of study	No. of cases	No. of controls	HPV type	Prevalence (%) Cases	Prevalence (%) Controls	OR (95%CI)
Eluf-Neto *et al.* [76]						
Brazil	199	225	HPV	84.0	17.0	37.1 (19.6–70.4) [a]
			HPV 16	53.8	5.3	74.9 (32.5–173)[a]
			16, 18, 31, 33	65.6	6.3	69.7 (28.65–169.6)[b]
Muñoz *et al.* [59, 108]						
Spain	250	238	HPV	69.6	4.6	46.2 (18.5–115.1)[c]
			HPV 16	45.8	3.1	14.2 (5.0–49.5)[d]
Colombia	186	149	HPV	72.4	13.3	15.6 (6.9–34.7)[c]
			HPV 16	50.6	9.2	5.5 (2.4–12.9)[d]

OR = odds ratio; CI = confidence interval.
[a]Adjusted for age and socioeconomic status.
[b]Adjusted for age, socioeconomic status, number of pap smears, parity, number of sexual partners, age at first intercourse, oral contraceptive use.
[c]Adjusted for age, study area, number of sexual partners, education, age at first birth, pap history.
[d]From data from pre-congress workshop of 14th International Papillomavirus Conference [108].

Table 9.2 Selected studies investigating the relationship between cervical inter-epithlial neoplasia (CIN) and human papillomavirus (HPV) that have used multivariate analysis and PCR-based methods for HPV detection

Reference source	Cases and their number	No. of controls	HPV type	Prevalence (%) Cases	Prevalence (%) Controls	OR (95%CI)
Liaw *et al.* [109] Taiwan	CIN I – 40 CIN II–III and Inv Ca – 48	261	HPV 16, 18, 31, 45	54.0	9.2	14.0 (6.1–32.0)
Bosch *et al.* [110, 108] Spain and Columbia	CIN III – 157	193	HPV 16	70.7 49.0	4.7 0.5	56.9 (24.8–130.6) 295.5 (44.8–1946.6)
	CIN III – 125	181	HPV 16	63.2 32.8	10.5 3.3	15.5 (8.2–29.4) 27.1 (10.6–69.5)
Olsen *et al.* [111] Norway	CIN II–III – 103	234	HPV 16	90.8 62.1	15.4 0.6	72.8 (27.6–191.9) 182.4 (54.0–616.1)
Kjaer [112] Denmark	low grade SIL – 120	1000	6, 11, X 16, 18, 31, 33	25.8 40.8	7.1 8.2	9.5 (5.6–16.1) 11.5 (7.0–18.7)
	high grade SIL – 79		6, 11, X 16, 18, 31, 33	15.2 62.0	7.1 8.2	8.3 (3.7–7.9) 32.9 (17.3–59.8)

OR = odds ratio; CI = confidence interval; SIL = squamous intra-epithelial lesions.

studies points towards a causal association between HPV infection and cervical carcinogenesis. Such a hypothesis assumes that HPV DNA is a marker for past or chronic persistent infection of the cervix. Other hypotheses that might explain such an association are increased susceptibility of cancer cells to HPV infection or enhanced detectability in cancer cells for some reason. Muñoz *et al.* (1992) showed a stable high rate of HPV DNA positivity amongst women with cervical cancer who reported being sexually active at the time of interview, as well as from those whose last sexual encounter was many years previous [59]. Assuming that sexual transmission is the major route of infection, this makes these other explanations unlikely.

COHORT STUDIES

Although cohort studies are regarded as one of the best means of obtaining evidence of a causal association, few are available for review and most are not completed.

Initial cohort studies based study entry on the presence of HPV infection by morphological criteria [60, 61]. These studies have methodological difficulties and should be interpreted with caution. One small study suggested a role for HPV 16 in the progression of CIN I to CIN III; however, it is likely that a significant percentage of subjects had higher grade disease at the outset [62]. This criticism can be directed at the larger cohort study by Syrjanen, who described an association between progression to more severe grades of CIN and HPV 16/18 infection [63, 64].

Caussy *et al.* (1990) used *in situ* hybridization to estimate the presence of HPV 16 DNA in cervical biopsy specimens of women who 2 years later developed cervical cancer [65]. He found no differences in HPV prevalence between women who subsequently developed disease as compared to controls. This may be related to the use of *in situ* hybridization, a relatively insensitive detection system, which may miss low copy numbers of the

virus. In contrast, Murthy *et al.* (1990) [66], also using *in situ* hybridization, compared the prevalence of HPV 16/18 infection in a group of 63 women with progression of low-grade CIN to CIN III, to that of 44 women with non-progressive dysplastic lesions. HPV 16/18 cervical infection was significantly associated with progression (OR of 5.9, 95% CI = 2.5–14.1).

Koutsky *et al.* (1992) have published the most comprehensive cohort study so far [67]. They described a cohort of 241 women aged between 16 and 50 years who at the outset of the study had negative cervical smears. These women were followed at 4-monthly intervals for 2 years with cytologic and colposcopic examinations of the cervix and were tested for cervical HPV DNA with dot-blot, Virapap or Southern transfer hybridization. On the basis of survival analysis, the cumulative incidence of CIN II or CIN III at 2 years was 28% for women positive for HPV DNA and 3% amongst those negative for HPV DNA. The risk was highest amongst those with HPV DNA type 16 or 18 infection as compared to women without evidence of infection (OR 11.0, 95% CI = 4.6–26).

In addition, several small follow-up studies have suggested that the presence or absence and type of HPV may influence prognosis in women with invasive cancer of the cervix [68–70].

In conclusion, the epidemiologic association of cervical neoplasia and human papillomavirus now fulfills all the established epidemiologic criteria for causality. There is strong evidence from case-control studies and preliminary data from cohort studies suggesting a causal link between specific high-risk HPV types and cervical carcinogenesis, with virtually no contrary studies. Although these viruses are strongly implicated, there is a consistent and significant background of cervical HPV infection in the healthy female population and it is obvious that other cofactors, in particular an individual's immune system, must play a large role in whether or not an infected individual actually goes on to develop cervical neoplasia.

MOLECULAR EPIDEMIOLOGY OF HPV INFECTIONS

In the late 1980s there was a realization that very little was known about the molecular epidemiology of cervical human papillomavirus infection, yet it had been recognized since biblical times that clinical genital wart infection was a sexually transmitted disease. The assumption was that cervical HPV infection was also sexually transmitted and would follow a similar history.

SEXUAL FACTORS

Although results from early correlational studies did not initially confirm that HPV was sexually transmitted [52, 71], it is now accepted that the majority of HPV infection is passed person-to-person by sexual contact. Several studies have confirmed that the lifetime number of sexual partners [72–74], especially number of most recent partners [75], is a significant risk factor for HPV infection. Also, when cervical HPV infection is included in multivariate analyses in case-control studies of cervical cancer, the apparent association of cervical cancer with increasing lifetime number of sexual partners is greatly weakened [59, 76].

Other investigators have searched for evidence of cervical or vulval HPV infection in virgins and initial findings have suggested that there is a background prevalence of infection with genital HPV types even in non-sexually active populations [71, 77]. Vertical transmission of infection at the time of birth has been suggested as an explanation. However, these studies were carried out in the early days of the use of the PCR, when problems with contamination were not fully recognized. More recent studies have not found HPV DNA in vaginal specimens collected from virginal women and children [78, 79]. There is no doubt that modes other than sexual transmission are possible, but rarely occur. The association between genital warts in the mother and the development of laryngeal papillomatosis in infants who were delivered by the vaginal route is well recognized [80]. Furthermore, HPV DNA has been identified at several sites (i.e. buccal mucosa, nasopharyngeal aspirates and penile foreskins) in infants both immediately after birth and for up to 2 years of age [81, 82]. Cason *et al.* (1995) have even gone so far as to suggest that there is a bimodal distribution of IgM antibodies to HPV 16 L1 and L2 proteins with peaks between the ages of 2 and 5 years of age and 13 and 16 years of age. These two peaks suggest that apart from sexual transmission in teenage years, a proportion of individuals may be infected much earlier in life [81].

AGE

The relationship of HPV with age may explain some of the widely varying prevalence data for HPV infection in cytologically normal women. This relationship was comprehensively studied in over 4000 women from district and hospital populations [83]. In this study general primer-mediated and type-specific PCR was carried out on cervical cell suspensions from women with inflammatory or normal smears. When combining all groups of their data, between which there was no significant difference, the prevalence rate of HPV infection (HPV types 6, 11, 16, 18, 31, 33 and unidentified HPV types) was 21% for women aged 20–24 years as compared to 6.1% for women aged 40–44 years. The prevalence rates for HPV types 16/18 for these age groups was 6.6 and 1.1%. Although there was no attempt to control for other risk factors such as sexual behaviour, this study clearly confirmed what had previously been suspected [84], that there is a strong inverse age-dependent relationship for cervical HPV infection, with higher rates being observed in women in their late teens and twenties and lower rates in women over the age of 35. This suggests that active HPV infection is often transient and not life-long (see Figure 9.3). This age distribution is

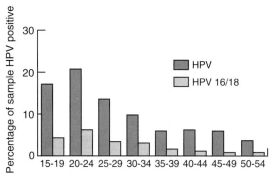

Figure 9.3 Suggested prevalence of HPV DNA in female populations from The Netherlands.

seen both for latent HPV infection and for low-grade CIN. There is a rapid rise in the prevalence of infection with HPV after initial sexual exposure, which is consistent with an epidemic curve. The subsequent dramatic decrease in prevalence is thought to be either due to clearance of infection by the host cellular immune mechanisms, or because of changes in sexual behaviour with age, or a combination of both.

OTHER RISK FACTORS FOR HPV INFECTION

Higher prevalence rates for cervical HPV DNA infection have been shown repeatedly in pregnant women (around 15–30%) compared to controls of similar age. Some studies describe a decrease in the postpartum period [85, 86]. It is well recognized that overt condyloma often increase in size and number during pregnancy and then regress postpartum. Data on the prevalence of specific HPV types in pregnancy are conflicting and confusing and no consistent opinion has been reached (for review see Schneider and Koutsky, 1992 [84]). Other risk factors that have been linked with genital HPV infection include smoking and oral contraceptive usage. Two cross-sectional studies have reported no significant association between smoking and cervical HPV infection after controlling for sexual history [71, 72]; however, Rohan *et al.* (1991) [87] found a significant relationship with a history of cigarette smoking

after controlling for potentially confounding variables, although their study had relatively small numbers.

In the study by Ley *et al.* (1991), a significant and independent risk of cervical HPV infection with oral contraceptive usage was found. Other studies have not confirmed this [71, 88]. Schiffman *et al.* (1993) confirmed an increased risk of CIN with oral contraceptive usage, but only in HPV DNA negative women [58]. In their case-control study it was suggested that the majority of CIN was attributable to cervical HPV infection; however, in HPV DNA-negative women, some of the commonly quoted risk factors appeared to persist, perhaps suggesting a further distinct epidemiology of disease in this small subset. Experimental evidence has also suggested a possible interactive effect of HPV and oral contraception. Oestrogen increases the transcription of the HPV type 16 transforming proteins in the SiHa cell line and the transcriptional regulatory regions of HPV DNA have been shown to contain hormone recognition elements [89]. Application of progestins used in oral contraceptives causes transformation of primary baby rat cells in the presence of HPV 16 DNA and the *ras* oncogene [90]. Such a hypothesis has not been confirmed and supporting evidence will have to be collected carefully in view of studies suggesting increased cervical HPV expression in oral contraceptive users [91].

PREVALENCE OF GENITAL HPV INFECTIONS AND FACTORS INFLUENCING DEVELOPMENT OF CERVICAL NEOPLASIA

Thus far, HPV infection has been defined by the presence or absence of HPV DNA. The morphology of genital HPV infections in the lower genital tract includes a spectrum of disorders (see Figure 9.4):

- Clinical HPV infection is usually defined as any lesion visible to the naked eye which may or may not cause symptoms, and is typically seen as condylomata accuminata.

Such infections are usually related to sexual activity, but may occasionally present in infants when vertical transmission at birth is a possible route of infection.

- Subclinical HPV infection does not usually cause symptoms and can only be diagnosed at colposcopic examination when flat condylomata are seen as acetowhite lesions after the application of acetic acid. Subclinical infection may sometimes be diagnosed histologically from biopsy material when HPV infection is not suspected.
- Latent infection is not associated with symptoms or morphological abnormality, but is confirmed by the presence of HPV DNA. The natural history of latent cervical infection is still uncertain.

Information on the prevalence and incidence of genital warts (condylomata accuminata) in the UK, USA, Sweden and Denmark, suggest a substantial increase in infection rates over the last two decades. In the UK there has been a 2–3-times increase in the reported incidence of genital warts in both sexes since 1971 (91.25 per 10^5 in 1982), with consistent male predominance of about 2:1 (male:female); while in the USA, a 5.5-times increase has been observed since 1966, with a female predominance, the sex ratio ranging from 0.4 to about 1. However, the published data come from different reporting systems, i.e. sexually transmitted

disease clinics in UK, National Disease and Therapeutic Index in USA, and not from population-based samples. The number of cases identified may well have been influenced by increased awareness among doctors and the general population, and changing consultation patterns. One epidemiological study in a defined population has been reported by Chuang *et al.* (1984) from Rochester, Minnesota [92]. They found that the annual incidence of genital warts rose from 13 per 10^5 individuals (age and sex adjusted to the 1970 US population) in 1950–1954 to 106 per 10^5 individuals in 1975–1978.

The prevalence of subclinical infection as determined by evidence of HPV infection in cervical cytology ranges from around 0.3 to 13%, depending upon the female population examined. As discussed earlier the high prevalence rates of cervical HPV DNA of 73–84%, which were initially reported with PCR, are likely to be an overestimate due to contamination. The prevalence of HPV DNA in healthy female populations (a percentage of whom have subclinical infection) is now consistently found to vary from around 4 to 40% depending upon the age and other characteristics of the women studied.

It is also well recognized that the identification of cervical HPV DNA will often fluctuate considerably within individuals over time. This may be due to sampling errors or even laboratory methodology; however, it may well reflect alterations in cervical viral load and even fluctuations in shedding. The transient nature of cervical HPV infections in adolescent women [93] has been demonstrated and persistent infection has been linked to older age and the presence of biopsy-proven CIN [94, 95]. This is in keeping with the idea that HPV is sexually transmitted and commonly acquired in many societies around the outset of sexual activity, either because women tend to be more promiscuous at this time, or (as is the case in Latin American countries), because of high carriage rates by men despite women's relative monogamy [96].

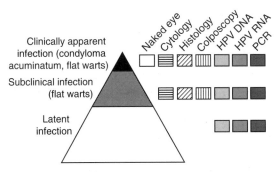

Figure 9.4 Spectrum and detection of HPV infection (HPV, human papillomavirus; PCR, polymerase chain reaction). Printed with permission of S. Garland.

Following acquisition of cervical HPV it is likely that the majority is cleared spontaneously depending upon viral load and the individual's immune mechanisms. Persistence is linked to the development of cervical neoplasia. Some of the risk factors for progression are recognized, i.e. high-risk HPV types, viral load, multiparity, nutritional status, certain HLA alleles (specifically genotypes of the HLA D loci [97]), smoking and oral contraceptive use (for review see Schiffman and Brinton [50]).

Except for older age no specific risk factors have been identified for the progression of CIN III to invasive cervical cancer. Because of obvious ethical difficulties with an investigational end point of invasive cancer the only study attempting to identify such risk factors used data from the four case-control studies of invasive cancer and of CIN III that were concurrently carried out in Spain and Colombia [98]. They concluded that both CIN III and invasive cancer had similar risk profiles.

A further complicating factor is that, until recently, CIN grades I to III have been regarded as a continuum with factors affecting disease progression and regression. Data from a study in Scotland [99] found that in a group of over 900 women with smears suggestive of CIN I or II, the proportion of women with histologic CIN III did not differ significantly between those who were given immediate treatment with loop excision and those who were under surveillance for 24 months and then treated. Most studies that discuss progression and regression of CIN have used colposcopically directed punch biopsy as their diagnostic tool. With the availability of large loop excision of the transformation zone punch biopsy has been shown to underestimate the prevalence of CIN II and III and overestimate the prevalence of CIN I [99, 100]. Such data challenge conventional studies which whilst using punch biopsy and cytology as their investigative tools, have suggested progression of a proportion of CIN I lesions to CIN II and III. The Scottish group also found that lower grades of CIN and HPV appeared to commonly revert to normal. This suggests that lower grades of cervical squamous neoplasia may have a very different natural history to those of a high grade and is in keeping with what is known about the transient nature of most cervical HPV infection.

THE IMMUNE RESPONSE TO PAPILLOMAVIRUSES

Our knowledge of the immunopathology of HPV is rudimentary. There is increasing evidence for immune effector mechanisms directed against HPV infections, despite the fact that these infections do not generalize, and are persistent in immunocompetent hosts. Work with cottontail rabbit papillomavirus (CRPV) confirms that cell-mediated immune responses appear of basic importance for the regression of papillomavirus-associated lesions. Suppression of cell-mediated immunity has been clearly shown to be associated with a high risk of skin warts, skin cancer and HPV-induced anogenital dysplasia and neoplasia in humans. Skin warts and skin carcinomas are frequently seen in sufferers of epidermodysplasia verruciformis (EV), a rare genetic disorder with known reduced cell-mediated immunity. Also, renal transplant recipients treated with immunosuppressive agents are at high risk of developing pre-invasive and invasive anogenital lesions, as are patients with acquired immunodeficiency related to Hodgkin's disease and HIV infection. *In vitro* experiments support such an hypothesis and it has been demonstrated that a delayed hypersensitivity reaction can be induced after injection of a recombinant HPV 16 L1 fusion protein into the skin of patients with CIN lesions [101].

There have now been several animal experiments involving the vaccination of calves with Bovine papilloma virus fusion proteins [102, 103]. Some of these experiments resulted in the production of neutralizing antibodies. The L1 and L2 fusion proteins of CRPV either expressed in bacteria or used as a recombinant vaccinia virus live vaccine have also been shown to protect animals from CRPV infections and induce neutralizing antibodies [104]. Unfortunately in humans, studies of humoral immunity against HPV infection have so far failed to confirm a significant role of antibodies in the immune response against HPV-induced lesions.

CONCLUSIONS

There is now evidence from both laboratory and epidemiological data supporting a strong, and probably, causal association between high-risk HPV types and cervical and anogenital carcinogenesis. This relationship is not that of simple infectious disease causation, but is complicated by a prolonged latent period and a relatively low prevalence of disease amongst infected individuals. It is obvious that other factors, such as the host immune response, play a major role. It has also previously been assumed that there is a continuum of disease from CIN I to CIN III and progression to invasive cervical cancer in a small percentage of individuals. This has been questioned by some who suggest that the majority of HPV infection and low-grade CIN may have a different natural history and that progression to CIN III or cancer is rare.

To date, cervical cytological screening has been shown to be the most effective means available for reduction in mortality from cervical cancer. Some groups advocate including a measure of cervical HPV DNA detection and genotyping into screening procedures [105]. Evidence is emerging that HPV typing may be useful in discriminating women who have borderline or low-grade cytology but who in fact have a high-grade CIN lesion warranting treatment [106]. This concept is explored in more detail in Chapter 6. Also HPV screening of women of older age groups could be used to supplement cytology as a means of defining those women at high risk. Such intervention is likely to benefit only a small number of women and studies so far suggest that HPV typing may be useful only as an adjunct to cytology and cost-effectiveness must be evaluated.

A further rapidly developing area of research is HPV vaccination. Investigators are working upon two approaches: a prophylactic vaccine which would elicit an immune response and prevent infection, and a therapeutic vaccine which would induce specific cell-mediated responses, thereby leading to the regression of existing lesions or even possibly malignant tumours. To date a successful animal model has been developed with the protection of cattle from bovine papillomavirus infection after vaccination [107]. Several trials of different therapeutic vaccines directed against HPV 16 and 18 have commenced in patients with CIN and cervical cancer. Recent prophylactic vaccine efforts have focused primarily upon structural protein based vaccines (i.e. L1 and L2) and results are awaited.

REFERENCES

1. Rous, P. (1911) A sarcoma of fowl transmissible by agent separable from tumour cells. *Journal of Experimental Medicine*, **13**, 397–411.
2. Muñoz, N. and Bosch, F.X. (1989) Epidemiology of cervix cancer, in *Human Papillomavirus and Cervical Cancer*, vol. 94 (eds N. Muñoz, F.X. Bosch and O.M. Jensen). IARC Scientific Publications, Lyon, pp. 9–40.
3. Bosch, F.X, Castellsague, X., Muñoz, N., de-Sanjose, S., *et al.* (1996) Male sexual behaviour and human papillomavirus: key risk factors for cervical cancer in Spain. *Journal of the National Cancer Institute*, **88**(15), 1060–7.
4. Alexander, E.R. (1973) Possible etiologies of cancer of the cervix other than herpes virus. *Cancer Research*, **33**, 1485–9.
5. Rojel, J. (1953) The interelation between uterine cancer and syphillis. *Acta Pathologica et Microbiologica Scandinavica*, **97** (suppl.), 1s–82s.
6. Clemmenson, J. (1965) Statistical studies in the aetiology of malignant neoplasms. *Acta Pathologica et Microbiologica Scandinavica*, **B1** (suppl.), 319.
7 zur Hausen, H. (1977) Human papillomaviruses and their possible role in squamous cell carcinomas. *Current Topics in Microbiology*, **78**, 1–30.

8. Mathews, R.E.F. (1982) Classification and nomenclature of viruses. *Intervirology*, **17**, 1–199.

9. Breitburd, F., Croissant, O. and Orth, G. (1987) Expression of human papillomavirus type 1 E4 gene products in warts in *Cancer Cells*, vol. 5 (eds B.M. Steinberg, J.L. Brandsma and L.B. Taichmann). Cold Spring Harbour, Cold Spring Harbour Laboratory, pp. 115–21.

10. Doorbar, J., Ely, S., Sterling, J., Mclean, C. and Crawford, L. (1991) Specific interaction between HPV 16 E1–E4 and cytolkeratins results in collapse of epithelial intermediate filament network. *Nature*, **352**, 824–7.

11. Pfister, H. (1987) Human papillomavirus and genital cancer. *Advances in Cancer Research*, **48**, 113–45.

12. Crawford, L.V. (1965) A study of human papillomavirus DNA. *Journal of Molecular Biology*, **13**, 362–72.

13. Bernard, H.U., Cahn, S.Y., Manos, M.M., *et al.* (1994) Identification and assessment of known and novel human papillomaviruses by polymerase chain reaction amplification, restriction fragment length polymorphisms, nucleotide sequence and phylogenetic algorithms. *Journal of Infectious Diseases*, **170**, 1077–85.

14. Van Ranst, M.A., Tachezy, R., Delius, H. and Burk, R.D. (1993) Taxonomy of human papillomaviruses. *Papillomavirus Report*, **4**(3), 61–5.

15. Vousden, K.H. and Jat, P.S. (1989) Functional similarity between HPV 16 E7, SV40 large T and adenovirus Ela proteins. *Oncogene*, **4**, 153–8.

16. Kanda, T., Furuno, A. and Yoshike, K. (1988) Human papillomavirus type 16 open reading frame E7 encodes a transforming gene for rat 3Y1 cells. *Journal of Virology*, **62**, 610–13.

17. Phelbs, B., Yee, C., Munger, C. and Howley, P. (1988) The human papillomavirus type 16 E7 gene encodes tranactivation and transformation functions similar to those of adenovirus ElA. *Cell*, **53**, 539–47.

18. Storey, A., Pim, D., Murray, A., Osborn, K., Banks, L. and Crawford, L. (1988) Comparison of the *in vitro* transforming activities of human papillomavirus types. *EMBO Journal*, **7**, 1815–20.

19. Pirisi, L., Yasumoto, S., Feller, M., Doniger, J. and DiPaolo, J.A. (1987) Transformation of human fibroblasts and keratinocytes with human papillomavirus type 16 DNA. *Journal of Virology*, **61**, 1061–6.

20. McCance, D.J., Kopan, R., Fuchs, E. and Laimins, L.A. (1988) Human papillomavirus type 16 alters human epithelial cell differentiation *in vitro*. *Proceedings of the National Academy of Sciences of the USA*, **85**, 7169–73.

21. Pecoraro, G., Lee, M., Morgan, D. and Defendi, V. (1989) Diffential effects of human papillomavirus type 6, 16 and 18 DNAs on immortalization and transformation of human cervical epithelial cells. *Proceedings of the National Academy of Sciences of the USA*, **86**, 563–7.

22. Woodworth, C.D., Doniger, J. and DiPaolo, J.A. (1989) Immortalization of human foreskin keratinocytes by various human papillomavirus DNAs corresponds to their association with cervical carcinoma. *Journal of Virology*, **63**(1), 159–64.

23. Hurlin, P.J., Kaur, P., Smith, P.S., Peres-Reyes, N., Blanton, R.A. and McDougall, J.K. (1991) Progression of human papilimavirus type 18 immortalized human keratinocytes to a malignant phenotype. *Proceedings of the National Academy of Sciences of the USA*, **88**, 570–4.

24. Hudson, J.B., Bedell, M.A., McCance, D. and Laimins, L.A. (1990) Immortalization and altered differentiation of human keratinocytes *in vitro* by the E6 and E7 open reading frames of human papillomavirus type 16. *Journal of Virology*, **64**, 519–26.

25. Hawley-Nelson, P., Vousden, K.H., Hubbert, N.L., Lowy, D.R. and Schiller, J.T. (1989) HPV 16 E6 and E7 proteins cooperate to immortalize human foreskin keratinocytes. *EMBO Journal*, **8**, 3905–10.

26. Vousden, K.H., Wrede, D. and Crook, T. (1991) HPV oncoprotein function: releasing the brakes on cell growth control. *Papillomavirus Report*, **2**(1), 1–3.

27. Cullen, A.P., Reid, R., Campion, M. and Lorincz, A.T. (1991) Analysis of the physical state of different human papillomavirus DNA in intraepithelial and invasive cervical neoplasm. *Journal of Virology*, **65**, 605–12.

28. Ayre, J.E. (1949) The vaginal smear. Precancer cell studies using a modified technique. *American Journal of Obstetrics and Gynecology*, **58**, 1205–19.

29. Meisels, A. and Fortin, R. (1976) Condylomatous lesions of the cervix and vagina. I. Cytological patterns. *Acta Cytologica (Baltimore)*, **20**, 505–9.

30. Muñoz, N., Bosch, X. and Kaldor, J.M. (1988) Does human papillomavirus cause cervical cancer? The state of the epidemiological evidence. *British Journal of Cancer*, **57**, 1–5.

31. Southern, E.M. (1975) Detection of specific sequences among DNA fragments separated by gel electrophoresis. *Journal of Molecular Biology*, **95**, 503–17.

32. Brandsma, J., Burk, R.D., Lancaster, W.D., Pflster, H. and Schiffman, M.H. (1989) Interlaboratory variation as an explanation for varying prevalence estimates of human papillomavirus infection. *International Journal of Cancer*, **43**, 260–2.

33. Schneider, A., Kraus, H., Schuhmann, R. and Gissmann, L. (1985) Papillomavirus infection of the lower genital tract: detection of viral DNA in gynecological swabs. *International Journal of Cancer*, **35**, 443–8.

34. Caussy, D., Orr, W., Daya, A.D., Roth, P., Reeves, W. and Rawls, W. (1988) Evaluation of methods for detecting human papillomavirus deoxyribonucleic acid sequences in clinical specimens. *Journal of Clinical Microbiology*, **26**(2), 236–43.

35. Cornelissen, M.T.E., van der Velden, K.J., Walboomers, J.M.M., *et al.* (1988) Evaluation of different DNA–DNA hybridization techniques in detection of HPV 16 DNA in cervical smears and biopsies. *Journal of Medical Virology*, **25**, 105–14.

36. Schiffman, M.H., Bauer, H.M. and Lorincz, A.T. *et al.* (1991) Comparison of Southern blot hybridization and polymerase chain reaction methods for detection of cervical human papillomavirus infection with types 6, 11, 16, 18.31, 33, and 35. *Journal of Clinical Microbiology*, **29**, 573–7.

37. Kiviat, N.B., Koutsky, L.A., Critchlow, C.W., *et al.* (1990) Comparison of Southern transfer hybridization and dot filter hybridization for detection of cervical human papillomavirus infection with types 6, 11, 16, 18, 31, 33, and 35. *American Journal of Clinical Pathology*, **94**, 561–5.

38. Saiki, R.K., Scharf, S., Faloona, F., *et al.* (1985) Enzymatic amplification of β globin genomic sequences and restriction site analysis for diagnosis of sickle cell anaemia. *Science*, **230**, 1350–4.

39. Kwok, S. and Higuchi, R. (1989) Avoiding false positives with the PCR. *Nature*, **339**, 237–8.

40. Gravitt, P.E. and Manos, M.M. (1992) Polymerase chain reaction based methods for the detection of human papillomavirus DNA. In *The Epidemiology of Human Papillomavirus and Cervical Cancer*, vol. 119 (eds N. Muñoz, F.X. Bosch, K.V. Shah and A. Meheus). IARC Scientific Publications, Lyon, pp. 121–34.

41. Chen, S., Tabrizi, S.N., Fairley, C.K., Borg, A.J. and Garland, S.M. (1994) Simultaneous detection and typing strategy for human papillomavirus based on PCR and restriction endonuclease mapping. *Biotechniques*, **17**(1), 138–41.

42. van den Brule, A.J.C., Claas, E.C.J., du Maine, M., *et al.* (1989) Use of anticontamination primers in the polymerase chain reaction for the detection of human papilloma virus genotypes in cervical scrapes and biopsies. *Journal of Medical Virology*, **29**, 20–7.

43. Tierney, R.J., Ellis, J.R.M., Winter, F.L., *et al.* (1993) PCR for the detection of cervical HPV infection: the need for standardization. *International Journal of Cancer*, **54**, 700–1.

44. Mansell, M.E., Ho, L., Terry, G., Singer, S. and Cuzick, J. (1994) Semi-quantiative human papillomavirus DNA detection in the management of women with minor cytological abnormalities. *British Journal of Obstetrics and Gynaecology*, **101**, 807–9.

45. Cox, J.T., Lorincz, A.T., Schiffman, M.H., Sherman, M.E., Cullen, A. and Kurman, R.J. (1995) Human papillomavirus testing by hybrid capture appears to be useful in triaging women with a cytological diagnosis of atypical squamous cells of undetermined significance. *American Journal of Obstetrics and Gynecology*, **172**, 946–54.

46. Wright, T.C., Sun, X.W., and Koulos, J. (1995) Comparison of management algorithms for the evaluation of women with low grade cytological abnormality. *Obstetrics and Gynecology*, **85**, 202–10

47. Galloway, D.A. (1992) Serological assays for the detection of HPV antibodies. In *The Epidemiology of Human Papillomavirus and Cervical Cancer*, vol. 119 (eds N. Muñoz, F.X. Bosch, K.V. Shah and A. Meheus). IARC Scientific Publications, Lyon, pp. 121–34.

48. Wideroff, L., Schiffman, M.H., Nonnenmacher, B., Hubbert, R., Kirnbauer, R. and Greer, C.E., *et al.* (1995) Evaluation of seroreactivity to human papillomavirus type 16 virus like particles in an incident case-control study of cervical neoplasia. *Journal of Infectious Diseases*, **172**, 1425–30.

49. Bosch, F.X., Manos, M.M., Muñoz, N. and Sherman, M., *et al.* (1995) Prevalence of human papillomavirus in cervical cancer: a world wide perspective. *Journal of the National Cancer Institute*, **87**(11), 796–800.

50. Schiffman, M.H. and Brinton, A.B. (1995) The epidemiology of cervical carcinogenesis. *Cancer*, **76**, 1888–901.

51. Donnan, S.P.B., Wong, F.W.S., Lau, E.M.C., Takashi, K. and Esteve, J. (1989) Reproductive and sexual risk factors and human papilloma virus infection among Hong Kong Chinese. *International Journal of Epidemiology*, **18**(1), 32–6.

52. Reeves, W.C., Brinton, L.A., Garcia, M., *et al.* (1989) Human papillomavirus infection and cervical cancer in Latin America. *New England Journal of Medicine*, **320**(22), 1437–41.

53. Peng, H.Q., Liu, S.L., Mann, V,. Rohan, T. and Rawls, W. (1991) Human papillomavirus type 16 and 33, herpes simplex virus type 2 and other risk factors for cervical neoplasia. *International Journal of Cancer*, **47**, 711–16.

54. Schmauz, R., Okong, P., de Villiers, E.-M., *et al.* (1989) Multiple infections in cases of cervical cancer from a high-incidence area of tropical Africa. *International Journal of Cancer*, **43**, 805–9.

55. Franco, E.L. (1992) Measurement errors in epidemiological studies of human papillomavirus and cervical cancer. In *The Epidemiology of Human Papillomavirus and Cervical Cancer*, vol. 119 (eds N. Muñoz, F.X. Bosch, K.V. Shah and A. Meheus). IARC Scientific Publications, Lyon, pp. 181–97.

56. Morrison, E.A.B., Ho, G.Y.F., Vermund, S.H., *et al.* (1991) Human papillomavirus infection and other risk factors for cervical neoplasia: a case control study. *International Journal of Cancer*, **49**, 6–13.

57. Manos, M.M., Schiffman, M.H., Bauer, H.M., *et al.* (1991) HPV and cervical neoplasia: a prevalence case control study. *1991 Papillomavirus Workshop*, Seattle (Abstracts: 108).

58. Schiffman, M.H., Bauer, H.M., Hoover, A.G., *et al.* (1993) Epidemiologic evidence showing that human papillomavirus infection causes most cervical intraepithelial neoplasia. *Journal of the National Cancer Institute*, **85**(12), 958–64.

59. Muñoz, N., Bosch, F.X., de Sanjose, S., *et al.* (1992) The causal link between human papillomavirus and invasive cervical cancer: a population-based case-control study in Columbia and Spain. *International Journal of Cancer*, **52**, 743–9.

60. Nash, J.D., Burke, T.W. and Hoskins, W. (1987) Biologic course of cervical human papillomavirus infection. *Obstetrics and Gynecology*, **69**(2), 160–2.

61. Mitchell, H., Drake, M. and Medley, G. (1986) Prospective evaluation of risk of cervical cancer after cytological evidence of human papillomavirus infection. *Lancet*, **i**, 573–5.

62. Campion, M.J., McCance, D.J., Cuzick, I. and Singer, A. (1986) Progressive potential of mild cervical atypia: Prospective cytological, colposcopic, and virological study. *Lancet*, **ii**, 237–40.

63. Syrjanen, K., Mantyjarvi, R., Saarikoski, S., *et al.* (1988) Factors associated with progression of cervical human papillomavirus (HPV) infections into carcinoma *in situ* during longterm prospective followup. *British Journal of Obstetrics and Gynaecology*, **95**, 1096–1102.

64. Syrjanen, K., Hakama, M., Saarkoski, S. *et al.* (1990) Prevalence, incidence, and estimated life time risk of cervical human papillomavirus infections in a non selected Finnish population. *Sexually Transmitted Diseases*, **17**, 15–19.

65. Caussy, D., Marrett, L.D., Worth, A.J., McBride, M. and Rawls, W.E. (1990) Human papillomavirus and cervical intraepithelial neoplasia in women who subsequently had invasive cancer. *Canadian Medical Association Journal*, **142**, 311–17.

66. Murthy, N.S., Sehgal, A., Satyanarayana, L., *et al.* (1990) Risk factors related to the biological behaviour of precancerous lesions of the uterine cervix. *British Journal of Cancer*, **61**, 732–6.

67. Koutsky, L.A., Holmes, K.K., Critchlow, C.W., *et al.* (1992) A cohort study of the risk of cervical intraepithelial neoplasia grade 2 or 3 in relation to human papillomavirus infection. *New England Journal of Medicine*, **327**, 1272–8.

68. Higgins, G.D., Davy, M. and Roder, D., *et al.* (1991) Increased age and mortality associated with cervical carcinomas negative for human papillomavirus RNA. *Lancet*, **338**, 910–13.

69. Walker, J., Bloss, J.D. and Liao, S.Y., *et al.* (1989) Human papillomavirus genotype as a prognostic indicator in carcinoma of the cervix. *Obstetrics and Gynecology*, **74**, 781–5.

70. Barnes, W., Delgado, G. and Kurman, R.J. *et al.* (1988) Possible prognostic significance of human papillomavirus type in cervical cancer. *Gynecologic Oncology*, **29**, 267–73.

71. Villa, L.L. and Franco, E.L.F. (1989) Epidemiological correlates of cervical neoplasia and risk of human papillomavirus infection in asymptomatic women in Brazil. *Journal of the National Cancer Institute*, **81**(5), 332–9.

72. Ley, C., Bauer, H.M., Reginold, A., *et al.* (1991) Determinants of genital human papillomavirus infection in young women. *Journal of the National Cancer Institute*, **83**(14), 997–1003.

73. Moscicki, A.-B., Palefsky, J., Gonzales, J. and Schoolnik, G.K. (1990) Human papillomavirus infection in sexually active adolescent females; prevalence and risk factors. *Pediatric Research*, **28**, 507–13.

74. Hildesheim, A., Gravitt, P., Schiffman, M.H., *et al.* (1993) Determinants of genital human papillomavirus infection in low-income women in Washington. *Sexually Transmitted Diseases*, **20**, 279–85.

75. Fairley, C.K., Chen, S., Ugoni, A., *et al.* (1994) Genital human papillomavirus infection is best predicted by the sexual partners in the last 5 years rather than life time number. *Obstetricsrics and Gynaecology*, **84**, 755–9.

76. Eluf-Neto, J., Booth, M., Muñoz, N., Bosch, F.X., Meijer, C.J.L.M. and Walboomers, J.M.M. (1994) Human papillomavirus and invasive cancer in Brazil. *British Journal of Cancer*, **69**, 114–19.

77. Jochmus-Kudielka, I., Schneider, A., Braun, R., *et al.* (1989) Antibodies against the human papillomavirus type 16 early proteins in human sera: correlation of anti-E7 reactivity with cervical cancer. *Journal of the National Cancer Institute*, **81**(22), 1698–703.

78. Fairley, C.K., Chen, S., Tabrizi, S.N., Quinn, M.A. and Garland, S.M. (1992) The absense of genital human papillomavirus DNA in virginal women. *International Journal of STD and AIDS*, **3**, 414–17.

79. Gutman, L.T., Claire, K.S., Herman-Giddens, M.E., *et al.* (1992) Evaluation of sexually abused and non abused young children for intravaginal human papillomaviruses. *American Journal of the Diseases of Childhood*, **146**, 694–9.

80. Shah, K., Kashima, H., Polk, B.F., *et al.* (1986) Rarity of caesarean delivery in cases of juvenile onset respiratory papillomatosis. *Obstetrics and Gynecology*, **68**, 795–9.

81. Cason, J., Kaye, J.N., Jewers, R.J., *et al.* (1995) Perinatal infection and persistance of human papillomavirus types 16 and 18 in infants. *Journal of Medical Virology*, **47**(3), 209–18.

82. Fairley, C.K., Tabrizi, S.N., McNeil, J.J., Chen, S., Borg, A.J. and Garland, S.M. (1993) Is HPV always sexually aquired? *The Medical Journal of Australia*, **159**, 724–6.

83. Melkert, P.W., Hopman, E., van de Brule, A.J., *et al.* (1993) Prevalence of HPV in cytolomorpholoigally normal cervical smears, as determined by the polymerase chain reaction is age dependent. *International Journal of Cancer*, **53**, 919–23.

84. Schneider, A. and Koutsky, L. (1992) Natural history and epidemiological features of genital HPV infection. In *The Epidemiology of Human Papillomavirus and Cervical Cancer*, vol. 119 (eds N. Muñoz, F.X. Bosch, K.V. Shah and A. Meheus). IARC Scientific Publications, Lyon, pp. 23–52.

85. Rando, R.F., Lindeheim, S., Hasty, L., Sedlacek, T.V., Woodland, M. and Eder, C. (1989) Increased frequency of detection of human papillomavirus deoxyribonucleic acid in exfoliated cervical cells during pregnancy. *American Journal of Obstetrics and Gynecology*, **161**, 50–5.

86. Czegledy, J., Gergley, L. and Endroedi, I. (1989) Detection of human papillomavirus deoxyribonucleic acid by filter *in situ* hybridization during pregnancy. *Journal of Medical Virology*, **28**, 250–4.

87. Rohan, T., Mann, V., McLaughlin, J., *et al.* (1991) PCR-detected genital papillomavirus infection: prevalence and associated risk factors for cervical cancer. *International Journal of Cancer*, **49**(6), 856–60.

88. Burkett, B.J., Peterson, C.M., Birch, L.M., *et al.* (1992) The realtionship between contraceptives, sexual practices and cervical human papillomavirus infection among a college population. *Journal of Clinical Epidemiology*, **45**(11), 1295–302.

89. Mitrani-Rosenbaum, S., Tsvieli, R. and Tur-Kaspa, R. (1989) Oestrogen stimulates differential transcription of human papillomavirus type 16 in SiHa cervical carcinoma cells. *Journal of General Virology*, **70**, 2227–32.

90. Pater, A., Bayapour, M. and Pater, M.M. (1990) Oncogenic transformation of human papillomavirus type 16 deoxyribonucleic acid in the presence of progesteron or progestins form the oral contraceptives. *American Journal of Obstetrics and Gynecology*, **1**, 1099–103.

91. Hildesheim, A., Reeves, W.C., Brinton, L.A., *et al.* (1990) Association of oral contraceptive use and human papillomavirus in invasive cervical cancers. *International Journal of Cancer*, **45**, 860–4.

92. Chuang, T.Y., Perry, H.O., Kurland, L.T. and Ilstrup, D.M. (1984) Condyloma accuminata in Rochester, Minn. *Archives of Dermatology*, **120**, 469–75.

93. Rosenfeld, W.D., Rose, E., Vermund, S.H., Schreiber, K. and Burk, R.D. (1992) Follow-up evaluation of cervicovaginal human papillomavirus infection in adolescents. *Journal of Paediatrics*, **121**, 301–11.

94. Ho, G.Y., Burk, R.D., Klein, S., *et al.* (1995) Persistance of genital human papillomavirus infection as a risk factor for persitant cervical dysplasia. *Journal of the National Cancer Institute*, **87**, 1365–71.

95. Hildesheim, A., Schiffmann, M.H., Gravitt, P.E., *et al.* (1994) Persistence of type specific human papillomavirus infection in cytologically normal women. *Journal of Infectious Diseases*, **169**, 235–40.

96. Muñoz, N., Castellague, X., Bosch, F.X., *et al.* (1996) Difficulty in elucidating the male role in cervical cancer in Columbia, a high risk area for the disease. *Journal of the National Cancer Institute*, **88**(15), 1068–70.

97. Odunsi, K., Terry, G., Ho, L., Bell, J., Cuzick, J. and Ganesan, T.S. (1995) Association between HLA DQB 1*03 and cervical-intraepithelial neoplasia. 1. *Molecular Medicine*, **1**(2), 161–71.

98. Moreno, V., Muñoz, N., Bosch, F.X., *et al.* (1995) Risk factors for progression of cervical intraepithelial neoplasia grade III to invasive cancer. *Cancer Epidemiology, Biomarkers & Prevention*, **4**(5), 459–67.

99. Flannelly, G., Anderson, D., Kitchener, H.C., *et al.* (1994) Mangement of women with mild and moderate cervical dyskaryosis. *British Medical Journal*, **308**, 1399–403.

100. Buxton, E.J., Luesley, D.M, Shafi, M.I., and Rollason, M. (1991) Colposcopically directed punch biopsy: a potentially misleading investigation. *British Journal of Obstetrics and Gynaecology*, **98**, 1273–6.

101. Hopfl, R., Sandbicher, M., Sepp, N., *et al.* (1991) Skin test for HPV type 16 proteins in cervical intraepithelial neoplasia. *Lancet*, **i**, 373–4.

102. Jin, X.W., Cowsert, L., Marshall, D., *et al.* (1990) Bovine serological response to a recombinant BPV-1 major capsid protein vacine. *Intervirology*, **31**, 345–54.

103. Jarrett, W.F., Smith, K.T., O'Neil, B.W., *et al.* (1991) Studies on vaccination against papillomavirus: prophylactic and theraputic vaccination with recombinant structural proteins. *Virology*, **184**, 33–42.

104. Lin, Y.L., Borenstein, L.A., Selvakumar, R., Ahmed, R. and Wettstein, F.O. (1992) Effective vaccination against papilloma development by immunization with L1 or L2 structural protein of cottentail rabbit papillomavirus. *Virology*, **187**, 612–19.

105. Cuzick, J., Szarewski, A., Terry, G., *et al.* (1995) Human papillomavirus testing in primary cervical screening. *Lancet*, **345**, 1533–6.

106. Herrington, C.S., Evans, M.F., Hallam, N.F., Charnock, F.M., Gray, W. and McGee, J.D. (1995) Human papillomavirus status in the prediction of high-grade cervical intraepithelial neoplasia. *British Journal of Cancer*, **71**, 206–9.

107. Campo, M.S., Grindlay, G.J., O'Neil, B.W., *et al.* (1993) Prophylactic and theraputic vaccination against mucosal papillomavirus. *Journal of General Virology*, **74**, 745–53.

108. Bosch, F.X., Muñoz, N. and de Sanjose, S. (1995) Epidemiology of HPV and cervical neoplasia. *Precongress Workshop of the 14th International Papillomavirus Congress*, Quebec City, 1995.

109. Liaw, K., Hsing, A.W., Chen, C.-J., Schiffman, M.H., Zhang, T.Y. and Hsieh, C.-Y., *et al.* (1995) Human papillomavirus and cervical neoplasia: a case-control study in Taiwan. *International Journal of Cancer*, **62**, 565–71.

110. Bosch, F.X., Muñoz, N., de Sanjose, S., *et al.* (1993) Human papillomavirus and cervical neoplasia grade III/carcinoma *in situ*: a case control study in Spain and Columbia. *Cancer Epidemiology, Biomarkers & Prevention*, **2**, 415–22.

111. Olsen, A.O., Gjoen, K., Sauer, T., *et al.* (1995) Human papillomavirus and cervical intraepithelial neoplasia grade II–III: a population-based case-control study. *International Journal of Cancer*, **61**, 312–15.

112. Kjaer, S.K., van den Brule, A.J.C., Bock, J.E., Poll, P.A. and Engholm, G., *et al.* (1996) Human papillomavirus – the most significant determinant of cervical intraepithelial neoplasia. *International Journal of Cancer*, **65**, 601–6.

Christine Bergeron

INTRODUCTION

Cytologic examinations are now an accepted, even necessary, component of surgical pathology. The cervical cytologic smear is one of the most successful diagnostic procedures in medicine, having reduced deaths from cervical cancer by more than 70%. It is a safe, easy and inexpensive method for screening large numbers of sexually active women. In countries where cervical screening programmes have been established, the incidence of invasive cervical cancer has markedly decreased [1, 2]. However, it is well recognized that despite repeated screening, a certain number of women continue to develop invasive cervical cancer [2]. This emphasizes the importance of identifying the potential failures of this detection system and of analysing the reasons for them.

THE FALSE-NEGATIVE DIAGNOSIS

The cervical smear to detect pre-cancerous lesions of the cervix was a method applied rapidly to the largest possible number of women without double-blind studies on the efficacy of cytology detection. The analysis of errors is therefore based on current practice or is retrospective. The quality of cytologic diagnoses is frequently debated [3, 4]. The rate of smear failure has been mostly studied in women with invasive carcinomas and the smear failures varies from 2% to 30% [5–16]. These false-negative smears have major medical, financial and occasionally legal implications. It has been suggested that a small number of cancers develop rapidly, during the interval between Pap smears [8, 16]. The sampling error rate represents a factor in false-negative cases. There are problems in obtaining diagnostic cells for examination, either due to low shedding rates, or the presence of cancers high in the endocervical canal. Laboratory errors can also result in incorrect diagnoses [3, 4]. No cytology laboratory is completely immune to screening errors and there is probably an irreducible error rate of at least 5–10% in routine screening. This means that cervical carcinoma will develop in some women despite appropriate screening.

CLASSIFICATIONS

In 1945, i.e. more than 50 years ago, the newly formed American Cancer Society endorsed the use of the vaginal smear as an effective cancer prevention test for carcinoma of the uterine cervix. Papanicolaou's sampling technique, initially developed to study the hormonal status of mice, was a vaginal pool smear, and this was the method originally used in clinical observations [17]. A few years later, J. Ayre proposed a wooden tongue depressor for a direct sampling of this organ [18]. The choice of device used in sampling is

Cancer and Pre-cancer of the Cervix
Edited by D.M. Luesley and R. Barrasso. Published in 1998 by Chapman & Hall, London. ISBN 0 412 56600 1.

influenced by the experience and preference of the person taking the sample. There are a variety of devices used to obtain the cervical samples: wooden and plastic spatulas, a cotton-tipped applicator, the cytobrush, the cervex brush. A meta-analysis of randomized and quasi-randomized studies performed by Buntinx and Brouwers (1996) [19], showed that there were no substantial differences in the yield of cytological abnormalities between the extended tip spatula, the Ayre spatula combined with the cytobrush or cotton swab, or the cervex brush. However, the Ayre spatula, cytobrush, or cotton swab used alone generally performed significantly worse than the combinations, the extended tip spatula, or the cervex brush.

For 25 years, the traditional approach to classification of the cervical smear results was based on the numerical Papanicolaou class designations from I to V according to the severity of cytological abnormalities [20]. Although this classification was endorsed by most laboratories and is still used by some, changes in medical behaviour modifying our understanding of cervical pre-cancerous lesions emphasized the need to revise the mode of classification: these changes were (1) a widespread use of cervical screening, (2) a colposcopic evaluation of the cervix, and (3) the introduction of new out-patient therapeutic procedures. Criticisms formulated to the Papanicolaou classification were: an absence of correlation between cytology and histology, a class II including benign and viral cytological modifications, and a class III and IV giving an imprecise diagnosis of malignancy. Reagan *et al.* [21] proposed a classification that could be used in cytology and histology. The term 'carcinoma *in situ*' was suggested for cytologic and histologic abnormalities associated with a true neoplastic lesion, and the term 'dysplasia' for cytologic abnormalities associated with a better differentiated lesion. The pre-cancerous potential of the latter was unknown. It was proposed to divide dysplasia into mild, moderate and severe, according to

the severity of abnormalities. The Richart classification (1967) was firstly an histologic classification and was later applied to cytology [22]. The term 'cervical intra-epithelial neoplasia' (CIN) was introduced to replace the terms dysplasia and carcinoma *in situ*. Grade I was proposed for cytological abnormalities confined to the lower third of the epithelium, grade II for cytological abnormalities confined to half and grade III for cytological abnormalities in more than two-thirds of the epithelium. The main objective of this classification was to define a morphologic and biologic continuum between dysplastic lesions and carcinoma *in situ*, leading to the concept that all these lesions were pre-cancerous lesions, including CIN I, and should be evaluated by colposcopy. Because of the difficulty in differentiating between severe dysplasia and carcinoma *in situ*, another objective of this classification was to combine these two entities under the term CIN III.

Until 1988, most laboratories used one of these three classifications, Papanicolaou, dysplasia–carcinoma *in situ* or CIN. The Bethesda System (TBS) for reporting cervical/vaginal cytologic diagnoses was proposed by a 1988 National Cancer Institute Workshop (NCIW) convened to consider the benefits of increased standardization in the diagnostic reports provided by cytology laboratories [23]. TBS was revised in 1991 in response to advances in the understanding of cervical neoplasia and to the needs of clinicians and cytopathologists. Criteria for the specific diagnostic entities used in TBS as well as criteria for specimen adequacy were formulated [24]. Interim guidelines for management of abnormal cervical cytology, focusing on low-grade squamous intra-epithelial lesions and atypical squamous cells of undetermined significance, were published in 1994 [25].

TBS

TBS was developed to reduce the confusion regarding diagnostic terminology in cytopathology, and thereby improve patient treatment by providing gynaecologists with clearly defined and standardized diagnoses [24]. This system requests three criteria for a pathologic report: (1) a statement regarding the adequacy of the specimen for diagnostic evaluation, (2) a general categorization of the diagnosis (normal or abnormal), and (3) a descriptive diagnosis (Table 10.1).

THE ADEQUACY OF THE SPECIMEN

Specimen adequacy has been a concern of the conscientious cytologist. Failure to sample a lesion accounts for one-half to two-thirds of false-negative cervico-vaginal cytopathology reports. There is evidence that both sampling and intra-laboratory causes of missed and underdiagnosed abnormalities could be reduced by careful attention to factors affecting the adequacy of the sample.

TBS for reporting cervical–vaginal cytological diagnoses incorporates the routine evaluation of adequacy as an integral part of the report, and the 1991 TBS update establishes adequacy criteria. If unsatisfactory or less than optimal, this should be noted in the report. Patient identification (age, date of last menstrual period), specimen identification and clinical information increase the sensitivity and the reliability of the evaluation. The specimen should contain both squamous cells and endocervical or squamous metaplastic cells proving that the transformation zone has been sampled. Blood, inflammation, thick areas, poor fixation, air-drying artifact that precludes interpretation of approximately 75% or more of the epithelial cells and a scant squamous epithelial component, defined as well-preserved and well-visualized squamous epithelial cells spread over less than 10% of the slide surface, should be considered as unsatisfactory for the detection of cervical epithelial abnormalities. In this case, the cytopathologist may suggest how a more diagnostic sample may be obtained. An adequate endocervical/transformation zone (EC/TZ) sample is defined as at least two

Table 10.1 Bethesda classification

Adequacy of the specimen

- Satisfactory for evaluation
- Satisfactory for evaluation but limited by ... (specify reason)
- Unsatisfactory for evaluation ... (specify reason)

Descriptive diagnoses

Benign cellular changes

Epithelial cell abnormalities

Squamous cell

- Atypical squamous cells of undetermined significance: qualify
- Low-grade squamous intra-epithelial lesion encompassing: HPV, mild dysplasia, CIN I
- High-grade squamous intra-epithelial lesion encompassing: moderate and severe dysplasia CIS/CIN II and CIN III

Glandular cell

- Atypical glandular cells of undetermined signifance: qualify
- Endocervical adenocarcinoma
- Endometrial carcinoma
- Extra-uterine adenocarcinoma

groups of at least five well-preserved endo-cervical glandular and/or metaplastic squamous cells. When an epithelial cell abnormality is identified, the specimen is never reported as unsatisfactory for evaluation. With these crite-ria, the unsatisfactory rate averages 1%. Scant cellularity is the reason for the unsatisfactory report in 70%, followed by obscuring inflam-mation and/or blood in 28%, and air drying, poor preservation, or degeneration in 2% [26].

The criteria for and importance of the EC/TZ component remain controversial [27]. Despite the association of endocervical and squamous metaplastic cells with increased detection of lesions, three longitudinal studies have found women, whose entry specimens lacked an EC/TZ component, actually to be at lower, not higher, risk of having a squamous intra-epithelial lesion on follow-up than women with EC/TZ cells on initial smears [28–30]. Therefore, TBS has taken a middle ground: EC glandular cells and/or squamous cells are accepted as evidence of the component in non-atrophic smears, and EC/TZ-negative smears are assigned to the 'satisfactory but limited' category.

The need to improve the quality of the slide has also encouraged the development of a number of devices aimed at automating slide-making. This can be done by using liquid-based cervical cytology. Rather than smearing the cervical cells onto a slide, the col-lection device is rinsed in a vial of preservative solution, capturing virtually all of the cell sam-ple. The patient's specimen is sent to a laboratory where the liquid is filtered to reduce blood, mucus and inflammation and a thin layer of the cells is applied to a microscope slide. Several recent studies have compared slide quality and detection of abnormalities in con-ventional smears and slides made using these types of monolayer preparation system [31–37]. These studies have shown an improvement in the adequacy of the specimen. They have not shown a statistically higher detection rate but a better definition of the abnormal cells with a lower rate of atypical squamous cells of under-mined significance (ASCUS) and a higher rate of low-grade squamous intra-epithelial lesion (LGSIL). From the only study that has been done, the monolayer slightly increases the sensitivity and the specificity of the Pap test in comparison with the conventional technique but with no statistical difference [38]. Due to the added cost of disposables, it is not clear if this method will be a substitute for the con-ventional technique read in a high-quality laboratory.

EPITHELIAL CELL ABNORMALITIES

Several investigators have studied inter-observer and intra-observer concordance in the evaluation of cervico-vaginal smears [14, 39–46]. The agreement among cytopathol-ogists is generally related to the complexity of the grading system: more complicated schemes with many diagnostic categories produce lower concordance than simpler sys-tems. TBS is the simplest system. Overall inter-observer and intra-observer reproducibility for the cytological SILs using a two-tiered system is fair to very good and the reproducibility is better for high grade squamous intra-epithelial lesion (HGSIL) than for LGSIL [42, 44]. Therefore, false-positives are rare in the HGSIL. The variability is lower in the same laboratory between several observers than in different sites [45], and between pathologists specializing in cytology than between general cytologists [47].

The original two-part categorization in normal and abnormal [23] has been modified into three parts: normal, benign cellular changes, and epithelial cell abnormalities. The latter is a category for potentially pre-malignant or malignant lesions. Human papil-lomavirus (HPV) DNA can be detected in the vast majority of squamous intra-epithelial lesions (SIL), whether or not cytomorphologic features of HPV are identified. The histo-logical and cytological separation of lesions into lower and higher grades is based on the increasing probability of association with high-risk HPV types and subsequent progression to invasive squamous carcinoma [48–50].

LGSIL

The better differentiated lesions where the neoplastic abnormalities are confined to the lower third of the epithelium are known as CIN I. These lesions are included in TBS among the LGSIL. Koilocytes, which are an expression of epithelial differentiation, are, by definition, present in LGSIL. Viral replication, which is intimately bound to differentiation of the squamous epithelium, is associated with a pathogenic effect, known as koilocytosis [51, 52]. This morphologic alteration of cells is common to all types of HPV infecting the anogenital region, whether or not the viruses are oncogenic. The koilocyte is a degenerated superficial or intermediate cell with lysis of the cytoplasm and a nucleus that contains a large number of viral particles. Koilocytosis is a descriptive, not diagnostic, term that should be applied to squamous cells showing both cytoplasmic vacuolization and nuclear abnormalities characterized by enlargement, hyperchromasia and wrinkling (Plate 11). Cytoplasmic vacuolization in the absence of nuclear alterations does not qualify as koilocytosis. This latter change can be found in normally glycogenated squamous epithelium or in association with various types of inflammation, notably that due to trichomonas. Koilocytes are associated with dyskeratosis, parakeratosis and hyperkeratosis. The latter are not specific of a neoplastic process but rather reflect a defect in keratin synthesis and may be found after laser therapy or even as a variant of a normal cervical epithelium. LGSIL do not contain abnormal basal or parabasal cells that do not exfoliate because they are present only in the lower third of the epithelium (Plate 12).

The lesions known as condylomata acuminata are associated with HPV types considered to be low-risk, most commonly types 6, 11, 42, 43 and 44, which with rare exceptions are not detected in invasive cancers of the uterine cervix. Cervical condylomas can be diagnosed by cervical smear and confirmed by colposcopic biopsy. The cervical lesions of this type are included in TBS among the squamous intra-epithelial lesions of low grade. Indeed, they are characterized by koilocytosis associated with parakeratosis.

TBS incorporates cellular changes associated with HPV (i.e. koilocytosis and CIN I) within the category of LGSIL because the natural history, distribution of HPV types and morphologic features of both of these lesions and thus their treatment are similar. Furthermore, studies evaluating the dysplasia/CIS or CIN terminology have repeatedly demonstrated a lack of inter-observer and intra-observer reproducibility [53]. The greatest lack of reproducibility is between koilocytosis and CIN I and pure condyloma and CIN I with koilocytosis, which cannot be distinguished on the basis of cytological features.

HGSIL

The less-differentiated lesions are known as CIN II and CIN III, and the neoplastic abnormalities are present in half or the lower two-thirds of the epithelium, respectively. These lesions are included together in TBS among the HGSIL. They are characterized by abnormal parabasal and basal cells with enlarged, hyperchromatic nuclei, often with smudged chromatin. Variations in nuclear size and cellular shapes are seen. The abnormal parabasal cells are associated with or without koilocytosis, depending on whether the surface is composed of poorly differentiated cells (Plates 13–15). Separating moderate dysplasia from severe dysplasia and severe dysplasia from CIS has been shown to be irreproducible [46, 53]. From a clinical standpoint, no useful purpose is served by this separation; moderate is not usually managed differently from severe dysplasia and CIS. Combining moderate dysplasia with severe dysplasia and CIS will reduce discordance between the cytology and the biopsy reports by reducing the possible categories from three to one. Since the management of cervical lesions is partly based on correlation of the cytologic and histologic findings, greater concordance will reduce the morbidity, anxiety and cost to the patient of repeated examinations.

ASCUS and AGCUS

TBS of 1988 for reporting cervical/vaginal cytologic diagnoses introduced the term 'atypical squamous cells of undetermined significance' (ASCUS). It was to be used as a category of squamous-cell abnormality for cases not diagnostic of a reactive/inflammatory, pre-neoplastic, or neoplastic condition. The 1991 Bethesda conference retained the ASCUS terminology and described criteria in a 1994 Bethesda System reference atlas. The term ASCUS has been reserved for cellular changes that are not clearly reactive but lack the nuclear criteria required for a definite diagnosis of SIL. The cellular changes observed in smears classified as ASCUS include a loss of the normal nuclear/cytoplasm ratio, associated with nuclear enlargement. Nuclear detail is often poorly defined but the nuclei may appear hyperchromatic (Plate 16). However, the criteria used to make this diagnosis vary significantly from laboratory to laboratory. Pathologists do not agree on what constitutes ASCUS. Communication between the cytopathologist and clinician is essential for determining the significance of a report of ASCUS. In our laboratory only about 1% of smears are given this designation. Follow-up of patients with ASCUS reveals a range of 10–25% with the diagnosis of SIL. Fewer than 5% of the cases, however, have a HGSIL. TBS recommends qualifying the diagnosis, when possible, to indicate whether a reactive process or pre-neoplastic condition is favoured. Interim guidelines concluded that an ASCUS diagnosis may be expected in about 5% of patients, and that a greater frequency

might constitute misuse of the term [25]. Cellular changes thought to be purely reactive in nature are described as such and are not included under the category of ASCUS.

The post-menopausal woman with severe atrophic vaginitis should be instructed to use systemic or topical oestrogen until the vaginal mucosa is oestrogenized, at which time the smear should be repeated. Occasionally, women receiving chemotherapy may demonstrate cytologic changes that appear abnormal. Invariably these changes will regress after cessation of treatment. Further follow-up in the above instances will depend on the findings noted in the repeat smear.

The clinical significance of a report of atypical glandular cells of undetermined significance (AGCUS) is poorly understood. In large measure this relates to the much lower frequency of glandular abnormalities of the cervix. Likewise, atypical glandular cells are encountered much less frequently than atypical squamous cells (0.2 vs 4.5%, respectively, see [54]). As a result, the experience of individual cytology laboratories with glandular abnormalities of the cervix has been limited. Additionally, until recently, the criteria for the diagnosis of glandular abnormalities have tended to be vague and subjective. This rate will vary somewhat depending upon cytologic criteria as well as the population screened but should apparently be less than 0.5%. AGCUS is associated with substantial underlying uterine pathology in 20% of the cases, half of them including squamous lesions. Cervical colposcopy, endocervical curettage and endometrial biopsy are recommended for the complete evaluation of AGCUS.

Lesions in the category of AGCUS include a morphologic spectrum from atypical-appearing, reactive processes (Plate 17) to adenocarcinoma *in situ* (Plate 18). The diagnosis should, if possible, indicate whether the cells are favoured to be of endocervical or endometrial origin. Potential pitfalls in the diagnosis of AGCUS have been identified and should be borne in mind by the gynecologists as well as the cytologist. Prior cervical conization has been reported to be associated with the subsequent reporting of atypical glandular cells. The presence of tubal metaplasia may be misleading.

THE FALSE-POSITIVE DIAGNOSIS

FALSE-POSITIVE FOR ASCUS AND LGSIL

False-positive diagnoses are not so threatening for the patient as false-negative, but they lead to unnecessary colposcopic examination, treatment and expense. One of the best methods of detecting the false-positive involves comparing cytologic findings with histologic findings. However, the false-positives are particularly frequent in the cytologic diagnosis of ASCUS and LGSIL. The lack of a clear cut-off between ASCUS and LGSIL and the lack of inter-observer agreement have lead to a non-specific overuse of the term ASCUS [55, 56]. It poses a major problem for the clinician and there is considerable controversy as to the best manner of handling poorly defined cytologic atypias and LGSIL. Some observers propose a conservative management approach by following-up the patient with repeat Pap smears, because they believe that the cytologic abnormalities correspond to the categories of benign repair or immature metaplasia. Further, even if these abnormalities correspond to true low-grade lesions of the squamous epithelium, the likelihood of regression is very high, whereas progression towards a high-grade epithelial lesion or invasive carcinoma would be unlikely [57, 58]. The slowly evolving lesions could be followed by cytology without any risk to the patient; the lesions could be treated if the abnormality persisted for 2 years. Other observers prefer to perform colposcopy. Their argument is that although the majority of smears with minor grade cytologic atypia spontaneously revert to normal, some persist and others may progress to high-grade lesions. Also, cytology is a screening test, it is not a diagnostic tool. As a result, 20% of women with minor grade cytologic atypia

may, in fact, have histologically high-grade cancer precursor [59, 60].

QUALITY ASSURANCE WITH HPV DNA TESTING (SEE ALSO CHAPTER 6)

Testing for HPV after an abnormal cytologic diagnosis has been proposed, particularly if the cytologic abnormality is poorly defined or of low grade, thus providing an objective marker of progression and help in the decision for or against treatment. The presence of HPV 16 or 18 associated with an atypical smear is probably a serious risk factor. In two studies using hybrid capture technology, between 93 and 100% of patients with high-grade CIN, but whose original Pap smear indicated low-grade lesions, were correctly identified [60, 61]. Conversely, a negative repeat Pap smear and HPV DNA test for high oncogenic HPV confers a negative predictive value of 95–100%. While the veritable contribution of the intermediate triage approach to reducing cost as compared to colposcopy with a biopsy still remains to be determined by additional clinical trials, it is suspected that cost-effectiveness could certainly be achieved if HPV DNA testing was performed without a repeat Pap smear. This can be done by using liquid-based cervical cytology [62]. HPV DNA testing is performed in the cellular residues of the collection kits used for preparing Thin-Prep smears. HPV testing is done in those women whose Thin-Prep smears contain minor-grade cytologic atypia including ASCUS and LGSIL.

The rate of HPV DNA positivity can be correlated with increasing severity of morphologic atypia in both cytologic and histologic specimens. In one study using a panel of five independent pathologists, a 100% correlation was achieved between the diagnosis of SIL and the concurrent detection of HPV DNA by Southern blot and PCR technology [56]. In contrast, in the absence of significant atypia, HPV DNA positivity was lower but increased, parallelling the increasing severity of atypia and probability of CIN. In cytopathology services in which positive rates do not conform to national or regional standards, HPV DNA testing may be used for quality control of diagnostic criteria. For example, HPV positivity is expected in between 15 and 25% of ASCUS smears, 65 and 75% of LGSIL smears, and 85 and 95% of HGSIL smears. When the results of viral testing deviate significantly (either toward higher or lower rates) from the above rates, the cytopathologists should readjust their criteria to diagnose HPV-related diseases.

THE QUALITY CONTROL FOR THE FALSE-NEGATIVE

Error rates of the cytologic smear show that quality control is necessary to improve false-negatives due to screening error. A study using strict quality-control measures in the cytologic laboratory, by rescreening all or at least a large number of smears using two or more observers, and by correlating the results with subsequent biopsy and clinical findings, would be the only way to know the screening error rate. However, this is impossible to realize on a routine basis.

Screening errors have led to quality assurance guidelines that are unique to cytology but are not universally accepted. The proposals are summarized in Table 10.2. The effectiveness of rescreening 10% of smears has been repeatedly questioned. Targeted rescreening [3] is probably the most efficient way of preventing false-negative results and has been used in our laboratory since 1992 [63]. High-risk smears selected from patients with previous abnormal smears and/or cervical biopsies, previous cervical treatment and colposcopic abnormalities are reviewed by another cytotechnician even if considered negative by the first cytotechnician. Besides being costly, the system requires co-operation of physicians in informing the laboratory about the clinical findings.

Other procedures are also used in our laboratory. Perceived non-negatives are given

Table 10.2   Cervical cytology quality assurance

- Rescreening of 10% of smears
- Rescreening high-risk patients
- Rescreening of all abnormal smears by a pathologist
- Review of previous negative slides from a case when abnormal smears are detected
- Comparison of abnormal cytologic and subsequent histologic findings
- Cytotechnologist participation in continuing education activities
- Laboratory participation in proficiency programmes

to the pathologist for confirmation, as well as ASCUS and the cases initially diagnosed by the cytotechnician. Furthermore, a program developed on the computer system provides a monthly print-out of each cytotechnician's performance. The computer program lists the breakdown of each cytotechnician's diagnoses by reporting the category according to TBS. These are compared with the laboratory averages for the month. Individual cytotechnician discrepancies according to the average number of abnormal cases and the number in each category are also provided. The individual cytotechnician negative rescreen discrepancy is compared to the overall negative rescreen discrepancy rate.

Within recent years, recognizing the nature of the problem of screening, several commercially sponsored systems have been developed that use partial automation of the reading process. The Papnet (Neuromedical Systems, Inc., New York, USA) testing system is a computer-assisted, interactive system that combines computerized image processing and neural network technology to recognize and record areas of abnormality on Papanicolaou-stained cervical smears [64]. It uses conventionally prepared smears. The system uses computer technology in tandem with human diagnostic judgement to reduce the false-negative rates. This system could be a complementary method to look at the abnormal cells but will not replace a final diagnosis made with the microscope.

External quality control is important to test the homogeneity of the diagnosis and test the inter-observer variability. The rate of a specific laboratory may be compared to the mean rate of the other laboratories. This type of quality control is particularly useful for a cytologist who is working alone and does not have the option of showing abnormal cases to his or her colleague.

CONCLUSION

Clinicians have to recognize the achievements and the limitations of the system and remember at all times that a cytologic result does not over-rule clinical findings, and that a persisting lesion must undergo biopsy, regardless of the results of the cervical smear. In spite of the quality control, a negative smear report carries a margin of error. Everybody would agree that quality control is necessary and that the different options that have been described could improve the quality of cytology. However, all these quality-control systems would increase the cost of cytology in a laboratory. There are cost offsets, for example, manual quality-control related savings, if automated apparatuses are used. Ultimately, savings will be derived by greater reliability of test results. The high error rate is indeed one of the strongest arguments against longer intervals between cytologic smears. Alternatively, an abnormal result has a very high positive predictive value, specifically the diagnosis of a HGSIL. The clinician is responsible for finding the lesion, whatever the method of diagnosis chosen, including diagnostic conization.

Finally, it is important when judging the Papanicolaou smear to realize that although it is imperfect, no other test has been as successful in preventing and detecting cancer. Women are more likely to have cervical cancer related to inadequate screening than to misdiagnosed Papanicolaou smears.

REFERENCES

1. International Association for Cancer Research Working Group on Valuation of Cervical Screening Programmes (IARC) (1986) Screening for squamous cervical cytology and its implications for screening policies. *British Medical Journal*, **293**, 659–80.
2. Laara, E., Day, N.E. and Hakama, M. (1987) Trends in mortality from cervical cancer in the Nordic countries: association with organised screening programmes. *Lancet*, **i**, 1247–9.
3. Koss, L.G. (1993) Cervical Pap smear. New directions. *Cancer*; **71**, 1406–12.
4. Melamed, M.R. and Flehinger, B.J. (1992) Reevaluation of quality assurance in the cytology laboratory. *Acta Cytologica*, **36**, 461–5.
5. Bergeron, C., Debaque, H., Ayivi, J., *et al.* (1997) Cervical smear histories of 585 women with biopsy-proven carcinoma *in situ*. *Acta Cytologica* **41**, 1676–80.
6. Gay, J.L., Donalson, L.D. and Goellner, J.R. (1985) False negative results in cervical cytologic studies. *Acta Cytologica*, **29**, 1043–6.
7. Kristensen, G.B., Skyggebjerg, K.D., Holund, B., *et al.* (1991) Analysis of cervical smears obtained within three years of the diagnosis of invasive cervical cancer. *Acta Cytologica*, **35**, 47–50.
8. Mitchell, H., Medley, G., Giles, G. (1990) Cervical cancers diagnosed after negative results on cervical cytology: perspective in the 1980s. *British Medical Journal*, **300**, 1622–6.
9. Morrell, N.D., Taylor, J.R., Snyder, R.N., *et al.* (1982) False-negative cytology rates in patients in whom invasive cervical cancer subsequently developed. *Obstetrics and Gynecology*, **60**, 41–5.
10. Ostergard, D.R. and Gondos, B. (1971) The incidence of false-negative cervical cytology as determined by colposcopically directed biopsies. *Acta Cytologica*, **15**: 292–3.
11. Pairwuti, S. (1991) False-negative Papanicolaou smears from women with cancerous and precancerous lesions of the uterine cervix. *Acta Cytologica*; **35**, 40–6.
12. Paterson, M.E.L., Peel, K.R. and Joslin, C.A.F. (1984) Cervical smear histories of 500 women with invasive cervical cancer in Yorkshire. *British Medical Journal*, **289**, 896–8.
13. Rylander, E. (1977) Negative smears in women developing invasive cervical cancer. *Acta Obstetrics and Gynecology Scandinavia*, **56**, 115–18.
14. Sherman, M.E. and Kelly, D. (1992) High grade squamous intraepithelial lesions and invasive carcinoma following the report of three negative Papanicolaou smears: screening failures or rapid progression. *Modern Pathology*, **5**, 337–42.

15. Van der Graaf, Y., Vooijs, G.P., Gaillard, H.J.L. and Go, D.M.D.S. (1987) Screening errors in cervical cytology screening. *Acta Cytologica*, **31**, 434–8.

16. Wain, G.V., Farnsworth, A. and Hacker, N.F. (1992) Cervical carcinoma after negative Pap smears: evidence against rapid-onset cancers. *International Journal of Gynecological Cancer*, **2**, 318–22.

17. Papanicolaou, G.N. and Traut, H.F. (1941) The diagnostic value of vaginal smears in carcinoma of the uterus. *American Journal of Obstetrics and Gynecology*, **42**, 193–206.

18. Ayre, J.E. (1947) Selective cytology smear for diagnosis of cancer. *American Journal of Obstetrics and Gynecology*, **53**, 609–17.

19. Buntinx, F. and Brouwers, M. (1996) Relation between sampling device and detection of abnormality in cervical smears: a meta-analysis of randomised and quasi-randomised studies. *British Medical Journal*, **313**, 1285–90.

20. Papanicolaou, G.N. and Traut, H.F. (1943) *Diagnosis of Uterine Cancer by the Vaginal Smear.* The Commonwealth Fund, New York.

21. Reagan, J.W., Seidemann, I.L. and Saracusa, Y. (1953) Cellular morphology of carcinoma *in situ* and dysplasia or atypical hyperplasia of uterine cervix. *Cancer*, **6**, 224–35.

22. Richart, R.M. (1967) The natural history of cervical intraepithelial neoplasia. *Clinical Obstetrics and Gynecology*, **10**, 748–84.

23. National Cancer Institute Workshop (NCIW) (1989) The 1988 Bethesda System for reporting cervical/vaginal cytological diagnoses. *Journal of the American Medical Association*, **262**, 931–4.

24. The Revised Bethesda System for Reporting Cervical/Vaginal Cytologic Diagnoses: Report of the 1991 Bethesda Workshop (1992) *Acta Cytologica*, **36**, 273–6.

25. Kurman, R.J., Henson, D.E., Herbst, A.L., *et al.* (1994) Interim guidelines of management of abnormal cervical cytology. *Journal of the American Medical Association*, **271**, 1866–9.

26. Nielsen, M.L., Davey, D.D. and Kline, T.S. (1993) Specimen adequacy evaluation in gynecologic cytopathology: current laboratory practice in the college of American pathologists, interlaboratory comparison program and tentative guidelines for future practice. *Diagnostics of Cytopathology*, **9**, 394–403.

27. Buntinx, F., Schouten, H.J., Knottnerus, J.A., *et al.* (1993) Interobserver variation in the assessment of the sampling quality of cervical smears. *Journal of Clinical Epidemiology*, **46**, 367–70.

28. Kivlahan, C. and Ingram, E. (1986) Papanicolaou smears without endocervical cells: are they inadequate? *Acta Cytologica*, **30**, 258–60.

29. Mitchell, H. (1994) Consistency of reporting endocervical cells. An intralaboratory and interlaboratory assessment. *Acta Cytologica*, **38**, 310–14.

30. Vooijs, G.P., Elias, A., Van der Graaf, Y. and Veling, S. (1985) Relationship between the diagnosis of epithelial abnormalities and the composition of cervical smears. *Acta Cytologica*, **29**, 323–8.

31. Awen, C., Hathway, S., Eddy, W., Voskuil, R. and Janes, C. (1994) Efficacy of ThinPrep preparation of cervical smears: a case 1000-case, investigator-sponsored study. *Diagnostics of Cytopathology*, **11**, 33–7.

32. Bur, M., Knowles, K., Pelow, P., Corral, O., *et al.* (1995) Comparison of ThinPrep preparations with conventional cervicovaginal smears. *Acta Cytologica*, **39**, 631–42.

33. Geyer, J.W., Hancock, F., Carrico, C. and Kirkpatrick, M. (1993) Preliminary evaluation of cyto-rich: an improved automated cytology preparation. *Diagnostics of Cytopathology*, **9**, 417–22.

34. Hutchinson, M.L., Agarwal, P., Denault, T., Berger, B. and Cibas, E.S. (1992) A new look at cervical cytology. ThinPrep multicenter trial results. *Acta Cytologica*, **36**, 499–504.

35. Laverty, C.R., Thurloe, J.K., Redman, N.L. and Farnsworth, A. (1995) An Australian trial of Thinprep: a new cytopreparatory technique. *Cytopathology*, **6**, 140–8.

36. Vassilakos, P., Cossali, D., Albe, X., *et al.* (1996) Efficacy of monolayer preparations for cervical cytology. Emphasis on suboptimal specimens. *Acta Cytopathologica*, **40**, 496–500.

37. Wilbur, D.C., Cibas, E.S., Meritt, S., *et al.* (1994) ThinPrep Processor: clinical trials demonstrate an increased detection rate of abnormal cervical cytologic specimens. *American Journal of Clinical Pathology*, **101**, 209–14.

38. Ferenczy, A., Robitaille, J., Franco, E., *et al.* (1996) Conventional cervical cytologic smears vs ThinPrep smears. A paired comparison study on cervical cytology. *Acta Cytologica*, **40**, 1136–42.

39. Confortini, M., Biggeri, A., Cariaggi, M.P., *et al.* (1993) Intralaboratory reproducibility in cervical cytology. Results of the application of a 100-slides set. *Acta Cytologica*, **37**, 49–54.

40. Fahey, M.T., Irwig, L. and Macaskill, P. (1995) Meta-analysis of Pap test accuracy. *American Journal of Epidemiology*, **141**, 680–9.

41. Jones, S., Thomas, G.D.H. and Williamson, P. (1996) Observer variation in the assessment of adequacy and neoplasia in cervical cytology. *Acta Cytologica*, **40**, 226–34.

42. Joste, N.J., Rushing, L., Granados, R., *et al.* (1996) Bethesda classification of cervicovaginal smears: reproducibility and viral correlates. *Human Pathology*, **27**, 581–5.

43. Kato, I., Santamaria, M., Alsonso de Ruiz, P. *et al.* (1995) Interobserver variation in cytological and histological diagnoses of cervical neoplasia and its epidemiologic implication. *Journal of Clinical Epidemiology*, **48**, 1167–74.

44. Klinkhamer, P.J.J.M., Voojis, G.P. and de Haan, A.F.J. (1988) Intraobserver and interobserver variability in the diagnosis of epithelial abnormalities in cervical smears. *Acta Cytologica*, **32**, 794–800.

45. Sama, D., Cotignoli, T., Guerrini, L., *et al.* (1996) Intralaboratory reproducibility of cervical cytology diagnoses in the external quality assurance scheme of the Emilia-Romagna region of Italy. *Gynecologic Oncology*, **60**, 404–8.

46. Yobs, A.R., Plott, A.E., Hicklin, M.D., *et al.* (1987) Retrospective evaluation of gynecologic cytodiagnosis. II. Interlaboratory reproducibility as shown in rescreening large consecutive samples of reported cases. *Acta Cytologica*, **31**, 900–10.

47. O'Sullivan, J.P., Ismail, S.M., Barnes, W.S.F., *et al.* (1994) Interobserver variation in the diagnosis and grading of dyskaryosis in cervical smears: specialist cytopathologists compared with non-specialists. *Journal of Clinical Pathology*; **47**, 515–8.

48. Bergeron, C., Barrasso, R., Beaudenon, S., *et al.* (1992) Human papillomaviruses associated with cervical intraepithelial neoplasia. Great diversity and distinct distribution in low-and high-grade lesions. *The American Journal of Surgical Pathology*, **16**, 641–9.

49. Lorincz, A.T., Reid, R., Jenson, A.B., *et al.* (1992) Human papillomavirus infection of the cervix: relative risk associations of 15 common anogenital types. *Obstetrics and Gynecology*, **79**, 328–37.

50. Schiffman, M.H., Bauer, H.M., Hoover, R.N., *et al.* (1993) Epidemiologic evidence showing that human papillomavirus infection causes most cervical intraepithelial neoplasia. *Journal of the National Cancer Institute*, **85**, 958–64.

51. Koss, L.G. and Durfee, G.R. (1956) Unusual patterns of squamous epithelium of uterine cervix; cytologic and pathologic study of koilocytotic atypia. *Annals New York Academy of Science*, **63**, 1245–61.

52. Meisels, A. and Fortin, R. (1976) Condylomatous lesions of the cervix and vagina. I. Cytologic patterns. *Acta Cytologica*, **20**, 505–9.

53. Ismail, S.M., Colclough, A.B., Dinnen, J.S., *et al.* (1990) Reporting cervical intra-epithelial neoplasia (CIN): intra and interpathologist variation and factors associated with disagreement. *Histopathology*, **16**, 371–6.

54. Kennedy, W.A., Salmieri, S.S., Wirth, S.L., *et al.* (1996) Results of the clinical evaluation of atypical glandular cells of undetermined significance (AGCUS) detected on cervical cytology screening. *Gynecologic Oncology*, **63**, 14–18.

55. Raffle, A.E., Alden, B. and Mackensie, E.F.D. (1995) Detection rates for abnormal cervical smears: what are we screening for? *Lancet*, **345**, 1469–73.

56. Sherman, M.E., Schiffman, M.H., Lorincz, A., *et al.* (1994) Toward objective quality assurance in cervical cytopathology. Correlation of cytopathologic diagnoses with detection of high-risk human papillomavirus types. *American Journal of Clinical Pathology*, **102**, 182–7.

57. Koutsky, L., Holmes, K.K., Critchlow, C.W., *et al.* (1992) A cohort study of the risk of cervical intraepithelial neoplasia grade 2 or 3 in relation to papillomavirus infection. *New England Journal of Medicine*, **327**, 1272–8.

58. Montz, F.J., Bradley, J.M., Fowler, J.M. and Nguyen, L. (1992) Natural history of the minimally abnormal Papanicolaou smear. *Obstetrics and Gynecology*, **80**, 385–8.

59. Mayeaux, E.J., Harper, M.B., Abreo, F., Pope, J.B. and Phillips, G.S. (1995) A comparison of the reliability of repeat cervical smears and colposcopy in patients with abnormal cervical cytology. *Journal of Family Practice*, **40**, 57–62.

60. Wright, T.C., Ferenczy, A., Sun, X.W., *et al.* (1995) Comparison of management algorithms for the evaluation of women with low-grade cytologic abnormalities. *Obstetrics and Gynecology*, **85**, 202–10.

61. Cox, J.T., Lorincz, A.T., Schiffman, M.H., *et al.* (1995) Human papillomavirus testing by hybrid capture appears to be useful in triaging women with a cytologic diagnosis of atypical squamous cells of undetermined significance. *American Journal of Obstetrics and Gynecology*, **172**, 946–54.

62. Ferenczy, A., Franco, E., Arseneau, J., *et al.* (1996) Diagnostic performance of hybrid capture human papillomavirus deoxyribonucleic acid assay combined with liquid-based cytologic study. *American Journal of Obstetrics and Gynecology*, **3**, 651–6.

63. Bergeron, C. and Fagnani, F. (1995) Individual cytotechnologist performance and quality control. *Acta Cytologica*, **39**, 274.

64. Mango, L.J. (1996) Reducing false negatives in clinical practice: the role of neural network technology. *American Journal of Obstetrics and Gynecology*, **4**, 1114–19.

11

THE DIAGNOSIS AND TREATMENT OF CERVICAL INTRA-EPITHELIAL NEOPLASIA

David M. Luesley and Renzo Barrasso

INTRODUCTION

The diagnosis of cervical intra-epithelial neo-plasia (CIN) requires either a directed or exci-sional biopsy. The treatment requires either destruction or excision. The simplicity of these statements does not reflect the complexity either may pose in clinical practice. Further-more, the concepts and applied technology applicable to the field have and will continue to evolve and change at great pace. It is there-fore timely to review the current status of the clinical management of intra-epithelial disease of the cervix.

APPROACHES TO DIAGNOSIS

The diagnostic process is initiated following the recognition of a cytological abnormality. As CIN is asymptomatic this rests on detection by screening. An abnormal cervical smear report should prompt further investigation and this will usually be a colposcopic exami-nation. Neither colposcopy nor cytology can be regarded as diagnostic procedures but both will prompt a biopsy in certain situations, which have become the object of critical review for several reasons:

- Not all abnormal cervical smear reports will reflect underlying CIN.
- Minor degrees of CIN may resolve without any form of intervention.
- The screening procedure, as applied to pop-ulations, whilst relatively sensitive, lacks specificity, particularly when minor cytolog-ical abnormalities are present.
- The interface between medicine and women in the screening context generates a consid-erable degree of stress and anxiety. This is true even in women who eventually receive normal smear reports [1–4, 5].

In an effort to minimize intervention and its economic and psychosocial sequelae, selective referral policies have been advocated. Such policies have not been derived from data generated from objective controlled clinical en-quiry but represent a consensus based on accu-mulated experience. There is certainly scope to develop referral policies based on proper trials.

BIOPSIES

The different types of biopsy are given in Table 11.1.

Representative histological material can be collected in two ways. Biopsies can be directed

Cancer and Pre-cancer of the Cervix
Edited by D.M. Luesley and R. Barrasso. Published in 1998 by Chapman & Hall, London. ISBN 0 412 56600 1.

Table 11.1   Types of biopsy

- Directed (punch) biopsies
- Excisional biopsies
- Loop excision
- Laser excision
- Knife cone biopsy
- Wedge biopsy
- Hysterectomy

Table 11.2   Colposcopic features suggestive of CIN

- Punctation
- Mosaicism
- Acetowhite
- Irregular surface contour
- Atypical vessels
- Lesion size
- Hyperkeratosis

using the colposcopic characteristics of CIN to guide the colposcopist to sample the most atypical area of the cervical transformation zone (TZ), or the whole cervical TZ can be removed as an excisional biopsy. The latter has become possible as a direct result of simple, cheap and rapid methods for TZ excision and has also generated some degree of controversy. The controversy centres on the possibility of overtreatment using this 'see and treat' approach [6–10].

DIRECTED BIOPSY

The colposcopist endeavours to take a directed biopsy or biopsies from what is considered to be the most atypical part of the cervical TZ. The colposcopic features suggestive of CIN that will guide the colposcopist are listed below. For this, the whole TZ should be easily visible and if any or all of this area is sequestered within the endocervical canal then the reliability of such biopsies could be questioned.

COLPOSCOPIC FEATURES SUGGESTIVE OF CIN

Table 11.2 contains a list of colposcopic features suggestive of CIN.

Punctation

Prior to the application of acetic acid (5%) punctation appears as small dots. These are subepithelial capillaries seen 'end on'. As the capillaries become more widely spaced, so do the punctate dots, and this appears as a coarsening of the punctate pattern and equates to more severe degrees of dysplasia. Punctation can be highlighted by using saline and a green filter, which makes the red capillaries appear black.

Mosaicism

Mosaicism is the subepithelial capillaries viewed side on and, like punctation, the wider or more coarse the mosaic pattern the greater the degree of dysplasia (Plate 19). Using the green filter after the application of normal saline also highlights this feature.

Acetowhite

Application of a 3–5% solution of acetic acid to the cervix will bring about changes if metaplasia, HPV infection or dysplasia is present. It may also emphasize an atypical capillary pattern as in Plate 19. The reaction takes between 30 and 60 seconds to develop. The degree of acetowhiteness has been correlated with the degree of atypicality, so that mild changes with fine capillary patterns may represent metaplasia whereas marked acetowhiteness with a sharp outline and coarse capillary patterns equates to high-grade intra-epithelial neoplasia. It can be difficult to distinguish between changes attributable to human papillomavirus (HPV) infection and low-grade lesions. Plate 20 illustrates the acetowhite change associated with CIN I.

Not all acetowhite lesions need be associated with abnormal capillary patterns. Acetowhite lesions may sometimes be seen outside the cervical TZ, these 'satellite' lesions are associated with HPV infection (Plate 21).

Irregular surface contour

High grades of CIN and early invasion may be seen in association with an irregular surface contour (Plate 22). When associated with contact bleeding as in this figure, invasion should be suspected.

Atypical vessels

Normally the subepithelial capillaries branch regularly and have an even calibre. As the changes become more coarse, irregular branching and unevenness in the vessel calibre occurs. Both of these features indicate possible invasive disease and should be biopsied.

Lesion size

If the atypical TZ is large, then the possibility of high-grade disease increases. The association between size and grade of lesion is not, however, absolute and the colposcopist should carefully assess all part of a large lesion to determine the areas within it that harbour the most atypical changes.

Hyperkeratosis

So-called leucoplakia can occur on the cervix as a result of chronic irritation as in procidentia, HPV infection and dysplasia. The thickened epithelium obscures the subepithelial vascular pattern and as such should always be suspect and the site of a biopsy.

Several investigators have now commented on the potential inaccuracies in directed punch biopsies [11–13]. Also, one must consider that in certain situations, the biopsy itself could influence the natural history of the condition, particularly if the original lesion was small [5, 14]. The hypothesis that trauma to the TZ could alter the natural history is controversial. Chenoy *et al.* [15] have recently published data from a prospective randomized trial where similar groups of women were subjected to either a central TZ biopsy, a peripheral TZ biopsy or no directed biopsy. All women were then treated by diathermy loop excision of the TZ 6 weeks following biopsy trauma. This study not only confirmed previous findings of the inaccuracy of punch biopsy but also showed that there was no significant effect in outcome histology. This, of course, may not hold true for very small lesions which might easily be removed by a diagnostic biopsy [15].

CONE BIOPSY

This procedure was primarily used when the upper limit of the squamo-columnar junction was not visible and therefore the degree of abnormality within the canal not assessable via the colposcope. There were other situations where formal cold knife conization was considered appropriate, but many of these have been replaced by alternative excisional procedures such as loop excision. It is still appropriate to consider conization as separate from TZ excision, as in the latter the intent is to excise as much tissue as one would normally have destroyed (if a destructive method had been employed): conization, however, is a procedure where there is deliberate intent to remove an additional segment of the endocervical canal. Thus the reasons for cone biopsy remain the same (see Table 11.3), it is only the technique that has altered.

Indications for cold knife conization that the authors feel are still valid are the woman with a smear suggestive of adenocarcinoma *in situ* (AIS) and the suspicion of Stage Ia_1 or Ia_2 cervical cancer.

In both of these situations the excision margins are of extreme importance. AIS is far less common than squamous intra-epithelial neoplasia and not infrequently found in association

Table 11.3   Indications for cold knife cone biopsy

- Suspected invasive disease
- Suspected adenocarcinoma *in situ*
- Upper limit of suspected high-grade lesion not seen[a]
- Previous treatment and persisting high-grade cytological abnormality[a]

[a]In these situations a large or deep loop or laser excisional cone may sometimes be used.

with its squamous counterpart [16]. The lesions may be small and situated deep in gland clefts. The whole endocervical canal may theoretically be at risk [17] although our own observations suggest that the majority of, if not all, lesions are sited at or very close to the TZ. We have also noted that in cases of AIS, subsequent persistence of abnormality is closely related to completeness of excision [18]. It is therefore important to have a specimen where the margins of excision can be accurately assessed.

All methods employing heat as a cutting tool will be associated with some thermal artefact at the margins and this could be of importance in interpretation [19]. The same is true for early invasive disease. Some authorities feel that local excision may be suitable treatment for these lesions [20], yet again, the diagnosis is wholly dependent upon the whole lesion being in the specimen and accurate assessment of the excision margins.

Unlike directed biopsies, cone biopsy and TZ excision are usually considered as therapeutic as well as diagnostic.

WEDGE BIOPSY

In this procedure, a wedge of tissue is removed from the most atypical part of the cervix. It is not considered as therapeutic but offers greater scope for histopathological interpretation than a directed biopsy. This is of particular importance when trying to either exclude or confirm invasion. Whilst it is possible to take representative wedge biopsies by either laser or diathermy, the problems with heat artefact remain. As accurate interpretation is of paramount importance in these situations, cold knife excision is to be preferred. The procedure would normally be conducted under general anaesthesia.

The major indication for wedge biopsy at present is in the exclusion of invasion during pregnancy. The morbidity of the procedure is considered to be less than that of formal cold knife conization and it offers superior diagnostic potential when compared with directed biopsies.

SELECTION FOR TREATMENT

This only becomes an issue should one accept that not all patients referred for investigation of their abnormal smears require treatment. There appears to be a growing body of opinion that feels that minor abnormalities (koilocytotic atypia and mild dysplasia) may be observed and here the issue is one of accurate initial diagnosis [21–25]. Between 20 and 40% of women with smears showing mild dyskaryosis will eventually be proven to have high-grade disease (CIN II and CIN III) [13, 26–28]. Most, if not all, practising colposcopists agree that high-grade disease merits treatment although the majority of these patients will not develop cervical cancer [29]. It would, based on currently available evidence, be ethically unacceptable to deny these women the opportunity of having their dysplastic condition treated. The issue at hand is therefore one of selecting those at most risk from a population where the majority will not be at high risk, in other words, improving the specificity of the whole process.

Some argue that as the process is inherently non-specific and that surveillance leads to cumulative default then all women suspected of having any grade of CIN should be offered treatment. This is the basis of an ongoing debate, selection for treatment versus 'see and treat'. Both have their advocates and critics and both cases can be supported or criticized using a variety of basic clinical, psychosocial and health economic arguments.

The selective approach is underpinned by the observations that minor degrees of cervical dysplasia can regress over time and that all forms of treatment have some degree of morbidity. The objective is to identify those with or who have progressed to high-grade disease (CIN II and III). Cytology, colposcopy and directed biopsy are all used to achieve this aim. Those without high-grade disease will be offered cytological and/or colposcopic surveillance and should understand the importance of their compliance with such a regimen. How long surveillance should be maintained, particularly if either cytological or colposcopic abnormalities persist, is unknown.

'SEE AND TREAT'

Prior to the advent of out-patient TZ excision several criteria had to be met prior to offering ablative treatment, perhaps the most important being to be able to exclude invasive disease and being able to visualize the whole lesion [30–32]. This usually necessitated the taking of a directed biopsy. Directed biopsies therefore formed the basis of management in all colposcopy clinics and a great deal of reliance was placed on the resultant histopathology.

When compared with whole TZ specimens one must also accept that the small directed biopsy is a specimen of lesser quality. It cannot be relied upon in situations of early invasive change (Stage Ia_1 and Ia_2), when AIS is present, and in our own experience there is a tendency to upgrade lesions when based on small biopsy specimens [11].

From a patient perspective, the directed biopsy prior to planned therapy will usually mean additional attendance for treatment purposes, if treatment is indicated. This will also have a cost effect. It seems, therefore, at least superficially, that there are major advantages to be gained by treating (by performing an excisional biopsy) at the primary colposcopic assessment. This will result in overtreatment. Between 10 and 30% of women may be overtreated using the treat-at-first-visit strategy if one accepts that patients with koilocytotic atypia alone do not require treatment. If one goes further and adopts a policy of not treating mild dysplasia, then the overtreatment rate will be higher (Table 11.4) [7].

Referral practice will inevitably impinge on the treatment strategy. If all first cytological abnormalities are referred, there will be a higher proportion of disease of doubtful significance and therefore treatment policies must take account of referral practice.

Overtreatment occurs in many areas of medical practice. The concern regarding overtreatment in women with abnormal smears is the potential for creating additional morbidity (both physical and emotional) in young women, many of whom will be nulliparous.

A review of published data on short- to medium-term physical morbidities such as cervical stenosis, secondary haemorrhage, etc., does not suggest very high rates [33]. However, if a universal 'see and treat' approach were adopted, the actual number of women developing morbidity would be quite substantial.

In an ideal situation, the decision to treat should be based upon proper selective strate-

Table 11.4 'See and treat': overtreatment

Outcome histology of loop excision	Number	Percent
Invasion and adeno-carcinoma *in situ*	15	1.3
CIN II and CIN III	392	61.0
CIN I	86	13.4
HPV only or no abnormality	146	24.3
Total	639	100

In this study there were 639 consecutive attenders to a colposcopy clinic. All grade of smear treated at the first attendence.

gies founded on sound prospective clinical enquiry, which would encompass those end-points felt to be of primary importance. As with referral policy, no such studies have yet been completed and reliance has therefore been placed on consensus based on accumulated experience.

WHAT TREATMENTS?

The 'see and treat' approach is only applicable to out-patient excisional procedures, but other treatment modalities are available and in widespread use (Table 11.5).

In a relatively short period of time the management of pre-invasive disease has shifted from the ultra-radical to the most conservative. Soon after the techniques for recognition became possible, i.e. cervical cytology, clinicians felt that they were looking at malignant disease and therefore radical treatment methods were employed. These included radical hysterectomy, occasionally with radiotherapy. Hysterectomy remained the standard treatment up until the 1960s, when knife conization became accepted. This acceptance was seen first in Europe where it was realized that a knife cone could offer more than just diagnostic capability. Cone biopsy was still morbid in terms of its immediate problems such as the requirement for general anaesthesia and haemorrhage. Also, the long-term problems of cervical stenosis and loss of a significant amount of endocervical epithelium carried separate morbidities of menstrual, fertility and pregnancy dysfunction.

Many women with confirmed CIN regarded as requiring treatment are young and not an insignificant proportion are nulliparous. Concern over subsequent cervical function is therefore very real and to a large extent has driven the trend toward conservative treatment.

DESTRUCTIVE (ABLATIVE) TREATMENTS

A basic principle of all local destructive techniques is that they should not be used if invasive disease cannot be excluded [30]. In all cases it is mandatory to fully visualize the cervical TZ and take directed biopsies from the most atypical areas to achieve this endpoint. Other criteria should also be met (Table 11.6).

All methods of local destruction are effective. Success rates of between 90 and 98% are the norm. Cryocautery may be less effective, although this may be because of poor case selection [34].

All grades of CIN, if treated, can be treated by the same method. The principles of treating CIN I do not differ from those applied to CIN III. The whole TZ should be regarded as suspicious even if only a small part appears

Table 11.5 Treatment methods for CIN

Excisional	Destructive (ablative)
Diathermy excision (LLETZ/LEEP)	Electrocoagulation
CO_2 laser excision	CO_2 laser vaporization
Cold knife cone	Cold coagulation
Hysterectomy	Cryocautery

Table 11.6 Criteria to be met prior to local destruction

- The whole transformation zone can be fully visualized
- The whole transformation zone can be assessed colposcopically
- Directed biopsy(ies) have excluded invasive disease
- Colposcopy does not suggest invasion
- Cytology does not suggest invasion
- There has been no previous treatment for CIN
- There is no suspicion of adenocarcinoma *in situ*
- The patient is not pregnant
- The patient has given consent for treatment

atypical colposcopically. The whole TZ, and not a part of it, should therefore be the target for treatment.

Treatment should ideally destroy tissue to a depth of 7 mm, as it has been demonstrated that CIN can and does involve gland crypts. Treating to a depth of 7 mm allows a suitable safety margin to account for this phenomenon [35].

When local ablative techniques were introduced they were either performed under a general anaesthetic or in an out-patient setting without any form of analgesia. Undoubtedly some discomfort was experienced and as a result of this there has been an increase in the use of local anaesthetic agents. These can be used alone or in conjunction with a vasoconstrictor to minimize blood loss. Haemorrhage is not a problem with electrodiathermy, cold coagulation or cryocautery but can be problematic during laser therapy. Care should be taken when injecting agents into the cervix to avoid any areas of atypical epithelium because of the theoretical risk of introducing malignant cells deep into the cervical stroma.

Electrodiathermy

The first destructive, as opposed to excisional, methods were also introduced in the 1960s and quite understandably employed electrodiathermy. This equipment was and is virtually universally available in most operating departments. The technique was initially performed under general anaesthetic and employs both diathermy needle (to achieve depth of destruction) and ball fulguration. The method was popularized by Chanen [36, 37]. Results using this technique are similar to other destructive methods and in excess of 95% of women so treated will be cytologically normal 6 months following treatment.

Cryocautery and cold coagulation

Cryocautery and 'cold coagulation' rapidly followed, and in contrast to the diathermy

techniques employed at that time were suitable for use in the unanaesthetized patient. Cold coagulation is really a misnomer, as it destroys through heat applied from an external source. This is similar to cryocautery, which destroys using cold temperatures applied directly to the cervix from an external source.

Carbon dioxide laser

In the 1970s and 1980s, the carbon dioxide laser rose to prominence as the most popular out-patient method for treating CIN. This technique had the advantage of precision and depth control. The necessary equipment is expensive and requires frequent maintenance. Nevertheless, the more powerful machines, especially those which can operate in superpulse mode, are versatile and allow for excisional use as well as vaporization.

EXCISIONAL TREATMENT

There are obvious advantages of an excisional method of management as opposed to a destructive method. The former allows a much more confident exclusion of early invasive disease, which may be overlooked even by highly trained colposcopists [38]. Excision also combines diagnosis with therapy, obviating the need for an initial assessment plus directed biopsy. Furthermore, if, at the time of follow-up, persistent colposcopic and or cytological abnormalities are found, a satisfactory archive specimen is available for review.

Laser excision

This method, first reported by Dorsey [39] used the carbon dioxide laser as a cutting rather than vaporizing tool. In order to achieve the desired outcome without producing excessive thermal damage, a powerful laser in superpulse mode is perhaps the best choice. Equipment for such excisional work tends to be more expensive than that normally employed for vaporization only.

Prior to treatment the whole TZ should be visualized and a suitable local anaesthetic and vasoconstricting agent infiltrated into the cervical stroma. The lesion is then identified and a preliminary furrow cut with the laser to circumscribe the ectocervical limit of the atypical TZ. This furrow is gradually deepened. The cylinder of tissue that will eventually form the specimen is then grasped anteriorly with specially designed grasping 'claws' and traction is applied, pulling the cylinder toward the operator. The furrow is progressively deepened as traction is maintained. The grasping claw is then transferred to the posterior and lateral aspects of the cylinder in turn so that the furrow is deepened symmetrically around the cylinder. Once a cylinder of the required depth has been cut, the bottom is transected either with sharp dissection or with the laser. The former method has an advantage in that there will be no thermal artefact at the point where the endocervical canal is transected. The residual cervical bed can then be cauterized with the laser defocused so as to ensure adequate haemostasis. Several authors have reported the results of this technique and while it takes longer and is associated with slightly increased blood loss when compared to vaporization techniques, it provides good material for histological assessment, is appropriate if the upper limit of the TZ is situated in the canal and can also be used instead of knife cone biopsy [27, 40–43].

Diathermy loop excision

Cartier first demonstrated that by utilizing fine wire loops, good quality cervical biopsies could be obtained without recourse to general anaesthesia [44]. Prendiville extended this concept by enlarging the size of the loop so that the whole TZ could be excised, usually using a local anaesthetic and vasoconstrictor directly injected into the cervical stroma [10]. This technique, known as large loop excision of the TZ (LLETZ) in Europe and loop electrosurgical excision procedure (LEEP) in the

USA, has since become one of the most popular methods of managing women suspected of having cervical intra-epithelial neoplasia.

The technique can be performed under local or general anaesthesia. The cervix is infiltrated with local anaesthetic and a vasoconstrictor as with laser excision. The cervix must be colposcopically assessed and the extent of the TZ identified. This is particularly important if the atypical area extends into the canal. Only if the extent of canal involvement can be gauged can a wire loop of the appropriate diameter be chosen to ensure adequate excision at the endocervical margin. In an ideal situation, the whole atypical TZ should be removed with one traverse of the loop (Figure 11.1).

Whether a lateral or antero-posterior traverse is chosen will depend on the size and shape of the cervix and TZ and the access to the cervix. If the TZ is large it may not be possible to remove this in one piece. The most atypical area should be removed as a large biopsy and the remainder sent as separate specimens. The objectives are identical to those of laser excision in that tissue of a minimum depth of 7 mm is excised. Haemostasis is achieved by ball fulguration.

When lesions are seen to extend into the endocervical canal it is often wise to take a second biopsy, removing a further piece of the endocervical canal. This allows greater scope for pathological interpretation and greater confidence in determining complete excision.

The technique combines the safety and precision of laser excision with the simplicity and cost benefits of ordinary diathermy equipment. As it is usually performed in an outpatient setting and is excisional, patients can, and often are, treated at their first attendance. Published results to date indicate that it is as effective as laser methods and is highly cost-effective [6, 7, 9, 10].

Cold knife conization

Knife cone biopsy has been practised for several decades and is still used in selected situa-

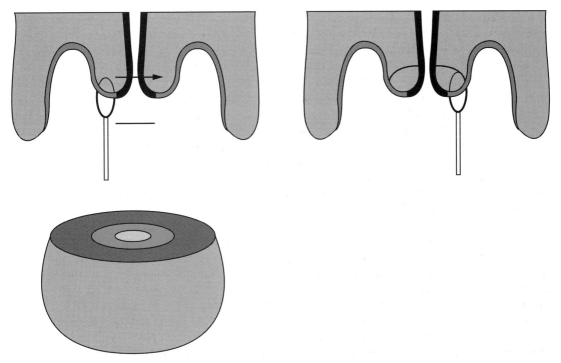

Figure 11.1 Diagram demonstrating the technique of loop excision of the cervical transformation zone. Ideally the transformation zone should be excised in one traverse of the loop although this may not be possible with large lesions.

tions. The indications for cone biopsy have, however, altered with time, partly as a result of increased use of excisional out-patient methods and partly as a result of an increased understanding of the natural history of the disease. The recognized morbidity of cone biopsy has also fostered a more conservative attitude, especially in younger women with mild degrees of cytological abnormality.

HYSTERECTOMY

Hysterectomy still retains a place in the management of women who have confirmed CIN. There should, however, usually be another indication for hysterectomy, such as symptomatic uterine fibroids or other known and symptomatic pelvic pathology. In this role, hysterectomy is not primarily being used to remove the CIN, the fact that it does is convenient.

It is important that these women undergo prior colposcopic assessment in order to exclude any vaginal extension of an atypical TZ. If this is inadvertently left in the vaginal vault at the time of surgery, (vaginal intra-epithelial neoplasia) VaIN may persist [45, 46].

Very rarely, hysterectomy may offer the least morbid and therefore safest method of removing the cervical TZ. The only situation where this may occur is in the post-menopausal patient who has a small, atrophic cervix flush with the vaginal vault. Local excision can be morbid in these situations. Hysterectomy on the other hand is usually not difficult and furthermore there is less chance of the procedure itself damaging the cervix. A cautionary note, however, is that in this group there will be an increased incidence of occult invasion and there is a theoretical risk of surgical undertreatment.

REASONS FOR EMPLOYING EXCISION RATHER THAN ABLATION

The concept underlying excision is the risk of either misdiagnosing invasive disease (a large biopsy is far more reliable than a directed biopsy) or not recognizing invasive disease because of its situation in the endocervical canal. It will be appreciated that even if a lesion can be seen in the endocervical canal, it will not be readily accessible to directed biopsy forceps. Furthermore, colposcopic recognition criteria rely very heavily on the subepithelial vascular pattern that is normally viewed at an angle of 90°. An increase in the viewing angle may alter the appearance of the microvasculature, resulting in misinterpretation.

An excisional method of treatment is always advised if, after local destruction, there are signs of persisting CIN. In this situation the colposcopist cannot be certain the residual malign epithelium has been sequestered beyond the view of the colposcope.

As alternative excisional methods have become available, the need for formal cold knife conization has become less obvious. Both laser and loop excision can fulfil the safety aspects of management that underpinned conization.

INDICATIONS FOR GENERAL ANAESTHESIA

Although the majority of women are now treated under local analgesia there are still situations where a general anaesthetic will be necessary. These are usually patient-orientated reasons such as intense anxiety, but occasionally it is not possible to achieve good access and the necessary instrumentation required to achieve good access can only be achieved with a relaxed and sedated patient.

TIMING OF TREATMENT

The timing of treatment can be viewed in relation to the timing of the first assessment. If an excisional method of treatment is to be used then the patient can be treated without waiting for the results of directed biopsies. This particular approach has been referred to already as 'see and treat'. The advantages and disadvantages of such a policy really relate to patient convenience and economics. If a biopsy/assessment is performed at the first visit, a subsequent visit will be required to administer treatment.

Treatment during pregnancy should be avoided, and this topic is dealt with later in this chapter.

Treatment should also be avoided if there is evidence of a vaginitis, this should be treated prior to either local destruction or local excision. Many clinicians also feel that treatment should not be performed during menstruation as this might increase the risk of haemorrhage. We have analysed our diathermy loop excision morbidities in relation to the time performed in the menstrual cycle and not found any correlation [47]. We would, however, suggest that if possible treatment should not be scheduled for this time as the view of the TZ may be obscured, particularly if bleeding is heavy. In practice, timing of treatment is not always easy. In situations where the clinician is more concerned about subsequent default it seems justifiable to offer treatment despite the fact that the woman may be menstruating.

Treatment produces short-term problems in the form of vaginal discharge and bleeding, and normally women are asked to refrain from intercourse and tampon insertion during the period of healing (4–6 weeks). It is the authors' practice to avoid treatment immediately prior to such events as holidays and weddings, where the short-term side effects may have considerable social impact.

COMPLICATIONS OF TREATMENT

PAIN

The immediate complications of both excision and ablation are pain and haemorrhage. The former can usually be avoided by judicious use of intracervical local anaesthetic, although even after adequate infiltration, some women experience some discomfort usually described as cramping or period-like pain. This may be due to a reflex effect causing uterine spasm. It generally passes off quite quickly. Pain was not found to be a major problem in our own series of loop excisions and this appeared to be related to the length of time that treatment took. Certainly it does appear that difficult and lengthy procedures are associated with more discomfort and indeed if at the time of initial assessment a lengthy procedure is anticipated, serious consideration should be given to electively performing treatment under a general anaesthetic.

HAEMORRHAGE

The risk of haemorrhage can be minimized by infiltrating the cervix with a suitable vasoconstricting agent: this is usually given as a combined injection with the local anaesthetic. Concurrent cervicitis or vaginitis increases the risk of bleeding and if these conditions are recognized, they should be treated prior to either excision or ablation. In the vast majority of situations haemorrhage can be controlled by either fulguration diathermy or by defocusing the laser beam. Delayed or secondary haemorrhage occurs in approximately 2% of all treated cases.

Some bleeding and discharge is almost universal following all forms of local treatment (with the exception of cryocautery) but bleeding sufficient to warrant further intervention is unusual. The most likely cause of secondary haemorrhage is infection of the exposed stroma with dislodgement of the slough that inevitably forms after thermal damage. Occasionally an arterial bleed will occur that might necessitate suturing, repeat diathermy or laser. The majority of secondary haemorrhages respond well to rest and antibiotics, however. Some clinicians routinely prescribe antibiotics following treatment. This practice has not yet been shown to reduce the risk of secondary haemorrhage or indeed pelvic infection.

GENITAL TRACT TRAUMA

Rare complications that can occur at the time of treatment are trauma to the vagina or even perforation of the cul-de-sac. This is only likely to occur with inexperienced operators or in situations where there has been previous cervical surgery, especially if the abnormality extends onto the vagina.

CERVICAL STENOSIS

One of the more worrying complications of treatment is cervical stenosis. There are no universally accepted definitions of cervical stenosis. Complete closure of the residual canal is undoubtedly a problem (Plate 23) and can result in haematometra or haematocervix (Plate 24) but symptoms such as dysmenorrhoea can arise even if the cervix is patent but the canal significantly narrowed.

We have adopted a standard approach to the diagnosis of stenosis based on the symptoms observed and the noted increased frequency of cytology samples not containing endocervical cells. A minimum canal diameter of 3 mm, estimated by the passage of a cervical sound, provides a cut-off by which we diagnose this problem. Between 2 and 3% of patients treated by loop excision will develop cervical stenosis based upon this definition. Not all of these patients are symptomatic but the frequency of dysmenorrhoea is double that of patients with less narrowing. Situations that may predispose to excessive cervical scarring and narrowing following treatment are the depth of the specimen removed [48] and the menstrual status of the patient. Women who are amenorrhoeic after treatment appear to be at an increased risk of developing cervical stenosis.

SUBFERTILLTY

Apart from the symptoms that may result from cervical distortion there is the theoretical risk of damaging fertility. This may arise through two basic mechanisms. The treatment may result in ascending pelvic infection with subsequent tubal damage. The second, and least understood process, relates to the role of the endocervical canal and cervical mucus in fertility. Whilst there have been documented pregnancies in women following large knife cone biopsies, there is almost certainly a relationship between cervical mucus production or lack of it and fertility. This has been very difficult to quantify and will probably remain a theoretical risk. It is virtually impossible to perform the ideal clinical study to assess the effect of treatment on subsequent fertility. One recent study, however, has suggested that the cumulative fertility rate of women treated by loop excision approximates to that of the background population [49].

PREGNANCY OUTCOME

Another outcome following cervical trauma is malfunction during pregnancy or labour. The risk of miscarriage, particularly mid-trimester abortion, is theoretically increased only if the integrity of the internal cervical os has been damaged. Two recent studies have indicated that whilst early pregnancy loss may not be increased, subsequent pregnancy performance may be altered in women following loop excision treatment of CIN. Two studies related to patients treated by loop excision only and noted that infants were born significantly lighter that those of matched controls [50, 51]. This was not an effect of gestation at the time of delivery and was felt to reflect social factors that may also relate to their having CIN, such as cigarette smoking. Another study noted that women treated for CIN by any conization method also had smaller infants than matched population controls. These workers also noted that similar differences were present prior to treatment, i.e. women who were to be treated by conization (of various types) had smaller infants than their controls even before they developed CIN and required treatment [52]. Both of these

studies suggest that factors related to the development of CIN may be of more importance than the actual treatment.

No effect on pregnancy performance has been reported associated with laser vaporization [53].

Cervical dystocia has also been cited as a possible sequel to local cervical treatment. This has been noted in the past but only after large knife cones. There have been no recent data relating to the more modern techniques of treatment.

FOLLOW-UP

The purpose of follow-up is to detect residual disease in the short term and recurrent disease in the long term. Even patients whose subsequent follow-up is normal remain at an increased relative risk of developing cervical cancer than those who have never had an abnormal smear (relative risk = 3). Those who have abnormal cytology during follow-up have 20 times the risk of developing cervical cancer than women who have never had an abnormal smear [29]. Thus, after treatment some sort of follow-up will be required. This may be a combination of colposcopy and cytology or cytology alone. The timing of the first follow-up visit varies somewhat but most clinicians will perform this between 4 and 6 months after treatment. If follow-up is performed too soon after treatment atypical cytology results will be seen with increasing frequency. This probably reflects an aspect of epithelial regeneration.

All patients who have had CIN diagnosed remain at an increased relative risk of developing cervical cancer [29]. This risk increases should there be an abnormal smear taken during the follow-up schedule. The schedule currently recommended in the UK with regard to patients treated for CIN is an annual smear for 5 years and then return to a 3-yearly recall schedule, should all follow-up smears remain negative. Any cytological abnormality in a treated patient should prompt an early colposcopic review with recourse to further biopsies if indicated.

Abnormal follow-up cytology is more likely in those who have had large lesions treated and in whom there is doubt with regard to the completeness of excision [54].

Our own data with regard to treatment failures following LLETZ identified the size of the colposcopically visible lesion and excision margin status of the loop specimen as the most important independent variables predicting failure of LLETZ treatment [54]. In this study, 58 cases of treatment failure as defined by a dyskaryotic smear 6 months after treatment were paired with two controls per case. Each control was matched for date of treatment. Lesion size was determined semiquantitatively by dividing the cervix into eight sectors. Incomplete excision was much more likely to have occurred with large lesions and these women were more likely to have persistent dyskaryosis following treatment. Furthermore, in the multivariate equation, each of these variables independently predicted for persistent dyskaryosis.

The exact role of colposcopy in the follow-up of treated patients is still unclear. It is well recognized that cytology alone can and does miss small lesions, although the significance of these remains uncertain [55]. Our practice is still to offer patients a colposcopic and cytological assessment 6 months following treatment. If both investigations are negative, patients are discharged to the community for cytological surveillance only. There may be a justification for following patients with cytology alone, particularly those found to have small, low-grade lesions that have been completely excised. Large, high-grade lesions even with demonstrable complete excision are probably best reassessed in the colposcopy clinic, not only because of the higher risk of residual disease but also to assess the possible side-effects, particularly cervical stenosis (these patients are more likely to have had larger areas of the cervix removed and are therefore more prone to develop cervical stenosis).

FOLLOW-UP AFTER HYSTERECTOMY

A minority of women will have undergone hysterectomy, either because they had associated gynaecological problems such as fibroids, menorrhagia, etc., or they were found to have CIN prior to or after primary treatment for an unrelated condition.

All women undergoing hysterectomy should have had a cervical smear performed within at least 3 years. If not, a smear should be performed and if any abnormality is found, colposcopy should be performed prior to hysterectomy. This is to try and avoid leaving atypical epithelium at the vaginal vault, i.e. to identify the minority who will have a vaginal extension of their atypical TZs. If an atypical vaginal extension is identified, this can either be treated by laser vaporization prior to the hysterectomy or the area of affected vagina can be excised along with the rest of the specimen. The latter is facilitated by a vaginal approach to the hysterectomy. Atypical epithelium if unrecognized might be buried in the vault suture line and result in VaIN or even carcinoma at the vaginal vault [45, 46]. The risk of burying atypical epithelium in the vaginal vault is small but still present. For this reason most authorities would recommend not closing the vaginal vault if CIN is thought to involve the cervix and upper vagina. Haemostasis can be achieved merely by oversewing the vagina.

All patients who have confirmed CIN on their hysterectomy specimens should have follow-up by vaginal vault cytology. This should be performed 6 months following the hysterectomy and, if normal, performed 12 months later. In situations where complete removal of the lesion can be confirmed and where both follow-up smears are negative, it is possible to discontinue further cytology follow-up. If there is any doubt with regard to lesion excision status, follow-up cytology should be as for those women who have had CIN treated and have an intact cervix.

Any woman having an abnormal vaginal vault smear should be assessed colposcopically. This is always more difficult following hysterectomy, particularly if the vaginal vault has been closed as this invariably results in the formation of 'pockets' in both angles (Plate 26). These can be difficult to visualize. Any atypical epithelium seen in the vagina will require a biopsy. If the area of abnormality is seen to involve the angles or vault suture line then the possibility of buried atypical epithelium should be considered and perhaps a wide excision biopsy under a general anaesthetic considered rather than a directed punch biopsy.

MANAGEMENT OF ABNORMAL SMEARS AFTER TREATMENT

Persisting cytological abnormality after treatment occurs in 10–15% of women (Tables 11.7a, b).

All cases with moderate dyskaryosis or worse regardless of colposcopic findings, and any cases with a colposcopic abnormality, were subjected to rebiopsy by repeat loop excision. These criteria were met in 43 cases with

Table 11.7a   Cytological and histological outcome after loop excision of CIN (*N* = 639)

Original histology	Percent normal cytology 6 months after treatment
No abnormality or HPV only	83%
CIN I	90%
CIN II and CIN III	92%

Table 11.7b   Subsequent biopsy results (*N* = 43)[a]

Histology	Number (%)
CIN (any grade)	12 (28)
HPV only	15 (35)
No abnormality	16 (37)

[a]Further biopsies were taken if the follow-up smear was moderate dyskaryosis or worse or colposcopy suggested dysplasia.

abnormal cytology. The histological outcome is detailed in Table 11.7(b).

Not all of these women will have residual CIN and our own experience is that the majority, on rebiopsy, will have no histological abnormality or changes consistent with wart virus infection only (Table 11.7b). The situation does demand careful assessment, particularly if the original treatment was complicated or the lesion felt to be large [54].

All of these women will require colposcopy. If the abnormal smear is moderate dyskaryosis or worse the likelihood is that residual disease is present and unless the whole TZ can be thoroughly scrutinized and seen to be normal, a further excisional biopsy will be required. Of women with an abnormal smear following a first loop excision, 98% can be expected to revert to normality following a second treatment [49].

Any colposcopic abnormality occurring in conjunction with a cytological abnormality should prompt a further excision unless both suggest HPV changes only. Referral to the original biopsy is often useful in these situations and is one of the advantages of having performed an initial excisional biopsy. If this demonstrated high-grade disease and was also incomplete then a further excision is mandatory. One of the problems, which again is theoretical, is the possibility of lesions lying outside the new TZ. This can occur once the original TZ has been treated and justifies recourse to further excision if cytological and/or colposcopic abnormalities persist.

A report of incomplete excision on the original specimen does increase the chances of there being abnormal cytology at a 6-month review [54] but does not automatically warrant further treatment. Careful surveillance is, however, mandatory [56].

Most abnormalities detected within 12 months of treatment will be residual disease [57], indeed the same might apply to those discovered even later. The rationale of following-up these patients is primarily to detect those with residual disease but also to detect the risk of them developing new disease. It is logical to assume that the circumstances that brought about dysplastic change in the first instance may continue to prevail, hence the accepted increased risk and heightened surveillance.

SPECIAL SITUATIONS

PREGNANCY

It is not surprising, given the age of the population of women presenting with abnormal smears, that a proportion will either be pregnant at the time or conceive in the interval between taking the smear and attending the colposcopy clinic. The diagnosis and treatment of CIN in pregnancy poses several problems. First, the cervix is altered quite considerably in the pregnant state. There is eversion, softening and widespread metaplasia, in addition to which the cervical mucus is both greater in volume and more tenacious. What might be a fairly straightforward colposcopic assessment in the non-pregnant state can become quite difficult.

The softening and increased vascularity make any biopsy and treatment procedures more hazardous in that haemorrhage is more likely. All this has to be set against the often highly charged emotional background that can accompany pregnancy, the fear of any procedure causing miscarriage, the fear that the disease process itself might either harm the unborn child or follow a more aggressive course. All of these issues have to be dealt with as well as reaching a reliable diagnosis.

Reaching the diagnosis will virtually always depend upon obtaining material for histopathological assessment and, as has already been stated, biopsy procedures may be more hazardous in the pregnant state. It has been our practice only to take biopsies if there is any suspicion of invasive disease, this includes Stages Ia_1 and Ia_2 cervical cancer. Most colposcopists are well aware that this is not always possible with colposcopy alone, therefore a fairly low threshold for intervention is usually accepted in pregnancy.

In early pregnancy (i.e. before 28 weeks), any suspicion of pathology greater than pre-invasive change should prompt a biopsy that will be sufficient to reliably exclude invasive disease. The authors feel that small directed biopsies are not adequate for this purpose and therefore either a limited cone or wedge of tissue is generally removed. This is performed under a general anaesthetic to enable good access and also allows the use of suture materials to secure haemostasis. In later pregnancy, if the foetus is considered viable and it is felt that biopsy is required, this may be delayed until delivery, unless of course there are signs of frank invasion. This course is justifiable on the grounds that achieving an adequate biopsy in late pregnancy is both difficult and hazardous and it has not been shown that delaying the definitive diagnosis of early invasive disease Ia_1/Ia_2 by a matter of a few weeks will result in a significant change in outcome.

If, at colposcopy, invasive disease is not suspected, a repeat colposcopic assessment may be scheduled later in pregnancy (28–32 weeks), or, if the cytological and colposcopic features are those of low-grade disease only, then the repeat assessment may be scheduled to take place 3 months post-partum.

One of the major problems to arise when assessing women who are pregnant is that of unsatisfactory colposcopy. In the non-pregnant state, women with high-grade cytological abnormalities who do not have fully visualized TZs are usually managed by some form of conization. This is not wholly appropriate in the pregnant state. In cases of doubt then perhaps a more experienced colposcopic opinion should be sought prior to a biopsy procedure that is likely to be morbid.

MENOPAUSE

The cervix undergoes both inversion, shrinkage and the epithelium becomes progressively atrophic once circulating oestrogen levels fall.

These changes can also result in difficulties in diagnosis. The risk of older women having early or occult invasion is somewhat higher than women in the reproductive age ranges, particularly if they have not been previously screened.

Colposcopy is far more likely to be unsatisfactory as the squamo-columnar junction appears to recede up the endocervical canal with the gradual shrinkage and inversion of the cervix. The thinned and atrophic squamous epithelium is easily traumatized and small areas of intra-epithelial neoplasia may be accidentally scraped off leaving nothing other than exposed stroma. Colposcopic examination may reveal small petechial haemorrhages and even contact bleeding suggestive of invasive disease. The acetic acid reaction is generally less marked and the hypo-oestrogenic state renders Schillers test unreliable.

To complicate matters further, the features mentioned above also combine to confuse the cytological picture. There is a much higher proportion of unsatisfactory smears in postmenopausal women and as the scrape is more likely to contain basal and parabasal cells these might give the cytological appearance of minor abnormalities. If interpretation is difficult, both colposcopically and cytologically, then there is justification in prescribing a local oestrogen preparation over a 6-week period and repeating both assessments. The delay in reaching a diagnosis is unlikely to affect outcome and the benefits are more reliable assessments and, for the majority, avoidance of needless surgical intervention.

Any suggestion of high-grade disease should, as in other situations, prompt appropriate biopsies. In this age group it is unlikely that a good view of the endocervical canal can be obtained and therefore some type of conization will be required [58].

Despite the fact that fertility is not an option in these women, cone biopsy is still morbid and in situations where the cervix is flush with the vaginal vault it may be impossible to resect a satisfactory specimen. Rather than risk un-acceptable morbidity and be left with a badly damaged histological specimen, simple hysterectomy should be considered as an option in selected cases. There is naturally a risk of finding invasive disease, although this is highly unlikely if an endocervical curettage is performed at the time of the initial colposcopic assessment. Even if a small occult lesion is discovered in the hysterectomy specimen, it is likely to be fully treatable with post-operative adjuvant pelvic radiotherapy or a planned parametrectomy.

VAGINAL EXTENSION

In 4% of women, colposcopy will identify a vaginally extended TZ. This may harbour dysplastic epithelium or it may just represent a congenital TZ. All such lesions should be biopsied in addition to biopsies from the most atypical areas of the cervical TZ. Taking biopsies from the vagina is slightly more difficult than from the cervix. There is no subepithelial stroma and the punch biopsy forceps traditionally employed to take biopsies from the cervix may cause a considerable crush artefact. This can only lead to difficulties and even errors in interpretation. A modified Keyes punch, as used extensively for skin and vulval biopsies, is preferable, enabling a selected disc of vaginal tissue to be removed and fixed flat on a piece of card. This avoids crushing and distortion during fixation. Should dysplasia be confirmed in the vagina then this, like the cervix, will require treatment. The vagina, however, is slightly more difficult to treat and certainly should not be treated to the same depth as the cervix. The carbon dioxide laser provides the best method for selective and precise destruction of atypical vaginal TZs [59] and can be combined with an excision procedure for the central cervical lesion. General anaesthesia is required. If lesions are large and involving both the anterior and posterior vaginal walls a two-stage approach is perhaps wiser than an attempt to destroy all of the atypical vaginal epithelium at one attempt. If both walls are treated there is a possibility of cross-vaginal adhesions forming, resulting in dyspareunia or apareunia.

Intravaginal applications of chemodestructive agents such as 5-fluorouracil have been advocated for the treatment of these vaginal lesions [59]. The procedure is protracted, can cause considerable discomfort, particularly if the vulva is inadvertently exposed, and the outcome is variable. There have not been, and are unlikely to be, any large-scale studies documenting the long-term outcome of women treated by this modality.

CONCLUSIONS

The diagnosis and management of pre-invasive disease of the cervix is often regarded as simple and non-challenging. Nothing could be further from the truth. A fine balance between prevention of invasive disease and avoiding morbidity must always be struck. The solution is not to destroy or excise every TZ, this is both irrational, unnecessary and will inevitably lead to unacceptable morbidity. As our understanding of both aetiology and natural history expands, the pressure for even more conservative approaches increases. This will demand an enhancement, not a reduction, of colposcopic skill. Screening programmes are now here and are unlikely to be abandoned. Any screening programme, particularly the cervical screening programme, will identify false-positive results. It is therefore mandatory that colposcopy and careful case selection are seen as part of, not an addition to, the cervical cancer prevention strategy.

REFERENCES

1. Campion, M.J., Brown, J.R., McCance, D.J., *et al.* (1988) Psychosexual trauma of an abnormal cervical smear. *British Journal Obstetrics and Gynaecology*, **95**, 175–81.
2. Marteau, T. (1993) Psychological effects of an abnormal smear result. In *Large Loop Excision of the Transformation Zone: A Practical Guide* (ed. W. Prendiville). Chapman & Hall, London.
3. Marteau, T.M., Walker, P., Giles, J. and Smail, M. (1990) Anxieties in women undergoing colposcopy. *British Journal Obstetrics and Gynaecology*, **97**, 859–61.
4. Posner, T. and Vessey, M. (1988) *Prevention of Cervical Cancer: The Patients' View.* King Edwards Hospital Fund for London, London.
5. Weinman, J. and Johnson, M. (1988) Stressful medical procedures: an analysis of the effects of psychological interventions and of the stressfulness of the procedures. In *Topics in Health Psychology* (eds S. Maes, P. Defares, I. Sarason, *et al.*). Wiley, Chichester.
6. Bigrigg, M.A., Codling, B.W., Pearson, P., Read, M.D. and Swingler, G.R. (1990) Colposcopic diagnosis and treatment of cervical dysplasia at a single clinic visit. Experience of low voltage diathermy loop in 1000 patients. *Lancet*, **336**, 229–31.
7. Luesley, D.M., Cullimore, J., Redman, C.W.E., *et al.* (1990) Loop diathermy excision of the cervical transformation zone in patients with abnormal cervical smears. *British Medical Journal*, **300**, 1690–3.
8. Mor-Yosef, S., Lopes, A., Pearson, S., and Monaghan, J.M. (1990) Loop diathermy cone biopsy. *Obstetrics Gynecology*, **75**, 884–6.
9. Murdoch, J.B., Grimshaw, R.N, and Monaghan, J.M. (1991) Loop diathermy excision of the abnormal cervical transformation zone. *International Journal of Gynecological Cancer*, **1**, 105–11.
10. Prendiville, W., Cullimore, J. and Norman, S. (1989) Large loop excision of the transformation zone (LLETZ). A new method of management for women with cervical intra-epithelial neoplasia. *British Journal of Obstetrics and Gynaecology*, **96**, 1054–60.
11. Buxton, E.J., Luesley, D.M., Shafi, M.I. and Rollason, T.P. (1991) Colposcopically directed biopsy: a potentially misleading investigation. *British Journal of Obstetrics and Gynaecology*, **98**, 1273–6.
12. Kirkup, W. and Hill, A.S. (1980) The accuracy of colposcopically directed biopsy in patients with suspected intra-epithelial neoplasia of the cervix. *British Journal of Obstetrics and Gynaecology*, **87**, 1.
13. Skehan, M., Soutter, W.P., Lim, K., Krausz, T. and Pryse-Davies, J. (1990) Reliability of colposcopy and directed punch biopsy. *British Journal of Obstetrics and Gynaecology* **97**, 811–16.
14. Richart, R.M. (1966) Influence of diagnostic and therapeutic procedures on the distribution of cervical intra-epithelial neoplasia. *Cancer*, **19**, 1635–8.

15. Chenoy, R., Billingham, L., Irani, S., Rollason, T.P., Luesley, D.M. and Jordan, J.A. (1996) The effect of directed biopsy on the atypical cervical transformation zone: assessed by digital imaging colposcopy. *British Journal of Obstetrics and Gynaecology*, **103**, 457–63.

16. Luesley, D.M., Jordan, J.A., Woodman, C.B.J., Watson, N., Williams, D.R. and Wadell, C. (1987) A retrospective review of adenocarcinoma-in-situ and glandular atypia of the uterine cervix. *British Journal of Obstetrics and Gynecology*, **94**, 699–703.

17. Jaworski, R.C., Pacey, N.F., Greenberg, M.L. and Osborn, R.A. (1988) The histologic diagnosis of adenocarcinoma-in-situ and related lesions of the cervix uteri. *Cancer*, **61**, 1171–81.

18. Cullimore, J.E., Luesley, D.M., Rollason, T.P., *et al.* (1992) A prospective study of conization of the cervix in the management of cervical intraepithelial glandular neoplasia (CIGN) – a preliminary report. *British Journal of Obstetrics and Gynaecology*, **99**, 314–18.

19. Montz, F.J., Holschneider, C.H. and Thompson, L.D.R. (1993) Large loop excision of the transformation zone: effect on the pathologic interpretation of the resection margins. *Obstetrics and Gynecology*, **81**, 976–82.

20. Morgan, P.R., Anderson, M.C., Buckley, C.H., *et al.* (1993) The Royal College of Obstetricians and Gynaecologists micro-invasive carcinoma of the cervix study: preliminary results. *British Journal of Obstetrics and Gynaecology*, **100**, 664–8.

21. Fletcher, A., Metaxas, N., Grubb, C. and Chamberlain J. (1990) Four and a half year follow-up of women with dyskaryotic cervical smears. *British Medical Journal*, **301**, 641–4.

22. Robertson, J.H., Woodend, B.E., Crozier, E.H. and Hutchinson, J. (1988) Risk of cervical cancer associated with mild dyskaryosis. *British Medical Journal*, **297**, 18–21.

23. Sedlacek, T.V. (1992) Clinical options in dealing with minor cytologic abnormalities. *The Colposcopist*, **14**, 1–2.

24. Duncan, I.D. (1992) NHS cervical screening programme: guidelines for clinical practice and programme management. National Co-ordinating Network.

25. Shafi, M.I. (1994) Cytological surveillance avoids overtreatment. *British Medical Journal*, **309**, 590.

26. Giles, J.A., Deery, A., Crow, J. and Walker, P. (1989) The accuracy of repeat cytology in women with mildly dyskaryotic smears. *British Journal of Obstetrics and Gynaecology*, **96**, 1067–70.

27. McIndoe, A.J., Robson, M.S., Tidy, J.A., Mason, P. and Anderson, M.C. (1989) Laser excision rather than vaporization: the treatment of choice for cervical intraepithelial neoplasia. *Obstetrics and Gynecology*, **74**, 165–8.

28. Shafi., M.I., Luesley, D.M. and Jordan J.A. (1992) Mild cervical cytological abnormalities. *British Medical Journal*, **305**, 1040.

29. McIndoe, W.A., McLean, M.R., Jones, R.W. and Mullius, P.W. (1984) The invasive potential of carcinoma *in situ* of the cervix. *Obstetrics and Gynecology*, **64**, 451–8.

30. Jordan, J.A. (1981) Treatment of CIN by destruction – Introduction. In *Pre-Clinical Neoplasia of the Cervix. Proceedings of the Ninth Study Group of the Royal College of Obstetricians and Gynaecologists* (eds J.A. Jordan, F. Sharpe and A. Singer), RCOG, London.

31. Luesley, D.M. (1992) Advances in colposcopy and management of cervical intraepithelial neoplasia. *Current Opinion in Obstetrics and Gynecology*, **4**, 102–8.

32. Luesley, D.M. and Cullimore, J. (1989) The treatment of cervical intra-epithelial neoplasia. *Cancer Surveys*, **7**, 529–45.

33. Hammond, R.H. and Edmonds, D.K. (1990) Does treatment for cervical intra-epithelial neoplasia affect fertility and pregnancy. *British Medical Journal*, **301**, 1344–5.

34. Van Lent, M., Trimbos, J.B., Heintz, A.P.M. and Van Hall, E.V. (1983) Cryosurgical treatment of cervical intra-epithelial neoplasia (CIN III) in 102 patients. *Gynecologic Oncology*, **16**, 240–5.

35. Anderson, M.C. and Hartley, R.B. (1980) Cervical crypt involvement by intraepithelial neoplasia. *Obstetrics and Gynecology*, **55**, 546–50.

36. Chanen, W. and Hollyock, V.E. (1971) Colposcopy and electrocoagulation diathermy for cervical dysplasia and carcinoma *in situ*. *Obstetrics and Gynecology*, **37**, 623–8.

37. Chanen, W. and Rome, R.M. (1983) Electrocoagulation diathermy for cervical dysplasia and carcinoma *in situ*: a 15 year survey. *Obstetrics and Gynecology*, **61**, 673–9.

38. Benedet, J.L., Anderson, G.H. and Boyes, D.A. (1985) Colposcopic accuracy in the diagnosis of microinvasive and occult invasive carcinoma of the cervix. *Obstetrics and Gynecology*, **65**, 557–62.

39. Dorsey, J.H. and Diggs, E.S. (1979) Microsurgical conization of the cervix by carbon dioxide laser. *Obstetrics and Gynecology*, **54**, 565–70.

40. Larsson, G., Alm, P. and Grundsell, H. (1982) Laser conization versus cold knife conization. *Surgery, Gynecology and Obstetrics*, **54**, 59–61.

41. Partington, C.K., Turner, M.J., Soutter, W.P., Griffiths, M. and Krausz, T. (1989) Laser vaporization versus laser excision conization in the treatment of cervical intra-epithelial neoplasia. *Obstetrics and Gynecology*, **73**, 775–8.

42. Schink, J.C. (1989) Laser vapourization versus laser excision conization in the treatment of cervical intra-epithelial neoplasia. *Obstetrics and Gynaecology*, **74**, 681.

43. Tabor, A. and Berget, A.B. (1990) Cold knife and laser conization for cervical intra-epithelial neoplasia. *Obstetrics and Gynecology*, **76**, 633–5.

44. Carrier, R. (1984) *Practical Colposcopy*. Laboratoire Carrier, Paris.

45. Woodman, C.B.J., Jordan, J.A. and Wade-Evans, T. (1984) The management of vaginal intra-epithelial neoplasia after hysterectomy. *British Journal of Obstetrics and Gynaecology*, **91**, 707–11.

46. Ireland, D. and Monaghan, J.M. (1988) The management of the patient with abnormal vaginal cytology following hysterectomy. *British Journal of Obstetrics and Gynaecology*, **95**, 973–5.

47. Luesley, D.M., Shafi, M., Finn, C. and Buxton, E.J. (1992) Haemorrhagic morbidity after diathermy loop excision: effect of multiple pretreatment variables including time of treatment in relation to menstruation. *British Journal of Obstetrics and Gynaecology*, **99**, 82.

48. Luesley, D.M., McCrumm, A., Terry, P.B., *et al.* (1985) Complications of cone biopsy related to the dimensions of the cone and the influence of prior colposcopic assessment. *British Journal of Obstetrics and Gynaecology*, **92**, 158–64.

49. Bigrigg, A., Haffenden, D.K., Sheehan, A.L., Codling, B.W. and Read, M.D. (1994) Efficacy and safety of large loop excision of the transformation zone. *Lancet*, **343**.

50. Blomfield, P.I., Buxton, J., Dunn, J. and Luesley, D.M. (1993) Pregnancy outcome after large loop excision of the cervical transformation zone. *American Journal of Obstetrics and Gynecology*, **169**, 620–5.

51. Braet, P.G., Peel, J.M. and Fenton, D.W. (1994) A case controlled study of the outcome of pregnancy following loop diathermy excision of the transformation zone. *Journal of Obstetrics and Gynaecology*, **14**, 79–82.

52. Eristensen, J., Langhoff-Roos, J. and Kristensen, F.B. (1993) Increased risk of preterm birth in women with cervical conization. *Obstetrics and Gynaecology*, **8**, 1005–8.

53. Saunders, N., Fenton, D.W. and Soutter, W.P. (1986) A case controlled study of the outcome of pregnancy following laser vaporization of the cervix. In *Reproductive and Perinatal medicine IV. Gynecological Laser Surgery* (eds F. Sharp and J.A. Jordan). Perinatology Press, Ithaca, NY.

54. Shafi, M.I., Dunn, J.A., Buxton, E.J., Finn, C.B., Jordan, J.A. and Luesley, D.M. (1993) Abnormal cervical cytology following large loop excision of the transformation zone: a case controlled study. *British Journal of Obstetrics and Gynaecology*, **100**, 145–8.

55. Giles, J.A., Hudson, E., Crow, J., Williams, D.R. and Walker, P. (1988) Colposcopic assessment of the accuracy of cervical cytology screening. *British Medical Journal*, **296**, 1099–102.

56. Murdoch, J.B., Morgan, P.R., Lopes, A. and Monaghan, J.M. (1992) Histological incomplete excision of CIN after large loop excision of the transformation zone (LLETZ) merits careful follow-up not retreatment. *British Journal of Obstetrics and Gynaecology*, **99**, 990–3.

57. Paraskevaidis, E., Jandial, L., Mann, E., *et al.* (1991) Pattern of treatment failure following laser for cervical intraepithelial neoplasia: implications for follow-up. *Obstetrics and Gynecology*, **78**, 883–7.

58. Constantine, G., Williams, D.R. and Luesley, D.M. 1987 The management of post-menopausal women with abnormal cervical cytology. *Colposcopy and Gynecologic Laser Surgery*, **3**, 93–7.

59. Krebs, H. (1989) Treatment of vaginal intra-epithelial neoplasia with laser and topical 5-fluorouracil. *Obstetrics and Gynaecology*, **73**, 657–60.

12

THE PSYCHOSOCIAL IMPACT OF CERVICAL INTRA-EPITHELIAL NEOPLASIA AND ITS MANAGEMENT

T. Natasha Posner

INTRODUCTION

For some time it was assumed that medical intervention after an abnormal smear was without untoward effects. Development of the technology for out-patient treatment of a large proportion of cases of cervical intra-epithelial neoplasia (CIN) reduced the morbidity associated with treatment when cone biopsy operation was used more widely. Investigation and treatment went ahead on the assumption that it was better to intervene earlier than later, and intervention itself would do no harm. Since the mid-1980s onwards, however, evidence has accumulated from one report after another spelling out the previously unacknowledged, mostly psychosocial, 'side-effects' of this intervention. This evidence added to the critical assessment of the cervical screening programme as a whole which was taking place. In weighing up the total costs and benefits, any cost to the screening programme participants in terms of associated morbidity, whether psychological or physical, would clearly need to be taken into account.

Most medical procedures have their unwanted side-effects it could be argued, and since it is considered generally worthwhile to have a cervical screening programme in operation, why pay too much attention to them, when the rationale for the intervention is to save lives? Various investigators have provided some compelling reasons for looking at, attempting to understand and reducing the negative aspects of the experience of an abnormal Pap smear and subsequent investigation and treatment. Lerman *et al.* [1] argued that not to do so could jeopardize the success of the medical intervention:

> The potential of cervical screening to produce psychologic sequelae cannot be ignored because of the possible impact of such sequelae on patient compliance with subsequent screening and diagnostic follow-up.

McDonald and associates [2] thought it important to address women's fears about abnormal smears, both because they could influence their receptivity to medical procedures, and because of the possibility of a harmful effect on the development of the condition. They suggested that 'if these fears and concerns can be studied and addressed, health care providers will be better able to manage and treat CIN patients'. The rationale for the research reported on by Posner and Vessey [3] was that the impact of the abnormal

Cancer and Pre-cancer of the Cervix
Edited by D.M. Luesley and R. Barrasso. Published in 1998 by Chapman & Hall, London. ISBN 0 412 56600 1.

smear finding from the woman's viewpoint had not been fully investigated, and the possibility of unnecessary morbidity in terms of psychosocial, psychosexual or physical distress resulting from the medical intervention for CIN had not been assessed. These authors argued that if the benefits of screening are to be maximized, 'any unintended negative consequences need to be reduced as far as possible; they first need to be acknowledged and understood' [3, p. 12].

A further argument for paying attention to the unintended negative consequences of medical intervention resulting from cervical screening is an ethical one. The screening programme mostly finds cervical abnormalities in well women who are symptomless. In a significant proportion of cases of CIN, there is the possibility of spontaneous regression of the condition without any medical intervention. If medical intervention goes ahead in this situation, it is clearly imperative in ethical terms to do whatever is possible to minimize any potentially harmful effects. Otherwise, while the screening programme may be beneficial for the population as a whole, there will be individual cases in which more harm than good may come of the intervention.

This chapter will review the evidence of the psychosocial impact on women of an abnormal Pap smear finding and the subsequent medical intervention, and will discuss psychological factors in the development of CIN and possible management strategies for alleviating the negative aspects of the impact of this process.

PSYCHOSOCIAL ASPECTS OF INVESTIGATION AND TREATMENT

Reelick *et al.* [4], reporting in 1984 on research carried out in Rotterdam in 1977–8, were the first to report the conclusion that 'mass screening for cervical cancer ... caused psychological side-effects to occur among women with a positive smear'. The research had a quasi-experimental design, comparing women with a negative Pap smear result with those who

had to be rescreened and those with a positive result, by carrying out a pre-test and two post-test structured interviews using a mixture of closed and open-ended questions. Two psychological measures were used: a six-point semantic differential of 10 contrast-pairs designed to indicate women's feelings before and after they were told the result of the screening test, and a four-point Likert scale asking their opinions about issues assumed to be of importance to them.

Most of the 99 women interviewed before and after getting a positive smear result reported that their first reaction was one of fright. The scores on the semantic differential indicated that women with a positive smear result became 'moodier' and felt more ill on the first post-test, whereas women with a negative smear (350) became more relaxed. A comparison of the semantic differential scores of women treated in hospital for a cervical abnormality indicated two significant shifts: after the operation and hospitalization, the women were both more relaxed and more cheerful than before, but they felt more ill than before the result of the test was known to them. Some women were treated by their GP (32) or with an out-patient procedure (23). In this study, it was only among women who had an operation requiring hospitalization (59) that there were lasting 'problems of a psychological or somatic nature' associated with the treatment (about 35%). Two variables were found to be significantly related to having an 'unfavourable reaction' towards a positive smear: worrying about the result of the screening test (before the result was known) and 'having a higher Pap classification', i.e. a more serious condition. Three variables were found to be related to an unfavourable reaction towards the operation and hospitalization: an unfavourable reaction to the positive smear, an attitude that the cervix and menstruation is important for feeling oneself to be a woman, and an attitude that the ability to become pregnant is important for feeling oneself to be a woman. The researchers concluded that:

Mass screening for cervical cancer in Rotterdam caused psychological side-effects to occur among women with a positive smear. However, for the majority of women with a positive smear they were not of a lasting or serious nature.

The paper by Beresford and Gervaize in 1986 [5] was the first to report a study of the 'emotional impact' of abnormal Pap smears. Fifty women were interviewed after colposcopy investigation in an Ottawa clinic to determine whether or not there were particular emotional concerns that accompany the diagnosis of CIN. A series of open-ended questions invited interviewees to describe their thoughts and experiences. They were subsequently asked to rate the intensity level of expressed concerns on a 10-point scale, and to suggest interventions which could have helped to reduce their anxiety.

The study results 'clearly indicated that women with abnormal Pap smears have identifiable and considerable fear'. Four major areas of concern were identified: these were fear of cancer, mentioned by 100%, fear of loss of reproductive and or sexual function (68%), fear of medical procedures (65%), and a concern about 'loss of bodily integrity', which was often expressed by interviewees as 'bodily betrayal' (62%). Over two-thirds of the interviewees rated their fears in the first two areas as severe (70% and 65%, respectively). Smaller proportions rated their fears in the other two areas as severe (44% and 32%, respectively); however, 52% rated their fear of loss of bodily integrity as moderate, and the researchers commented that this resulted from 'the first realization in young, typically healthy women that their bodies might not be functionally perfect and under their direct control'. The researchers also recorded 'behavioural symptoms of distress', which were described by the colposcopy clinic patients. Over half (52%) mentioned sleep disturbances, 44% irritability, 30% crying episodes, 26% outbursts of anger, 26% difficulties in interpersonal relationships and 22% difficulties in sexual relationships.

They found no relationship between 'the degree of difficulty experienced' and the level of disease (28% were found to be 'negative', 12% had CIN I, and 60% had CIN II or III), the woman's age, parity or marital status. Results similar to these in terms of areas of concern and indications of distress were to be repeated in many subsequent studies.

Early in 1988, the results of a study designed to assess the psychosexual effects of the diagnosis and treatment of CIN were published [6]. The research was conducted in colposcopy, laser and genitourinary clinics in London, and compared three main groups of patients, aged between 17 and 26 years of age. Group 1 consisted of 30 women referred to a colposcopy clinic. Group 2 comprised 50 women called to a genitourinary medicine clinic because their regular partner had been found to have evidence of human papilloma virus (HPV) infection. This group was divided into group (a) the 52% who had cervical HPV infection with or without CIN, and group (b) who had no evidence of cervical abnormalities. Group 3 was made up of 25 women who had been referred to a genitourinary medicine clinic as the regular sexual partner of a male with non-specific urethritis. Structured questionnaires were used at the initial clinic visit and at a review appointment some months after any treatment necessary had been carried out. The questionnaires asked about six aspects of sexual behaviour and response in the previous 6 months for the initial questionnaire, at 5–6 months for the second administration: frequency of spontaneous sexual interest, frequency of intercourse, frequency of adequate vaginal lubrication and sexual arousal with intercourse, frequency of orgasm with intercourse, frequency of dyspareunia, and frequency of negative feelings towards intercourse. The women were shown a scale relating to each question and asked to choose a point on the scale descriptive of their response.

There was no indication that the initial responses of the four groups were different.

The results from the second administration of the questionnaire, however, indicated a dampening of sexual interest in groups 1 and 2(a) after diagnosis and treatment of cervical pre-invasive atypia, the scale values being significantly depressed by comparison with initial values ($p < 0.001$ for each group). Diagnosis and treatment of pre-invasive cervical atypia was also associated with significant decrease from the first to the second administration of the questionnaire in the variables intended to be measures of sexual function and response, frequency of intercourse, adequacy of vaginal lubrication and sexual arousal and frequency of orgasm. There was also a tendency to increased discomfort with intercourse for groups 1 and 2(a), and a significant increase in the frequency of negative feelings towards sexual intercourse ($p < 0.001$ for each group). There were no significant differences in the before and after measurements on these variables for groups 2(b) and 3. The responses of groups 1 and 2(a), and 2(b) and 3 were found to be similar throughout and the differences between these two groupings highly significant.

The study authors, Campion *et al.* [6] concluded that it had demonstrated a significant alteration in sexual attitudes, behaviour and response in the women after diagnosis and treatment of pre-invasive cervical atypia (mostly CIN), which had important social implications pointing to 'a need for supportive and informative counselling'. They suggested that the findings of decreased spontaneous sexual interest and increased negative feelings towards sexual intercourse with a regular partner were particularly significant. In discussing the reasons for this apparent trauma, they focused on the negative feelings associated with anxiety and hostility, and the tendency towards discomfort associated with intercourse after treatment with CO_2 laser, which appeared to be related to the anxiety and diminished sexual response. Given that the nature of the treatment was 'conservative' in comparison with major surgery for gynaecological malignancy with its accompanying genital change and readily understood psychosexual sequelae, they felt the altered attitudes towards sexuality and sexual behaviour were not fully explained.

A comprehensive investigation of the impact on women of the medical process following an abnormal Pap smear finding and referral for colposcopy was carried out in 1983–1984 by the present author [3]. This study followed a series of 153 women in two different hospitals in England through the course of their investigation and treatment. The women were interviewed at least twice, using a semi-structured in-depth interview, focusing on their experiences and conceptualizations. Altogether 359 interviews were conducted. The first interview took place at the first visit to the colposcopy clinic, the last interview was conducted after any treatment needed had been carried out and reviewed, 6–9 months after the first interview, mostly in the patient's home. The ages of the women in the study ranged from under 25 years (16%) to 45 years and over (7%). Thirty-nine per cent of the sample were nulliparous; the educational and marital status, and social class of the sample covered the whole range. Thirteen per cent had no treatment to the cervix after the colposcopy examination; the rest of the sample, who had varying degrees of CIN (all except five or six who had invasive carcinoma) was roughly divided between those who had out-patient treatment (27% cryocautery, 14% laser therapy), and those who had in-patient surgery (35% cone biopsy, 8% hysterectomy). The data were analysed quantitively and qualitatively and produced a description and analysis of the experiences of the women, published in book form in 1988 [3].

The study found that news of the abnormality on the smear and need for further investigation was often a considerable shock, plunging the woman into a period of uncertainty about its implications. When they were asked how they felt at this time, interviewees described feelings of shock, horror, alarm, extreme anxiety, anger, helplessness and some disbelief: 65%

said they were 'worried', 'alarmed', 'concerned', 'anxious' or 'nervous'; 27% described themselves as being 'shocked', 'stunned', 'surprised', 'devastated' or 'very upset'. The feelings of alarm, anxiety and horror related to beliefs that the abnormal smear implied cancer, necessitating a hysterectomy, an end to childbearing, and possible death. There was also much fear of the unknown, particularly in regard to the colposcopy examination.

Mapping out this anxiety, the time which was most often cited as the time of greatest anxiety (by 46%) of the women was prior to the colposcopy examination; for a third (33%), however, the most anxious time was when they were waiting to have treatment. Asked for how long the thought of the abnormal cells was 'on their mind a lot of the time', nearly a quarter (24%) replied until they had the colposcopy examination, just over a fifth (22%) said until they had treatment, for 9% it was all the time, but 21% said it was not on their mind at all. Many women came for the colposcopy examination not knowing what to expect and fearing the worst about their condition. Over half the sample (52%) felt better after colposcopy, reassured, relieved and more relaxed, but 19% felt worse at the prospect of treatment or in reaction to the stress of the examination.

The findings of this study suggested that the sensitivity of the cervix and the stressfulness of out-patient treatment procedures had previously been underestimated. Among the patients having cryocautery or laser therapy, 42% experienced pain or other symptoms during treatment that they described as severe; post-treatment, 30% experienced pain and 26% described feelings of trauma, depression, vulnerability and/or a sense of violation for a period of time. Nulliparous women were significantly more likely to experience severe symptoms than parous women ($p < 0.01$).

In the last interview (6 months after treatment), respondents were asked what had been their worst fear initially. The answers reflected what they said about their feelings when they first heard there was something wrong. The dominating fear was of cancer and its assumed consequences, hysterectomy, infertility and possible death. The women's images of cancer were investigated at this point, by asking them how they would describe it as a disease. Most of the descriptions were in extremely negative and often metaphorical language, equating the disease with destruction of the body and a horrible death (examples of such descriptions were: 'Imagine a black fungus, creeping, mouldy ...'; 'A very painful, lingering, very nasty, dreadful death ... worst thing you could possibly have'; 'Just a terrible disease. It just seems to eat people away – to eat at the body' [3, p. 56]); only a small minority talking in technical terms, or distinguishing between different types of cancer. At the same time, however, these women thought that cervical cancer was largely curable: no one thought that it was incurable, over 80% saying it was curable if caught early, a further 11% that it was usually curable. Although there was knowledge and optimism about a specific cancer relevant to study participants, it appeared that their image of the disease was, for the most part, dominated by the prevailing cultural image of cancer. There was thus an ambivalent response to a question about the likelihood of recurrence – 63% could not be confident that recurrence was unlikely; and to a question that asked whether they still felt 'threatened by the possibility of cancer', 45% replied negatively and 47% positively.

Women in this sample were asked whether knowing about the abnormal smear had made them feel differently about their bodies. About 45% said it had not. Among those who did feel differently, the most common feelings were of being detached from, or out of control of the body, and defilement expressed as feeling 'unclean' or 'dirty'. This research also attempted to assess the impact of the abnormal smear finding and subsequent medical process on interviewees' sexual relations. In the first interview women were asked if the knowing that they had an abnormal smear finding had changed how they felt about sexual relations, and 42% replied that it had. In the last interview women were asked if the abnormal smear finding had in any way disturbed their sexual relations (in addition to the period of post-treatment abstention from intercourse): 43% replied that it had done so. This disturbance continued for a proportion of women for at least 6 months after colposcopy and treatment; when asked if their sex life was 'back to normal', 14% said that it was not. The upset appeared to be related to resentment or guilt about the sexually transmitted nature of the condition, to the changes in body image involving a sense of contamination, defilement or vulnerability, and to the unavoidable invasion of 'private space' involved in the medical intervention.

There were often feelings of embarrassment and sometimes guilt about their condition among the women interviewed. Some of the embarrassment was due to the *implied* guilt – the stigma associated with having an abnormal smear. This stigma resulted from the public image of the condition as presented in the media at that time causing the women to feel 'tainted' with a 'spoiled identity' [7]. Associated with this, some women felt resentful about, and victimized by, the stereotype of the promiscuous woman who brought the condition upon herself by her sexual behaviour. Feelings of 'why me' were common and led to a search for an explanation of its development, often couched in terms of other gynaecological conditions experienced, such as childbirth or infections, or associated factors such as contraception. For most women, however, the abnormal smear finding came 'out of the blue'. There was some awareness of the contraceptive pill and smoking as relevant factors, very much less knowledge of the protective effect of barrier methods of contraception, and almost no mention of the role of the male partner's previous sexual behaviour.

The majority of women in this sample (73%) felt that they were 'back to normal in all respects' 6–9 months after the medical inter-vention, and that they had got over the episode (80% felt this), but that they would not readily forget the stress and anxiety it caused them at the time. One of the main conclusions of the study was that there was a large gap between the medical and lay interpretations of CIN: what is seen by the doctor as minor and requiring a preventive procedure may be experienced by the patient as traumatic and life-threatening. Most of the women in the study were well and in the prime of life. The positive smear, experienced as the threat of cancer, faced them with evidence of their body's imperfection and thoughts of their mortality. It was an existential jolt. Posner and Vessey [10] concluded that:

Whenever cancer is a possible threat, lack of information will allow the patient to assume the worst. In the absence of a fuller understanding, the framework of lay interpretation is the curative rather than the preventive medical mode, and the meaning attached to the condition is strongly influenced by the cultural image of cancer which presents the disease as inevitably destructive. The merest hint of cancer can be overlaid with the same negative metaphorical image as the fully developed disease and this may affect the way a woman with abnormal cervical cells thinks about her body, her health status and her future prospects ... Healing requires attention to the personal and social meaning of medical intervention as much as to the physical state of the cervix.

In the University of Tennessee Medical Centre, McDonald and colleagues [2] investigated the impact of CIN diagnosis and treatment on self-esteem and body image. They gave colposcopy clinic patients a questionnaire to complete on four different occasions during the process of investigation and treatment for varying degrees of CIN. They were able to report on the data for 20 patients only, demonstrating variations in levels of self-esteem and various aspects of body image using an adjective generation technique and scores across three categories. The mean score for self-esteem was lowest at the time of surgery and highest at the postsurgery visit, and significantly higher then than on the first visit. Positive body image was greatest when the results of the colposcopic biopsy were received and at the postsurgery visit. Patients were asked to rank 'potential concerns' presented to them at each visit. Concern about cancer overrode all other concerns except at the postsurgery visit, when a slightly higher proportion of patients had anxieties about loss of attractiveness than expressed concerns about the recurrence of cancer. 'Loss of sexual functioning' was also high on the list of patients' concerns at each visit. The researchers felt that health care providers' awareness of the impact of the process on self-esteem and body image could help to provide more 'holistic care' to patients.

The following year (1990), there were reports of further studies. In the UK, the *British Journal of Obstetrics and Gynaecology* carried a report of a study designed to measure levels of distress in women having a colposcopy examination after an abnormal cervical smear, and to examine factors contributing to the distress [8]. Thirty women were asked to complete a questionnaire before, and 28 days after, their first colposcopy examination. General levels of anxiety were assessed using the Spielberger State–Trait Anxiety Inventory [9]; specific anxieties were assessed using a seven-point rating scale.

This study demonstrated extremely high levels of anxiety in this group of women, with a mean score on the Spielberger Anxiety Inventory of 51.2. This score is similar to anxiety levels in women following abnormal results on Alpha-Foeto Protein screening for foetal abnormality [10], and higher than anxiety levels in women the night before surgery; (the norm for adult women is 35). The anxiety was as much about undergoing the colposcopy as it was about what might be found wrong on examination. However, the general level of anxiety was significantly associated with anxiety about the colposcopy examination, and women who were uncertain about what would happen had higher mean anxiety scores. The three most frequently mentioned concerns about colposcopy were that it would be painful, that it would be uncomfortable and uncertainty over what would happen during the procedure. No relationship between anxiety scores and the seriousness of the referred problem, the perceived seriousness of the problem, parity or the time since referral, was found. The women were significantly less anxious overall after the consultation than beforehand. These researchers concluded that the distress experienced was 'more strongly related to anticipation of the procedure than the outcome'.

An article by Quilliam in the *Health Education Journal* [11] summarized the emotional issues relating to a positive smear finding, which had been most commonly identified by women she had interviewed at length for material for her book based on her own experience [12]. Self-blame and feelings of guilt about their own sexual history were common. Many women made links between their current condition and past gynaecological or sexual trauma such as abortion or rape, and this obviously added greatly to their distress. Fears about infertility and death were often acknowledged. Quilliam also found anxiety coming from feeling dependent on health professionals for survival, and guilty feelings about expressing fears and emotions, lest it be seen as 'making a fuss' about the situation. In her assessment of why reactions to a positive smear finding could be so strong, Quilliam suggested that 'mixed messages' concerning the meaning of a positive smear result and equally muddled and sometimes judgmental messages about the causes, were part of the reason. The emotional upset of episodes relating to cancer or sexual trauma in a woman's own history could be retriggered by a positive smear result, she suggested. Furthermore, the current situation could provoke family or partnership upset and conflict because of the implied threat of cancer, or the implications of sexually transmitted disease, which might result in contraceptive changes or the investigation of a male partner.

A study in the USA by Lauver and Rubin [13] attempted to document the 'breadth of women's responses to abnormal Pap results' among clients attending hospital clinics providing gynaecology and family planning services to a largely socioeconomically disadvantaged and black population. The researchers initially contacted the participants (118 women, 94% black) by telephone and informed them of their abnormal Pap smear results (indicating CIN or HPV) and the need for colposcopic investigation. They were then asked if they had any questions and these questions were recorded as an index of women's concerns. Answers were given to their questions and a follow-up appointment made. Subsequently, at the time of colposcopy, subjects were interviewed (107 women) about their reactions to the abnormal Pap results, using an open-ended interview question as a second index of the women's concerns. It is not clear whether these interviews took place before or after the colposcopy examination.

The most commonly asked questions reflected concerns about the further evaluation (36%) or about the causes of the abnormality (33%). Not surprisingly, however, 39% of the study participants asked no questions at this point, having just been told of the abnormal results. At the time of the colposcopy examination, the most common reaction to the abnormal Pap result reported was concern about the seriousness and future implications of the abnormality (54%). Forty-three per cent reported feeling nervous, fearful or upset; and 36% said they were uncertain about the meaning of the abnormal Pap test. The researchers noted some discrepancy between their findings and those in earlier research they cited [5, 6] in regard to the amount of fear and worry expressed. They acknowledged that their results might not be applicable to other groups of women with abnormal Pap results.

An investigation carried out in Stockport, Manchester [14] aimed to assess levels of dissatisfaction with information received, and levels of knowledge about the colposcopy procedure, as well as the state anxiety (again using the Spielberger State–Trait Anxiety Inventory [9]) experienced by women undergoing colposcopy in a hospital out-patient clinic. Thirty women were interviewed prior to the colposcopy examination.

The study findings revealed significant levels of dissatisfaction with the information that had been available to the women. All but two of those interviewed thought it would have been helpful to have received a leaflet explaining colposcopy before they came to the clinic. Just over half the sample felt they did not know enough about colposcopy and why they were having the examination, nor did they have sufficient understanding of the relevant anatomy or the procedure; however, nearly everyone considered the procedure 'very important'. The researchers thus concluded there was good evidence that

> women undergoing colposcopy may not have an objectively high level of knowledge about the procedure, despite considering it important, and that they recognise they are lacking in information.

The mean score on the Spielberger Anxiety Inventory for the sample was 47.1 (SD 12.8), indicating a high level of state anxiety in these women at the time immediately preceding colposcopy.

Further evaluation of the psychological effects associated with receiving a positive Pap test result was conducted by Lerman and colleagues [1] among a hospital family planning and colposcopy clinic population drawn predominantly from the lower-income community of North Philadelphia. The investigation compared 106 women with normal Pap test results with 118 women who were referred for colposcopy examination to follow-up a positive Pap smear result. A brief structured interview administered over the telephone about 3 months after the women had received their Pap test results was designed to have respondents rate themselves on the following variables: current frequency of worry about cervical cancer, current impairment in daily activities as a result of cervical cancer worries, tension and mood in the past month, sexual interest and sleep patterns.

This research found 'substantial and generalized negative psychologic consequences of a positive Pap test result'. Thirty per cent of the women with positive results said that they often worried about cervical cancer, as compared with 14% of women with negative results. For 25% of women with positive results, this worry resulted in some impairment of their daily activities, as compared with 11% of women with negative results. Fifty per cent of women with positive results reported impairment in sexual interest, 40% had sleep disturbance and 24% described their mood as 'very bad or mostly bad', as compared with 31, 26 and 6%, respectively, among those with negative results. The researchers found these differences were significant, and were independent of all demographic and sexual screening history confounding variables examined. However, when the positive smear group was divided into those who had complied with colposcopy and those who had not, it was found that the effects depended on the women's follow-up status. Women who had undergone colposcopy did not have increased worry, disturbance of mood or sexual interest when compared with women who had negative results.

The researchers suggested that failure to attend for colposcopy examination might 'create a vicious cycle in which failure to comply sustains uncertainty about disease and accompanying distress which in turn contributes to delay in follow-up'. They recognized that it was not possible to determine from their study assessing distress at one point in time, 3 months after notification of results, whether psychological distress among non-compliers resulted from the continuing uncertainty about their diagnoses or whether excessive psychological distress had been a contributory factor in the non-compliance. They also recog-

nized that as their study population was largely young, black and less well-educated, their findings might not apply equally to the larger cervical screening population.

Greimel and colleagues in Graz carried out an investigation of 'the subjective reactions and mental state' (researchers' translation) of 50 patients with CIN [15]. They found that patients with CIN reacted to the diagnosis in a similar manner to patients confronted with the diagnosis of cancer, and that the subjective experience of the condition was largely independent of its objective severity. The possibility that CIN can develop into cancer produces 'an existential threat' (present author's translation). These researchers suggest that illness behaviour was determined more by social factors than medical, irrespective of personality factors. 'Effective coping' (researchers' translation) with the condition depended on adequate medical information being received by the women. Patients who were satisfied with the medical information they received had a more positive attitude towards the condition and found it less life-threatening than those who were dissatisfied with the information received. The degree of empathy between doctor and patient significantly affected the experience of the illness episode, and the quality of the explanation given was of crucial importance in reducing anxiety about the future course of the condition and in minimizing negative changes in subjective quality of life.

Palmer and colleagues saw their research [16] as building on the findings of Posner and Vessey [3], using psychological measures of the degree of trauma experienced and a control group to investigate the impact of diagnosis as opposed to treatment. Their study design compared two groups of 20 subjects each. The control group had received a negative smear result, and the other had been diagnosed with CIN in the colposcopy clinic. The group with CIN were also compared after diagnosis and after treatment. The subjects were visited at home and asked to complete a questionnaire using an impact of events scale [17], the Spielberger *et al.* State–Trait Anger Scale [18], and the Walston and Walston Multi-dimensional Health Locus of Control Scale [19]. The women diagnosed with CIN were found to have significantly higher scores than those with a negative smear result for intrusive thoughts, avoidance and state anger (p <0.001); there was no difference in the trait anger scores between the two groups. On the assumption that intrusive thoughts and avoidance occur when an individual is processing information associated with a traumatic event [20, 21], the researchers concluded that this result supported the hypothesis that the diagnosis of CIN was a traumatic event. There was no difference between the two groups in their health locus of control belief scores. To investigate the effect of treatment, the group with CIN completed the questionnaire again after treatment. There were no significant differences in scores of intrusive thoughts, avoidance, total impact of events or state anger, and thus no evidence that the treatment had an additional traumatic effect over and above diagnosis.

At the second visit to the treated group, they also answered interview questions about their understanding and experience of their condition, and beliefs about its causes. These questions covered some of the same ground as those in the interviews in the study by Posner and Vessey [3], and the findings were along similar lines to those of the earlier study. Asked about their feelings about their bodies, women in this study also spoke about feelings of defilement (43%), and detachment or being out of control (36%). Sixty per cent reported a change in their feelings about sexual activity related to feelings of contamination (50%) or fear of discomfort (10%). The subjects in this study were more aware of the causal role of HPV and of smoking (42 and 10%, respectively, mentioned these factors), but were similarly unaware, it seemed, of the role of male promiscuity. Asked about their experience of treatment, the researchers felt that the responses of 45% implied a need for more anticipatory information.

A study carried out in the USA [22] was designed to investigate the relationship between relevant health beliefs, health locus of control (using the Multi-dimensional Health Locus of Control Scale [19]), and diagnosis of HPV and compliance with medical regimens for an abnormal Pap test. Two hundred and seventy-two women were surveyed, of whom 11% failed to comply with medical recommendations for follow-up. The study found there were no statistically significant relationships between compliance and locus of control, HPV diagnosis or demographic variables. There were also no significant relationships with the three subscales from the health belief scale used; however, there were statistically significant relationships with two individual items: 'the uncertainty about my Pap test makes me nervous', and 'I have not been able to cope with my abnormal Pap test'. These researchers concluded that women who were nervous about receiving an abnormal Pap test were more likely to comply with medical recommendations than women who were not nervous about the test. Along with the finding that women who indicated that they could not cope with their abnormal Pap test were much less likely to comply with medical recommendations, this has implications for counselling situations since, as Funke and Nicholson, concluded, 'it appears there is an effective medium between not being concerned enough to act and being so fearful that they take no action at all'.

Investigators in a genitourinary medicine clinic in London [23], found that women who had previously undergone laser therapy and were subsequently referred again for colposcopic investigation of a cervical abnormality were at increased risk of psychological morbidity. This study demonstrated higher levels of social dysfunction, anxiety and somatic symptoms (using the GHQ-28 [24]) in this group of women and suggested that psychological evaluation and counselling should be offered.

In a recent study [25] of the emotional reactions of women attending a colposcopy clinic in Oxford, Gath *et al.* assessed a series of 102 new attenders 4 weeks before, and then again 4 and 32 weeks after their first colposcopy clinic appointment. The aim of the study was to assess the frequency, nature, severity and duration of emotional symptoms in women and to elicit their views. A battery of measures was used in this assessment: Present State Examination (PSE) (9th edition) [26], General Health Questionnaire (GHQ) [24], Beck depression inventory [27], Leeds depression and anxiety scale [28], Spielberger State–Trait Anxiety Inventory [9], a social adjustment scale [29], Eysenck's Personality Inventory [30], and a semi-structured interview about gynaecological health and psychosexual functioning and women's experience of attending the clinic.

It was found that the PSE mean score was significantly lower at the last assessment than at the first, and there was a statistically significant decrease in the frequency of psychiatric syndromes. Compared with a community sample, the colposcopy clinic patients at the first assessment were significantly more likely to be suffering from situational anxiety, tension, impaired concentration, and somatic features of depression.

At the first assessment, most women reported that the emotional impact of the abnormal smear had been severe in the first week after notification, with just over half the sample (52 women, 51%) reporting shock, panic and horror. These women were subsequently referred to in the report of the study as the 'shock subgroup'. When compared with the other 50 patients in the sample, they had significantly higher mean total scores on most of the measures at assessment 1 and some significant differences at later assessments, although these differences diminished over time. At the first assessment 90% of the whole sample expressed fear and worry, 67% reported depressed mood, 65% pessimism, and smaller percentages reported other indicators of stress such as irritability, sleep disturbance, headaches.

Asked about their experience of colposcopy at least half the sample (51%) reported pain, embarrassment, shock or distress. There was some evidence of an impact on psychosexual functioning, with about one-third of women with a sexual partner reporting that their sexual functioning was impaired after they had been informed about the abnormal smear. At the last assessment, however, only 4% of those with a sexual partner reported that their sexual relationship had deteriorated.

The authors reported that in the sample as a whole, psychiatric morbidity was found to be 'transient and relatively minor' and concluded that 'after an abnormal cervical smear, further investigation by colposcopy is generally associated with low levels of anxiety and depression'. This conclusion appears at first reading to be somewhat at variance with other studies (particularly in relation to anxiety levels), but it is clearly a summative assessment of the patients in the study using psychiatric case criteria. The findings of this investigation, both in terms of the initial reaction to the abnormal result and referral for investigation, the nature of that reaction, the experience of the colposcopy examination, the reduction of anxiety and distress after treatment, and the desire for more initial information, were very much in line with other reported studies. There was evidence of considerable distress in the initial stages, with over half the patients in the study forming a 'shock subgroup'. It seems unlikely that these women would consider the anxiety and upset involved in the episode 'relatively minor'. It may be that in trying to standardize the measurement of distress using psychiatric tools, we are in danger of pathologizing an understandable human reaction to a perceived threat and missing the existential point from the viewpoint of the patients. The study authors made the helpful suggestion that GPs could offer supportive counselling to prevent persistent anxiety and depression and if necessary further support could be provided by a nurse counsellor at the colposcopy clinic. Such intervention might also prevent transient but unnecessary anxiety and distress.

Another recently reported study [31] carried out in Aberdeen linked anxiety with indicators of 'social adjustment' and compared the psychological responses of women waiting for colposcopy examination with women being kept under surveillance and with a control group of women who had had recent negative cytology results. The investigation again demonstrated that anxiety levels (as measured by the Hospital Anxiety and Depression Scale [32]) were significantly higher in women with abnormal smears than women with negative results. Those waiting for colposcopy examination showed the highest mean anxiety, which declined significantly following the colposcopy clinic visit; however, one in five women remained very anxious. Raised levels of anxiety were found to be associated with overall 'social maladjustment' and with reports of negative feelings about the self, feeling less attractive, tarnished, let down by their bodies and with psychosexual problems. Problems of 'social adjustment' (work in the home, social lives, relationships with partners, children and families), were overall more evident in the surveillance group. The investigators concluded that a positive Pap smear may be psychologically traumatic for a significant minority of women irrespective of management strategy. If this finding is repeated in other investigations, it has implications for the assessment of the relative costs of immediate intervention versus surveillance for CIN I results on the Pap smear.

A logical development of the psychological measurement of the distress reported in the literature reviewed above was the development of a standardized specific measurement tool, which could be used to compare the distress of patients in one clinical situation with those in another, and to evaluate the efficacy of intervention programs to alleviate the distress. Doherty and colleagues in London

reported in 1991 [33] on their development, the 'Abnormal Smears Questionnaire' (ASQ), a standardized questionnaire to assess feelings and cognitions relating to an abnormal smear result and subsequent medical procedures. Alongside the Spielberger State–Trait Anxiety Inventory [9] and the GHQ [24], it was used on four categories of patients (pre- and post-colposcopy and pre- and post-laser treatment), 80 patients altogether. These investigators demonstrated concurrent internal consistency and validity for the questionnaire, and reported that the ASQ is 'a reliable, valid and conveniently brief questionnaire for measuring both the affective and cognitive dimensions of distress relating to an abnormal smear result'.

This study found the cognitions associated with distress were related to causality, outcome of treatment, perceived severity and possible consequences of the condition, and, to a lesser extent, concern about the sexually transmitted nature of the disease. Analysis of the ASQ affective data indicated that the abnormal smear result had some degree of negative emotional effect on, and the medical procedures caused some distress to, almost all the women (97%). The abnormal smear result was a source of 'severe distress' for a small minority (8%), and for the most part the distress was reduced in the post-treatment group. These investigators suggested that an intervention including both information giving and counselling based on cognitive techniques would be beneficial, and noted that there had been no published studies evaluating the effectiveness of a combined information and counselling service in reducing distress.

Another standardized measure was developed by Bennetts and colleagues in New South Wales [34]. The questionnaire, which they called PEAPS-Q (Psychosocial Effects of Abnormal Pap Smears Questionnaire), was intended to measure distress associated with the notification of an abnormal Pap smear result and subsequent management, enabling the dimensions to be examined separately.

Focus group discussions identified five major dimensions of distress, which were then translated into 200 potential questionnaire items (with the deletion of any item that could not be formatted in terms of a five-point Likert scale), which became the 41 core items forming the basis of the PEAPS-Q. This questionnaire was then used in 1993 to measure distress in two groups of women attending a New South Wales family planning clinic: a group of 93 women with abnormal Pap test results and a group of 257 women being followed-up after treatment for the first or second time. A four-factor, 14-item exploratory model of psychosocial distress was tested on the data from the three groups. Concurrent validity, internal consistency and high repeatability was demonstrated for the questionnaire and it was suggested that PEAPS-Q was a reliable and valid measure of the psychosocial impact of management of a cervical abnormality. Acknowledging an overlap with the ASQ [33], these authors claimed that the psychometric properties of the PEAPS-Q had been more rigorously scrutinized, thus providing researchers 'with a valid method for quantification of distress, both generally and within dimensions'. They further commented that the distress experienced by women is multi-dimensional and different at the various stages, aspects which had indeed been amply demonstrated by the range of earlier studies, but provide a challenge in terms of making a standardized measure comprehensive enough and appropriate for a particular stage of the process.

SUMMARY

In the past decade, there has been increasing recognition and understanding of the psychosocial aspects of women's experience of the investigation and treatment of abnormal cervical shears. This experience has been given various labels: 'psychological side-effects' [4], 'emotional impact' resulting in 'behavioural symptoms of distress' [5], 'unin-

tended negative effects' of 'the impact of the medical process' [3], 'the non-medical, emotional and psychosocial adverse effects' [35], 'substantial and generalized negative psychological consequences' [1]. There is a large measure of agreement about the main features of this experience, though when attempts have been made to measure the extent of the distress it has varied between different patient populations. In descriptions and assessments of the degree of upset caused, it has perhaps not been emphasized enough how varied people's reactions can be to the potentially stressful and possibly distressing process following a referral for an abnormal smear. There are women who do not experience distress in any form, and for whom the process is not traumatic, though they might experience some passing discomfort, and for a time, be somewhat concerned about the outcome.

What has been described in the literature is a frequent reaction of shock, alarm, fear and anxiety to the news of an abnormal Pap smear requiring further investigation, which amounts to a traumatic experience [3, 5, 8, 14, 16, 25, 36]. The anxiety is about cancer and its possible implications, loss of reproductive function, loss of sexual function and attractiveness, bodily 'betrayal', the medical process itself, particularly colposcopy and treatment, and possible future recurrence of the condition. Although the investigation and treatment of the condition may be relatively unproblematic in medical terms, colposcopy is very often a stressful medical investigation for patients [3, 8], and out-patient treatment can be experienced as distressing and may have side-effects [3]. The episode may result in adverse changes in the woman's body image [2, 3, 5, 31], psychosexual upset [1, 3, 6, 12], lowered self-esteem [2] and feelings of embarrassment, guilt [3, 9, 12], or anger [16] about the condition, and these effects may impact on personal relationships. Several researchers commented on the finding that the severity of a woman's experiential distress was not determined by the objective seriousness of her condition [3, 8, 15]. Many investigators have suggested that the provision of increased amounts of information at an earlier stage of the process and supportive counselling in some form would be beneficial in reducing anxiety and distress, and several research studies (reviewed below) have demonstrated such benefits.

The recognition, identification, description and analysis of different components of women's experience has been followed by measurement of distress using standardized tools and then the development of measurement instruments specific to the situation. In the attempt to quantify this distress, however, it is important to bear in mind that it relates to a process taking place over a period of time. The meaning of this process, from the patient's point of view, can change rapidly, along with accompanying emotions, when more information is received or a certain point is passed. The fact that acute distress is transitory for most women, does not lessen its existential significance. The episode is not **experienced** as routine and is still something which has to be got through.

PSYCHOLOGICAL FACTORS IN THE DEVELOPMENT OF CIN

Both psychological and psychophysiological factors are now thought to play important roles in disease processes. There is considerable evidence, as Lambley [37] has documented, from research and clinical studies of other sexually transmitted and viral-borne disease to suggest the need for and effectiveness of an approach to the treatment of cervical cancer and its precursors, which takes account of the role of such factors in the promotion of the condition.

The work of Schmale and Iker in the 1960s and 1970s investigated the association between negative psychological factors, particularly feelings of hopelessness, and the progression of CIN to carcinoma of the cervix, ending one report [38] with the conclusion:

The psychological state of the organism represented by the experiencing and reporting of hopelessness may in some as yet undetermined way provide a permissive atmosphere or a facilitating role which allows those who are already biologically predisposed to cancer to develop the clinical manifestations of the disease.

In 1971 Schmale and Iker published the results of a semi-prospective study [39] amongst women awaiting the outcome of abnormal Pap smears using video-taped interviews and a self-report questionnaire. Neither the researchers nor the women knew the results of the smear tests at the time of the study, but the researchers were able to predict whether or not the subject had a 'cancerous' lesion in nearly 75% of cases on the basis of an assessment of the degree of 'hopelessness' expressed by the subject in the interview and questionnaire.

The work of Antoni and Goodkin has developed this understanding further. In 1986 they reported on a study of stress and hopelessness in the promotion of CIN to invasive disease [40]. This work was followed by two papers describing 'host moderator variables' in the promotion of CIN: one relating to personality facets and the other to dimensions of life stress. The contribution of life stress to the development of different types of cancer has been widely investigated, and the psychosocial literature generally supports a relationship between stressful experiences and neoplastic growth, mediated by various factors including the controllability and predictability of the negative life events. Antoni and Goodkin [41] examined the relationships between controllability, predictability and coping style in 75 patients awaiting the results of punch biopsies taken during colposcopy examination. A modified form of the Life Experience Scale [42] and a semi-structured interview were administered. Those subjects defined as susceptible according to their earlier research (i.e. more passive, pessimistic, conforming, avoiding and somatically anxious), had positive although mostly non-significant correlations between life events and promotion, while resilient subjects (more optimistic and likely to employ more active coping strategies in dealing with stressors) had negative correlations. To explain this finding these researchers speculated that for certain individuals stressful events mobilized effective coping, which is accompanied by improved health strategies. A sociable and confident style was a beneficial moderator in the relationship between controllability of life events and CIN; an inhibited style and pessimistic attitude detrimentally affected this relationship. These researchers also found that the predictability of life events did not contribute to CIN promotion beyond the effects of controllability, and concluded that their study provided some support for the idea that a specific host interpersonal characteristic (coping style) moderates the association between life event and CIN promotion.

Reviewing previous psychological and psychosocial work in relation to the aetiology of cervical cancer and its precursors, Lambley [37] proposed a new model for the development of cervical cancer, which incorporated existing epidemiological and medical formulations into a new multi-factor framework, and examined the implications of the model for treatment. He stressed that 'effective broad-based treatment programmes require broad-based, multifactor theories'. The risk factors included in this model were: sexually transmitted viruses, which women who have had many sexual partners or whose partners have had many sexual contacts are more likely to acquire; early interpersonal experiences which 'appear to play a significant role in helping to establish these risks both in terms of sexual behaviour choice, and in terms of important other health-care behaviours such as the failure to use contraceptives'; the experience of interpersonal stress and particularly how an individual copes with stress, which appears to play a significant role in creating immune vulnerability: those women with a hopeless–depressed coping style in response

to stress, or with a high degree of dependency and lack of autonomy, appear to be most at risk; diet and smoking, which appear to act as important co-carcinogenic factors helping to exacerbate unhealthy developments. The progression of dysplasia to malignancy, the model suggests, may require several risk factors to be present simultaneously, with viral factors plus reduced immunity being the most significant physiological factors necessary.

Lambley [37] argued that,

> It is only by studying the broader bio-psychological context in which the disease is contracted that effective preventive and treatment programmes will be developed to cope with the problem.

The multi-factor model he proposed implies that greater attention should be given to the 'risk matrix' in which the patient lives, and consideration given to the possibility of psychological interventions to enhance the effectiveness of physical treatments. Lambley suggested, in particular, that the sexual habits of the patient and her partner, and the health-care patterns of the patient, should be assessed for their contribution to her risk status, with advice about modification. The patient's interpersonal circumstances and stress-coping style could also be examined with her in order to facilitate improvements that could enhance immune functioning. Besides this, the distress caused by diagnosis of the condition and 'the emotional after-effects of treatment ... can be severe and may have an important bearing on immune functioning', as Lambley noted, but 'relatively little attention is given to providing psychological advice or help to patients ...'.

MANAGEMENT OF THE PSYCHOSOCIAL IMPACT

THE PROVISION OF INFORMATION

There is general agreement in the literature that timely and appropriate information is required above all in order to reduce the amount of distress, particularly anxiety, experienced after a woman is informed of an abnormal Pap smear result. 'Accurate information is needed not only to make sense of what is happening, to provide some framework of interpretation, but also to prepare women for the different stages of the medical process' [3, p. 97]. Wilkinson *et al.* [36] provided a brief summary of an overall management strategy by writing 'good education and counselling from the onset may prevent future psychiatric and psychological morbidity and improve compliance with treatment'.

Whenever they have been asked whether they have received enough information about their condition, its investigation and treatment, some women have invariably said that they would have liked more information. The investigation of Palmer *et al.* [16] found that:

> despite fairly lengthy consultation, [the interviewees] still felt that they had not been given enough information ... [and] wanted to know more about the nature of their disease, its cause, what would happen at and after treatment, and the outcome of treatment.

Kincey *et al.* [14] asked women in their study about satisfaction with knowledge and communications. The responses led to their conclusion that there was 'significant dissatisfaction with the information available at the time'. Nearly half their sample answered negatively to the question 'Overall, do you feel satisfied that you know enough about your colposcopy and why you are having it?' Lauver and Rubin [13] concluded from their research:

> Overall, these findings suggest that women at risk for cervical cancer often need more information upon receiving abnormal Pap results. Women may want information not only about the causes and diagnostic implications of abnormal Pap results but also about follow-up procedures.

In the investigation reported by Bennetts *et al.* in 1995 [34], there was also evidence of a need for the provision of more information before arrival at the colposcopy clinic. During the initial assessment 47% indicated that they had felt the need for more information about screening at the time of the initial Pap test. At the last assessment, most women (85%) agreed that a detailed and illustrated pamphlet should be provided before the first clinic attendance.

There is a question about how this information is best given, when it should be given, who is best placed to give it, and what it should cover. A World Health Organization (WHO) consultation in 1990 on psychological implications of mass screening for cancer [35] recommended that the aim of the screening test and the meaning of the possible results and terms used need to be properly explained to potential participants; that all health professionals involved in screening programmes need to receive training for their communication with programme participants, and that the communications should be 'honest, concise, understandable, and take account of the growing body of knowledge about health messages'. The provision of adequate and accurate information was seen as an essential measure to reduce adverse effects of screening and as an 'ethical imperative' if there was to be informed choice. It was recognized that the risks of intervention and the various options available are currently not always discussed with screening participants, and suggested that there were 'ethical and psychological reasons for such open discussion in appropriate cases'.

Lauver and Rubin [13] suggested that nurses are in ideal positions to provide information, and that they could assess the concerns of their clients and intervene appropriately. Such intervention would include explaining the reason for obtaining a Pap test in the first place; explaining the causes, seriousness, and meaning of an abnormal Pap test; providing information about evaluation and treatment; and recognizing and discussing any negative emotions the client may have regarding normal or abnormal Pap results. Informational intervention about Pap results need not, in their opinion, be limited to women with abnormal results. Providing such information as part of the initial preparation for undergoing a smear test could help to lessen the shock and panic that may be experienced if an abnormal result is understood to mean cancer is present.

Very often it is the general practitioner (GP), family physician or family planning clinic doctor who is the most likely to receive the abnormal smear result and to be in a position to inform the woman. This is best done in an interactive way in a consultation in the surgery or clinic, or over the phone, so that the woman can express any worries or ask any questions that may immediately occur to her. The doctor's explanation of the meaning of the abnormal smear finding and the medical process following referral can be extremely important in reducing anxiety levels and preparing the woman for subsequent investigation and possible treatment. Unless the appointment for colposcopy is very soon, it can be helpful if the doctor invites the woman, and possibly her partner, to visit again or to telephone to discuss any concerns she may have. It would be difficult for her to think of all the questions she might want to ask about the condition shortly after hearing about the abnormality. She is likely to need some reassurance about her future prospects in terms of her future health, fertility and sexuality. As Tayler and Cherry [43] suggested,

> Family physicians need to remain aware of a woman's likely physical and emotional responses and their possible effects on her personal well-being and relationships ... The woman needs to be given the opportunity to express her fears and concerns and to receive support.

There may be an opportunity at this time to raise the question of sexual relations in order to reassure the woman (and her partner) that continuing to have sexual intercourse will not worsen the woman's condition, or harm the partner.

It is essential for the woman to have some prior explanation of the purpose of the colposcopy examination and what it entails, in order to reduce the degree of anticipatory anxiety. This explanation can be given by the doctor informing the woman of her abnormal result, or by the colposcopy clinic when an appointment is made, either over the phone or with a leaflet through the post. There is much evidence in the psychological literature that supports the wisdom of providing patients with prior explanations and as much detailed information as they wish of likely experiences during medical procedures [44, 45]. Once in the colposcopy clinic, the colposcopist may need to give the patient some explanation of the procedure as she goes along. In some clinics a closed circuit television (CCT) facility is available, and some patients may welcome a chance to 'look for themselves'. However, it should not be assumed that every patient will benefit from viewing a magnified and undeniably real image of the lesion. After the examination, the colposcopist is in a position to say something about the nature of the condition of the cervix and to discuss treatment options or likely treatment options with the patient. Very often, this provides an opportunity to reassure the patient that there is no cancer present, and that treatment is to remove the abnormal cells in the surface layer of the cervix which constitute a risk factor, in order to prevent any possible adverse development.

It is not sufficient, however, simply to reassure patients that they do not have cancer, Posner and Vessey [3] argued, because this leaves a conceptual gap to be filled only by the vague term 'abnormal cells'. These authors suggested that 'patients need a name for their condition and some way of locating it, both in terms of the degree of abnormality and in terms of its size and position' and that 'the idea of CIN as a stage between normality and disease, and the acknowledgement that abnormal cells may not necessarily progress to become a cancer, are keys to changing the black and white conception of the cervical smear test' [3, pp. 96–7, 101]. A diagrammatic presentation of the cervix and the CIN lesion can be helpful. If the patient is shown a photograph of her cervix or a CCT picture, it is important to explain the degree of magnification involved. 'The challenge to health education in this field is to make the black and white convincingly grey, the invisible and threatening, visible and less threatening ...' [3, p. 96]. The information and reassurance given to the patient by the colposcopist can do most to change a woman's concept of her condition from an uncontained threat to her life and her body, or at least her sexuality and fertility, to a located, containable and temporary risk.

Patients coming for treatment, whether as out-patients, day surgery patients or in-patients, need information about the procedures they will undergo and any possible after-effects such as discharge or pain. Prior explanation of the treatment in a setting that allows the patient to ask questions is the most helpful. Careful reassurance about resumption of sexual intercourse after the recommended abstention is also required. Women having cryocautery need to be prepared for the subsequent watery discharge; those having cone biopsy operations tend to overestimate the size of the cone to be excised, and an indication of the actual size can be reassuring.

Carefully worded written information can be provided at all stages of the process to anticipate the patient's concerns. Posner and Vessey [3] recommended written information as a way of reinforcing and supplementing verbal explanation of the meaning of the abnormal smear result and the subsequent medical process given by medical personnel involved. They suggested that leaflets can help to familiarize women with words they might not have come across before, such as 'colposcopy', and that such written information can be shared with family and friends, whose understanding can help to reinforce the patient's own grasp of new concepts. Quilliam [11] suggested that patients be allowed to tape record consultation sessions so that they could listen to them afterwards at a time of less stress.

Several research studies have specifically looked at the effect of providing written information. Wilkinson and colleagues [36] conducted a controlled trial in Cardiff looking at differences between two randomly selected groups, one sent the standard computerized letter (31 patients) to inform them of the abnormal smear result (showing dyskaryosis), the other an explanatory leaflet and more personalized letter (29 patients). When they attended the hospital colposcopy clinic, the differences between the two groups were measured by comparing the women's beliefs about their health as assessed by a structured questionnaire administered in the clinic, and their anxiety levels measured before and after consultation by the Spielberger State–Trait Anxiety Inventory [9].

There were significant differences between the two groups in their beliefs about their health status. Among the group of women who did not receive a leaflet, 19 thought that they had cancer and 12 that their health had deteriorated, compared with only one who thought she had cancer and two who thought their health had deteriorated from the group who did receive a leaflet. The group not sent a leaflet had a significantly higher initial state anxiety level (50) than the group who were (39) ($p<0.0001$). After consultation, however, state and trait anxiety levels were not significantly different between the groups. The researchers commented that the mean level of anxiety in the group of women not sent a leaflet was similar to that provoked by very stressful situations, while the mean in the group sent a leaflet was similar to that expected for general medical and surgical outpatients.

In Toronto, Stewart *et al.* [46] carried out a similar investigation to assess the effect of the receiving written information prior to attending the colposcopy clinic. One hundred and twenty-five consecutive women with a Pap test finding of dysplasia were alternatively assigned to receive a mailed educational brochure or to receive no brochure. When they arrived in the colposcopy clinic, they were given a questionnaire which included the Brief Symptom Inventory [47], specific questions about their feelings relating to an abnormal smear, as well as a test of their knowledge of dysplasia, colposcopy and the recommended follow-up. It was found that women who had received the brochure, compared with those who did not, showed a reduction in psychological distress, which was statistically significant for the psychoticism subscale, and significant reductions in specific concerns about having an abnormal smear, fear of cancer and worries about future health. The women in the group who had received the brochure were also significantly more knowledgeable about the meaning of dysplasia, the definition of colposcopy, and the recommended duration of follow-up of an abnormal smear. All but one of the women who received the brochure said they had read it and found it helpful in understanding their condition. The authors of this report concluded that:

> Given the usefulness of the brochure in reducing psychological distress and increasing knowledge, the ease and low cost of this intervention, there is good reason to recommend that an educational brochure routinely accompany or immediately follow notification of the woman about an abnormal Pap smear.

There is some evidence of differences between individuals in the amount of information about health care procedures that is optimal for their coping strategies. Miller and Mangan [48] investigated the effects of providing high and low amounts of information about colposcopy on negative emotions of women with individual differences in preference for information and involvement. Based on self-reported preference to seek or avoid information and involvement in stressful situations, these researchers identified 'monitors' and 'blunters' (information avoiders). Information avoiders who received a high amount of information reported a considerable increase in anxiety before colposcopy, whereas 'monitors' did not.

Barsevick and Johnson [49] investigated individual differences in preferences for information and involvement in health care, information-seeking behaviours and emotional responses in 36 women undergoing colposcopy. The study findings were consistent with the ideas of Krantz and associates [50] that information seeking in stressful health care situations is related to individual differences in preference for information and behavioural involvement, but were not in accordance with the claim of Miller and Mangan [48] that information-seeking behaviour is related to being a 'monitor' or a 'blunter'. Women who asked more questions reported a higher preference for information than women who asked fewer questions; and women who asked for an information sheet about colposcopy preferred involvement in their health care to a greater extent than women who did not seek an information sheet. Women who requested information reported greater confidence than women who did not. However, there was only a moderate correlation between information seeking and preference for information, leading these researchers to conclude that women who ask questions or request written information do not always have a characteristic preference for information or involvement in health care, so that 'assessment of these characteristics will not be useful in guiding the clinicians responses to women's information-seeking efforts'. However, they reassuringly suggested:

> The clinician who responds to women's questions regardless of their characteristic preference for information or involvement probably will not affect emotional responses negatively and may have positive effects on women's confidence.

SUPPORT AND COUNSELLING

One of the potential measures to reduce adverse effects of cancer screening programmes recommended by the WHO consultation mentioned above, was that:

Individual attention for screening programme participants suffering distress should be available over the telephone or in person from an appropriately trained source [35].

The Women's Nationwide Cancer Control Campaign in London* has for several years run a Helpline using trained counsellors to provide support, counselling and information to women concerned about breast and cervical changes detected by screening programmes.

Women's need for support after an abnormal cervical smear has been recognized by a number of investigators:

> Reassurance, warmth and understanding from doctors and nurses and the patient's family and friends have all been shown to be important in providing support for women as they go through the medical process [3, p. 94].

Sensitive care, personal attention and kindly treatment on the part of nurses and doctors in the clinic can be immensely supportive and helpful in reducing stressful feelings and some of the negative impact of the procedures. McDonald *et al.* [2] noted that 'a flippant comment at this time of patient vulnerability may have widespread repercussions', and argued that a professional manner that was 'caring, understanding and empathetic' could help bolster self-esteem and body image. 'If possible, a patient should have the full attention of a nurse or friend in the room with her during treatment', Quilliam [14] suggested. Recognizing the degree of conflict between the detachment and objectivity required for professional internal examination of the body, and the attention to the person required for a comfortable and supportive relationship, Posner and Vessey [3, p. 31] wrote that:

*The Women's Nationwide Cancer Control Campaign, Suma House, 128 Curtain Road, London EC2A 3AR, UK. Helpline: 0171 729 2229 (Monday to Friday, 9.30 am–4.30 pm); recorded general information on cervical screening: 0171 729 5061.

Recognition of the patients as people, and reassurance wherever possible, helps, at least to some extent, to resolve the stress and heal the emotional hurt of the medical investigation.

In both the Posner and Vessey study [3] and in the study by Palmer *et al.* [16], respondents reported that the manner in which they were treated in the colposcopy clinic was a positive factor in an episode which was overall distressing for many. Palmer *et al.* [16] observed that the clinic 'environment' was experienced as 'emotionally supportive' and 28% of all experiences mentioned entailed 'the comfort provided by nurses, which was singled out as immensely supportive'. Suggesting that health professionals needed to beware of biased or judgmental attitudes which will be hurtful to the patient, Quilliam [11] wrote that 'it makes a great deal of difference to a woman how she is treated by health professionals, and in particular how her cervical condition is seen'.

If the colposcopy clinic nurse can devote most of her attention to the patient while she is having a colposcopy examination or outpatient treatment, this can be very helpful in providing distraction, reassurance and support. Rickert and colleagues [51] have demonstrated the effectiveness of another means of distraction during the colposcopy examination. They found that allowing female adolescents undergoing colposcopy to watch music videos reduced bodily movements indicative of discomfort or pain. These investigators carried out two studies randomly allocating patients to a control or experimental group, and then observing patient behaviour across ten dimensions using multiple measures to determine behavioural, physiological and cognitive components of anxiety, self-reports of pain and discomfort, satisfaction with and feelings about the procedure. The first study examined the effect of allowing the experimental group to watch the procedure on a television monitor. It found no difference in bodily movements or anxiety rating in the experimental group. In the second study patients watched a music video and demonstrated significantly fewer bodily movements indicative of pain and discomfort, requiring less reassurance from the doctor and fewer procedural explanations. These investigators concluded that including music videos in the colposcopy protocol was 'a simple and easily used strategy to minimize anxiety, distress, and related bodily movements for those adolescents undergoing colposcopy'.

It can be reassuring for the patient to see the same nurse and doctor on each visit to the clinic, where possible. This allows a more personal relationship to develop, which provides some counterbalance to the depersonalizing aspects of investigation and treatment. Out-patient treatment for CIN, however uncomplicated and brief, may still be experienced as a shock to the system, physically, psychologically and emotionally, and patients deserve some nursing care and attention for a time afterwards, particularly if they have come to the clinic on their own. Women have described understandable feelings of vulnerability and being unbalanced after treatment. Preparation for out-patient treatment should include advice to the patient to take the day off work, and to bring someone supportive with her to the clinic and to accompany her home.

McDonald *et al.* [2] wrote that 'in coping with cancer or even the potential of cancer, a support system can be a tremendous advantage'. In her book *Positive Smear*, Quilliam [12] suggests that women need to have a support network to help them through this time, and that there are ways to find other women to confide in, ways of communicating and negotiating with partners, and of organizing practical help if needed. In the UK there have been two telephone support lines for women with concerns about abnormal smears, one based in Liverpool where there is also a self-help group, and one in Nottingham, both run by women with their own experiences to draw

on.* Such peer counselling 'in which a woman who had experienced an abnormal Pap test result in the past would spend time listening to the fears and concerns of another woman currently facing the same problem, may be a very effective strategy to promote effective coping' [22]. Mutual support can be a very powerful way of dissolving feelings of stigma, isolation and other forms of emotional pain.

A possible need for counselling, particularly in relation to psychosexual upset, has been acknowledged by a few writers, for example, Quilliam [12]. A recommendation in the report by Posner and Vessey was:

> There should be greater awareness of a possible need for psychosexual counselling, particularly for women who are not in stable relationships, and for women who are treated by hysterectomy. The offer of such counselling needs to be made sensitively in order not to intrude further on a private sphere in a way which is unwelcome [3, p. 100].

Greater awareness of the role of HPV in the aetiology of CIN has added the element of infectivity with all its complications and connotations to the psychosexual experience of this process. Initial anxiety and morbid feelings about the possibility of cancer may be replaced by anxiety and negative feelings about having a sexually transmitted infection. Women need accurate information about the nature, prevalence and significance of HPV and they may need counselling support in order to come to terms with the implications of acquiring the infection.

As a result of their investigation of 'the psychosexual trauma of an abnormal cervical smear', Campion *et al.* [6] suggested there was a need for 'informative and supportive counselling' and that a research study was required

---

*CERCAN, 6 Landford Avenue, Liverpool L9 6BR, UK, tel. 0151 525 2848 (Lily); Abnormal Smear Care, c/o Long Eaton and District Volunteer Bureau, Community House, 173 Derby Road, Long Eaton, Nottingham, UK, tel. 01602 731 778 (Linda, Monday to Friday, 9.15 am–1.15 pm).

to 'assess the effects of counselling in decreasing anxiety associated with the diagnosis and treatment of pre-invasive cervical disease'. Four years later in 1992, Wolfe and colleagues reported an evaluation of the acceptability of counselling in a study also designed to compare several different information leaflets [52]. The offer of a chance to discuss concerns was acceptable to 80% of the patients to whom it was made. The response to the counsellor's approach and to various components of the interview was a very high level of satisfaction. The study did not evaluate the effect of the counselling by itself, but did demonstrate lower levels of anxiety, less distress related to the abnormal smear result and medical procedures and improved mood after medical intervention had taken place.

PRACTICAL ASPECTS OF MANAGEMENT

Apart from offering sufficient information and support to allay unnecessary anxiety and provide reassurance during the medical process after referral for an abnormal cervical smear, there are some practical aspects to management that can reduce distress. In the first place, any delay in the investigation of abnormal results and carrying out any subsequent necessary treatment, is an avoidable cause of distress: 'the provision of adequate facilities for investigation and treatment with the minimum of delay is thus a necessary measure' [32]. The time of greatest anxiety is likely to be prior to the colposcopy examination, so that if there is any delay at this time, women should have access to a source of counselling.

In the colposcopy clinic, it is important to avoid anything which is an unnecessary invasion of a woman's privacy or cause for embarrassment, and thus it should not be possible for people to come into the clinic during examination and treatment, and observers should not be allowed, except where absolutely necessary and the woman has freely given her consent. Additional people in the clinic makes it less likely that the patient will feel able to discuss

her concerns fully, and more likely that she will feel embarrassed and exposed. Questions about aspects of a woman's sexual history may also cause embarrassment and be experienced as a further intrusion into her privacy [53], so that care needs to be taken to collect only that information which is necessary in clinical terms.

The very nature of the technical aspects of management involve a degree of exposure and invasion for the patient, and sensitivity to this aspect of her experience on the part of health professionals is important.

Interviewees in the study by Posner and Vessey [3] were asked if they had objections to any questions they had been asked in the clinic. In answer to this, they frequently mentioned being asked at what age they first had sexual intercourse. Posner and Vessey suggested:

> Further distress is caused by lack of awareness of the meaning for women of medical intervention which involves a transmutation of private space. The hurtfulness of it may not be confined simply to physical pain. In the cause of preserving the physical integrity of her cervix, a woman may feel that her personal integrity is violated and her moral integrity impugned.

CONCLUSION

In an area of preventive medical care, intervention in the form of colposcopy and treatment for CIN has been found to result in a considerable degree of psychosocial morbidity. The possible adverse effects include a period of extreme anxiety and stress, behavioural symptoms of distress, negative changes in the woman's body image and view of her health status, psychosexual upset, and lowered self-esteem and feelings of embarrassment, guilt or anger, which may impact on personal relationships. This morbidity relates to the experience of the woman as a person rather than to the physical state of her cervix, and results from unhelpful conceptualizations of the condition and negative reactions to the processes involved in medical intervention. The threat of cancer in an area of the body physically and psychologically vital to a woman's reproductive and sexual functions, combined with the intrusive and uncomfortable nature of the examination and treatment, has the potential to cause distress. The meaning of the episode for the woman in the context of her life can be very different from the medical view of it. This divergence of view was expressed by an interviewee in the study by Posner and Vessey [3]: 'They know it's common; they know it's only a small thing – to you, it's the end of the world'. The suggestion that counselling be offered has come very belatedly, but in recognition of the nature of women's distress. In order to prevent unnecessary anxiety and morbidity, it is essential to provide adequate information from the time the Pap test is taken. Understanding, reassurance and support from the health professionals involved, and others, is important in helping to alleviate any negative aspects of a woman's experience of this intervention. The task for health professionals in the colposcopy clinic is to be technically accurate, effective and efficient at the level of the cervix in the cause of prevention, while at the same time relating in a positive way to the embodied persons at the end of the colposcope, in the cause of healing.

REFERENCES

1. Lerman, C., Miller, S., Scarborough, R., *et al.* (1991) Adverse psychologic consequences of positive cytologic cervical screening. *American Journal of Obstetrics and Gynecology*, **163**(3), 658–62.
2. McDonald, T., Neutens, J., Fischer, L. and Jessee, D. (1989) Impact of cervical intraepithelial neoplasia diagnosis and treatment on self-esteem and body image. *Gynecologic Oncology*, **34**, 345–9.
3. Posner, T. and Vessey, M. (1988) *Prevention of Cervical Cancer: the Patient's View*. King Edward's Hospital Fund for London, London.
4. Reelick, N., Haes, W. and Schuurman, J. (1984) Psychological side-effects of the mass screening on cervical cancer. *Social Science and Medicine*, **18**(12), 1089–93.
5. Beresford, J. and Gervaize, P. (1986) The emotional impact of abnormal Pap smears on patients referred for colposcopy. *Colposcopy and Gynecologic Laser Surgery*, **2**(2), 83–7.
6. Campion, M., Brown, J., McCance, D., *et al.* (1988) Psychosexual trauma of an abnormal cervical smear. *British Journal of Obstetrics and Gynaecology*, **95**, 175–81.
7. Goffman, E. (1964) *Stigma: Notes on the Management of Spoiled Identity*. Penguin, Harmondsworth.
8. Marteau, T., Walker, P., Giles, J. and Smail, M. (1990) Anxieties in women undergoing colposcopy. *British Journal of Obstetrics and Gynaecology*, **97**, 859–61.
9. Spielberger, C., Gorsuch, R. and Lushene, R. (1970) *Manual for the State–Trait Anxiety Inventory*. Consulting Psychologists Press, Palo Alto, CA.
10. Marteau, M., Kidd, J., Cook, R., *et al.* (1988) Screening for Down's syndrome. *British Medical Journal*, **297**, 1469.
11. Quilliam, S. (1990) Positive smear: the emotional issues and what can be done. *Health Education Journal*, **49**(1), 19–20.
12. Quilliam, S. (1989) *Positive Smear*. Penguin, Harmondsworth.
13. Lauver, D. and Rubin, M. (1991) Women's concerns about abnormal Papanicolaou test results. *Journal of Obstetrics, Gynecology and Neonatal Nursing*, **20**(2), 154–9.
14. Kincey, J., Statham, S. and McFarlane, T. (1991) Women undergoing colposcopy: their satisfaction with communication, health knowledge and level of anxiety. *Health Education Journal*, **50**(2), 70–2.
15. Greimel, E., Girardi, F., Freidl, W., *et al.* (1992) Vorstufen des zervixkarzinoms (CIN) – erleben und kranheitsverarbeitung. *Wiener klinische Wochenschrift*, **104**(13), 396–8. (The author thanks F. de Looze for his help with the translation of this paper).
16. Palmer, A., Tucker, S., Warren, R. and Adams, M. (1993) Understanding women's responses to treatment for cervical intra-epithelial neoplasia. *British Journal of Clinical Psychology*, **32**, 101–12.
17. Horowitz, M., Wilner, N. and Alvarez, W. (1979) Impact of events scale: a measure of subjective stress. *Psychosomatic Medicine*, **41**(3), 209–18.
18. Spielberger, C., Johnson, E., Russel, S., *et al.* (1983) Assessment of anger: the stait trait anger scale. In *Advances in Personality Assessment* (eds C. Spielberger and J. Butcher). Erlbaum, Hillsdale, NJ.

19. Walston, K. and Walston, B. (1976) Development of the multi-dimensional Health Locus of Control Scales. *Journal of Consulting and Clinical Psychology*, **44**, 580–5.

20. Chemtrob, C., Roitblat, H., Hamada, R., *et al.* (1988) A cognitive action theory of post traumatic stress disorder. *Journal of Anxiety Disorders*, **2**, 253–75.

21. Horowitz, M. (1979) Psychological responses to serious life events. In *Human Stress and Cognition* (eds V. Hamilton and D. Warburton). Wiley, Chichester.

22. Funke, B. and Nicholson, M. (1993) Factors affecting compliance among women with abnormal Pap smears. *Patient Education and Counselling*, **20**, 5–15.

23. Boag, F., Dillon, A., Catalan, J., *et al.* (1991) Assessment of psychiatric morbidity in patients attending a colposcopy clinic situated in a genitourinary medicine clinic. *Genitourinary Medicine*, **67**, 481–4.

24. Goldberg, D. and Hillier, V. (1979) A scaled version of the General Health Questionnaire. *Psychiatric Medicine*, **9**, 139–45.

25. Gath, D., Hallam, N., Mynors-Wallis, L., *et al.* (1995) Emotional reactions in women attending a UK colposcopy clinic. *Journal of Epidemiology and Community Health*, **49**, 79–83.

26. Wing, J., Cooper, J. and Sartorius, N. (1974) *The Measurement and Classification of Psychiatric Symptoms.* Cambridge University Press, London.

27. Beck, A., Ward, C., Mendelson, M., *et al.* (1961) An inventory for measuring depression. *Archives of General Psychiatry*, **4**, 561–71.

28. Snaith, R., Bridge, G. and Hamilton, M. (1976) The Leeds scale for the self-assessment of anxiety and depression. *British Journal of Psychiatry*, **128**, 156–65.

29. Cooper, P., Osborn, M., Gath, D. and Feggetter, G. (1982) Evaluation of a modified self-report measure of social adjustment. *British Journal of Psychiatry*, **141**, 68–75.

30. Eysenck, H. and Eysenck, S. (1963) *The Eysenck Personality Inventory.* University of London Press, London.

31. Bell, S., Porter, M., Kitchener, H., *et al.* (1995) Psychological response to cervical screening. *Preventive Medicine*, **24**, 610–16.

32. Zigmond, A. and Snaith, R. (1983) The Hospital Anxiety and Depression Scale. *Acta Psychiatrica Scandinavica*, **67**, 361–70.

33. Doherty, I., Richardson, P., Wolfe, C. and Raju, K. (1991) The assessment of the psychological effects of an abnormal cervical smear result and subsequent medical procedures. *Journal of*
Psychosomatic Obstetrics and Gynaecology, **12**, 319–24.

34. Bennetts, A., Irwig, L., Oldenburg, B., *et al.* (1995) PEAPS–Q: A questionnaire to measure the psychosocial effects of having an abnormal Pap smear. *Journal of Clinical Epidemiology*, **48**, 1235–43.

35. WHO Report (1990) *Consultation on Psychological Implications of Mass Screening for Cancer.* WHO Regional Office for Europe, Geneva.

36. Wilkinson, C., Jones, M. and McBride, J. (1990) Anxiety caused by abnormal result of cervical smear test: a controlled trial. *British Medical Journal*, **300**, 440.

37. Lambley, P. (1993) The role of psychological processes in the aetiogy and treatment of cervical cancer: a biopsychological perspective. *British Journal of Medical Psychology*, **66**, 43–60.

38. Schmale, A. and Iker, H. (1966) The psychological setting of uterine cervical cancer. *Annals of the New York Academy of Sciences*, **125**, 807–13.

39. Schmale, A. and Iker, H. (1971) Hopelessness as a predictor of cervical cancer. *Social Science and Medicine*, **5**, 95–100.

40. Goodkin, K., Antoni, M. and Blaney, P. (1986) Stress and hopelessness in the promotion of cervical intraepithelial neoplasia to invasive squamous cell carcinoma of the cervix. *Journal of Psychosomatic Research*, **30**, 67–76.

41. Antoni, M. and Goodkin, K. (1989) Host moderator variables in the promotion of cervical neoplasia – II. Dimensions of life stress. *Journal of Psychosomatic Research*, **33**(4), 457–67.

42. Sarason, I., Johnson J. and Siegel, J. (1978) Assessing the impact of life changes: development of the Life Experience Survey. *Journal of Consulting and Clinical Psychology*, **46**, 932–46.

43. Tayler, S. and Cherry, S. (1989) Cervical atypia. *Australian Family Physician*, **18**(3), 223.

44. Johnston, M. (1980) Anxiety in surgical patients. *Psychological Medicine*, **10**, 145–52.

45. Ley, P. (1988) *Communicating with Patients: Improving Communication, Satisfaction and Compliance.* Croom Helm, London.

46. Stewart, D., Lickrish, G., Sierra, S. and Parkin, H. (1993) The effect of educational brochures on knowledge and emotional distress in women with abnormal Papanicolaou smears. *Obstetrics and Gynecology*, **81**(2), 280–2.

47. Derogatis, L. and Melisaratos, N. (1983) The brief symptom inventory: an introductory report. *Psychological Medicine*, **13**, 595–605.

48. Miller, S. and Mangan, C. (1983) Interacting effects of information and coping style in adapting to gynecologic stress: should the doctor tell all? *Journal of Personality and Social Psychology*, **45**, 223–36.

49. Barsevick, A. and Johnson, J. (1990) Preference for information and involvement, information seeking and emotional responses of women undergoing colposcopy. *Research in Nursing and Health*, **13**, 1–7.

50. Krantz, D., Baum, A. and Wideman, M. (1980) Assessment of preference for self-treatment and information in health care. *Journal of Personality and Social Psychology*, **39**, 977–90.

51. Rickert, V., Kozlowski, K., Warren, A., *et al.* (1994) Adolescents and colposcopy: the use of different procedures to reduce anxiety. *American Journal of Obstetrics and Gynecology*, **170**, 504–8.

52. Wolfe, C., Doherty, I., Raju, K., *et al.* (1992) First steps in the development of an information and counselling service for women with an abnormal smear result. *European Journal of Obstetrics and Gynecology and Reproductive Biology*, **45**, 201–6.

13

PRE-MALIGNANT GLANDULAR LESIONS

A.D. Blackett, J. Nevin and F. Sharp

INTRODUCTION

As is shown elsewhere in this volume, adenocarcinomas of the uterine cervix are becoming an increasingly important problem. It has been noted that in the 1970s the proportion of cervical cancers reported as adenocarcinomas doubled [1]. This could either be because of a genuine increase in incidence, improved detection or more effective management of the majority of squamous pre-malignancies. A particularly worrying feature of the trend is that the greatest increase in frequency seems to be occurring in the younger (<35 years) age groups, where the doubling in incidence between 1973 and 1982 was equivalent to a 10% annual increase in incidence [2]. Similar observations have been made using routine data from UK regional cancer registries [3].

An understanding of the natural history of the malignancy is a necessary foundation on which to build strategies for improved detection and treatment, but as a consequence of the lower incidence than squamous lesions coupled with harder detection, comparatively little is known of adenocarcinoma. Whereas with cervical squamous carcinomas the spectrum of conditions through the grades of intra-epithelial neoplasia to frank invasive malignancy is well characterized, it is by no means universally accepted that there is a similar progression from glandular dysplasias, through adenocarcinoma *in situ* (AIS) to invasive adenocarcinomas. Similarly for squamous lesions, the role of human papillomavirus (HPV) in providing one step on the pathway of oncogenesis and the implication of other risk factors in its aetiology are widely accepted, yet for glandular lesions they remain matters of debate.

Those who seek to prevent adenocarcinoma of the cervix must also be mindful of the long and arduous road that was travelled to reach the current day management of squamous precursors. After Papanicolaou developed his cytological techniques, a generation of women with cervical smear abnormalities were subjected to hysterectomy, sacrificing fertility. The introduction of the cone biopsy improved fertility for the next generation, but this operation carries a high complication rate [4, 5]. The introduction of colposcopy permitted treatment tailored to the location and extent of the transformation zone (TZ) and the evolution of locally destructive therapy for squamous intra-epithelial abnormalities. Cone biopsy remains integral to the management of suspect glandular lesions and at the end of this century, such women are facing the same complications as those who had squamous abnormalities in the middle of the century.

This chapter emphasizes the shortcomings in our knowledge of cervical intra-epithelial

Cancer and Pre-cancer of the Cervix
Edited by D.M. Luesley and R. Barrasso. Published in 1998 by Chapman & Hall, London. ISBN 0 412 56600 1.

glandular lesions and how these are being addressed.

HISTOPATHOLOGY

Glandular pre-maligant lesions will be located within the field of cervical columnar epithelium. This includes columnar epithelium lying in the depths of the complex arrangement of crypts and tunnels of the endocervix (Figure 13.1 and [6]), as well as the columnar epithelium lying beneath squamous metaplastic epithelium in the glands of the TZ. This target area for glandular neoplasia extends from the outer limit of the acquired TZ (the line of the original squamo-columnar junction) to the upper limit of the endocervical canal at the level of the internal cervical os, where the glandular epithelium meets the endometrium.

CLASSIFICATION SYSTEMS

There is a lack of uniformity both in nomenclature and classification criteria for pre-malignant glandular lesions adopted by various authors. It is unfortunate that data from a number of small studies have to be pooled to reach a consensus on what is still a comparatively infrequently met condition. As with squamous intra-epithelial lesions, the terminology for glandular lesions is in the process of evolution and we will attempt to highlight and explain the various terminologies.

AIS was first described 40 years ago [7]. It can be defined as an intra-epithelial, non-invasive stage of endocervical adenocarcinoma [8]. The only problem with this definition is the interpretation of invasion, which in glandular lesions may not be so obvious and allows differences of interpretation to be manifest.

Glandular dysplasia as a term was coined almost 25 years ago [9] and is still used, although other terminology is now current. Confusingly, in addition to endocervical glandular dysplasia (EGD), reference will also be

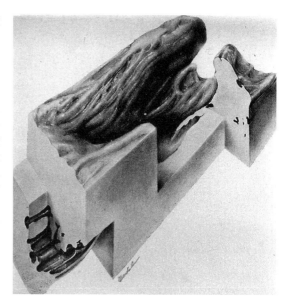

Figure 13.1   Three-dimensional reconstruction of a segment of the endocervical canal showing the complex arrangement of crypts and tunnels, lined by columnar epithelium and liable to neoplastic transformation (reproduced with permission from plate I in Fluhmann [6]).

found to endocervical glandular atypia (EGA), cervical glandular atypia (CGA), cervical glandular intra-epithelial neoplasia (CIGN) and glandular intra-epithelial neoplasia (GIN). All of these terms describe, in slightly different ways, the same conditions.

Van Roon *et al.* [10] were the first to attempt a three-step sub-division of the pre-malignant glandular spectrum, using the terms mild/moderate atypia, severe atypia and AIS. Brown and Wells [11] derived another three-step classification system, with those areas showing less marked dysplasia than AIS being termed either mild or severe glandular atypia (see also [12]). These three grades are approximately equal to CIGN grades I–III as classified by Gloor and Hurlimann in the same year [13]. A more recent classification considers it best not to attempt sub-division of the atypias [14]. The differentiating factors employed by these classification systems are shown in Table 13.1.

Table 13.1 Classification systems for premalignant glandular lesions

After Brown and Wells [11] and Wells and Brown [12]	Low-grade glandular atypias	Nuclear stratification to < 2/3 epithelial height Enlarged, elongated nuclei Increased nuclear/cytoplasmic ratio Nuclear pleomorphism or abnormal chromatin pattern Hyperchromatic nuclei Abnormal glandular profile, possibly irregular branching and budding Some intraluminal tufts Low papillary projections with stromal cores Abnormalities of mucin production Subnuclear vacuolation
	High-grade glandular atypias	Most characteristics shared with low-grade atypias but more pronounced Nuclei occasionally vesicular Nuclear stratification > 2/3 epithelial height Increased mitosis, but still rare (< 1 per profile) More intraluminal tufts Papillary projections lacking stromal cores
	Adenocarcinoma *in situ* (AIS)	Greater degree of aberrant glandular morphology Numerous mitoses (>5 in some glands) Glands smaller and more crowded Back to backing and cribriform patterns (not seen in atypias)
After Gloor and Hurlimann [13]	Cervical intra-epithelial glandular neoplasia (CIGN) I	Nuclei slightly hyperchromatic Single row of nuclei at base of cells Only occasional mitoses visible
	CIGN II	Nuclei oval with greater hyperchromia Nuclear crowding with some pseudostratification More mitoses visible Reduced cytoplasmic mucin
	CIGN III	Thicker epithelium with elongated nuclei Dense crowding with marked pseudostratification Many more mitoses visible Much reduced or absent cytoplasmic mucin
After Jaworski [14]	Endocervical glandular dysplasia (EGD)	Nucleus large and oval Chromatin fine to moderately granular Nucleoli inconspicuous Nuclear pseudostratification minimal Apoptosis present Mitosis occasional (1–2/gland) Glandular budding possible Tunnel clusters possible Papillary processes possible Cribriform glands absent
	AIS	Nucleus large, oval or irregular Chromatin moderate to coarsely granular Nucleoli variable, often larger (but not in all cases) Nuclear pseudostratification moderate to marked Apoptosis present, possibly prominent Mitosis frequent and possibly abnormal (>2/gland) Glandular budding possible Tunnel clusters possible Papillary processes possible Cribriform glands possible

It must be acknowledged that the differences between glandular atypias, AIS and malignancy are for the most part merely matters of degree. Demarcations are necessarily artificial and to an extent arbitrary. Accurate definitions of each category, of whichever classification scheme is adopted, are therefore vital to minimize the subjectivity of the appraisers.

Histological features that need to be examined fall into two distinct groupings – the cellular and the architectural. Cellular features include variation in nuclear size and shape (pleomorphism/anisokaryosis), increased nuclear staining (hyperchromasia), increased nuclear/cytoplasmic ratio, loss of nuclear polarity, nuclear stratification, increased mitotic activity and reduction of cytoplasmic mucin production (Figure 13.2). Architectural features include crowding and variation in size and shape of glands.

The overall architectural pattern is often retained in pre-malignant conditions and if the changes observed are more than slight then a diagnosis of invasive adenocarcinoma should be contemplated. This dividing line between AIS and overt malignancy is one of the most contentious. Some argue that AIS only affects surface and superficial crypts and that if crypt architecture is in any way perturbed then the condition is no longer merely *in situ*. An acceptable depth limit of 4 mm has been proposed, beyond which the adenocarcinoma can no longer be considered *in situ* [15]. Others accept deeper lying AIS and a degree of architectural perturbation including outpouchings, complex foldings and intraluminal papillary projections [14, 16]. This obviously has ramifications when considering experimental data relating to particular degrees of abnormality.

Invasion is particularly hard to spot in adenocarcinomas as it is performed by means of newly formed glands pushing out into the stroma and not by individual cells or groups of cells invading surrounding tissues, as in the squamous condition. Budding and reduplication of crypts leads to back-to-back or cribriform patterns of glands, but by this stage the condition is probably no longer AIS.

Staining patterns of basement membrane seem to be a productive way of identifying early invasive adenocarcinomas. Yavner *et al.* [17] have shown intact basement membranes in AIS but fragmented and irregular basement membranes on some gland buds and outpouchings of early invasive cancer.

The concept of microinvasive squamous cell carcinoma has been under consideration since the introduction of the term almost three decades ago. In contrast, until recently little effort had been devoted to the study of small adenocarcinomas, and to whether such well-defined tumours might carry the same favourable prognosis as their squamous counterparts, and be treated in a similar less radical way. As to the size that might be considered as a microinvasive adenocarcinoma, Jaworski suggested the term be applied to a tumour which invades to a depth of less than 5 mm [14]. Burghardt [18] suggested a volume limit of less than 500 mm³, whilst coining the term endocervical microcarcinoma. Recent important work in this area has been that of Östör who studied a series of 77 microinvasive adenocarcinomas over a 25-year period [19]. He concluded that this is a clinicopathological entity having the same prognosis as its squamous counterpart and should be treated in the same way, multicentricity or 'skip' lesions not being seen to be a problem.

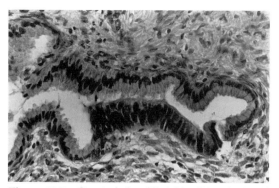

Figure 13.2 Cervical gland lined in the majority of its surface by normal columnar epithelium. The middle third of the lower aspect is replaced by adenocarcinoma *in situ* (H & E, ×128).

Cervical glandular 'deviations' are a multitude of benign conditions unrelated to the process of oncogenesis, which can nevertheless produce signs suggestive of dysplasia and therefore need consideration. They include inappropriately located Müllerian tissues, papillary endocervicitis and other causes of inflammation, Arias–Stella reaction and other changes associated with pregnancy, nabothian cysts, Gärtner's duct hyperplasia, reserve cell hyperplasia, mesonephric hyperplasia, diffuse laminar endocervical glandular hyperplasia, endometriosis, tubal metaplasia, intestinal metaplasia, tunnel clusters, isthmic mucosa and microglandular hyperplasia.

Microglandular hyperplasia is a benign lesion sometimes occurring in women taking oral contraceptives and during pregnancy. It has been shown not to be causally related to the development of endocervical carcinoma [20]. For a review of the literature on microglandular hyperplasia, see [21].

Sub-categorization of AIS

As discussed elsewhere in this volume, invasive adenocarcinomas are classified into various categories based on their histological appearance. It might reasonably be expected that if all adenocarcinomas arise from AIS, then a similar subdivision of the latter into the same categories might be possible. Whilst some have attempted to do this it is not a straightforward task, bearing in mind the comparatively low numbers of specimens to examine and the nature of a precursor lesion tending towards a more homogeneous morphology.

Gloor and Ruzicka [22] saw two distinct types of AIS: type I being most common with columnar cells and type II more uncommon with ballooned cells, structural disorder and large clear nuclei. Jaworski *et al.* subdivided AIS into four types analogous to some of the categories of adenocarcinomas. They were able to recognize endocervical, endometrioid and intestinal groups, plus a miscellaneous group including adenosquamous and clear

cell types [23]. Kudo stayed with two types: endocervical and endometrioid, the latter lacking mucin production [16]. Clear cell adenocarcinomas with an *in situ* component have been reported [24].

RELATIONSHIP BETWEEN PRE-MALIGNANT LESIONS AND ADENOCARCINOMA

Luesley commented on the disparity between the incidence of adenocarcinomas and that of corresponding pre-invasive lesions [25]. This would be the result if some tumours arose *de novo* without passing through a pre-malignant phase, but the accepted multi-hit model of carcinogenesis makes this unlikely. It is much more likely that this disparity is merely a function of the difficulty in detecting pre-malignant lesions, most of which are still discovered fortuitously.

Between areas of AIS and normal cells may lie regions of atypia [11]. This could be variously interpreted as an area of atypia in which a cell or cells have undergone the next genetic lesion in the oncogenic pathway and progressed to AIS, or an area of AIS exerting its influence on surrounding cells (perhaps by secretion of growth factors?). Because it is not possible to observe cells *in situ* on a longitudinal basis it remains speculation as to whether the cells showing such an intermediate morphology are observed in the process of evolving into AIS or are to be subsequently subsumed by the clonal expansion of AIS cells.

Brown and Wells reported that they observed atypias 16-times as frequently as AIS, which may indicate the ratio of lesions progressing from this first stage away from normality [11]. It may, however, be simplistic to interpret such data in this fashion, as once again we are making inferences based on snapshots at a point in time. As we know nothing of the rate of progress that lesions might make we cannot discount the possibility that cells exist as atypias for, on average, 16-times as long as they do as AIS. We assume that a series of genetic lesions must be accumulated to progress along the path to invasive cancer but we cannot expect this accumulation to be linear with time – indeed, there are grounds to expect a mutation to result in a cell which, being less fit, would then be at greater risk of subsequent mutation. Another trap to fall into is the assumption that the grades imposed by histology are equivalent to genetic events, whereas one is inherently discontinuous and the other has discontinuity artifactually imposed upon it.

The ages of patients showing particular grades of lesion might present us with an indication of the rate of progress to be expected, although it must be stressed at the outset that when comparing results from various studies, the different populations and classification by different pathologists in each case render such interpretations tenuous. Nevertheless the mean ages reported for the various conditions do appear to fall into a pattern, as shown in Table 13.2.

Table 13.2 Mean ages of patients reported with glandular lesions

Condition	Mean age (years)	Reference
Atypia	32.0	35
Atypia	36.9	11
Atypia + AIS	34.7	35
AIS	35.9	99
AIS	36.0	100
AIS	36.4	29
AIS	37.0	14
AIS	38.4	41
AIS	39.0	27
AIS	39.0	26
AIS	40.5	35
Microinvasive adenocarcinoma	39.0	19
Microinvasive adenocarcinoma	44.0	26
Invasive adenocarcinoma	40.8	29
Invasive adenocarcinoma	47.0	100
Invasive adenocarcinoma	49.0	26
Invasive adenocarcinoma	57.0	27

The condition in each case is as classified by the authors concerned.

As we are dealing with a biological system we should dismiss any temptation to expect individual cases to show strict adherence to this model with progression from atypias to AIS in 2 years and to invasive cancer in another 10, but it is nevertheless a useful indication of a temporal trend.

Does AIS progress to adenocarcinoma? The answer seems to be that it can and that the time period taken to do so is anywhere from 3 to 14 years (Table 13.2). The average age of diagnosis of AIS is several years earlier than that of adenocarcinoma, which is suggestive of AIS being a precursor lesion. Also, adenocarcinomas typically will show areas of AIS at their periphery.

When previous samples taken from adenocarcinoma patients were reviewed with the benefit of hindsight, evidence was found of AIS that was not reported at the time. Bousfield *et al.* [26] showed signs of high-grade glandular atypia on smears taken 5–10 years prior to diagnosis of cancer; and Boon *et al.* [27] found AIS in five of 18 biopsies taken 3–7 years prior to cancer being discovered. These, coupled with isolated reports of individuals known to have AIS which progressed eventually to invasive adenocarcinoma [16], make powerful evidence for the acceptance of AIS as a real precursor lesion.

Whilst there is convincing evidence of the pre-malignant nature of AIS and even severe glandular atypia, the milder lesions are usually self-limiting. In one report of the follow-up of smears showing glandular atypia [28], 167 out of 251 patients with low-grade glandular atypias regressed cytologically. Only two patients developed cytological evidence of progressive disease. This is a compelling argument for restricting the use of cone biopsy in these patients and limiting the follow-up to cytology, including an endocervical sampling method.

THE LOCALIZATION OF AIS

The localization of AIS has caused some debate. It can doubtless be found anywhere that columnar epithelium exists, including ectopies and throughout the length of the endocervical canal [11]. Whether this means that dysplasia can arise amongst columnar cells anywhere is uncertain, as some maintain that its seat of origin is always the squamo-columnar junction [16, 29] and that when more far-flung sites are discovered, they are merely distant ramifications of the original dysplastic area rather than discrete foci. Teshima *et al.* [30] found that all eight cases of AIS they examined (four endocervical, four endometrioid) were situated just proximal to the squamo-columnar junction and Colgan and Lickrish [31] reported that three-quarters of their AIS samples were beneath the TZ and thereby covered by squamous epithelium, as earlier reported by Christopherson *et al.* [32]. Colgan and Lickrish [31] also reported that 16% of their cases were multifocal, which provides grounds for examining multiple sections and for dismissing the utility of punch biopsies as a means of diagnosing AIS [33]. Given a genetic predisposition or a widely spread oncogenic agent there is no reason why multifocal sites should not arise, although it is never easy to prove that all foci are absolutely discrete.

RELATIONSHIP BETWEEN PRE-MALIGNANT GLANDULAR AND SQUAMOUS LESIONS

Areas of co-existing cervical squamous intra-epithelial neoplasia (CIN) have been found adjacent to areas of AIS in between 48 and 89% of cases [23, 29, 31, 33–39] just as they have been in adenocarcinomas [40]. As CIN is more likely to be detected colposcopically, the possibility of co-existent AIS provides grounds for favouring excisional (with histological analysis) rather than destructive procedures for the removal of CIN.

Just as CIN can be found adjacent to AIS, so can areas of glandular atypia be found in samples diagnosed as CIN. Brown and Wells found 16 glandular atypias and one AIS in 105 CIN III cases, compared to only two glandular atypias in 100 normal controls [11].

These areas of adjacent dysplastic squamous and glandular cells suggest either a common causative agent or derivation from a common cell lineage as proposed by Qizilbash [41] who saw the reserve cell as being the origin for both CIN and glandular atypias. It is quite possible that these areas are developed by clonal expansion from a single mutated (or infected?) reserve cell [42], with some progeny subsequently undergoing metaplasia due to the pH changes at the TZ. Cytogenetic evidence has been put forward to show the clonal nature of invasive cervical cancers [43].

AETIOLOGY

Parity and/or number of sexual partners seem to be risk factors that may not apply to glandular lesions to the same extent as they do to their squamous counterparts, a greater proportion of adenocarcinomas than squamous cancers occurring in single women. The evidence pointing to a sexually transmitted factor is nevertheless strong and much work has gone into identifying papillomaviruses in cervical glandular cancers and pre-cancers. Whilst attempts have been made to implicate oral contraceptives [2], it is hard to separate their input from that of sexually transmitted elements as in any epidemiological study the two will inevitably co-segregate.

HPV

HPV presence has been sought by various methods, the most frequently used one having been *in situ* hybridization. Not surprisingly, the same technique has produced different levels of positivity in the hands of different workers, a summary of findings being presented in Table 13.3. Those authors discriminating between HPV types found slightly more HPV 18 than HPV 16.

For glandular atypias the figures for HPV positivity reported have usually been much lower, as shown in Table 13.3. For microglandular endocervical hyperplasia, in accord with the

Table 13.3   HPV positivity in glandular atypias and AIS

Condition	HPV positive/total cases (%)	Method	Reference
Glandular atypia	0/10 (0)	ISH[a]	49
	2/36 (6)	ISH	101
	2/36 (6)	ISH	44
	1/5 (20)	PCR[b]	102
	6/13 (46)	ISH	103
Total	11/100 (11)		
AIS	7/28 (25)	ISH	88
	10/37 (27)	Dot Blot	100
	15/36 (42)	PCR	102
	9/21 (43)	ISH	49
	4/7 (57)	PCR	37
	31/47 (66)	Dot blot + PCR	36
	12/21 (67)	ISH	44
	14/21 (67)	ISH	48
	21/30 (68)	ISH	103
	4/4 (100)	ISH	104
Total	127/252 (50.4)		

[a]ISH = *in situ* hybridizatio; [b]PCR = polymerase chain reaction.

benign nature of the condition, no HPV positivity was detected in a series of 16 cases [44].

Higgins *et al.* [35] used a variation on the *in situ* hybridization protocol by adopting ^{125}I-labelled riboprobes to detect transcriptionally active HPV. In their 42 cases, 37 containing AIS and five lower grades of atypia, 13 showed positive for HPV 16 and 28 positive for HPV 18. Collectively this gave 40 out of 42 (95%) positive for HPV.

The different sensitivities of detection techniques was amply demonstrated by Griffin *et al.* [45] and Alejo *et al.* [37], who compared the results of PCR and *in situ* hybridization. By combining the results from these series, PCR found nine out of 15 cases of AIS (60%) to be HPV-positive (one HPV 18, all others HPV 16), whereas *in situ* hybridization could show positivity in only two out of 15 cases (13%). As many of the previously mentioned studies detected higher incidences than these using *in situ* hybridization one wonders how much greater these would have been found in their cases by PCR. This also raises the question of whether the extreme sensitivity of the most recent techniques gives meaningful results, or

whether in fact a less sensitive technique might actually screen out those levels too low to be biologically relevant and detect only those where the level of virus is high enough to exert an effect on the system. This question is far from being answered at present.

Young *et al.* [46] were unable to detect HPV intranuclear signals in 21 cases of adenocarcinoma and suggest that techniques not allowing accurate localization of signal may give positive results because of contamination. They consider that HPV infection may not have a major role in the aetiology of adenocarcinoma of the cervix, although this view is not in accordance with the majority of authors. It may take the widespread adoption of the next advance in detection technology, *in situ* PCR [47], to clarify the presence and location of HPV within abnormal tissues.

An interesting observation on HPV in areas of CIN co-existing with AIS [48] was that not only were the incidences comparable, but the HPV type was the same in each coincident CIN/AIS pair. Whilst all the CIN-only samples examined by these authors contained HPV 16, those co-existing with AIS contained predominantly HPV 18, as per the AIS regions. This is strong evidence in support of the common origin of such co-existent dysplasias, with a metaplastic process or bilateral differentiation of reserve cells occurring subsequent to the initial infection.

HPV type 18 is generally the prevalent type in glandular cell abnormalities, whilst type 16 predominates in squamous cell abnormalities. Neither is unique to one condition or another and the distribution seems not to vary appreciably in different grades of lesion. Where variation is seen, however, it is in geographical distribution [49]. Type 18 is particularly prevalent in Africa and is more common than type 16 in Indonesia. Other HPV types are locally common and this should be borne in mind when comparing data prepared in different areas of the world.

Work on transfecting human cells *in vitro* with HPV has produced evidence in favour of HPV being responsible for one step in the multi-step oncogenic process. Human cervical squamous cells have been immortalized with both HPV types 16 and 18 and implanted under the skin of athymic mice, producing no evidence of tumour formation or invasion. Similar experiments were performed with cells passaged for varying numbers of population doublings after transfection. Early passage HPV 18-transformed cells produced dysplasia in 65% cases whilst HPV 16-transformed cells produced none. With late passage cells, however, dysplasia was produced 80% of the time by both [50, 51]. The more aggressive effects of HPV 18 compared with HPV 16 are as expected. The greater effect of late, as opposed to early, passage cells is of more interest and shows the action of additional mutations that accumulate over time in the presence of HPV genes. The mechanism operating here may well be that of the E6 gene of HPV repressing the action of the host *TP53* tumour suppressor gene. The expression of this gene can normally block entry to the cell cycle until any DNA damage is repaired. When it is repressed there is no longer control over the accumulation of

mutations, some of which will, by chance, play a role in oncogenesis.

Regarding the stage at which HPV exerts its effect, the highest rates of HPV detection are in AIS samples, with much lower levels in atypias. It seems that it is infection of already somewhat disturbed cells that stimulates progression to AIS. Further progression then depends upon genetic events, precipitated presumably by the deleterious effects of HPV E6 and E7 genes on the host cell DNA repair mechanisms, allowing accumulation of mutations. Once this process is initiated it may or may not require the maintenance of HPV sequences within the cell. If this scenario is accepted as a working hypothesis the question remains as to what causes the initial pre-HPV perturbations manifest as glandular atypia. Other aetiological factors, perhaps acting through disturbances of the local immune system, may be implicated. Alternatively, this early stage may also be caused by HPV but, being still in the process of developing into AIS, the full phenotypic picture has not yet evolved and the viral load has not had the chance to increase to detectable levels. Perhaps the situation is analogous to that in squamous epithelia, where HPV resides at low copy number in dividing cells and only multiplies as the cells differentiate on their passage to the epithelial surface.

Investigations of HPV incidence in adenocarcinomas [52–55] suggest a significant difference between the younger, pre-menopausal women, who are more likely to have detectable HPV, and the older, post-menopausal women, who are less likely to have detectable HPV. This age effect has not been reported for AIS, presumably because the age distribution is altogether lower, but it is nevertheless interesting to speculate on its significance as we assume that all such adenocarcinomas at one time passed through an AIS stage. It may be that in older age groups the likelihood of mutations in the *TP53* and *RB* tumour suppressor genes is higher, and the presence of the HPV E6 and E7 genes (which exert a direct effect on the host cell tumour suppressor systems) is thereby not obligatory. On the other hand it may be that most, if not all, tumours originate from an HPV infection giving rise to AIS, but once the process of adenocarcinogenesis has been initiated the virus is no longer necessary. Those tumours found in older subjects may have merely been there longer and given the virus time to fall back to undetectable levels. A corollary of this is that tumours in older subjects, which have taken longer to develop, will have had slower cell turnover and thus been given less scope for concomitant viral replication.

BIOLOGICAL INDICATORS OF DISEASE

In even the earliest stages of neoplastic change, alterations to normal cellular functioning are brought about that are manifest at a variety of functional levels. Any of these modifications that can be monitored will provide insight into the nature of the neoplasia and potentially the likelihood of its further progression. Investigations have been carried out to this end on several cellular systems.

ANTIGEN MARKERS

Human milk fat globulin (HMFG1) shows cytoplasmic staining in glandular atypias, as with endocervical neoplasia and not merely the luminal border staining seen in normal endocervical epithelia and microglandular hyperplasia [56]. HMFG1 is thought to be a marker of neoplastic change brought about by a disorder of the cells' glycocalyx and is thus a means of differentiating the benign nature of microglandular hyperplasia from that of the glandular atypias.

Carcinoembryonic antigen (CEA) has been examined in AIS with variable results, one paper reporting positivity in 93% [57], another found 67% positive [58], whilst a third found only 9% positive [59].

A monoclonal antibody, known as 1C5, has been produced [60], which appears specific for adenocarcinomas of the endocervical type. It stained positive in three out of four adenocarcinomas and six of six adenosquamous carcinomas, whilst only two-thirds stained positive for CEA.

Cytokeratins may be useful in diagnosing AIS [58] as a bank of antibodies against the various species of molecule are now available and their distribution is known to be highly specific to cell type.

Mutations in the *TP53* gene giving rise to altered expression of the p53 tumour suppressor protein are regarded as the commonest genetic alterations in cancer. A study looking at p53 overexpression in a number of cervical conditions [61] found a remarkably low positivity of only 11% in invasive adenocarcinomas and none at all in the samples of AIS they examined. These mutations are often seen as late events in tumorigenesis and so would not necessarily be expected in pre-malignant conditions such as AIS.

McCluggage *et al.* [62] used antibodies against the protein products of two oncogenes, Ki-6 and Mib 1, in an attempt to differentiate between adenocarcinomas, AIS and tubo-endometrial metaplasia. Interestingly the staining indices they derived enabled them to distinguish the tubo-endometrial metaplasia specimens from the others but could not distinguish AIS from adenocarcinoma. This might be seen as evidence suggestive of AIS and adenocarcinoma being on the same pathological continuum and separable from benign conditions not subjected to the same form of mitotic up-regulation.

The *c-myc* oncogene protein product has been shown by immunohistochemistry to be present within the nucleus of some normal cells at moderate levels. Polacarz *et al.* [63] looked at expression of this gene in glandular atypias, AIS and invasive adenocarcinomas. They reported nuclear and cytoplasmic staining in 17 out of 17 invasive adenocarcinomas, 10 out of 10 AIS and three out of four glandular atypias (Figure 13.3). In comparison only three out of 14 normal samples showed staining. In addition to the staining pattern

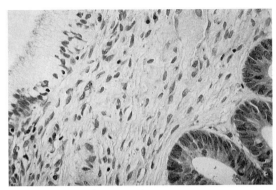

Figure 13.3   *c-myc* Oncogene expression in human cervix, demonstrated using antibody to p62^{c-myc} oncoprotein. Normal columnar epithelium (above left) and AIS (below right). The cell nuclei and cytoplasm in AIS stain darkly due to the final reaction product. Immunoperoxidase stain (reproduced with permission from Anderson *et al.* [73]).

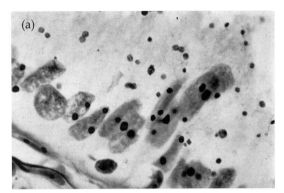

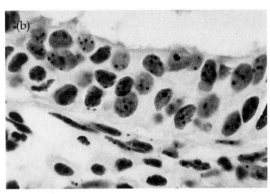

Figure 13.4   Silver staining nucleolar organizer regions (AgNORs) in nuclei. (a) Normal columnar epithelium (low dot counts); (b) AIS (high dot counts). Silver staining technique; final reaction product black (reproduced with permission from Anderson *et al.* [73]).

observed in histologically abnormal areas, they also reported staining in areas of normal morphology adjacent to AIS or atypias in nine out of 12 cases and adjacent to adenocarcinomas in nine of nine cases. This field change effect has close parallels with that observed with AgNORs (see below), as might be expected with both assays measuring aspects of cellular activity.

SILVER STAINING NUCLEOLAR ORGANIZER REGIONS (AgNORs)

As a measure of transcriptional activity, AgNORs have shown AIS to lie between the baseline levels of normal tissue and the increased activity of adenocarcinomas. Darne *et al.* studied a series of 20 AIS, 30 invasive adenocarcinomas and 15 normals [64]. They found that the AIS and adenocarcinomas were easily separable from the normals (Figure 13.4) and whilst there was overlap between the two disease groups, the mean figure for the adenocarcinomas was 150% that of the figure for AIS. As in so many ways AIS gives results intermediate between normals and cancer but its transcription level of more than twice that

of normal tissue makes it hard to see it as a benign condition. Another paper using the same technique [65] found AIS and invasive adenocarcinomas to show higher AgNOR levels than normals and microglandular hyperplasias, but the authors were unable to differentiate between the disease groups.

Another interesting observation has been the reporting of a field change effect in AgNORs [64]. In six out of 17 AIS cases examined it was noted that elevated AgNORs were visible in adjacent, morphologically normal glandular tissue. If AIS spreads by growth promoting substances secreted by cells exerting an effect on neighbouring cells, then this is precisely the sort of observation that might be expected.

Elevated transcription is almost certainly an early sign, which would be visible prior to the subsequent morphological changes. The reason that not all cases exhibited this phenomenon and other workers have failed to see it at all [66] may well be that it is a function of the rate of expansion of the area of AIS. If it is only a minority that make fast enough progress to show such a field change effect then it is surely this minority that is likely to progress to invasion.

ULTRASTRUCTURAL ANALYSIS

Ultrastructural analysis of AIS has shown the presence of tonofibres and secretary granules, which, whilst found in reserve cells, are not normally found in columnar epithelium [16]. This suggests the origin of the dysplastic change in reserve cells prior to differentiation.

PLOIDY

Little work has been done on the state of the genetic material in AIS. Jaworski and Jones [67] reported analysis of three AIS specimens, which showed two to be tetraploid and one diploid. Work performed in our laboratories has also found aberrant ploidy in AIS, where two of six AIS specimens showed aneuploid peaks on flow cytometry compared to 12 of 24 adenocarcinomas (M.E.S. Flynn, personal communication). Such an apparently high incidence of aneuploidy in AIS (albeit with small sample numbers) suggests that major genetic perturbations may be an earlier stage in adenocarcinogenesis than would be imagined. More work needs to be done on ploidy analysis as a means of diagnosing AIS or more particularly those AIS cases likely to progress to malignancy.

CLINICAL MANIFESTATIONS

The disease is usually asymptomatic although the occasional patient presents with abnormal bleeding [29, 32–34, 68, 69] or a discharge [33, 34].

More commonly, but not inevitably, the possibility of AIS is indicated by a smear abnormality [22, 33, 68, 70–72]. It may be unsuspected [22, 25, 32, 34, 70–73], and discovered during the investigation of a squamous abnormality, or a chance finding in a hysterectomy specimen. As awareness increases and cytological skills improve, the incidental discovery of glandular pre-malignancies is likely to diminish [26, 74].

CYTOLOGY

The investigation of cervical glandular abnormalities is most often instigated by laboratory findings rather than symptoms and signs, and under these circumstances, patients and clinicians are guided entirely by cytopathologists [69, 75]. Since the ensuing procedures have significant complication rates (see below), it is pertinent for laboratory personnel to be acutely aware of the implications of their recommendations. When laboratory observations, in isolation, direct clinical practice, there is added responsibility for those who report them. The cytology of pre-malignant glandular lesions of the cervix has yet to achieve acceptable levels of accuracy, a reality requiring repeated emphasis.

The cytological features of epithelial repair and regeneration in the cervix have been shown to mimic neoplasia [76, 77] and it is recognized that there are a number of benign and non-neoplastic entities which mimic glandular atypias [28, 73, 77] – inflammation, changes associated with IUCDs and hormonal contraception, ectropions, Müllerian remnants and microglandular hyperplasia. Given their generally self-limiting nature [28], minor glandular derangements should be followed up cytologically and only be an indication for biopsy if persistent, progressive and not without treatment of any co-existing infection.

Of all the cervical glandular abnormalities and their cytological manifestations, most attention has been given to AIS. Despite this, the cytological inaccuracies of the late 1970s [32] remain a reality of practice in the nineties. A recent study [78] retrospectively reviewed biopsy-based diagnoses subsequent to referral for smear results showing endocervical glandular atypia. Of those patients whose smears were seen as severe atypia suggestive of AIS, only 26% were diagnosed as AIS histologically, the others having squamous lesions (53%) or benign conditions (21%).

One common error, distressing since it may precipitate unnecessary cone biopsy, is the false-positive report [28, 33, 74, 77, 79–81]. Few studies have quantified this problem. Whittaker reports that three of 11 patients with index smears suggesting AIS had no neoplasia of any description on final histology. Nasu *et al.* [80] found AIS or adenocarcinoma in 82% of women with smears suggesting high-grade abnormalities. However, in an Australian centre only one in 47 cytological reports suggesting AIS was falsely positive [75]. Contributing factors to false positivity are the considerable overlap with squamous lesions [28, 68, 73, 77]; cervical endometriosis [74, 77]; and normal endometrial cells may be misinterpreted as atypical cervical glandular cells [33, 72, 74]. Smears from women with a history of cone biopsy are more likely to result in an incorrect diagnosis of atypical glandular cells [33, 74, 77, 82]. Other entities which create confusion are mesonephric remnants, tubal metaplasia [74, 77, 81] and microglandular hyperplasia [28]. Finally, it is reiterated that inflammatory and reparative changes can mimic neoplasia [69, 76, 77, 79].

Lee *et al.* [77], after scrutinizing and scoring 41 parameters in the cytology of 30 patients with AIS, are closest to solving these problems and they identify several major (feathering, cellular crowding and nuclear/cytoplasmic ratio >50%) and supplementary prerequisites (rosettes, mitotic figures and cellular strips) which, in combination, contribute to an accurate diagnosis of AIS. Contrary to earlier opinion [83], they feel that ciliated cells are more

likely to be benign, especially if macronucleoli are absent. Where doubt exists, a repeat smear may be useful in distinguishing patients less likely to need a cone biopsy. Recently van Aspert-van Erp *et al.* [84] used a bank of 46 architectural, cellular and nuclear features, 17 scored at three levels of expression, to evaluate a series of smears known to exhibit endocervical columnar cell atypias. They concluded that their accuracy in detecting grades of endocervical cell abnormalities at least equalled that of detecting squamous lesions.

Undercalling glandular abnormalities is potentially catastrophic and made possible by the considerable overlap on cytology between *in situ*, micro- and frankly invasive adenocarcinoma. Although progress has recently been made in identifying the cytological features associated with microinvasive adenocarcinoma, the accuracy is at best 50% [85]. Despite their success in limiting false-positivity, Laverty *et al.* [75] found histological evidence of invasion in 38% of 16 patients with smears suggesting AIS. Attempts at such *cytological* accuracy lack credence whilst confusion exists about the *histological* distinction between intra-epithelial and microinvasive adenocarcinoma [33]. Nguyen and Jeannot [72] and Betsill and Clark [83] describe the cytological features of AIS and microinvasive adenocarcinoma as being similar. Some authors claim that the cytological distinction between AIS and frank invasion is possible [22] but others are less successful – Bousfield *et al.* [26] undercalled (as AIS) five of 11 cases with deeply invasive adenocarcinoma and overcalled five of 22 cases of AIS/endocervical dysplasia. Lee *et al.* [77] admit that the distinction between AIS and invasive disease is difficult and logistical regression analysis required seven variables in 12 different combinations to fully separate these cases. Keyhani-Rofagha *et al.* [86] emphasize the importance of a necrotic background and macronucleoli as markers of invasion.

Hysterectomy should *never* be undertaken on the strength of a cytological report suggesting AIS, as this ignores the possibility of un-suspected invasion. An index cone biopsy is mandatory.

At present, cytology as a screening tool for cervical glandular pre-cancer has its limita-tions. The detection of glandular lesions is reliant on special skills and cytologists are less familiar with the relevant features [77, 83]. This has been shown by several authors [68, 69, 72] who have reviewed the cytology of patients with histologically confirmed AIS and found previously unidentified atypical glan-dular cells in as many as 20–25%. Poor sensi-tivity (illustrated in Table 13.4) may relate to lesions in the bases of glands being sampled haphazardly [23, 32, 77, 83] and, without the utilization of sampling techniques specific to the endocervix, foci high in the canal may be missed by cytological screening [72, 77].

The detection of glandular lesions is improved by the cytological identification of associated squamous abnormalities. In three reports, a smear suggestive of a squamous abnormality *alone* resulted in the diagnosis of 12 out of 36, five out of 17 and seven out of 15 patients with AIS, respectively [33, 68, 72].

Unfortunately, squamous abnormalities may predominate and result in the associated glandular abnormality being overlooked [22, 32, 69, 71, 73, 86] or mistaken as being of squa-mous origin [11, 73]. They may also obscure the grade of glandular abnormality [26].

Sensitivity may be improved by more wide-spread use of sampling techniques specific to

Table 13.4 Patients with histologically confirmed AIS and cervical cytology negative for glandular abnormality

Author	False-negative (%)
Muntz *et al.* [34]	55
Nguyen and Jeannot [72]	47
Cullimore *et al.* [70]	47[a]
Andersen and Arfman [68]	33
Östör *et al.* [33]	29
Luesley *et al.* [71]	26
Ayer *et al.* [85]	7

[a]Included CIGN.

the endocervix, such as brush smears [70, 72, 75], and by improvements in the cytological recognition of these abnormalities [70]. How-ever, one must beware of resulting overdiag-nosis [77, 79].

Special skills and experience are required for the accurate interpretation of glandular cyto-logical abnormalities [28, 77, 79]. It is a rapidly evolving field [69] and it seems obvious that careful review of any possible glandular lesion, by a senior cytologist, is mandatory before recourse to cone biopsy. Where there is co-existing infection, this should be treated appropriately and the smear repeated.

ENDOCERVICAL CURETTAGE

This procedure is a source of debate in the context of management of squamous lesions and although theoretically attractive, its role in the diagnosis of AIS is not established. As an ancillary procedure in suspect glandular dis-ease, it has been shown to have limitations and will miss lesions, particularly if small [41, 68, 87]. Nguyen and Jeannot report that endo-cervical curettage (ECC) was performed in four patients with AIS and was falsely nega-tive in three [72]. Poynor *et al.* see AIS in only 33% of ECCs done prior to diagnostic cone biopsy [39].

When diagnosing AIS, it is also important to exclude occult and unsuspected invasion [33, 68, 73]. For this, an adequate margin of sub-epithelial stroma is essential in any biopsy, as the epithelium concerned may be at the base of glands which may extend 6 mm [88]. ECC, especially in an out-patient setting, is extreme-ly unlikely to sample to this depth. ECC has also been known to interfere with the diagno-sis of AIS in cone biopsy specimens because of the resultant loss of surface epithelium [74–88], but may have a role if cytology is equivocal and suggestive of a lesser grade of glandular atypia. A negative ECC may justify a 'wait and see' policy and reduce the occur-rence of unnecessary cone biopsies [77]. This proposal awaits proper, prospective analysis.

COLPOSCOPY

The colposcopic characteristics of AIS are not well established [33, 68, 71, 73, 75, 81, 89], which consequently limits the diagnostic role of punch biopsy [32, 71, 83]. Considering its possible endocervical location, any cervical AIS is usually hidden within the canal and will not be visible colposcopically [33, 72]. If located ectocervically, the surface configuration and vascular pattern usually remain unchanged [33, 68]. Occasionally, AIS is suggested by dense acetowhite changes of the villous projections [68, 90] or epithelium that is densely actowhite, fragile and easily stripped [73]. However, any foci identified may only be part of a greater abnormality which is hidden beneath the surface [33, 73, 91] and is not usually colposcopically apparent [14, 71].

Despite being insensitive, colposcopy remains an essential adjunct in the investigation of AIS. As much as 50% of AIS is associated with CIN [22, 23, 31, 32, 34, 41, 68, 69, 72, 73, 83, 88, 92] and it is mandatory to define the extent of any squamous lesion [73], so that the appropriate biopsy approach may be selected. A recent case report [93] documents recurrent AIS at the vaginal vault after hysterectomy and emphasizes the importance of vaginal colposcopy during primary investigation, since co-existing vaginal intra-epithelial neoplasia would require a concomitant colpectomy. It is also essential to map the TZ preoperatively so that cone biopsy may be appropriately tailored.

DEFINITIVE DIAGNOSIS

The definitive diagnosis is ultimately dependent on a cone biopsy specimen [16, 25, 38, 68, 73]. Since the glandular epithelium lines the canal rather than the intravaginal portion of the cervix, the length of the cone theoretically should be greater than for squamous lesions.

The technique of conization is important. The width should be greater than the TZ as defined colposcopically. Since AIS is usually buried beneath the TZ and may involve the

bases of endocervical glands, it is recommended that the cone be cylindrical, which will ensure that the depths of TZ and glands within the canal are not 'undercut' [4, 29, 94] (Figure 13.5). The depth of gland involvement has been quantified and confirms that glands involved by AIS may extend as far as 6 mm into the underlying stroma [88]. This should be taken into account when fashioning the initial circumferential incision. If 25 mm in length, such a cylinder will excise the majority of AIS (>90%) and result in normal follow-up cytology in most patients (>90%) [70].

Since the cone biopsy may have deleterious effects in proportion to its size on subsequent pregnancies [95], young patients who wish to

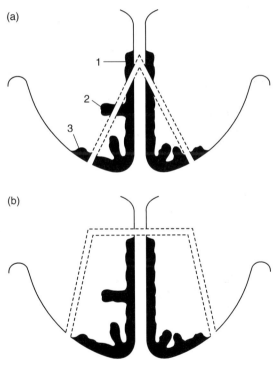

Figure 13.5 (a) Possible sites for incomplete excision of intra-epithelial neoplasia (glandular or squamous) during cone biopsy. (1) Cone apex; (2) glands in the stroma near the apex; (3) ectocervical limit of the transformation zone. (b) Colposcopically directed truncated cone or cylinder reduces these problems.

retain their fertility deserve special consideration. Bertrand *et al.* [29] localized more than 75% of AIS within 15 mm of the point of maximal cervical convexity and Nicklin *et al.* [88] found that 13 out of 14 patients less than 36 years of age had lesions within 10 mm of the squamo-columnar junction. Thus if the squamo-columnar junction (SCJ) is situated on the ectocervix, a cone of 15–20 mm in length would seem a reasonable compromise in the young patient.

No contemporary discussion of cone biopsy is complete without some reference to the type of knife utilized. Modern electrosurgical techniques are well established in the management of squamous neoplasia but their application to glandular lesions is open to debate. On first principles such practice would seem illogical – the ideal length of a cone biopsy specimen is 25 mm with most loops being only 20 mm in diameter – and we have witnessed the uncertainties created by the thermal artefact and fragmentation that accompanies this new technique. More compelling is the evidence [96] that cold knife cone biopsies are associated with less recurrences than those cut with electrosurgical loops. The same group has reported the occurrence of invasive disease after a 'hot knife' cone [97]. Opinions in favour of these new methods [98] are based on anecdote and we propose that cone biopsies for glandular abnormalities be fashioned with standard scalpels or the Beaver blade.

THERAPY

Radical hysterectomy is unnecessary if cervical glandular neoplasia is intra-epithelial. Any debate is between cone biopsy and simple hysterectomy for AIS. Early opinion suggesting the need for hysterectomy in all cases [32] was followed by appeals for greater conservatism [34, 70, 71, 73]. Crucial to this issue are the status of excision margins in the diagnostic cone specimen and the woman's desire for future fertility.

If excision lines are involved, there is often residual disease [33, 34, 38, 68, 71, 88, 96].

Further surgery, in the form of repeat cone biopsy or simple hysterectomy, is required [14, 33, 34, 73] to exclude any occult, undetected invasion [39] and to prevent progression of any remaining disease.

Many authors [33, 34, 68, 88] have demonstrated a reduction in residual AIS if histological review suggests complete excision. These findings have resulted in the opinion that, under such circumstances, cone biopsy is adequate treatment if future childbearing is desired. Recently, these conclusions have been prospectively tested and tentatively confirmed in 35 patients, where at a median follow-up interval of 12 months, cytology has been normal in 83% and no glandular lesion has been confirmed in the remainder [70]. More substantive follow-up on this cohort is awaited since, despite seemingly adequate excision, residual disease (so-called 'skip lesions') in the order of 40% has been reported [38, 39] and there is undisputed evidence that recurrences and invasive disease do indeed follow conservative management [39, 96]. None of this is surprising for an entity which is multifocal in about 15% of patients [14].

Attention should also be given to the doubts expressed about the reliability of post-cone cytology especially in the presence of cervical stenosis [34, 73]. Admittedly the interim results of the above-mentioned prospective study [70] suggest that cytological follow-up after conization for AIS may predict residual disease more accurately than observed involvement of excision margins, but more substantive data are needed before this can be applied to all women who have had cone biopsy for AIS.

With all of the above considerations in mind it is therefore reasonable to recommend a simple hysterectomy to the patient who has a completed family or attendent benign gynaecological disease.

Mindful of the frequency of associated squamous disease, intra-operative attention must be paid to the vaginal vault and adequate treatment of any co-existing vaginal intra-epithelial neoplasia or vaginal adenosis

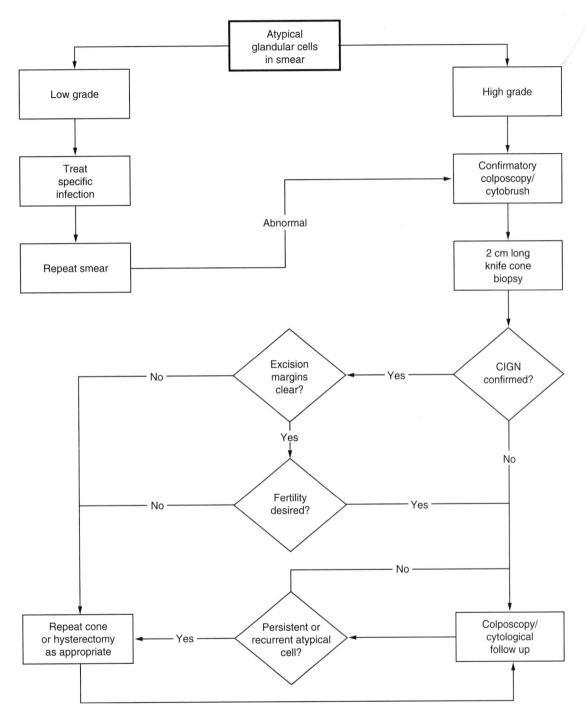

Figure 13.6 Recommended strategy for management of a woman presenting with atypical glandular cells in the cervical smear. (CIGN = cervical intraglandular neoplasia. Strategy assumes appropriate management of invasive disease uncovered by biopsy or apparent clinically.)

ensured. The importance of preoperative colposcopy is emphasized and Lugol's Iodine should be applied routinely as part of the operative preparation of the patient.

All treated patients require diligent cytological and colposcopic follow-up. The latter is necessary because of the occurrence of subsequent squamous abnormalities. This does not only apply to patients treated by cone biopsy as there have been reports of post-hysterectomy recurrence [93].

Figure 13.6 summarizes in flow chart form a reasonable management strategy for the woman who presents with atypical glandular cells in her cervical smear.

CONCLUSIONS

There is reasonable evidence of intra-epithelial precursors of cervical adenocarcinoma, namely AIS. At the present time, our knowledge of these precursor lesions is limited in many important aspects including their recognition (cytological, colposcopic, histological), their natural history (particularly their potential to progress to invasion), and when and how best to treat them when they are uncovered. There is an urgent need to refine all aspects of current knowledge so that less women are subjected to unnecessary cone biopsies and more are saved from invasive adenocarcinoma of the cervix.

As long as cone biopsy, with all its sequelae, remains integral to the investigation of this disease, management regimens should be under continuous scrutiny by centres for gynaecological oncology. Only then will the co-ordinated, integrated and multicentre prospective studies that the disease demands evolve into more appropriate and effective interventions.

REFERENCES

1. Shingleton, H.M., Gore, H., Bradley, D.H. and Soong, S.-J. (1981) Adenocarcinoma of the cervix. I. Clinical evaluation and pathological features. *Obstetrics and Gynecology*, **139**, 799–812.

2. Schwartz, S.M. and Weiss, N.S. (1986) Increased incidence of adenocarcinoma of the cervix in young women in the United States. *American Journal of Epidemiology*, **124**, 1045–7.

3. Silcocks, P.B., Thornton-Jones, H. and Murphy, M. (1987) Squamous and adenocarcinoma of the uterine cervix: a comparison using routine data. *British Journal of Cancer*, **55**, 321–5.

4. Sharp, F. and Cordiner, J.W. (1985) The treatment of CIN: cone biopsy and hysterectomy. *Clinical Obstetrics and Gynaecology*, **12**, 133–48.

5. Monaghan, J.M. (Ed.) (1989) Complications of cervical biopsy and cone biopsy. In *Complications in Surgery: Complications in the Surgical Management of Gynecological and Obstetric Malignancy*. Balliére Tyndall, London, pp. 42–61.

6. Fluhmann, C.F. (1961) *The Cervix Uteri and its Diseases*. W.B. Saunders, Philadelphia, PA.

7. Friedell, G.H. and McKay, D.G. (1953) Adenocarcinoma *in situ* of the endocervix. *Cancer*, **6**, 887–97.

8. Fox, H.H. and Buckley, C.H. (1991) *Pathology for Gynaecologists*, 2nd edn. Edward Arnold, London.

9. Barter, R.A. and Waters, E.D. (1970) Cyto- and histomorphology of cervical adenocarcinoma-in-situ. *Pathology*, **2**, 33–40.

10. Van Roon, E., Boon, M.E., Kurver, P.J.H. and Baak, J.P.A. (1983) The association between precancerous columnar and squamous lesions of the cervix: a morphometric study. *Histopathology*, **9**, 887–96.

11. Brown L.J.R. and Wells, M. (1986) Cervical glandular atypia associated with squamous intraepithelial neoplasia: a premalignant lesion? *Journal of Clinical Pathology*, **39**, 22–8.

12. Wells, M. and Brown, L.J.R. (1986) Glandular lesions of the uterine cervix: the present state of our knowledge. *Histopathology*, **10**, 777–92.

13. Gloor, E. and Hurlimann, J. (1986) Cervical intraepithelial glandular neoplasia (adenocarcinoma *in situ* and glandular dysplasia). *Cancer*, **58**, 1272–80.

14. Jaworski, R.C. (1990) Endocervical glandular dysplasia, adenocarcinoma *in situ*, and early invasive (microinvasive) adenocarcinoma of the uterine cervix. *Seminars in Diagnostic Pathology*, **7**, 190–204.

15. Anderson, M.C. and Hartley, R.B. (1980) Cervical crypt involvement by intraepithelial neoplasia. *Obstetrics and Gynecology*, **55**, 546–50.

16. Kudo, R. (1992) Cervical adenocarcinoma. *Current Topics in Pathology*, **85**, 81–111.

17. Yavner, D.L., Dwyer, I.M., Hancock, W.W. and Ehrmann, R.L. (1990) Basement membrane of cervical adenocarcinoma: an immunoperoxidase study of laminin and type IV collagen. *Obstetrics and Gynecology*, **76**, 1014–19.

18. Burghardt, E. (1984) Microinvasive carcinoma in gynecological pathology. *Clinical Obstetrics and Gynecology*, **11**, 239–57.

19. Östör, A.G. (1997) Microinvasive adenocarcinoma of the cervix: a clinicopathological study of 77 women. *Obstetrics and Gynecology*, **89**, 88–93.

20. Jones, M.W. and Silverberg, S.G. (1989) Cervical adenocarcinoma in young women: possible relationship to microglandular hyperplasia and use of oral contraceptives. *Obstetrics and Gynecology*, **73**, 984–9.

21. Daniel, E., Nuara, R., Morello, V., Nagar, C., Tralongo, V. and Tomasino, R.M. (1993) Microglandular hyperplasia of the uterine cervix. Histo-cytopathological evaluation, differential diagnosis and review of literature. *Pathologica*, **85**, 607–35.

22. Gloor, E. and Ruzicka, J. (1982) Morphology of adenocarcinoma *in situ* of the uterine cervix: a study of 14 cases. *Cancer*, **49**, 294–302.

23. Jaworski, R., Pacey, N., Greenberg, M. and Osborn, R. (1988) The histological diagnosis of adenocarcinoma *in situ* and related lesions of the cervix uteri. *Cancer*, **61**, 1171–81.

24. Hasumi, K. and Ehrmann, R.L. (1978) Clear cell carcinoma of the uterine cervix with an *in situ* component. *Cancer*, **42**, 2435–8.

25. Luesley, D. (1989) Cervical adenocarcinoma. In *Progress in Obstetrics and Gynecology* (ed. J. Studd). Churchill Livingstone, Edinburgh, pp. 369–87.

26. Bousfield, L., Pacey, F., Young, Q., Krumins, I. and Osborn, R. (1980) Expanded cytologic criteria for the diagnosis of adenocarcinoma *in situ* of the cervix and related lesions. *Acta Cytologica*, **24**, 283–96.

27. Boon, M.E., Baak, J.P.A. and Kurver, P.J.H., Overdiep, S.H. and Verdonk, G.W. (1981) Adenocarcinoma *in situ* of the cervix: an underdiagnosed lesion. *Cancer*, **48**, 768–73.

28. Whittaker, J. (1993) A critical analysis of the grading system at present in use at Groote Schuur Hospital Cytology Department for endocervical glandular atypical changes with recommendations for improved criteria and terminology. M. Med. (Anat. Path.) thesis, University of Cape Town, 1993.

29. Bertrand, M., Lickrish, G.M. and Colgan, T.J. (1987) The anatomic distribution of cervical adenocarcinomas *in situ:* implications for treatment. *American Journal of Obstetrics and Gynecology*, **157**, 21–5.

30. Teshima, S., Shimosato, Y., Kishi, K., Kasamatsu T., Ohmi, K. and Uei, Y. (1985) Early stage adenocarcinoma of the uterine cervix. Histopathologic analysis with consideration of histogenesis. *Cancer*, **56**, 167–72.

31. Colgan, T.J. and Lickrish, G.M. (1990) The topography and invasive potential of cervical adenocarcinoma *in situ*, with and without associated squamous dysplasia. *Gynecologic Oncology*, **36**, 246–9.

32. Christopherson, W.M., Nealon, N. and Gray, L.A. (1979) Noninvasive precursor lesions of adenocarcinoma and mixed adenosquamous carcinoma of the cervix uteri. *Cancer*, **44**, 975–83.

33. Östör, A.G., Pagano, R., Davoran, R.A.N., Fortune, D.W., Channen, W. and Rome, R. (1984) Adenocarcinoma-in-situ of the cervix. *International Journal of Gynaecological Pathology*, **3**, 179–90.

34. Muntz, H.G., Bell, D.A., Lage, J.M., Goff, B.A., Feldman, S. and Rice, L.W. (1992) Adenocarcinoma *in situ* of the uterine cervix. *Obstetrics and Gynecology*, **80**, 935–9.

35. Higgins, G.D., Phillips, G.E., Smith, L.A., Uzelin, D.M. and Burrell, C.J. (1992) High prevalence of human papillomavirus transcripts in all grades of cervical intraepithelial glandular neoplasia. *Cancer*, **70**, 136–46.

36. Duggan, M.A., Benoit, J.L., McGregor, S.E., Inoue, M., Nation, J.G. and Stuart, G.C. (1994) Adenocarcinoma *in situ* of the endocervix: human papillomavirus determination by dot blot hybridization and polymerase chain reaction amplification. *International Journal of Gynecological Pathology*, **13**, 143–9.

37. Alejo, M., Macado, I., Matias-Guiu, X. and Prat, J. (1993) Adenocarcinoma *in situ* of the uterine cervix: clinicopathalogical study of nine cases with detection of human papillomavirus DNA by *in situ* hybridizationa and polymerase chain reaction. *International Journal of Gynecological Pathology*, **12**, 219–23.

38. Im, D.D., Duska, L.R. and Rosenheim, N.B. (1995) Adequacy of conization margins in adenocarcinoma *in situ* of the cervix as a predictor of residual disease. *Gynecologic Oncology*, **59**, 179–82.

39. Poynor, E.A., Barakat, R.R. and Hoskins, W.J. (1995) Management and follow-up of patients with adenocarcinoma *in situ* of the uterine cervix. *Gynecologic Oncology*, **57**, 158–64.

40. Maier, R.C. and Norris, H.J. (1980) Co-existence of cervical intraepithelial neoplasia with

primary adenocarcinoma of the endocervix. *Obstetrics and Gynecology*, **56**, 361–4.

41. Qizilbash, A. (1975) In-situ and microinvasive adenocarcinoma of the uterine cervix. A clinical, cytologic and histologic study of 14 cases. *American Journal of Clinical Pathology*, **64**, 155–70.

42. Alva, J. and Lauchlan, S. (1975) The histogenesis of mixed carcinoma of the cervix: the concept of endocervical columnar cell dysplasia. *American Journal of Clinical Pathology*, **64**, 20–5.

43. Spriggs, A.I. (1984) Precancerous states of the cervix uteri. In *Precancerous States* (ed. R.L. Carter). Oxford University Press, London, pp. 317–55.

44. Okagaki, T., Tase, T., Twiggs, L.B. and Carson, L.F. (1989) Histogenesis of cervical adenocarcinoma with reference to human papillomavirus-18 as a carcinogen. *Journal of Reproductive Medicine*, **34**, 639–44.

45. Griffin, N.R., Dockey, D., Lewis, F.A. and Wells, M. (1991) Demonstration of low frequency of human papillomavirus DNA in cervical adenocarcinoma and adenocarcinoma *in situ* by the polymerase chain reaction and in situ hybridization. *International Journal of Gynecological Pathology*, **10**, 36–43.

46. Young, F.I., Ward, L.M. and Brown, L.J.R. (1991) Absence of human papilloma virus in cervical adenocarcinoma determined by *in situ* hybridization. *Journal of Clinical Pathology*, **44**, 340–1.

47. Nuovo, G.J. (1992) *PCR In Situ Hybridization: Protocols and Applications*. Raven Press, New York.

48. Tase, T., Okagaki, T., Clark, B.A., Twiggs, L.B., Ostrow, R.S. and Faras, A.J. (1989) Human papillomavirus DNA in adenocarcinoma *in situ*, microinvasive adenocarcinoma of the uterine cervix, and coexisting cervical squamous intraepithelial neoplasia. *International Journal of Gynecological Pathology*, **8**, 8–17.

49. Samaratunga, H., Cox, N. and Wright, R.G. (1993) Human papillomavirus DNA in glandular lesions of the uterine cervix. *Journal of Clinical Pathology*, **46**, 718–21.

50. Barnes, W.A., Woodworth, C.D., Waggoner, S.E., *et al.* (1990) Rapid dysplastic transformation of human genital cells by HPV 18. *Gynecologic Oncology*, **38**, 343–6.

51. Waggoner, S.E., Woodworth, C.D., Stoler, M.H., Barnes, W.A., Delgado, G. and DiPaulo, J.A. (1990) Human cervical cells immortalised *in vitro* with oncogenic human papillomavirus DNA differentiate dysplastically *in vivo*. *Gynecologic Oncology*, **38**, 407–12.

52. Wilczynski, S.P., Walker, J., Liao, S.-Y., Bergen, S. and Berman, M. (1988) Adenocarcinoma of the cervix associated with human papillomavirus. *Cancer*, **62**, 1331–6.

53. Bjersing, L., Rogo, K., Evander, M., Gerdes, U., Stendahl, U. and Wadell, G. (1991) HPV 18 and cervical adenocarcinomas. *Anticancer Research*, **11**, 123–7.

54. Matsuo, N., Iwasaka, T., Hayashi, Y., Hara, K., Mvula, M. and Sugimori, H. (1993) Polymerase chain reaction analysis of human papillomavirus in adenocarcinoma and adenosquamous carcinoma of the uterine cervix. *International Journal Gynaecological Obstetrics*, **41**, 251–6.

55. Milde-Langosch, K., Schreiber, C., Becker, G., Loning, T. and Stegner, H.-E. (1993) Human papillomavirus detection in cervical adenocarcinomas by polymerase chain reaction. *Human Pathology*, **24**, 590–4.

56. Brown, L.J.R., Griffin, N.R. and Wells, M. (1987) Cytoplasmic reactivity with the monoclonal antibody HMFG1 as a marker of cervical glandular atypia. *Journal of Pathology*, **151**, 203–8.

57. Kluzack, T. and Kraus, F. (1986) Adenocarcinoma of the uterine cervix: immunohistochemical staining for carcinoembryonic antigen and human papillomavirus. *Laboratory Investigations*, **54**, 32A.

58. Hurlimann, J. and Gloor, E. (1984) Adenocarcinoma *in situ* and invasive adenocarcinoma of the uterine cervix. *Cancer*, **54**, 103–9.

59. Tobon, H. and Dave, H. (1988) Adenocarcinomas *in situ* of the cervix. Clinicopathological observations of 11 cases. *International Journal of Gynaecology and Obstetrics*, **7**, 139–51.

60. Kudo, R., Sasano, H., Koizumi, M., Orenstein, J.M. and Silverberg, S.G. (1990) Immunohistochemical comparison of new monoclonal antibody 1C5 and carcinoembryonic antigen in the differentiatial diagnosis of adenocarcinoma of the uterine cervix. *International Journal of Gynecological Pathology*, **9**, 325–36.

61. Holm, R., Skomedal, H., Helland, A., Kristensen, G., Borresen, A.L. and Nesland, J.M. (1993) Immunohistochemical analysis of p53 protein overexpression in normal, premalignant and malignant tissues of the cervix uteri. *Journal of Pathology*, **169**, 21–6.

62. McCluggage, W.G., Maxwell, P., McBride, H.A., Hamilton, P.W. and Bharucha, H. (1995) Monoclonal antibodies Ki-67 and Mib1 in the distinction of tuboendometrial metaplasia from endocervical adenocarcinoma and ade-

nocarcinoma *in situ* in formalin fixed material. *International Journal of Gynecological Pathology*, **14**, 209–16.

63. Polacartz, S.V., Darne, J., Sheridan, E.G., Ginsberg, R. and Sharp, F. (1991) Endocervical carcinoma and precursor lesions: c-myc expression and the demonstration of field changes. *Journal of Clinical Pathology*, **44**, 896–9.

64. Darne, J.F., Polacarz, S.V., Sheridan, E., Anderson, D., Ginsberg, R. and Sharp, F. (1990) Nucleolar organiser regions in adenocarcinoma *in situ* and invasive adenocarcinoma of the cervix. *Journal of Clinical Pathology*, **43**, 657–60.

65. Allen, J.P. and Gallimore, A.P. (1992) Nucleolar organizer regions in benign and malignant glandular lesions of the cervix. *Journal of Pathology*, **166**, 153–6.

66. Cullimore, J.E., Rollason, T.P. and Marshall, T. (1989) Nucleolar organiser regions in adenocarcinoma *in situ* of the endocervix. *Journal of Clinical Pathology*, **42**, 1276–80.

67. Jaworski, R.C. and Jones, A. (1990) DNA ploidy studies in adenocarcinoma *in situ* of the uterine cervix. *Journal of Clinical Pathology*, **43**, 435–6.

68. Andersen, E.S. and Arffmann, E. (1989) Adenocarcinoma *in situ* of the uterine cervix: a clinico-pathologic study of 36 cases. *Gynecologic Oncology*, **35**, 1–7.

69. Ayer, B., Pacey, F., Greenberg, M. and Bousfield, L. (1987) The cytologic diagnosis of adenocarcinoma-in-situ of the cervix uteri and related lesions. I. Adenocarcinoma-in-situ. *Acta Cytologica*, **31**, 397–411.

70. Cullimore, J.E., Luesley, D.M., Rollason, T.P., *et al.*, (1992) A prospective study of conization of the cervix in the management of cervical intraepithelial glandular neoplasia (CIGN) – a preliminary report. *British Journal of Obstetrics and Gynaecology*, **99**, 314–18.

71. Luesley, D.M., Jordan, J.A., Woodman, C.V.J., Watson, N., Williams, D.R. and Waddel, C. (1987) A retrospective review of adenocarcinoma-in-situ and glandular atypia of the uterine cervix. *British Journal of Obstetrics and Gynaecology*, **94**, 699–703.

72. Nguyen, G. and Jeannot, A.B. (1984) Exfoliative cytology of *in situ* and microinvasive adenocarcinoma of the uterine cervix. *Acta Cytologica*, **28**, 461–7.

73. Anderson, M., Jordan, J., Morse, A. and Sharp, F. (1992) *Integrated Colposcopy*. Chapman & Hall Medical, London.

74. Pacey, F., Ayer, B. and Greenberg, M. (1988) The cytologic diagnosis of adenocarcinoma *in*

situ of the cervix uteri and related lesions. III. Pitfalls in diagnosis. *Acta Cytologica*, **32**, 325–30.

75. Laverty, C.R., Farnsworth, A., Thurloe, J. and Bowditch, R. (1988) The reliability of a cytological prediction of cervical adenocarcinoma *in situ*. *Australia and New Zealand Journal of Obstetrics and Gynaecology*, **28**, 307–12.

76. Giersson, G., Woodworth, F.E., Patten, S.F. and Bonfiglio, T.A. (1977) Epithelial repair and regeneration in the uterine cervix. I. An analysis of the cells. *Acta Cytologica*, **21**, 371–8.

77. Lee, K.R., Manna, E.A. and Jones, M.A. (1991) Comparative cytologic features of adenocarcinoma *in situ* of the uterine cervix. *Acta Cytologica*, **35**, 117–26.

78. Lee, K.R., Manna, E.A. and St. John, T. (1995) Atypical endocervical glandular cells: accuracy of cytological diagnosis. *Diagnostic Cytopathology*, **13**, 202–8.

79. Lee, K.R. (1988) False-positive diagnosis of adenocarcinoma *in situ* of the cervix. *Acta Cytologica*, **32**, 276–7.

80. Nasu, I., Meurer, W. and Fu, Y.S. (1993) Endocervical glandular atypia and adenocarcinoma: a correlation of cytology and histology. *International Journal of Gynecological Pathology*, **12**, 208–18.

81. Novotny, D.B., Maygarden, S.J., Johnson, D.E. and Frable, W.J. (1992) Tubal metaplasia. A frequent potential pitfall in the cytologic diagnosis of endocervical glandular dysplasia on cervical smears. *Acta Cytologica*, **36**, 1–10.

82. Lee, K.R. (1993) Atypical glandular cells in cervical smears from women who have undergone cone biopsy. A potential diagnostic pitfall. *Acta Cytologica*, **37**, 705–9.

83. Betsill, W.L. and Clark H. (1986) Early endocervical glandular neoplasia. I. Histomorphology and cytomorphology. *Acta Cytologica*, **30**, 115–26.

84. van Aspert-van Erp, A.J., van't Hof-Grootenboer, A.B., Brugal, G. and Vooijs, G.P. (1995) Endocervical columnar cell intraepithelial neoplasia. II. Grades of expression of cytomorphologic criteria. *Acta Cytologica*, **39**, 1216–32.

85. Ayer, B., Pacey, F. and Greenberg, M. (1988) The cytologic diagnosis of adenocarcinoma-in-situ of the cervix uteri and related lesions. II. Microinvasive adenocarcinoma. *Acta Cytologica*, **32**, 318–24.

86. Keyhani-Rofagha, S., Brewer, J. and Prokorym, P. (1995) Comparative cytologic findings of *in situ* and invasive adenocarcinoma of the uterine cervix. *Diagnostic Cytopathology*, **12**, 120–5.

87. Weisbrot, I.M., Stabinsky, C. and Davis, M. (1972) Adenocarcinoma-in-situ of the uterine cervix. *Cancer*, **29**, 1179–87.

88. Nicklin, J.L., Wright, R.G., Bell, J.R., Samaratunga, H., Cox, N.C. and Ward, B.G. (1991) A clinicopathological study of adenocarcinoma *in situ* of the cervix. The influence of cervical HPV infection and other factors, and the role of conservative surgery. *Australia and New Zealand Journal of Obstetrics and Gynaecology*, **31**, 179–83.

89. Lickrish, G.M., Colgan, T.J. and Wright, V.C. (1993) Colposcopy of adenocarcinoma *in situ* and invasive adenocarcinoma of the cervix. *Obstetrics and Gynecology Clinics of North America*, **20**, 111–22.

90. Coppleson, M. and Pixley, E.C. (1981) Colposcopy of the cervix. In *Gynecologic Oncology* (ed. M. Coppleson). Churchill Livingstone, Edinburgh, pp. 205–24.

91. Ueki, M. (1984) Clinical characteristics and colposcopic findings. In *Cervical Adenocarcinoma*. Ishiyaku EuroAmerica, St. Louis, MO, pp. 22–6.

92. Burghardt, E. (1981) Pathology of preclinical invasive carcinoma of cervix. In *Gynecologic Oncology* (ed. M. Coppleson). Churchill Livingstone, Edinburgh, pp. 434–50.

93. Cullimore, J.E., Luesley, D.M., Rollason, T.P., Waddell, C. and Williams, D.R. (1989) A case of glandular intraepithelial neoplasia involving the cervix and vagina. *Gynecologic Oncology*, **34**, 249–52.

94. Nevin, J., Denny, L., Megevand, E., van der Merwe, E. and Hendricks, S. (1996) The role of vasoconstrictor therapy in cold knife cone biopsy. *Journal of Gynecologic Techniques*, **2**, 129–33.

95. Leiman, G., Harrison, M.A. and Rubin, A. (1979) Pregnancy following conization of the cervix: complications related to cone size. *American Journal of Obstetrics and Gynecology*, **136**, 14–8.

96. Widrich, T., Kennedy, A., Myers, T.M., Hart, W.R. and Wirth, S. (1996) Adenocarcinoma *in situ* of the uterine cervix: management and outcome. *Gynecologic Oncology*, **61**, 304–8.

97. Kennedy, A.W., El Tabbakh, G.H., Biscotti, C.V. and Sirth, S. (1995) Invasive adenocarcinoma of the cervix following LLETZ (large loop excision of the transformation zone) for adenocarcinoma *in situ*. *Gynecologic Oncology*, **58**, 274–7.

98. Townsend, D.E. (1996) In correspondence. *Gynecologic Oncology*, **60**, 332.

99. Wolf, J.K., Levenback, C.M., Malpica, A., Morris, M., Burke, T. and Mitchell, M.F. (1996) Adenocarcinoma *in situ* of the cervix: significance of cone biopsy margins. *Obstetrics and Gynecology*, **88**, 82–6.

100. Duggan, M.A. (1993) The human papillomavirus status of 114 endocervical adenocarcinoma cases by dot blot hybridization. *Human Pathology*, **24**, 121–5.

101. Tase, T., Okagaki, T., Clark, B.A., Twiggs, L.B., Ostrow, R.S. and Faras, A.J. (1989) Human papillomavirus DNA in glandular dysplasia and microglandular hyperplasia: presumed precursors of adenocarcinoma of the uterine cervix. *Obstetrics and Gynecology*, **73**, 1005–8.

102. Lee, K.R., Howard, P., Heintz, N.H. and Collins, C.C. (1993) Low prevalence of human papillomavirus type 16 and type 18 in cervical adenocarcinoma *in situ*, invasive adenocarcinoma and glandular dysplasia by polymerase chain reaction. *Modern Pathology*, **6**, 433–7.

103. Leary, J., Jaworski, R. and Houghton R. (1991) In-situ hybridization using biotinylated DNA probes to human papillomavirus in adenocarcinoma-in-situ and endocervical glandular dysplasia of the uterine cervix. *Pathology*, **23**, 85–9.

104. Nielsen, A.L. (1990) Human papillomavirus type 16/18 in uterine cervical adenocarcinoma *in situ* and adenocarcinoma. A study by *in situ* hybridization with biotinylated DNA probes. *Cancer*, **65**, 2588–93.

14

LOWER GENITAL TRACT INTRA-EPITHELIAL NEOPLASIA IN IMMUNOSUPPRESSED PATIENTS AND PATIENTS INFECTED WITH HUMAN IMMUNODEFICIENCY VIRUS

Ann M. Wright and Michelline Byrne

INTRODUCTION

As the year 2000 approaches, doctors will certainly be encountering an increasing number of women in their everyday practice who are immunocompromised. There are two main reasons for this: first is the recognition of human immunodeficiency virus (HIV), the causative agent of the acquired immunodeficiency syndrome (AIDS), characterized by a profound defect in cell-mediated immunity and which is now estimated to affect at least 9 million women world-wide. Second is the increasing number of women receiving renal and other organ transplants whose long-term prognosis of survival is greatly improved but who need lifelong iatrogenic immunosuppression with drugs to control organ rejection.

Immunosuppressed women have special problems and needs related to their sexual health; in particular, they seem to be at increased risk of developing cervical and other lower genital tract neoplasia. It is not yet clear, however, whether this will lead to an increased incidence of lower genital tract cancers.

After the British government's white paper of 1992 on the health of the nation [1], which targeted, amongst other areas, a reduction in both cervical carcinoma and HIV infection, this chapter aims to address and review the issues surrounding the occurrence of lower genital neoplasia specific to immunocompromised women, particularly those infected with HIV.

IMMUNOSUPPRESSION

WHAT IS IT?

Immunity can be defined as resistance of the body to the effects of a deleterious agent. Anything which compromises this resistance can therefore give rise to immunodeficiency. The mechanism and level of immunosuppression conferred by a disease or drug varies depending on its aetiology or the particular drug. The diagnosis may be descriptive or it may be made on more objective grounds. History of certain illnesses, intercurrent infections or medications may be suggestive of

Cancer and Pre-cancer of the Cervix
Edited by D.M. Luesley and R. Barrasso. Published in 1998 by Chapman & Hall, London. ISBN 0 412 56600 1.

immunosuppression, but a more objective measure which is reproducible and can allow comparisons of immunocompromise is a measure of T-cell lymphocyte counts, more specifically, the CD4 subset.

WHO DOES IT AFFECT?

Women can be immunocompromised for a number of reasons. In brief, immunosuppression may arise secondary to illness which may be congenital, i.e. hereditary immune deficiency, or acquired, e.g. renal disease, or it may be produced iatrogenically by a variety of drugs, e.g. cytotoxic agents used for the treatment of cancer, steroids or azathioprine.

Table 14.1 lists some of the diseases treated with immunosuppressive drugs. Women with insulin-dependent diabetes or asthma are not considered immunocompromised even if they require steroids.

Table 14.1 Some of the diseases treated with immunosuppressive drugs

- Rheumatoid arthritis
- Polymyositis
- Scleroderma
- Dermatomyositis
- Polyarthritis
- Chronic juvenile arthritis
- Behçet's syndrome
- Pemphigus
- Psoriasis
- Glomerulonephritis
- Goodpasture's syndrome
- Necrotizing vasculitis
- Wegener's granulomatosis
- Lymphoid granulomatosis
- Multiple sclerosis
- Pure red cell aplasia
- Chronic uveitis
- Chronic active hepatitis

WHERE ARE THEY?

As already mentioned, immunodeficiency can be a difficult condition to recognize, especially in the absence of a history of any of the associated conditions or drugs, and because of the widely differing aetiologies these patients may present via a number of sources. Accurate data are available for the prevalence and incidence of HIV infection in women but there are few such data in relation to the other causes of immunosuppression. As awareness increases, data surrounding other causes of immunosuppression would be expected to increase also. When the academic department of public health medicine at St Mary's Hospital undertook a needs assessment of sexual health services for immunocompromised women in Parkside health district [2], data to assess the number of immunocompromised women, both new and existing, were obtained in many ways, which included reporting by medical staff treating them. This was unwieldy and not easily reproducible as reporting depended, to some extent, on the interest of the carers in such data and highlighted the need for better information systems to allow identification of immunocompromised women with accurate statistics to target prevention and treatment strategies.

HIV

INTRODUCTION

AIDS has been recognized for over 15 years now and although the text aims to cover all aspects of immunosuppression it will concentrate on particular aspects relating to HIV/AIDS, as this is anticipated to be the commonest cause of immunodeficiency in women in the next few years.

EPIDEMIOLOGY

The current picture

HIV has spread to almost every population on the globe. For the year ending 1995 the joint

United Nations programme on HIV/AIDS (UNAIDS) estimated that 20.1 million infections (over 11 million men and about 9 million women) had occurred since the start of the epidemic in the early 1980s. Women are increasingly becoming infected with HIV and at a significantly younger age than men. By the year 2000 more than 14 million women are expected to have been infected [3]. Infection tends to dominate urban rather than rural areas [4]. The global programme on AIDS (GPA) of the World Health Organization (WHO) collects and provides information on the number of HIV infections and AIDS cases recorded throughout the world. Table 14.2 shows the number of reported AIDS cases and estimated adult AIDS cases by region at the end of 1994 [5]. Table 14.3 shows the estimates of the numbers of adults in major areas of the world who had been infected with HIV by late 1994 since the start of the pandemic.

Table 14.2 Number of reported and estimated adult AIDS cases by region at the end of 1994

	Reported		*Estimated*	
	(%)	*Number*	*(%)*	*Number*
USA	39	401,789	9	405,000
Africa	34	347,713	70	>3,150,000
Europe	12.5	127,886	4	180,000
Americas	12	124,893	>9	>450,000
Asia	2	17,057	<6	<270,000
Oceania	0.5	5,735	<1	<45,000
Total		1,025,073		>4,500,000

Table 14.3 Estimates of the numbers of adults in major areas of the world who had been infected with HIV by late 1994 since the start of the pandemic

North America	>750,000
Latin America/Caribbean	>1,500,000
Western Europe	450,000
North Africa/Middle East	>100,000
Sub-Saharan Africa	>8,000,000
Eastern Europe/central Asia	>50,000
East Asia/Pacific	>50,000
South and south-east Asia	2,500,000
Australasia	>20,000

As can be seen from the tables the highest estimated prevalence of HIV infection is in sub-Saharan Africa, where more than 8 million adults had been infected by the end of 1994. By the end of 1995 this figure had risen to 12 million, with over 4 million cases in south and south-east Asia and 1.5 million cases in Latin America and the Caribbean. By the end of 1995 approximately 750,000 individuals were infected in the USA, and in western Europe about half a million infections were estimated to have occurred [6]. By the end of 1996 a total of 28,447 HIV infections had been reported in England and Wales, with about 10% being acquired through heterosexual transmission [7].

Future trends

There is little evidence yet that the global HIV epidemic is abating [4]. In the USA and western Europe it is thought that the annual incidence of HIV infection peaked in the first half of the 1980s with the relatively explosive spread of HIV among promiscuous gay men and injecting drug users sharing needles. Subsequently the annual incidence of infections in these countries has decreased. In sub-Saharan Africa, extensive spread of HIV is thought to have started in the late 1970s and early 1980s and in this region the annual incidence is still rising. The rate of spread within Africa itself, however, is not uniform and while the incidence of HIV infection appears to have reached a steady state in Central Africa, there has been a huge increase in the number of cases in East Africa in the last few years, which is thought to be related to migration [8]. In East Asia and China the incidence is also increasing. In Asia extensive HIV transmission was initially documented in the late 1980s in only a few countries in the south and south-east. Since then, spread in these countries, notably Vietnam, has been rapid [9]. More recently it has been observed that the incidence of sexually transmitted disease has been rising in eastern Europe and this is expected to lead to a corresponding increase in HIV infection in these countries over the next few years [10]. Because of the interval between HIV infection and development of AIDS, the annual incidence of adult AIDS mirrors the annual incidence of adult HIV infection by about 10 years. In England and Wales, 2986 cases of HIV infection were reported in 1996, the highest annual total to date and there were 1862 new cases of AIDS [7]. The incidence of HIV infection due to heterosexual transmission is expected to increase steadily.

MALE-TO-FEMALE RATIO

In developed countries (North America and Western Europe), HIV has tended to largely affect men, although the proportion of affected women is increasing on a yearly basis. By contrast, in Africa, Asia and Latin America the male-to-female sex ratio is gradually decreasing and is now almost equal [4]. In Uganda three population-based HIV serosurveys have shown women to have a higher infection rate than men [11]. Thus of the 20 million infected adults world-wide nearly half of them are women. As the number of women affected increases the number of immunosuppressed women will also be expected to rise, including women with 'full blown' AIDS.

TRANSMISSION

Heterosexual transmission is the predominant mode of spread of HIV in the developing world and, as it now exceeds transmission due to intravenous drug use (IVDU) in the western world, it is responsible for the majority of HIV infections world-wide [3].

In conclusion, HIV is affecting an increasing number of women, with over 9 million women affected world-wide to date. In addition, approximately one million women are being infected on an annual basis, mainly in the developing countries but also elsewhere. As a substantial number of these women will go on to develop AIDS in the next few years HIV infection is rapidly becoming the single most commonly identified cause of immunosuppression in women.

IMMUNOSUPPRESSION AND CANCER

INTRODUCTION

There is an increased prevalence of malignancy among all immunocompromised patients, whatever the aetiology of their immunosuppression [12]. Patients with congenital immunodeficiency have a 2–15% incidence of carcinoma which is about 10,000-times that of the normal population [13]. In one study of patients with Wiskott–Aldrich syndrome the prevalence of malignancy was 15.4% [14]. Interestingly, certain types of carcinoma predominate with lymphomas, leukaemias and cutaneous malignancies being the most common [15]. Cervical neoplasia is more common in immunocompromised women [12].

Immunosuppressed patients also show an increased incidence of certain viral infections, most notably human papillomavirus (HPV) and herpes simplex virus (HSV), both of which have been associated with the development of certain cancers [16].

Table 14.4 shows the relative risk for specific malignancies of patients immunocompromised following renal transplantation.

Table 14.4 The relative risk for specific malignancies of patients immunocompromised following renal transplantation

Author	Cancer type	Relative risk
Hostell *et al.* [17]	Squamous cell (SCC)	36.4
Penn [12]	Kaposi's sarcoma (KS)	500
Porreco *et al.* [18]	Cervix	13

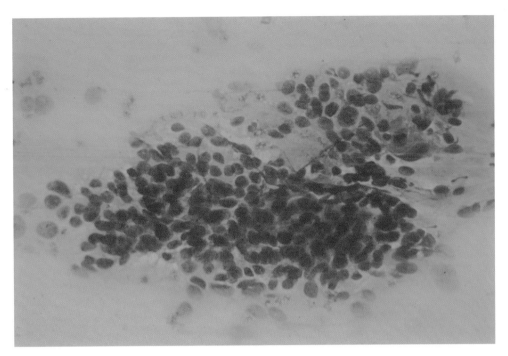

Plate 17 AGUS – the columnar cells have an increasing nuclear/cytoplasmic ratio.

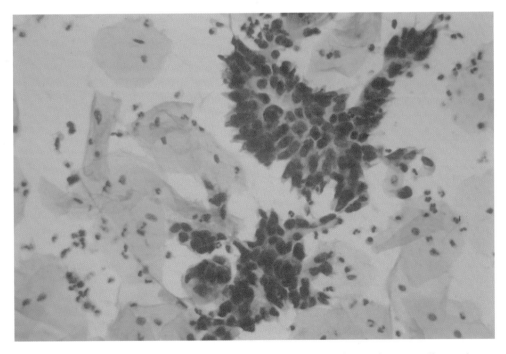

Plate 18 ACIS – the architecture and the nuclear/cytoplasmic ratio of the columnar cells are characteristic of a neoplastic process.

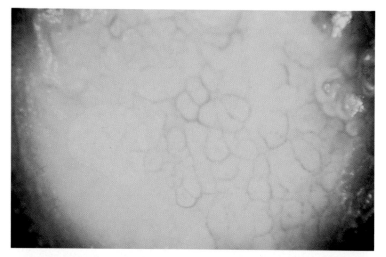

Plate 19 Cervix after the application of 5% acetic acid demonstrating a coarse mosaic pattern consistent with a diagnosis of high-grade intraepithelial neoplasia.

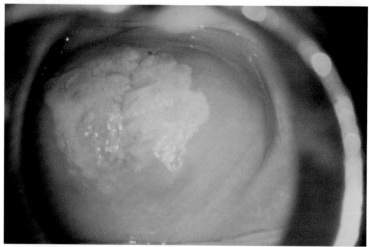

Plate 20 Cervix after the application of 5% acetic acid demonstrating acetowhite changes but no marked capillary patterns. The edge of the lesion is well defined and subsequent biopsy confirmed CIN 2.

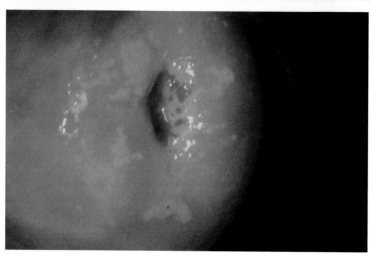

Plate 21 Cervix after the application of 5% acetic acid demonstrating acetowhite changes outside the transformation zone. These changes are consistent with HPV infection.

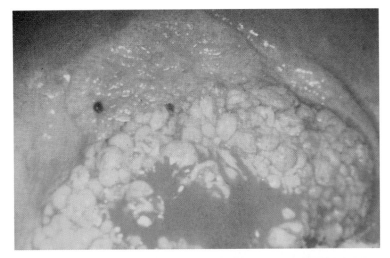

Plate 22 Cervix after the application of 5% acetic acid. This shows columnar epithelium (stained white) and a raised area peripheral to it. Contact bleeding is also apparent. Subsequent cone biopsy confirmed a stage Ib adenosquamous cervical cancer.

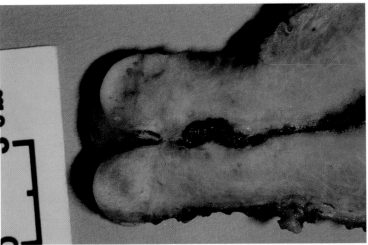

Plate 23 Hysterectomy specimen performed after a deep loop excisional procedure. This bisected view of the cervix shows complete stenosis of the residual endocervical canal with old blood in the residual canal and endometrial cavity.

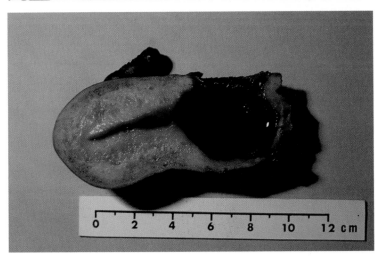

Plate 24 Another case of complete stenosis showing ballooning of the residual cervix as a result of menstrual obstruction.

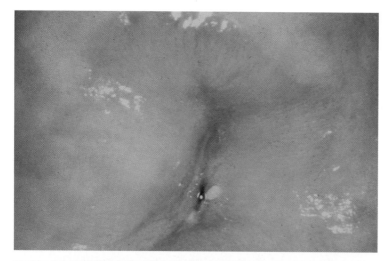

Plate 25 Colposcopy of a cervix 6 months following treatment by loop excision. The canal is patent but very narrow and it is impossible to accurately assess the new transformation zone.

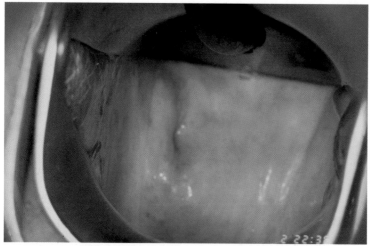

Plate 26 Vaginal vault after hysterectomy demonstrating a pocket in the right vaginal angle.

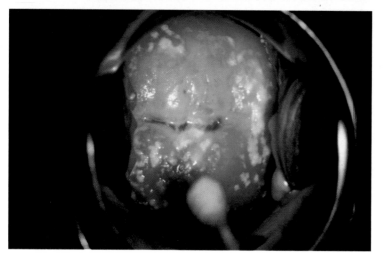

Plate 27 Vaginal candidiasis in an HIV-infected woman.

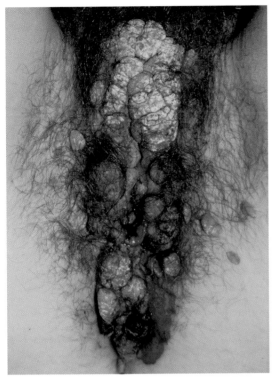

Plate 28 Extensive anogenital warts in an HIV-infected woman.

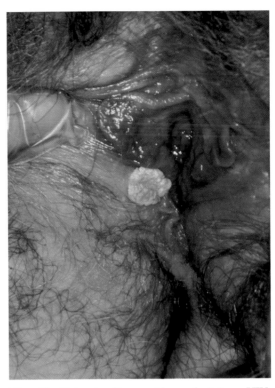

Plate 30 Extensive Bowenoid papulosis in an HIV-infected woman.

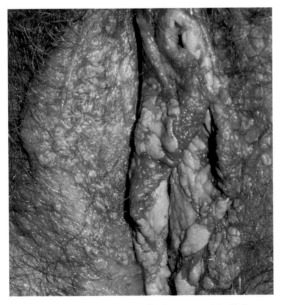

Plate 29 Concomitant vulval intra-epithelial neoplasia and genital herpes following renal transplantation.

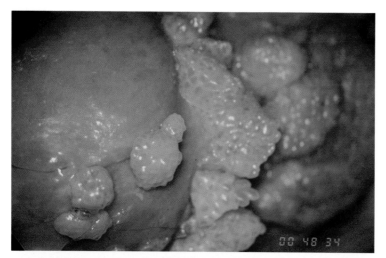

Plate 31 Typical accuminate warts.

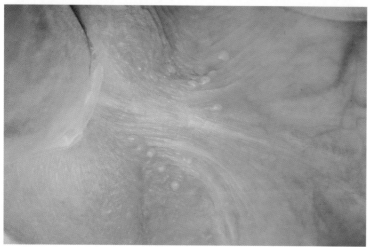

Plate 32 Pearly papules.

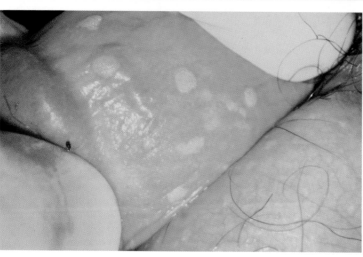

Plate 33 Flat warts, apparent after the application of acetic acid.

Plate 34 Non-specific white reaction during yeast infection.

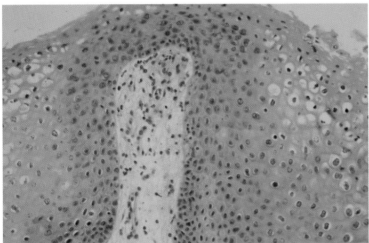

Plate 35 Histology of an endophytic wart.

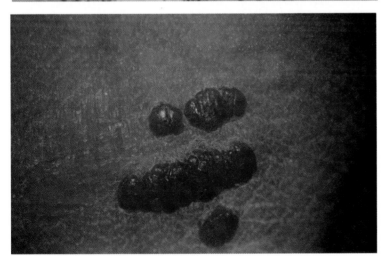

Plate 36 Pigmented papules, showing histology of intra-epithelial neoplasia.

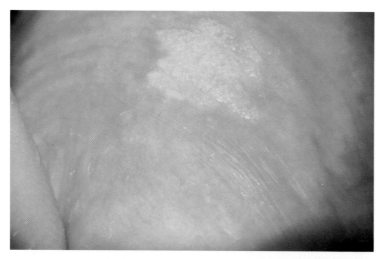

Plate 37 Spiked, leucoplastic area, showing histology of intra-epithelial neoplasia.

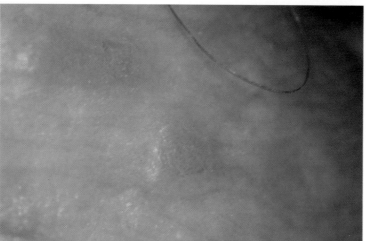

(a)

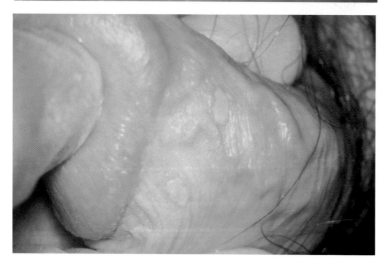

(b)

Plate 38 (a) Red, well-delimited areas, (b) strongly whitening after the acetic acid test. Histology will show intra-epithelial neoplasia.

HIV AND CARCINOMA

General

At least one in six people with AIDS in Europe and the USA develops a carcinoma during the course of their illness [19, 20]. The commonest types are: Kaposi's sarcoma, non-Hodgkin's lymphoma (NHL) and squamous cell carcinoma of the mouth and anogenital tract (homosexual men) [19, 20]. In fact, it was an epidemic of Kaposi's sarcoma associated with profound immunosuppression among homosexual men in 1981 that lead to the recognition of AIDS [21]. Malignancy may present atypically in these patients and be of high grade, with a tendency to run a more aggressive course. Although the development of carcinoma has been related to HIV-induced immunosuppression, it is also found in relatively immune-competent HIV-infected patients, suggesting other cofactors are involved.

Cervical carcinoma

In 1993, based on both the epidemiological evidence emerging of an association between HIV infection and cervical intra-epithelial neoplasia (CIN), and the adverse effects HIV infection appeared to have on the clinical course and treatment of both CIN and cervical carcinoma, biopsy-proven invasive cervical cancer (ICC) was added to the AIDS surveillance definition by both the centre for disease control (CDC) and the UK expert advisory board [22].

To date there has, however, been little hard evidence of an actual increased incidence of ICC in immunocompromised women, and evidence for this tends to be limited to anecdotal reporting [23, 24]. Although Maiman did find a 20% HIV-positive seroprevalence rate among women under 50 years presenting with carcinoma of the cervix, leading him to suggest that women diagnosed with cervical cancer at a young age should be offered HIV testing [25, 26], the converse was true in Africa, with HIV infection being found less frequently among women with cervical cancer

[27]. In addition, although HIV infection and cervical carcinoma are both common in Africa, there has been no increase in the incidence of invasive cervical cancer associated with the increasing incidence of HIV [28].

In St Mary's Hospital, of approximately 380 women with HIV infection who have attended the clinic between 1988 and 1997, not one is known to have developed ICC. One possible reason for this apparent lack of impact of HIV infection on the incidence of ICC is that many of the recently infected women may not have survived sufficient time to develop invasive disease, given that it is generally accepted that progression of CIN to ICC takes, on average, 10–20 years in immune-competent patients. As the number of HIV-infected women increases and as antiretroviral therapy improves longevity, ICC may emerge as a significant complication of HIV infection. The presence of ICC as an AIDS defining illness (ADI) does, however, highlight the need for HIV infected women to have regular gynaecological assessment.

IMMUNOSUPPRESSION AND CIN

GENERAL

Intra-epithelial neoplasia (IN) of the lower genital tract is found with increased frequency among immunosuppressed women [29]. The dominant lesion is CIN, although IN may affect the vulva and vagina also. Other conditions associated with an increased incidence of CIN include renal transplantation including post-operative drugs [18, 30–32], glomerulonephritis [33], chronic active hepatitis [34], breast cancer [35] and chemotherapy [36, 37] regardless of indication.

HIV AND CIN

There is strong evidence for an increased prevalence of CIN in HIV infected women ($\times$ 10–100 that of the general population). Many reports have shown that HIV and CIN co-exist. Bradbeer [38] was the first to report an increased incidence of CIN in HIV-seropositive women and since then several authors have noted a high prevalence of CIN in HIV-positive women, with a range between 22 and 70% (see Table 14.5).

Looking at the reverse relationship Maiman found that 33% of women referred with a cervical abnormality were HIV-seropositive [25]. Although the early studies were uncontrolled the findings have been corroborated by later controlled studies which also linked the occurrence of CIN to the degree of immunodepletion (see Table 14.6).

Studies on the natural history suggest degree of immunosuppression is related to the duration of HIV infection, which would agree with the finding that onset of malignancy in renal transplant patients is related to the length and degree of immunosuppression.

IMMUNOSUPPRESSION: MECHANISMS

There are several possible explanations for the association between neoplasia and immunosuppression [28, 37].

HIV-negative immunosuppression almost certainly allows development of carcinoma types that would normally be prevented by immunosurveillance [19]. The role of HIV in the pathogenesis of carcinoma, however, is not yet fully understood and it is not clear whether the mechanism is similar to that for HIV-negative immunosuppression. HIV itself is probably not carcinogenic. The main target of the virus is the CD4 subset of T-lymphocytes.

As the CD4 cell count drops the individual's immunocompetence also falls [45].

Other factors linking CIN and immunodepression in HIV-seropositive women include: an inverted CD4/CD8 ratio; functional immunosuppression; symptomatic HIV disease; CDC 4; p24 antigenaemia; increased microglobulin.

Some cancer types associated with immunosuppression can be linked to a viral aetiology, e.g. HPV and HIV, and at a molecular level the Tat protein of HIV can modulate HPV gene expression in cell culture [46].

HPV has been strongly implicated in the aetiology of cervical cancer and in HIV-positive patients [47, 48], who may incidentally, through behavioural factors, be at increased risk of acquiring HPV and other sexually transmitted diseases (STDs), it has been suggested that interaction between HPV and HIV at a cellular level may be significant in the development of carcinoma [46]. Similarly HIV may interact with

Table 14.5 Reports noting high prevalence of CIN in HIV-positive women

Author	Group	Prevalence (%)
Byrne *et al.* [39]	GUM Pts SMH	32
Vermund *et al.* [40]	GUM Pts New York	52
Laga *et al.* [41]	Prostitutes Zaire	27

Table 14.6 Studies linking occurrence of CIN to the degree of immunodepletion

Author	Group	HIV +ve	HIV −ve	CIN (%) +ve	CIN (%) −ve	p value
Smith *et al.* [42]	IVDU/Hetero	43	43	35	19	NS
Conti *et al.* [43]	IVDU	224	160	36	9	<0.01
Schafer *et al.* [44]	IVDU/Hetero	111	76	41	9	<0.01

other potential oncogenic viruses, e.g. hepatitis B virus, cytomegalovirus, herpes simplex virus and Epstein–Barr virus [49].

DIAGNOSTIC DIFFICULTIES

Much evidence suggests that the cytological diagnosis of CIN in immunocompromised women may be more unreliable than in those who are immunocompetent. Several studies have shown that there is a high false-negative rate for smears among immunosuppressed women, making cytology a poor predictor for the presence of CIN.

Evidence for the correlation of cytological examination and CIN is conflicting for HIV-positive patients. Some groups have shown poor correlation between cervical cytological results, colposcopy and colpobiopsy in HIV-positive patients (see Table 14.7). Fink reported low sensitivity (0.46) for cytology even though the inter-reader variability was minimal [50].

There is other evidence for cervical smears being positively predictive for disease (see Table 14.8).

Table 14.7 Studies showing false-negative rates for smears among immunosuppressed women

	Pt gp	*CIN(colp)*	*+ve smear*
Cordiner [31]	Renal tx	5	2
Alloub [32]	Renal tx	24	11
Hughes [34]	Breast ca	8	0

Table 14.8 Studies showing evidence for the correlation of cytological examination and CIN in HIV-positive patients

	HIV+ve	*+ve cytology*	*CIN*
Byrne [39]	19	6	4
Fink [50]	51	22	18
Khalsa [52]	13	2	7
Maiman [51]	32	1	13

Korn *et al.* found that the sensitivity of the smear was not reduced in HIV-positive women [53].

Because of the increased risk of CIN and because of the possible inaccuracy of cytology it has been suggested that immunosuppressed patients should be evaluated colposcopically as well as with regular smear tests. The frequency at which this is done needs to be balanced against the cost, discomfort and anxiety evoked by routine colposcopy.

One possible plausible explanation for the high false-negative cytology rate in HIV-positive patients is the presence of concomitant cervico-vaginal infection (also common in immunosuppressed patients) when the smear is taken, which could lead to difficulties with cytological interpretation (see Plate 27). The usual reasons for cytological insensitivity, e.g. small volume disease or sampling technique [54] are unlikely to explain the false-negative results, as the lesions in these patients tend to be large and high-grade and the smears are generally taken at colposcopy by experienced personnel under optimal conditions for recognition of the transformation zone. The optimum method for screening for dysplasia in HIV-positive women is not completely clear and would require a randomized trial with long-term follow-up outcomes. Most practices currently depend on the resources available.

NATURAL HISTORY

As the incidence of HIV infection increases the number of immunosuppressed women attending any one colposcopy clinic would be expected to rise. The effect of immunosuppression on the natural history of CIN has not yet been fully elucidated. Evidence to date for an increasing rate of disease progress with earlier development of ICC is conflicting. Klein *et al.* [55] failed to show rapid progression of CIN in women with HIV infection, whereas Conti *et al.* [56] prospectively studying women with untreated low-grade lesions, found HIV-positive women were more likely to progress.

Follow-up in the former study was short. Only large-scale prospective studies will elucidate the natural history of CIN and help in understanding the aetiology of carcinoma.

Other studies have shown that CIN lesions tend to be higher grade, more extensive and to persist or recur after treatment at a higher rate in immunosuppressed women than in immunocompetent women [57–60].

IMMUNOSUPPRESSION AND HPV

There is a strong association between immunosuppression and HPV in that many immunosuppressed patients have warts, especially anogenital warts. This might imply that immunosuppression enhances HPV entry in cervical cells and facilitates the development of infection. In one study of immunodepressed renal transplant patients the prevalence of skin warts was 40% after 1–4 years of immunosuppressive therapy, the incidence of warts increasing with the duration of immunosuppression [16]. Understandably this may represent a significant problem for long-term survivors. In a study of 61 immunosuppressed patients of varying aetiologies on suppressive therapy, genital HPV types seem to predominate whatever the site of the warts [61]. The incidence of cervical HPV infection is significantly higher in women who are immunosuppressed, or at risk of immunosuppression, than in the general population [62]. High rates of clinical, subclinical and latent infection have been reported [29, 32, 40]. Byrne *et al.* found evidence of HPV infection in 18 of 19 HIV-positive women, 10 (i.e. over half) of whom had infections which were subclinical and were only discovered at colposcopy [39]. Icenogle *et al.*, looking at cervico-vaginal lavage samples obtained from prostitutes in Kinshasa, found mixed HPV types present [63].

The frequency of HPV has been related to T-cell count and degree of immunosuppression. Women with a CD4 count less than 200 have five times the incidence of HPV, suggesting that the severity of HPV-induced cervical infection is related to the degree of immunosuppression [64]. In the same study there was a high incidence of high-risk oncogenic HPV types (16 and 18) but no corresponding increase in high-grade CIN.

Clinical warts tend to be multiple, widespread and large (see Plate 28). HPV infection tends to be persistent, recurrent and possibly progressive in this situation [29] and women presenting such a pattern of warts should be considered for HIV testing if their status is unknown.

In a study of HPV types found in recurrent warts affecting HIV-positive, long-term immunosuppressed women following ablative laser or cryotherapy, nine of 12 subjects were found to have the same HPV types [65], this was in stark contrast to the findings in a similar study by the same author of immunocompetent women, in whom 13 out of 14 had different HPV types pre- and post-treatment. Byrne *et al.* found HPV type 16 disappeared after laser destruction of the transition zone in seven immunocompetent women, two of whom later became infected with a different HPV type [67]. One possible explanation for this might be that immunocompetent women develop a type resistance to HPV preventing reinfection, which is lost with immunosuppression.

HPV is strongly implicated in the aetiology of cervical cancer and pre-cancer [48–50] and it is likely that HPV infection is a major contributory factor in the development of both intra-epithelial and invasive carcinoma. The mechanism for this is not fully understood as yet but appears to be complex. With immunocompromise the functions of the normal regulatory mechanisms are also compromised and modified and fail to keep HPV expression in check. It remains unclear, however, whether it is the virus or the host or a combination of both that promotes the development of progression.

The role of sexual partners has not yet been evaluated. As they might be responsible for treatment failure through reinfection, early assessment of the partner is desirable to improve treatment efficacy.

NON-CERVICAL LOWER GENITAL TRACT INTRA-EPITHELIAL NEOPLASIA AND IMMUNOSUPPRESSION

GENERAL

Cancers are often multifocal in immunodeficiency. Immunosuppressed women with CIN are also at risk of IN at other sites of the lower genital tract, namely vulva and vagina (VIN and VAIN). Disease represents a field change and may be synchronous or sequential.

VULVAL AND VAGINAL INTRAEPITHELIAL NEOPLASIA

In contrast to CIN, the data relating the occurrence of immunosuppression and VIN and VAIN are limited. After the cervix the commonest site for neoplastic change is the perineum (vulva) followed by the vagina. Penn, looking at occurrence of anogenital cancer in renal transplant patients, found that these neoplasms were often multiple, extensive and occurred later than all other post-transplant malignancies [12]. In Penn's cohort of 777 renal transplant patients with malignancy 49 (6%) were anogenital, 41% being *in situ* and 59% invasive, the majority being squamous cell.

Sillman *et al.* [29] found 8/20 women, immunosuppressed for various reasons, had anogenital neoplasia; most had concomitant involvement of the cervix and in 55% the lesions were multifocal. As with cervical neoplasia, immunosuppressed patients with anogenital neoplasia at other sites have aggressive lesions recalcitrant to treatment. They also tend to suffer from recurrent/persistent viral infections (see Plates 29 and 30).

HIV infection has also been associated with malignant and pre-malignant vulval and perineal tumours that may be difficult to recognize because of the presence of multiple, large genital warts (see Plate 28). Conversely in women with genital neoplasia–papilloma syndrome (characterized by the occurrence of HPV-associated lesions, one of which includes an *in situ* and one an invasive neoplastic tumour in multiple genital sites), Carson *et al.* found a reduced suppresser/cytotoxic T-cell ratio and reduced immunocompetence compared with a control group [68]. It is possible that the presence of other infections in the lower genital tract may enhance neoplasia (especially in the presence of some immunocompromise) and eradication might be a valuable prophylactic measure.

SUMMARY AND CONCLUSION

The purpose of this chapter was to highlight and discuss issues and problems relating to the commonly occurring lower genital intra-epithelial neoplasia associated with immunosuppression and HIV infection. As numbers of immunocompromised women increase guidelines for appropriate management and treatment will be required. As the mechanism for carcinogenesis in HIV infection becomes clearer a more fundamental approach to treatment might be possible. It may be that responses to standard treatments are altered and rather than treating the presenting complaint treatment of the underlying disease by lowering the viral load, e.g. with antiretrovirals, may either reverse the carcinogenic process or make it more responsive to conventional treatments. It is becoming clear that a less aggressive approach might be more appropriate to prevent multiple treatment courses being required.

REFERENCES

1. *The Health of the Nation: a Strategy for Health in England* (1992) Presented to Parliament by the Secretary of State for Health, HMSO, London.
2. *Needs Assessment of Sexual Health Services for Immunocompromised Women in Parkside* (1992) Presented to the Academic Department of Public Health Medicine, St Mary's Hospital Medical School, London.
3. *HIV/AIDS: Figures and Trends*. Joint United Nations Programme on HIV/AIDS, March 1996.
4. Chin, J. (1991) Global estimates of HIV infections and AIDS cases: 1991. *AIDS*, **5** (Suppl. 2), S57–61.

5. Mertens, T.E., Burton, A., Stoneburner, R., *et al.* (1994) Global estimates and epidemiology of HIV infection and AIDS. *AIDS,* **8**(Suppl. 1), S361–72.

6. Mertens, T.E. (1995) HIV/AIDS: trends of the Epidemic. Plenary Speech Presented at the *IXth International Conference on AIDS and STDs in Africa,* Kampala, December 1995.

7. *Communicable Disease Report,* Vol. 7, No. 4, January 1997.

8. Buve, A., Carael, M., Hayes, R. and Robinson, N.J. (1995) Variation in HIV prevalence between urban areas in sub-Saharan Africa: do we understand them? *AIDS,* **9** (Suppl. 1), S103–9.

9. Trong Thi Xuan, L., Giang, L.T., Lap, V.D., *et al.* (1996) Rising HIV infection rates and STD prevalence herald AIDS epidemic in Vietnam. *XIth International Conference on AIDS,* Vancouver, July 1996 (abstract Mo C.455).

10. Gromyko, A. (1996) Epidemiological trends of AIDS and other sexually transmitted diseases in the eastern part of Europe. *XIth International Conference on AIDS,* Vancouver, July 1996 (abstract Tu C.205).

11. Berkley, S., Namara, W. and Okware, S. (1990) AIDS and HIV infection in Uganda – are more women infected than men? *AIDS,* **4**, 1237–42.

12. Penn, I. (1996) Cancers of the anogenital region in renal transplant recipients. *Cancer,* **58**, 611–6.

13. Gatti, R.A. and Good, R.A. (1971) Occurrence of malignancy in immunodeficiency diseases. *Cancer,* **28**, 89–98.

14. Filipovich, A.H., Spector, B.D. and Kersey, J. (1980) Immunodeficiency in humans as a risk factor in the development of malignancy. *Preventative Medicine,* **9**, 252–9.

15. Penn, I. (1978) Malignancies associated with immunosuppressive or cytotoxic therapy. *Surgery,* **83**, 492–502.

16. Spencer, E.S. and Andersen, H.K. (1970) Clinically evident, non terminal infections with Herpes viruses and the Wart Virus in immunosuppressed renal allograft recipients. *British Medical Journal,* **3**, 251–4.

17. Hostell, E.O., Mandel, J.S., Murray, S.S., *et al.* (1977) Incidence of skin carcinoma after renal transplantation. *Archives of Dermatology,* **113**, 436–40.

18. Porreco, R., Penn, I., Droegemueller, W., *et al.* (1975) Gynaecological malignancies in immuno-suppressed organ homograft recipients. *Journal of Obstetrics and Gynaecology,* **45**, 359–63.

19. Esplin, J.A. and Levine, A.M. (1991) HIV – related neoplatic disease. *AIDS,* **5** (Suppl. 2), S203–10.

20. Beral, V. (1991) The epidemiology of cancer in AIDS patients. *AIDS,* **5** (Suppl. 2), S99–103.

21. Centers for Disease Control (1981) Kaposi's sarcoma and pneumocystis pneumonia among homosexual men: New York and California. *Morbidity and Mortality Weekly Report,* **30**, 305–8.

22. Centers for Disease Control (1987) Revision of the CDC surveillance case definition for acquired immunodeficiency syndrome. Reported by council of state and territorial epidemiologists: AIDS programme, Center for Infectious Diseases, CDC. *Morbidity and Mortality Weekly Report,* **36** (Suppl. 1S), 15–115.

23. Rellihan, M.A., Dooley, D.P., Burke, T.W., *et al.* (1990) Rapidly progressing cervical cancer in a patient with Human Immunodeficiency Virus infection. *Gynecologic Oncology,* **36**, 435–8.

24. Monfardini, S., Vaccher, E., Pizzocaro, G., *et al.* (1989) Unusual malignant tumours in 49 patients with HIV infection. *AIDS,* **3**, 449–52.

25. Maiman, M., Fruchter, R.G., Serur, E., *et al.* (1990) Human Immunodeficiency Virus infection and cervical neoplasia. *Gynecologic Oncology,* **38**, 377–82.

26. Maiman, M., Fruchter, R.G., Guy, L., *et al.* (1993) Human Immunodeficiency Virus infection and invasive cervical carcinoma. *Cancer,* **71**, 402–6.

27. Schmauz, S., Okong, P., de Villiers, E–M., *et al.* (1989) Multiple infections in cases of cervical cancer from a high incidence area in tropical Africa. *International Journal of Cancer,* **43**, 805–9.

28. Rabkin, C.S. and Blattner, W.A. (1991) HIV infection and cancers other than nonHodgkins lymphoma and Kaposi's sarcoma. *Cancer Surveys,* **10**, 151–60.

29. Sillman, F., Stanek, A., Sedlis, A., *et al.* (1984) The relationships between human papillomavirus and lower genital intraepithelial neoplasia in immunosuppressed women. *American Journal of Obstetrics and Gynecology,* **150**, 300–8.

30. Kay, S., Frable, W.J. and Hume, D.M. (1970) Cervical dysplasia and cancer developing in women on immunosuppression therapy for renal homotransplantation. *Cancer,* **26**, 1048–52.

31. Cordiner, J.W., Sharp, F. and Briggs, J.D. (1980) Cervical intraepithelial neoplasia in immunosuppressed women after renal transplantation. *Scottish Medical Journal,* **25**, 275–7.

32. Alloub, M.I., Barr, B.B., Mclaren, K.M., *et al.* (1989) Human papillomavirus infection and cervical intraepithelial neoplasia in women with renal allografts. *British Medical Journal,* **298**, 153–6.

33. Hartveit, F., Bartelsen, B., Thunold, S., *et al.* (1991) Risk of cervical intraepithelial neoplasia in women with glomerulonephritis. *British Medical Journal,* **302**, 375–7.

34. Norfleet, R.G. and Samson, C.E. (1978) Carcinoma of the cervix after treatment with Prednisone and Azathioprine for chronic active hepatitis. *American Journal of Gastroenterology*, **70**, 383–4.

35. Hughes, R.G., Colquhoun, M., Alloub, M., *et al.* (1993) Cervical intraepithelial neoplasia in patients with breast cancer: a cytological and colposcopic study. *British Journal of Cancer*, **67**, 1082–5.

36. Schramm, G. (1970) Development of severe cervical dysplasia under treatment with Azathioprine. *Acta Cytologica*, **14**, 507–9.

37. Nyberg, G., Erickson, O. and Westberg, N.G. (1981) increased incidence of cervical atypia in women with systemic lupus erythematosus treated with chemotherapy. *Arthritis and Rheumatism*, **24**, 648–50.

38. Bradbeer, C. (1987) Is infection with HIV a risk factor for cervical intraepithelial neoplasia? *Lancet*, **ii**, 1277–8.

39. Byrne, M.A., Taylor–Robinson, D., Munday, P.E., *et al.* (1989) The common occurrence of human papillomavirus infection and intraepithelial neoplasia in women infected with HIV. *AIDS*, **3**, 379–82.

40. Vermund, S.H., Kelley, K.F., Klein, R.S., *et al.* (1991) High risk of human papillomavirus infection and cervical squamous intraepithelial lesions among women with symptomatic human immunodeficiency virus infection. *American Journal of Obstetrics and Gynaecology*, **165**, 392–400.

41. Laga, M., Icenogle, J.P., Marsella, R., *et al.* (1992) Genital papillomavirus infection and cervical dysplasia – opportunistic complications of HIV infection. *International Journal of Cancer*, **50**, 45–8.

42. Smith, J.R., Kitchen, V.S., Botcherby, M., *et al.* (1993) Is HIV infection associated with an increase in the prevalence of cervical neoplasia? *British Journal of Obstetrics and Gynaecology*, **100**, 149–53.

43. Conti, M., Agarossi, A., Muggiesca, M.L., *et al.* (1991) Risk of genital HPVi and CIN in HIV positive women. *The VIIth International Conference on AIDS*, Florence (abstract M B 2408).

44. Schafer, A., Friedmann, W., Mielke, M., *et al.* (1991) The increased frequency of cervical dysplasia–neoplasia in women infected with the human immunodeficiency virus is related to the degree of immunosuppression. *American Journal of Obstetrics and Gynaecology*, **164**, 593–9.

45. Jeffries, D.J. (1986) Virological aspects of AIDS. In *Clinics in Immunology and Allergy, AIDS and HIV infection* (ed. A.J. Pinching). W.B. Saunders, London, pp. 627–44.

46. Vernon, S.D., Hart, C.E. and Reeves, W.C. (1993) The HIV–1 tat protein enhances E2-dependent human papillomavirus 16 transcription. *Virus Research*, **27**, 135–45.

47. Singer, A. and McCance, D. (1985) The wart virus and genital neoplasia; a casual or causal association. *British Journal of Obstetrics and Gynaecology*, **92**, 1083–5.

48. Schwarz, E., Freese, U.K., Gissman, L., *et al.* (1985) Structure and transcription of human papillomavirus sequences in cervical carcinoma cells. *Nature*, **314**, 111–4.

49. Zur Hausen, H., Gissman, L. and Schlehofer, J.R. (1984) Viruses in the etiology of human genital cancer. *Progress in Medical Virology*, **30**, 170–86.

50. Fink, M.J., Fruchter, R., Maiman, M., *et al.* (1993) Cytology, colposcopy and histology in HIV positive women. *The IXth International Conference on AIDS*, Berlin, 1993 (abstract P0-B23-1947).

51. Maiman, M., Tarricone, N., Vieira, J., *et al.* (1991) Colposcopic evaluation of human immuno-deficiency virus – seropositive woman. *Obstetrics and Gynecology*, **78**, 84–8.

52. Khalsa, S.K., Salmon, L.E. and Sandler, K.R. (1993) Gynecological manifestations in early HIV disease, and screening pap. smears versus colposcopy. *The IXth International Conference on AIDS*, Berlin, 1993 (abstract P0-B14-1647).

53. Korn, A.P., Autrey, M., Deremer, P.A., *et al.* (1994) Sensitivity of the Papanicolaou smear in human immunodeficiency virus-infected women. *Obstetrics and Gynecology*, **83**, 401–4.

54. Giles, J.A., Hudson, E., Crow, J., *et al.* (1988) Colposcopic assessment of the accuracy of cervical cytology screening. *British Medical Journal*, **296**, 1099–102.

55. Klein, R., Adachi, A., Fleming, I., *et al.* (1992) A prospective study of genital neoplasia and human papillomavirus (HPV) in HIV infected women. *The VIIIth International Conference on AIDS*, Amsterdam, 1992 (abstract Tu.B. 0527).

56. Conti, M., Agarossi, A., Muggiesca, M.L., *et al.* (1992) High progression rate of HPV and CIN in HIV infected women. *The VIIIth International Conference on AIDS*, Amsterdam, 1992 (abstract POB 3050).

57. Provencher, J.B. (1988) HIV status and positive Papanicolaou screening: identification of a high risk population. *Gynecologic Oncology*, **31**, 184–6.

58. Sillman, F.H., Boyce, J.G., Mecasaet, M.A., *et al.* (1981) 5-Fluorouracil/Chemosurgery for intra-epithelial neoplasia of the lower genital tract. *Obstetrics and Gynecology*, **58**, 356–60.

59. Fruchter, R., Maiman, M., Serur, E., *et al.* (1992) Cervical intraepithelial neoplasia in

HIV-infected women. *The VIIIth International Conference on AIDS*, Amsterdam, 1992 (abstract MOB 0057).

60. Agarossi, A., Muggiasca, M.L., Ciminera, N., *et al.* (1993) Follow–up of HIV 1 positive women treated for cervical neoplasia. *The IXth International Conference on AIDS*, Berlin, 1993 (abstract P0-B14-1634).

61. Bakir, T.M., Shuttleworth, D., Mckenna, D., *et al.* (1992) Detection of human papillomavirus DNA in skin warts from immunocompromised patients but not in semen from men whose wives have abnormal cervical cytology. *Journal of Hygiene, Epidemiology and Immunology*, **36**, 279–91.

62. Schneider, V., Kay, S. and Lee, M.H. (1983) Immunosuppression as a high-risk factor in the development of condyloma acuminatum and squamous neoplasia of the cervix. *Acta Cytologica*, **27**, 220–4.

63. Icenogle, J.P., Laga, M., Miller, D., *et al.* (1992) Genotypes and sequence variants of human papillomavirus DNAs from human immuno-deficiency virus type 1-infected women with cervical intraepithelial neoplasia. *Journal of Infectious Diseases*, **166**, 1210–6.

64. Johnson, J.C., Burnett, A.F., Willet, G.D., *et al.* (1992) High frequency of latent and clinical human papillomavirus cervical infections in immunocompromised human immunodefi-ciency virus infected women. *Obstetrics and Gynecology*, **79**, 321–7.

65. Nuovo, G.J., Babury, R. and Calayag, P.T. (1991) Human papillomavirus types and recurrent cervical warts in immunocompromised women. *Modern Pathology*, **4**, 632–6.

66. Nuovo, G.J. and Pedemonte, B.M. (1990) Human papillomavirus types and recurrent cervical warts. *Journal of the American Medical Association*, **263**, 1223–6.

67. Byrne, M.A., Taylor-Robinson, D., Wickenden, C., *et al.* (1988) Prevalence of human papillo-mavirus types in the cervices of women before and after laser ablation. *British Journal of Obstetrics and Gynaecology*, **95**, 201–2.

68. Carson, L.F., Twiggs, L.B., Fukushumi, M., *et al.* (1986) Human genital papilloma infections: an evaluation of immunologic competence in the genital neoplasia–papilloma syndrome. *American Journal of Obstetrics and Gynecology*, **155**, 784–9.

Alan S. Evans

INTRODUCTION

Invasive cancer of the cervix is a preventable disease. Early detection of asymptomatic intra-epithelial pre-invasive lesions through mass screening programmes using exfoliative cervical (Papanicolaou) cytology with appropriate treatment, subsequently returning the woman to normal cytology is the aim of all national screening programmes [1, 2]. Wide coverage reaching women at high risk of developing invasive cancer will not only detect pre-invasive disease, but will also allow earlier intervention in the natural history of the developing tumour.

At present it is accepted that cervical intra-epithelial neoplasia (CIN) III is a precursor of cervical cancer, with 30–70% of women developing invasive cancer within a 10-year follow-up period [3–5]. For some women in this group (less than 10% of patients), lesions can progress from intra-epithelial neoplasia to invasive tumour in under 1 year. In the continuum from CIN 3 to early invasive squamous cell cancer, there is often a difficulty in diagnosis: either expert colposcopically directed biopsies or cone biopsy are required to exclude invasive disease before treatment is undertaken. In recent years, with electrosurgical excision of the transformation zone of the cervix (LEEP or LETZ) in the management of CIN, it has been demonstrated that there can be errors in diagnosis of microinvasive cancer of up to 15.9%,

and also that as many as 25% of patients at colposcopy may have a more severe lesion than a directed biopsy suggests, including unsuspected Stage Ia_1 cancer [6–9].

The prevalence of primary adenocarcinoma has been variously assessed at 5–8% of all epithelial malignancies of the cervix. CIN often co-exists with adenocarcinoma *in situ* (AIS) on the cervix. The criteria for diagnosing AIS or glandular intra-epithelial neoplasia (GIN) of the endocervix is ill-defined. GIN may be focal or superficial and is consequently under-diagnosed [10, 11]. In contrast, in the more extensive multicentric lesions of GIN, early invasive adenocarcinomas are difficult to diagnose. The differentiation of GIN from adenocarcinoma should be made only on adequate conization or hysterectomy specimens and may only be suspected in directed punch biopsy specimens at colposcopy [12].

STAGING DEFINITIONS IN EARLY INVASIVE DISEASE

For decades early invasion of cervical cancer was associated with confusion. Investigators used differing terms to describe microinvasion of the cervix with understandable confusion and conflicting treatment outcomes. In July 1973 the committee of the International Federation of Gynaecology and Obstetrics (FIGO) reclassified cervical cancer. but there was no recommendation concerning the size or depth

Cancer and Pre-cancer of the Cervix
Edited by D.M. Luesley and R. Barrasso. Published in 1998 by Chapman & Hall, London. ISBN 0 412 56600 1.

of invasion that should be considered micro-invasive (Stage Ia). The 1985 FIGO definitions subdivided Stage I cancer into Ia and Ib, and in an attempt to define tumour volume in early disease, in September 1994 recommendations were made to modify these definitions further. These changes take into account the depth of invasion in early stage disease (see Table 15.1).

DIAGNOSIS OF EARLY INVASION

The diagnosis of early invasion may be made at colposcopy, but is usually based on microscopic examination of the removed tissue. Features in the directed biopsy histology that may be associated with early invasion are extensive involvement of the epithelial surface and endocervical crypts, luminal necrosis, and marked cellular pleiomorphism. Serial

sections should be performed to exclude microinvasion [13]. The diagnostic features at colposcopy, when present, are of atypical vessels in a background of acetowhite epithelium and are characteristic, but these vessels are less effusive and less atypical than those associated with invasive cancer; atypical vessels are of a more narrow calibre, they show asperites and are 'tufted' or 'sausage-like' (see Figures 15.1 and 15.2), however, these changes may be subtle, focal, or absent. Kolstad (1989) reported that the diagnosis of microinvasive cancer was made colposcopically in only 32.6% of patients [14]. This is confirmed in other reports [15], and in every large

Table 15.1 FIGO staging of early invasive carcinoma of the cervix

Stage 0	Intra-epithelial neoplasia, CIN, carcinoma *in situ*
Stage I	The carcinoma is strictly confined to the cervix: extension to the uterine corpus should be disregarded
Stage Ia	Invasive cancer identified only microscopically
	All gross lesions, even with superficial invasion, are Stage Ib cancers
	Invasion is limited to measured stromal invasion with a maximum depth of 5 mm and no wider than 7 mm diameter
Stage Ia$_1$	Measured invasion of stroma no greater than 3 mm in depth and no wider than 7 mm diameter
Stage Ia$_2$	Measured invasion of stroma greater than 3 mm and no greater than 5 mm in depth and no wider than 7 mm in diameter
Stage Ib	Clinical lesions confined to the cervix, or preclinical lesions greater than Stage Ia

Figure 15.1 Colposcopic features suggestive of microinvasion; atypical vessels of narrow calibre in a non-arboreal pattern.

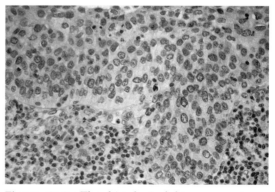

Figure 15.2 The histological biopsy specimen showing early microinvasion from squamous inter-epithelial neoplasia.

series of loop excision biopsies of the transformation zone (LEEP or LETZ), there are two or more patients with unsuspected microinvasive cancer in the cone specimen.

Although the morphological and clinical characteristics of microinvasive squamous carcinoma are now well defined, those of its endocervical counterpart are not. The main problem is recognizing features of microinvasion. Unlike squamous tumours, where invasion develops as a breakthrough of single cells or tongues of cells, many adenocarcinomas invade by entire glands without an associated stromal response. It is therefore impossible to assess with any degree of certainty the depth of invasion, or where the origin of the tumour is, given a limited biopsy.

Lack of correlation between cytology and biopsy, or extension of disease into the endocervical canal, makes a laser, loop or cold knife conization mandatory [16, 17]. A cone biopsy is an excisional biopsy of the cervix with its base on the ectocervix, which should include all the visible transformation zone and extend up the cervix to include a variable portion of the endocervical canal. A cone biopsy is also performed where there is a suspicion of squamous carcinoma microinvasion, AIS, tumour invasion, or there is an unsatisfactory colposcopy [18].

The aims of a cone biopsy may be both diagnostic and therapeutic. Cone biopsy is a relatively safe procedure; however, there are possible complications of:

- excessive haemorrhage, especially if carried out in pregnancy;
- general anaesthetic risks;
- subsequent cervical stenosis or cervical incompetence with future pregnancy losses.

The reason for defining microinvasion is to identify those women who are not at risk of metastasis to the pelvic or paracervical lymph nodes, and who are appropriately treated by more conservative measures such as cone biopsy alone. It is important to have adequate histological information; good quality cone biopsies are obtained using a technique that does not handle the mucosa and is performed after the mucosa has had time to regenerate following the colposcopic investigations.

The subsequent pathology should be described in terms of:

- depth of invasion;
- width and breadth of invasion;
- lymphatic space involvement;
- type and grade of tumour;
- conization clearance margins, at the endocervical and ectocervical margins.

Although electrosurgical loop excision of the transformation zone has been used by many clinicians as an alternative to formal cone biopsy, the limitations of the techniques are the same; persistent disease is related to positive surgical margins and lesion size [19–22].

Inadequate reporting may be given in up to half of patient reports [23].

MANAGEMENT OF EARLY INVASION

Where the depth of invasion is less than 3.0 mm (Stage Ia_1), as shown at cone biopsy or its equivalent, with the excision margins clear and no vascular lymphatic channel invasion identified, the frequency of residual disease or pelvic lymph node metastasis is sufficiently low that further surgery can be avoided. In those women where preservation of fertility is not an issue, total hysterectomy, either vaginally or abdominally, may also be undertaken [24–26]. For women with tumour invasion between 3 and 5 mm (Stage Ia_2), the information available suggests that the risk of lymphatic spread may be of the order of 8–10% [27]. Radical hysterectomy with pelvic lymph node dissection is recommended in this situation, and may also logically be undertaken for patients where the depth of invasion is uncertain due to artefact or where there is invasive tumour to the cone margins [28].

An alternative approach to Stage Ia_2 disease is either cone biopsy followed by endoscopic pelvic lymphadenectomy or radical trachealectomy, again with endoscopic removal of the nodes. Both techniques require further evaluation.

Intracavity radiation, with one or two insertions with tandem and ovoids, is reserved for patients who are not suited to surgery; 10,000–12,000 cGy vaginal surface dose is optimum [29].

REFERENCES

1. Anderson, G., Boyes, D.A., Benedet, J.L., *et al.* (1988) Organisation and results of the cervical cytology screening program in British Columbia, 1955–85. *British Medical Journal*, **296**, 975–8.
2. Austoker, J. (1994) Screening for cervical cancer. *British Medical Journal*, **3**, 61–7.
3. Richart, R.M. (1968) Natural history of cervical intraepithelial neoplasia. *Clinical Obstetrics and Gynecology*, **10**, 748.
4. Peterson, O. (1956) Spontaneous course of cervical precancerous conditions. *American Journal of Obstetrics and Gynecology*, **72**, 1063–71.
5. McIndoe, W.A., McLean, M.R., Jones, R.W., *et al.* (1984) The invasive potential of carcinoma *in situ* of the cervix. *Obstetrics and Gynecology*, **64**, 634–8.
6. Benedet, J.L., Anderson, G. and Boyes, D.A. (1985) Colposcopic accuracy in the diagnosis of microinvasive and occult invasive carcinoma of the cervix. *Obstetrics and Gynecology*, **65**, 557–62.
7. Buxton, E.J., Luesley, D.M., Shafi, M.I. and Rollason, M. (1991) Colposcopically directed punch biopsy: a potentially misleading investigation. *British Journal of Obstetrics and Gynaecology*, **98**, 1273–6.
8. Choo, Y.C., Chan, O.L., Hsu, C. and Ma, H.K. (1984) Colposcopy in microinvasive carcinoma of the cervix – an enigma of diagnosis. *British Journal of Obstetrics and Gynaecology*, **92**, 1156–60.
9. Anderson, E.S., Neilson, K. and Pederson, B. (1995) The reliability of preconization diagnostic evaluation in patients with cervical intraepithelial neoplasia and microinvasive carcinoma. *Gynecologic Oncology*, **59**, 143–7.
10. Brown, L. and Wells, M. (1986) Cervical glandular atypia associated with squamous intraepithelial neoplasia: a premalignant lesion? *Journal of Clinical Pathology*, **39**, 22–8.
11. Laverty, C.R. (1988) The reliability of a cytological prediction of cervical adenocarcinoma *in situ*. *Australia and New Zealand Journal of Obstetrics and Gynecology*, **28**, 307–12.
12. Jaworski, R.C. (1988) The histological diagnosis of adenocarcinoma *in situ* and related lesions of the cervix uteri. *Cancer*, **61**, 1171–81.
13. Al-Nafussi, A. and Hughes, D.E. (1994) Histological features of CIN predictive of microinvasive carcinoma. *British Journal of Obstetrics and Gynaecology*, **101**, 821.
14. Kolstad, P. (1989) Long term follow up of 1121 cases of carcinoma *in situ*. *Gynecologic Oncology*, **33**, 265–9.
15. Evangelos, P., Kitchener, H., Miller, I., *et al.* (1992) A population based survey of microinvasive disease of the cervix – a colposcopic and cytologic analysis. *Gynecologic Oncology*, **45**, 9–12.
16. Bloss, J.D. (1993) The use of electrosurgical techniques in the management of premalignant diseases of the vulva vagina and cervix; an excisional rather than ablative approach. *American Journal of Obstetrics and Gynecology*, **169**, 1081–85.
17. Wright, T.C., Gagnon, S., Richart, R.M., *et al.* (1992) Treatment of cervical intraepithelial neoplasia using the loop electrosurgical excision procedure. *Obstetrics and Gynecology*, **79**, 173–8.
18. Townsend, D.E., Ostergard, D.R., Mishel. D.R., *et al.* (1970) Abnormal papanicolaou smears. Evaluation by colposcopy, biopsies and endocervical currettage. *American Journal of Obstetrics and Gynecology*, **108**, 429–34.
19. Naumann, R.W., Bell, M.C., Alvarez, R.D., *et al.* (1994) LLETZ is an acceptable alternative to Diagnostic Cold-Knife Conization. *Gynecologic Oncology*, **55**, 224–8.
20. Shafi, M.I., Dunn, J.A., Buxton, E.J., *et al.* (1993) Abnormal cervical cytology following large loop excision of the transformation zone: a case controlled study. *British Journal of Obstetrics and Gynaecology*, **100**, 145–8.
21. Howe, D.T. and Vincenti, A.C. (1991) Is large loop excision of the transformation zone (LLETZ) more accurate than colposcopically directed punch biopsy in the diagnosis of cervical intra-epithelial neoplasia? *British Journal of Obstetrics and Gynaecology*, **98**, 588–91.
22. Golbang, P., Scurry, J., de Jong, S., *et al.* (1997) Investigation of 100 consecutive cone biopsies. *British Journal of Obstetrics and Gynaecology*, **104**, 100–5.
23. Kennedy, A.W., Belinson, J.L., Wirth, S., *et al.* (1995) The role of the loop electrosurgical excision procedure in the diagnosis and management of early invasive cervical cancer. *International Journal of Gynecological Cancer*, **5**, 117–20.

24. Sevin, B.U., Nadji, M., Averette, H., *et al.* (1992) Microinvasive carcinoma of the cervix. *Cancer,* **70**, 2121–8.
25. Jones, W.B., Mercer, G.O., Lewis, J.L., *et al.* (1993) Early invasive carcinoma of the cervix, *Gynecologic Oncology,* **51**, 26–32.
26. Creasman, W.T., Fetter, B.F., Clarke, D.L., *et al.* (1985) Management of stage 1A carcinoma of the cervix. *American Journal of Obstetrics and Gynecology,* **153**, 164–72.
27. Simon, N.L., Gore, H., Shingleton, H.M., *et al.* (1986) Study of superficially invasive carcinoma of the cervix, **68**, 19–24,

28. Hoskins, W.J., Ford, J.H., Lutz, M.H., *et al.* (1976) Radical hysterectomy and pelvic lymphadenectomy for the management of early invasive cancer of the cervix. *Gynecologic Oncology,* **4**, 278–90.
29. Grigsby, P.W. and Perez, C.A. (1991) Radiotherapy alone for medically inoperable carcinoma of the cervix: Stage 1A. *International Journal of Radiation Oncology, Biology and Physiology,* **21**, 375–8.
30. (1996) Cervical and vulva cancer: changes in FIGO definitions of staging. *British Journal of Obstetrics and Gynaecology,* **103**, 405–6.

Susan J. Houghton and Mahmood I. Shafi

INTRODUCTION

Since Hans Hinselmann first described the principle of colposcopy in 1925 [1], cervical imaging techniques have been used by gynaecologists to investigate the uterine cervix. Its initial usage was for benign and pre-cancerous lesions. The technique was used in many European countries, then exported to Latin America, where wide experience was gained in recognition of benign pathology as it was used as a primary screen. The colposcope has become widely accepted as an appropriate tool for assessing the cervix of women with suspected cervical intra-epithelial neoplasia (CIN) and for obtaining directed biopsies from abnormal areas. The basic optical system of the colposcope has not changed since its introduction, apart from the use of a green filter to enhance vascular appearances at the time of colposcopic assessment. Colposcopy uses a stereoscopic microscope to provide a three-dimensional image of the cervix (or other area of the female lower genital tract), magnified 6–40 times. The colposcopist manipulates the cervix or changes viewing angles to enhance the exposure of the portio or endocervical canal as the examination proceeds.

In 1956, Wied suggested the use of stereoscopic colpophotography for cervical cancer screening [2]. The focusing in colpophotography is done through the oculars of the colposcope and minor changes in the examiners'

refraction can result in blurred photographs. Also, as the magnification in colpophotography is relatively high, only a small portion of the cervix is visible in one picture. Colpophotography proved impractical as a screening procedure due to excessive cost, limitation of depth of focus and need for high magnification and precise focusing.

Alone, colposcopy is considered an inefficient screening method for cervical cancer. The combined use of cervical cytology and colposcopy enhances the detection of CIN [3]. However, colposcopy is neither a practical nor an economically feasible adjunct to large-scale cervical cancer screening due to the relatively high cost of the procedure and the absence of sufficient, adequately trained colposcopists [4].

Cervicography was introduced by Adolf Stafl as a method of photographing the cervix after the application of acetic acid [5], and was initially developed for remote areas with poor access to colposcopy. It is a less expensive 'snap-shot' version of colposcopy that allows permanent, objective documentation of cervical findings. The main objective in its development was that cervicography would overcome the limitations of colposcopic screening and that its use would improve the sensitivity of screening methods for pre-malignant disease of the cervix. Table 16.1 explains the differences between colposcopy, colpophotography and cervicography.

Cancer and Pre-cancer of the Cervix
Edited by D.M. Luesley and R. Barrasso. Published in 1998 by Chapman & Hall, London. ISBN 0 412 56600 1.

Table 16.1   Differences between colposcopy, colpophotography and cervicography

Colposcopy
    Visual examination of the cervix in ×6–40 magnification

Colpophotography
    Photographical documentation of cervical findings in ×6–40 magnification

Cervicography
    Permanent photographical documentation of the entire cervix in magnification of ×0.7 on the original slide. Slide is projected and observed from a short distance, magnification being dependent upon this distance

In cervicography, the image on the slide is actually reduced (×0.7 magnification), so that the entire cervix is visible on a single slide. The magnification is achieved by projection onto a large screen. Critical in the success of cervicography is the expert evaluation of the photograph or cervigram. Colposcopic findings are currently recorded in a written description of the vascular and tissue pattern of the cervix, with no objective documentation of the findings. With cervicography, there is permanent documentation of cervical findings and it is possible to evaluate objectively the expertise of the colposcopist by comparing the cervicographic findings of a case with the colposcopic interpretation.

Future directions in cervical imaging will undoubtedly involve the computerized digital imaging colposcope, which has been developed for the enhancement and analysis of colposcopic images. This system has many applications including image archiving, image enhancement, processing and quantitative measurements of various features in the images [6]. A wide range of digital filters may also be applied to the system to accentuate different aspects of the image. It is hoped that analysis of these images with regard to histopathology will result in the development of a new, more diagnostically accurate colposcopic methodology.

CERVICOGRAPHY

Cervicography is a diagnostic method that enables permanent objective documentation of normal and abnormal cervical findings; the projected cervigram is comparable to direct visual colposcopic magnification and resolution. The cervicograph apparatus is inexpensive and consists of a 35 mm camera body, a 50 mm extension ring, a 100 mm macrolens and a ring strobe light [7]. The cervicogram can be obtained by a technician and sent to an expert for evaluation, thus obviating the need for a trained colposcopist to assess the cervix prior to cervicography.

THE TECHNIQUE OF CERVICOGRAPHY

The cervix is visualized with a self-retaining speculum and a smear is taken. The cervix is then cleaned with a dry gauze and moistened with acetic acid. The cervicograph is hand held and is focused on the cervix by moving the system back and forth. The cervicograph should be held in prolongation of the axis of the vagina so both the anterior and posterior lips are fully visible. In rare situations where the cervix is exceptionally large, e.g. a pregnant cervix or grand multipara, separate pictures of the anterior and posterior lip may be taken. After the cervix is inspected with the cervicograph, it is again moistened with acetic acid. Immediately after the second application, when the cervix is in focus, a picture is taken. The cervicogram can be obtained by a physician, nurse or technician after a short period of training and involves no risk as the camera does not come into contact with the patient. It can be included in the patient's medical records and can also be sent to experts for consultations or second opinions. The basis of cervicographic evaluation is identical to that for colposcopy and therefore the results of cervicography depend on the expertise of the evaluators. The evaluation of cervicograms should be performed by experts in colposcopy who have been properly trained in the technique of cervicography. The cervicograph is projected onto a screen 10 feet or greater in width and observed from a distance of 3 feet in a totally darkened room. The apparent magnification of the projected slide is comparable to direct visual colposcopic magnification. The cervicograph findings are divided into four groups:

- negative – the entire squamo-columnar junction is fully visible, no abnormal feature is present;
- suspicious – an abnormal feature (acetowhite epithelium, punctation, mosaic, atypical vessels) is present;
- unsatisfactory – the squamo-columnar junction is not visible; and
- technically defective – picture out of focus, or the entire cervix is not visible.

INTERPRETATION OF CERVICOGRAMS

Abnormal colposcopic lesions are often compatible histologically with changes ranging from immature squamous metaplasia to CIN III. However, it can be difficult to distinguish colposcopically between immature metaplasia and true dysplasia or even neoplasia. The severity of the histological changes depends on the following factors: intercapillary distance, surface patterns, colour tone and the borders with normal tissue [8]. The evaluation of these parameters with colposcopy is highly subjective and cervicography allows their objective measurement. To date it is not known which parameters are more important in the prediction of histological changes, and it may be the case that lesion surface area, which is not objectively assessed using cervicography, is important as a predictor of histological grade of abnormality [9].

ROLE OF CERVICOGRAPHY IN CLINICAL PRACTICE

Cervicography has many potential clinical applications in gynaecology. These include:

cervicography screening; cervicography follow-up; evaluation of benign cervical lesions; documentation of visualization of squamo-columnar junction; evaluation of atypical Pap smears not suspicious for CIN or cervical cancer; documentation prior to out-patient treatment of CIN; photographic documentation of vulvar or vaginal lesions; evaluation of a patient with a suspicious or positive Pap smear where colposcopy is not available; for expert colposcopic consultation and for teaching and research purposes.

Cervicography has comparable accuracy in diagnosing cervical abnormality when compared to cervical cytology and colposcopy [10]. However, the rate of unsatisfactory findings in cervicography is slightly higher (17.6%) than with colposcopy (12%) [11], mainly due to a failure to visualize the whole transformation zone and problems with mucous or blood obscuring the view. Despite these limitations, cervicography provides a simple and inexpensive dimension in the investigation and diagnosis of cervical neoplastic disease.

SCREENING FOR CERVICAL CANCER

Since the observation by Papanicolaou and Traut that vaginal vault aspirates could be used to detect abnormal epithelial cells from a clinically normal cervix [12], cervical cytology has been widely adopted as a screening test for CIN.

Screening tests are relatively simple procedures designed to separate healthy persons from those with a high probability of having the disease under study, and thus permit early diagnosis and treatment. Since criteria for interpreting Pap smears were formulated at a time when a positive result led to a cold knife conization and possible hysterectomy, cytological cut-off points have favoured specificity over sensitivity [13]. In addition to being valid and acceptable, screening tests must be justifiable economically. The Pap smear test has not succeeded, however, in the eradication of this theoretically preventable disease and there are several reports in the literature of invasive cervical cancer occurring a short time after a negative (normal) or atypical (Class II) smear [14–16], attributable mainly to the false-negative rate of cytology. Cytological case finding may fail because of inadequate samples, insufficient time devoted to screening or human fatigue in sample analysis [17]. Formal colposcopic examination is impractical for use as a rapid screening test; however, cervicography solves the logistic difficulties confronting screening colposcopy and has been reported to detect cervical carcinoma in patients with negative cervical cytology [18]. The initiation of cervical neoplasia occurs almost exclusively in puberty and early adolescence, and during the first pregnancy [19, 20]. Therefore, Stafl suggested that a single cervicographic examination any time after a woman's first pregnancy could be used in screening for cervical cancer [10].

THE ROLE OF CERVICOGRAPHY IN SCREENING FOR CERVICAL CANCER

Several studies have investigated the use of cervicography as a screening procedure for cervical cancer [21, 22]. Tawa *et al.* [23] compared cervical smear cytology to cervicography in the routine screening of 3271 patients who had history of cervical abnormalities. The accuracy of each screening test was evaluated independently and the results compared with colposcopically directed biopsy results. The cervigram was significantly more sensitive than the Pap smear ($p < 0.001$), whereas the Pap smear was significantly more specific than the cervicogram in detecting CIN ($p < 0.001$).

Cervicography is more sensitive than cytology but has a high false-positive rate when a reporting system of 'any abnormality' results in referral for colposcopy. This is usually due to the presence of metaplastic or condylomatous changes without dysplasia. In an attempt to improve the false-positive rate, Szarewski *et al.* [24] used the revised reporting criteria for

cervicograms as modified by Stafl in conjunction with the marketing company National Testing Laboratories (Table 16.2) and found that this reporting system reduced false-positive rates for cervicography whilst maintaining sensitivity.

Cecchini *et al.* [25] evaluated the possible advantages of combining cervicography with cervical cytology in a screened population. They also studied the feasibility of screening for cervical cancer by cervicoscopy (naked eye examination of the cervix after acetic acid lavage). The three screening methods were compared according to positivity rate, CIN II–III detection rate and positive predictive value. Positivity rate was 3.8, 15.3 and 25.4% for cytology, cervicography and cervicoscopy, respectively. This study confirmed the limited sensitivity of cytology for high-grade CIN [23, 26], and the need for alternative or adjunctive screening tests; the use of the Pap smear and cervigram together increasing the detection of cervical dysplasia.

Cost analysis is important when evaluating any new screening method, as it must be proven to be cost-effective before it will be widely accepted as an alternative screening test. Cecchini *et al.* considered cytology as the basic screening method and estimated the costs of the combination of cervicography or cervicoscopy with cytology. Detecting one CIN II–III lesion at cytology cost $5543. The cost per each additional cytologically negative CIN II–III lesion detected at cervicography or cervicoscopy was $12,947 or $3916, respectively. The final cost of cervicography per cytologically negative CIN II–III detected, therefore, is unacceptably high. The specificity of cervicography is poor and the final cost heavily influenced by the cost of unnecessary colposcopies and biopsies in false-positive cases. However, cervicoscopy increased the detection rate of high-grade CIN at relatively low cost, with a higher sensitivity and despite poor specificity, and therefore warrants further investigation in its own right as a screening test for cervical cancer.

Table 16.2 Structured cervicography report, as revised by Stafl

---

Cervicography result:

Negative

(a) The squamo-columnar junction and transformation zone are fully visible. No abnormal lesion present

(b) The squamo-columnar junction is not fully visible, but components of the transformation zone are visible on both lips. No abnormal lesion present

(c) No abnormal lesion on the visible portion of the cervix, but squamo-columnar junction and transformation zone not visible lesion in endocervical canal cannot be excluded and endocervical smear is recommended

(d) Immature squamous metaplasia. Repeat cervigram in one year

Suspicious	Compatible with	
Abnormal lesion:	Trivial change of doubtful significance, repeat cervigram in 6–12 months	
acetowhite epithelium		
punctation	Papilloma, condyloma (HPV)	
mosaic	CIN I (mild dysplasia)	
atypical vessels	CIN II (moderate dysplasia)	Colposcopy recommended
other	CIN III (severe dysplasia/CIN)	
	microinvasive cancer/invasive cancer	

Technically defective – retake cervicogram!

Out of focus	Insufficient acetic acid or cervicogram taken too late after second application of acetic acid
Overexposed	Mucous or blood obscuring view
Underexposed	Speculum obscuring view
	Vaginal wall obscuring view
	Other

Other (vagina, vulva, penis, anus)

USE OF CERVICOGRAPHY IN SCREENING HIGH-RISK POPULATIONS

Cervicography has proven to be a very sensitive screening tool in the general population for cervical abnormality but should also be considered for use in screening populations at high risk of developing cervical dysplasia. Schauberger *et al.* concluded that cervicography in addition to cervical cytology should be considered for all women with vulvar condylomata and therefore at greater risk of cervical disease [27].

CERVICOGRAPHY AS A TRIAGE PROCEDURE

Recently, there has been a significant increase in the number of smears being reported as 'borderline nuclear abnormality' or 'ASCUS' (atypical squamous cells of uncertain significance), i.e. those without a definite cytological diagnosis of an intra-epithelial lesion. Previous studies evaluating the atypical smear have shown that a significant percentage were related to both condylomata without CIN and to CIN [28, 29, 30]. Management of these atypical smears has included: repeat smear in 6–12 months [31, 32], empiric treatment for inflammatory disease with a repeat smear [33, 34] and immediate colposcopy [35, 36]. It has been well documented that a repeat smear (for confirmation of or clarification of a previous ASCUS) may not detect significant lesions, especially in populations where follow-up is poor [37]. Cervicography has been proposed as a triage method for women with an inconclusive cytological diagnosis of 'atypia', in order to identify those in need of further evaluation with colposcopy and/or biopsy [34, 39]. This would not apply to patients with a definitive diagnosis of an intra-epithelial lesion (dysplasia or CIN), as these women would presumably be referred directly for colposcopy.

Its high sensitivity and high negative predictive value (important attributes for any screening procedure) and simplicity of technique and training, appear to justify the use of cervicography as a triage procedure in certain settings.

THE USE OF SPECULOSCOPY AS A TRIAGE TECHNIQUE

A further technique involving visualization of the cervix, speculoscopy, may have a clinical role in the evaluation of women with atypical cervical cytology. Speculoscopic examination of the cervix and vagina using chemiluminescent light and low-power magnification is a simple, inexpensive and rapid test, making it potentially acceptable as an adjunctive screening tool [40].

In a study by Massad *et al.*, the use of speculoscopy in selecting women with atypical Pap smears who would most benefit from referral for colposcopy was examined [41]. This study suggests that speculoscopy performed at the time of initial screening can accurately select women with atypical Pap smears who require colposcopy for diagnostic biopsy in a cost-effective manner. Speculoscopy warrants further investigation as a possible adjunctive test in the assessment of patients with atypical cervical cytology.

CERVICOGRAPHY FOLLOW-UP

Often, an abnormality detected by colposcopy does not require immediate treatment and follow-up is recommended. Follow-up colposcopy could easily be replaced by a cervicographic assessment by the referring gynaecologist, general practitioner or specialist nurse, at a much reduced cost and without the need for a further colposcopic appointment. Follow-up cervicography may also be appropriate in the management of women who have had local ablative or excisional treatment and follow-up colposcopy may be replaced by a combination of cytology and cervicography.

CERVICOGRAPHY IN THE EVALUATION OF BENIGN CERVICAL LESIONS

Local destructive methods, such as cryocautery, are often employed in the treatment of benign cervical lesions, for example ectropions or chronic cervicitis. It has been demonstrated that 50% of patients who develop invasive cancer after out-patient treatment were inappropriately treated for so-called benign cervical lesions [42]. Pre-treatment cervicographic assessment of these cases may offer a cost-effective method of evaluation in order to attempt to reduce the number of such cases. In cases where cervical cancer develops after out-patient treatment, it is far easier to assess the visual findings at the time of treatment from assessment of a cervicogram than from a subjective written description when trying to determine the reason for treatment failure.

CERVICOGRAPHY AND ITS USE IN TEACHING AND RESEARCH

With cervicography, a permanent, objective documentation of interesting and abnormal cervical, vulvar and vaginal lesions can be obtained. Teaching the methods of colposcopy on a one-to-one basis is time-consuming and has many limitations. The projection of cervicograms for teaching enables the discussion of cases in correlation with the cytologic, colposcopic and biopsy results to a far wider audience, without any added inconvenience or embarrassment to the patient. Permanent documentation by cervicography and objective evaluation of the cervicogram obtained also allows collection of data for research involving many aspects of cervical pathology and physiology. It may be particularly useful in long-term follow-up studies, where the cervical image may change over the study period.

CONCLUSION

Cervicography offers an inexpensive, technically simple and relatively accurate diagnostic dimension in the investigation of cervical disease. It replaces subjective colposcopic evaluation with objective documentation of cervical findings and enables critical assessment of colposcopic expertise. It is a technique that allows reproducibility in most clinical settings and can easily be undertaken by specially trained paramedical personnel. It may, therefore have many clinical roles in the evaluation and follow-up of patients with abnormal cytology, but will continue to be hampered by its high sensitivity and low specificity in the clinical setting.

FUTURE DIRECTIONS: DIGITAL IMAGING COLPOSCOPY

The static state of the present colposcope used in most clinics limits its use to primary diagnosis of the abnormal smear as a single event. However, the technique of computerized digital imaging colposcopy, as described by Craine *et al.* [43], allows several modes of image enhancement and manipulation. This system incorporates computerized enhanced image analysis, viewed and measured through a range of light sources, and can result in much more detail of the atypical transformation zone than traditional colposcopy. This is of particular relevance given the increasing popularity of large loop excision of the transformation zone (LLETZ) [44, 45], and the trend for a 'see and treat' management strategy for abnormal cervical cytology, which is often associated with overtreatment and its attendant physical and psychological morbidity. Computerized digital imaging colposcopy may offer the opportunity to improve our

understanding of both benign and malignant cervical disease and has potential as a teaching and research tool.

THE COMPUTERIZED DIGITAL IMAGING COLPOSCOPE

The digital imaging colposcope consists of a photocolposcope, to which is attached an optical interface, and a video format, solid state two-dimensional charge-couple image device (CCD). The incoming image is transferred to the CCD by means of an optical beam-splitter in the colposcope and, with the adjustment of transfer optics, the scene viewed through the binocular eyepiece is identically transferred to the CCD. The CCD camera is silicone-based and offers a broad range of spectoral response from the ultraviolet cut-off to the optical infrared. It gives a high signal to noise ratio, excellent geometric stability, linear photometric response, a large dynamic range, high detective quantum efficiency, high spatial resolution and a high speed of data acquisition.

The video signal is carried through a coaxial cable to a digitizing board set in the microcomputer. This board allows two images to be instantly available at all times for processing. The incoming video images are currently digitized at a spatial resolution of 512×480 px. Following digitization the images are displayed on the video monitor. The image may then be downloaded onto computer and either archived or printed (Figure 16.1).

The system is completed by an application specific software package in a menu-driven format. This package allows real-time viewing, instant image capture and a range of file manipulation functions, including image-processing (Figure 16.2) and analysis. Areas of interest to the clinician may be enlarged and the image processed by introducing digital filters to accentuate differences in cervical morphology, allowing enhancement of colposcopic detail (e.g. vessel patterns). Measurement facilities are available, allowing automated measurements and quantification

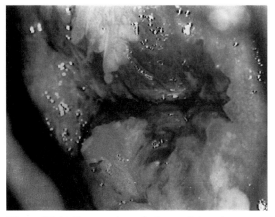

Figure 16.1   Image of atypical transformation zone.

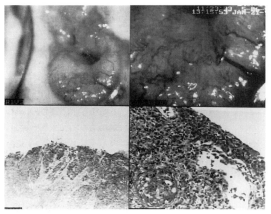

Figure 16.2   Images stored on Denvu using multi-image viewing facility.

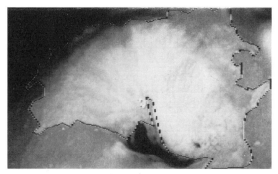

Figure 16.3   Lesion circumscribed using Denvu automeasure facility.

TECHNIQUE OF DIGITAL IMAGE CAPTURE

In digital image colposcopy the desired magnification for viewing is selected at the colposcope and a glass optical filter, if needed, is inserted in the fibre-optic illumination system. The area of interest is selected, either by direct visual observation through the colposcope or by monitoring the video display. At this stage, the colposcopist may utilize the contrast enhancement and probe the range of the current live image. When a satisfactory image is displayed, it may be digitized immediately by a keystroke. The digital image is then displayed and, if not satisfactory or desired for any reason, it may be simply overwritten by a new digitized image. This process is immediate and the digital image may then be processed and studied, or stored on the hard disk.

It is sometimes desirable to capture two or more digital images that are almost exactly identical in terms of scale, rotation and translation (i.e. the images precisely overlay each other). This situation may arise if one wanted to perform blink comparison of images obtained at different times or perform processing operations involving the subtraction from or division of one image by another. In order to obtain multiple digital images, when the desired scene is displayed in the live video format, it is digitized into one image buffer (e.g. image 1). The second scene is then established on the live video and the two images, image 1

of features such as intercapillary distance and lesion surface area (Figure 16.3). The multi-image viewing facility allows comparison either between serial colposcopies or with reference images for comparison. A blink comparison of input and digital images or just digital images is possible, allowing comparison between serial colposcopic assessments. The controlling computer image storing capacity allows storage of information, including patient details, measurement graphics and other text. The images may be individually downloaded and stored on disk, capacity being dependent upon type of disk.

and input video, are sequentially displayed at moderately high frequency, with the blink comparison capability. Any errors in registration between the two images are readily apparent and easily corrected by moving the colposcope, thus changing the orientation of the live input image. When satisfactory registration between the two scenes is achieved, the input image is then digitized as image 2. Therefore, it is possible to compare an archived image with the live image from the current examination. This allows direct comparison between the colposcopic appearances in consecutive colposcopic visits and more accurate identification of progression or change in cervical disease in that particular patient.

PROCESSING AND VASCULAR ENHANCE-MENT OF THE DIGITAL COLPOSCOPIC IMAGE

One of the primary goals of the processing of colposcopic images is to accentuate the pattern of cervical blood vessels, which is a diagnostic feature of CIN and cervical cancer. A common feature of colposcopes is the presence of a green filter through which to observe the cervix and increase the contrast of the vessels. However, this results in a decrease in the overall light intensity of the image. An image processing algorithm has been developed for use in digital image colposcopy to take advantage of the ability to obtain vascular images with the green filter.

This involves a pixel-by-pixel comparison of the green filter image with the normal white light image. Those pixels that are accentuated with the green filter are multiplied by a factor to emphasize them and the specific areas of vessels are highlighted and seen as smoother, broader regions of enhanced intensity. This type of enhancement may prove particularly useful in the extraction of vascular features in colposcopic images. This may assist in the identification of CIN and result in more accurate colposcopic diagnosis of the severity of the cervical lesion. This would have a clinical role in those patients with deferred treatment of their cytological and colposcopic abnormality, especially in the younger age groups who have low-grade lesions, where there has been an awareness of the need for damage limitation to the cervix.

CLINICAL USE OF DIGITAL IMAGING COLPOSCOPY

Digital imaging colposcopy has been used to investigate colposcopic features that may predict histological outcome as defined by LLETZ [46]. In our unit, 50 consecutive patients having cytological and colposcopic abnormality treated with LLETZ were studied. At colposcopy, a detailed history was obtained from the patient and index cytology noted. After colposcopic examination, the colposcopist gave an opinion as to the likely histological abnormality, knowing the cytology report. LLETZ was then performed if the lesion satisfied the criteria for local excisional treatment. The digital images were analysed blind of patient details. Univariate analysis was undertaken of the patient and colposcopic variables in relation to histological diagnosis.

The percentage correct diagnosis at colposcopy was 76%; however, the specificity for predicting high-grade lesions was high (false-positive rate 13%), but was relatively insensitive (false-negative rate of 42%). Index cytology and smoking were strongly associated with histological diagnosis. Age, parity and contraception were not significant. Of the colposcopic variables, focality, surface pattern, degree of acetowhitening, punctation or mosaic pattern and intercapillary distance (measured as a continuous variable) were significantly associated with histological diagnosis. Surface area also showed a positive trend for larger lesions to be high-grade, while edge parameters of definition and regularity were not significant.

Multivariate analysis identified three factors of independent prognostic importance – index cytology, current smoking status and surface pattern. Using a derived logistic model incorporating all independent variables, the percentage correct diagnosis increased to 88%. The specificity for predicting high-grade lesions was higher (false-positive rate of 6%), and the false-negative rate was reduced to 23%, compared with colposcopic diagnosis.

There is a need to objectively characterize colposcopic abnormalities and particularly to develop models differentiating high- and low-grade lesions. It is postulated that these have different behavioural and progressive potentials [47], and thereby treatment may be tailored to differing clinical situations. Index cytology and lesion size have previously been shown to be important variables in relation to histological grade of CIN and in the formulation of a colposcopic opinion [48]. Characterization has been attempted in other studies, but the technology used has been suboptimal and clinically not applicable. The statistical models derived from the analysis used by Shafi *et al.* provide useful data for comparison with the colposcopic opinion. There appear to be advantages in trying to introduce objectivity into colposcopic assessment, which may be translated into better diagnostic potential. These models need validation in a prospective cohort of patients. Use of the derived models in a prospective cohort study would test its stability in a clinical situation. Of note, patients wrongly classified by the model as being low-grade lesions were mostly small lesions, and probably would have been treated on grounds of age of the women concerned. These lesions may have a different progressive potential in comparison to the large high-grade lesions.

Digital imaging colposcopy has also been used to assess the effect of directed punch biopsy trauma on the natural history of atypical cervical transformation zones [49]. One hundred and sixty-one patients referred for colposcopic evaluation of their abnormal cervical cytology screening test were prospectively randomized into three groups: (i) no biopsy, (ii) central biopsy and (iii) peripheral/junctional biopsy. Quantitative assessment of the change in surface area of lesions of CIN over a 6-week period was performed using digital imaging colposcopy for all patients. The initial colposcopic or histologic diagnosis was compared to the histological outcome at LLETZ at the 6-week review. Chenoy *et al.* concluded that punch biopsy trauma has no significant effect on the regression rates of CIN regardless of the site of biopsy or severity of the lesion [49].

VIDEO COLPOGRAPHY

Recently, a new technique termed video colpography, which uses a video camera to capture a digital image of the cervix on Hi8 video tape, has been described [50]. The images can be viewed on a colour monitor or interfaced to a computer for enhancement or manipulation as outlined previously [46]. Fifty consecutive women referred for colposcopy had video colpography performed to determine the diagnostic accuracy of the technique and assess its potential as a secondary cervical screening method. The portable video camera (Sony CCD-TR780E) equipped with a macrolens, an autofocus mechanism and a narrow beam lighting system is mounted on a pan/tilt device attached to a portable stand allowing 'hands-free' operation. All women had video colpography of the cervix and vaginal fornices successfully performed in less than 2 minutes, after the application of acetic acid, prior to colposcopy. Standard colposcopic assessment, including directed punch biopsy and LLETZ where necessary, was then performed and the video tape subsequently reviewed independently by a different colposcopist.

The video colpographic images were satisfactory or good in 94% of cases and there was a high level of agreement on severity of abnormality between colposcopist and screener. Of the 29 cases where a histological diagnosis was obtained, video colpography correctly diagnosed 76% of cases, and more significantly 83% of cases of high-grade CIN. If video colpography had been used as a secondary screen for women with low-grade cytological abnormalities, 61% would have avoided referral for colposcopy.

CONCLUSION

Computerized digital imaging colposcopy allows for storage of the image, serial observations and analysis of the lesion, with accurate measurement of the lesion and manipulation of the digital image with the use of digital filters. The enhanced colposcopic image analysed with any or all of the available functions provides an invaluable tool for basic and clinical research into the surface structure of the cervix and for quantitative and natural history studies. It provides the means by which to build a substantial database of previously unobtainable colposcopic views of the cervix, which can be used for patient education and for teaching junior medical and nursing staff both in a clinical setting or by developing an archive of teaching material. The system can also act as an internal quality control for colposcopy units and as a tool for clinical audit.

The system allows multicentre comparisons to be conducted, allowing larger studies to be undertaken, thereby raising the power of these studies. In the digital imaging systems ports are also provided to interface with a network or modem allowing remote consultation with another colposcopist using conventional telephone lines. Satellite colposcopy may be useful to obtain a second opinion from an expert colposcopist.

Video colpography is a portable and relatively inexpensive technique that combines the simplicity of a video camera with the versatility of computerized digital imaging. It has potential for use in a primary health care setting as a secondary screening for women with minor cytological abnormalities and undoubtedly will be used for teaching, research and quality assurance.

With digital images being analysed in the context of histopathology, the goal of digital image analysis of colposcopic or video colpographic images in the future is to develop a new, dynamic cervical imaging methodology to replace existing, standard colposcopic techniques.

REFERENCES

1. Hinselmann, H. (1925) Verbesserung der inspektionsmoglichkeit von vulva, vagina und portio. *Münchener Medizinische Wochenschrifte*, **77**, 1733.
2. Wied, G.L. (1956) Stereocolpophotography: a new method for centralized screening for cervical carcinoma. *American Journal of Obstetrics and Gynecology*, **71**, 1301.
3. Ferenezy, A., Hilgarth, M., Jenny, J., *et al.* (1988) The place of colposcopy and related systems in gynecological practice and research. *Journal of Reproductive Medicine*, **33**, 737.
4. Limburg, H. (1958) Comparison between cytology and colposcopy in the diagnosis of early cervical carcinoma. *American Journal of Obstetrics and Gynecology*, **75**, 1298.
5. Stafl, A. (1981) Cervicography: a new method for cervical cancer detection. *American Journal of Obstetrics and Gynecology*, **139**, 815–25.
6. Crisp, W.E., Craine, B.L. and Craine, E.A. (1990) The computerized digital imaging colposcope: future directions. *American Journal of Obstetrics and Gynecology*, **62**, 1491–8.
7. Stafl, A. (1983) Cervicography. *Clinical Obstetrics and Gynecology*, **26**, 1007–16.
8. Kolstad, P. and Stafl, A. (1982) *Atlas of Colposcopy*, University Park Press, Baltimore, MD.
9. Shafi, M.I., Dunn, J.A., Finn, C.B., *et al.* (1993) Characterization of high- and low-grade cervical intraepithelial neoplasia. *International Journal of Gynecological Cancer*, **3**, 203–7.
10. Stafl, A. (1981) Cervicography: a new approach to cervical cancer detection. *Gynecologic Oncology*, **12**, 293–301.
11. Stafl, A. and Mattingly, R.F. (1973) Colposcopic diagnosis of cervical neoplasia. *Obstetrics and Gynecology*, **41**, 168–76.
12. Papanicolaou, N. and Traut, H.F. (1941) The diagnostic value of vaginal smears in carcinoma of the uterus. *American Journal of Obstetrics and Gynecology*, **42**, 193–205.
13. Reid, R. (1984) What does an abnormal Papanicolaou smear really mean? *Colposcopy and Gynecologic Laser Surgery*, **1**, 199–203.
14. Morell, M.D., Taylor, J.R., Snyder, R.N., *et al.* (1982) False-negative cytology rates in patients in whom invasive cervical cancer subsequently developed. *Obstetrics and Gynecology*, **60**, 41–5.
15. Figge, D.C., Bennington, J.L. and Schweid, A.I. (1970) Cervical cancer after initial negative and atypical vaginal cytology. *American Journal of Obstetrics and Gynecology*, **108**, 422–8.

16. Fruchter, R.G., Boyce, J. and Hart, M. (1980) Invasive cancer of the cervix: failures in prevention – Part 1. Previous Papanicolaou smear tests and opportunities for screening. *New York Journal of Medicine*, **80**, 743.

17. Koss, L.G. (1989) The Papanicolaou test for cervical cancer detection: a triumph and a tragedy. *Journal of the American Medical Association*, **261**(5), 737–43.

18. Spitzer, M.D., Burton, A., Krumholz, B., *et al.* (1986) Cervical cancer detection by cervicography in a patient with negative cervical cytology. *Obstetrics and Gynecology*, **68**, 685–705.

19. Coppleson, M. and Reid, B.M. (1967) *Preclinical Carcinoma of the Cervix Uteri*. Pergamon Press, London.

20. Stall, A. and Mattingly, R.F. (1974) Vaginal adenosis: a precancerous lesion? *American Journal of Obstetrics and Gynecology*, **120**, 666–77.

21. Blythe, J.G. (1985) Cervicography: a preliminary report. *American Journal of Obstetrics and Gynecology*, **152**, 192–7.

22. Gunderson, J.H., Schauberger, C.W. and Rave, N.R. (1988) The Papanicalaou smear and the cervigram: a preliminary report. *Journal of Reproductive Medicine*, **33**, 46–8.

23. Tawa, K., Forsythe, A., Cove, J.K., *et al.* (1988) A comparison of the Papanicolaou smear and the cervigram: sensitivity, specificity and cost analysis. *Obstetrics and Gynecology*, **71**, 229–35.

24. Szarewski, A., Cuziek, J., Edwards, R., *et al.* (1991) The use of cervicography in a primary screening service. *British Journal of Obstetrics and Gynaecology*, **98**, 313–17.

25. Cecchini, S., Bonardi, R., Mazzotta, A., *et al.* (1993) Testing cervicography and cervicoscopy as screening test for cervical cancer. *Tumori*, **79**, 22–5.

26. Spitzer, M., Burton, A., Krumholz, B., *et al.* (1987) Comparative utility of repeat Papanicolaou smears, cervicography and colposcopy in the evaluation of atypical Papanicolaou smears. *Obstetrics and Gynecology*, **69**, 731–5.

27. Schauberger, C.W., Rowe, N., Gunderson, J.H., *et al.* (1991) Cervical screening with cervicography and the Papanicalaou smear in women with genital condylomata. *Journal of Reproductive Medicine*, **36**, 100–2.

28. Evan, A.S. and Menaghan, J.M. (1985) Spontaneous resolution of cervical warty atypia. The relevance of clinical and nuclear DNA features: a prospective study. *British Journal of Obstetrics and Gynaecology*, **92**, 165–9.

29. Selvaggi, S.M. (1986) Cytological detection of condylomata and cervical intraepithelial neoplasia of the uterine cervix with histologic correlation. *Cancer*, **58**, 2076–81.

30. Maier, R.C. and Shultenover, S.J. (1986) Evaluation of the atypical squamous cell Papanicolaou smear. *International Journal of Gynecological Pathology*, **5**, 282.

31. Ridgley, R., Hernandez, E., Cruz⁺, *et al.* (1988) Abnormal Papanicolaou smears after earlier smears with atypical squamous cells. *Journal of Reproductive Medicine*, **33**, 383.

32. Reiter, R.C. (1986) Management of initial atypical cervical cytology: a randomised prospective study. *Obstetrics and Gynecology*, **68**, 237–40.

33. Kohan, S., Noumoff, J., Beckmann, E., *et al.* (1985) Colposcopic screening of women with atypical Papanicolaou smears. *Journal of Reproductive Medicine*, **30**, 383.

34. Himmelstein, L.R. (1989) Evaluation of inflammatory atypia. *Journal of Reproductive Medicine*, **34**, 634.

35. Andrews, S., Hernandez, E. and Miyazawa, K. (1989) Atypical squamous cells in Papanicolaou smears. *Obstetrics and Gynecology*, **73**, 747–50.

36. Koss, L.G. (1989) The Papanicolaou test for cervical cancer: a triumph and a tragedy. *Journal of the American Medical Association*, **267**, 737.

37. Lozowski, M.S., Yousri, M., Tulebain, F. and Solitaire, G. (1982) The combined use of cytology and colposcopy in enhancing diagnostic accuracy in preclinical lesions of the uterine cervix. *Acta Cytologica*, **26**, 285–91.

38. Solomon, D. and Weid, G.L. (1989) Cervicography: an assessment. *Journal of Reproductive Medicine*, **34**, 321–3.

39. August, N. (1991) Cervicography for evaluating the 'atypical' Papanicolaou smear. *Journal of Reproductive Medicine*, **36**, 89–94.

40. Lanky, N.M. and Edwards, G. (1992) Cemiluminescent light improves observer ability to visualize acetowhite epithelium. *American Journal of Gynecological Health*, **6**, 6.

41. Massad, L.S., Lanky, N.M., Mutch, D.G., *et al.* (1993) Use of speculoscopy in the evaluation of women with atypical Papanicolaou smears. *Journal of Reproductive Medicine*, **38**, 163–9.

42. Townsend, D.E. and Richart, R.M. (1982) Diagnostic errors in colposcopy. In *Preclinical Neoplasia of the Cervix* (eds J.A. Jordan, F. Sharp and A. Singer). Royal College of Gynaecologists, London, p. 145.

43. Craine, E.R., Engel, J.R. and Craine, B.L. (1990) Role of color and spatial resolution in digital imaging colposcopy. *SPIE Medical Imaging IV*, 1232.

44. Prendiville, W., Cullimore, J. and Nouon, S. (1989) Large loop excision of the cervical transformation zone (LLETZ): a new method of management for women with CIN. *British Journal of Obstetrics and Gynaecology*, **96**, 1054–60.

45. Luesley, D.M., Cullimore, J., Redman, C.W.E., *et al.* (1990) Loop diathermy excision of the cervical transformation zone in patients with abnormal cervical smears. *British Medical Journal*, **300**, 1690–3.

46. Shafi, M.I., Dunn, J.A., Chenoy, R., *et al.* (1994) Digital imaging colposcopy, image analysis and quantification of the colposcopic image. *British Journal of Obstetrics and Gynaecology*, **101**, 234–8.

47. Richart, R.M. (1990) A modified terminology for cervical intraepithelial neoplasia. *Obstetrics and Gynecology*, **75**, 131–5.

48. Tidbury, P., Singer, A., Jenkins, D., *et al.* (1992) CIN3: The role of lesion size in invasion. *British Journal of Obstetrics and Gynecology*, **99**, 583–6.

49. Chenoy, R., Billingham, L., Irani, S., *et al.* (1996) The effect of directed biopsy on the atypical transformation zone: assessed by digital imaging colposcopy. *British Journal of Obstetrics and Gynaecology*, **103**, 457–62.

50. Etherington, I.J., Dunn, J., Shafi, M.I., *et al.* (1997) Video colpography: a new technique for secondary cervical screening. *British Journal of Obstetrics and Gynaecology*, **104**, 150–3.

17

HUMAN PAPILLOMAVIRUS INFECTION IN THE MALE

Renzo Barrasso

EPIDEMIOLOGY

Condylomata acuminata were known to the ancient Greeks, who already supposed their sexual transmission. These lesions were associated with a viral cytopathic effect in 1968 and with the presence of human papillomaviruses (HPV) in 1981 (for review, see [1, 2]).

The identification of DNA of HPV in genital intra-epithelial and invasive neoplasia in women and in men [1, 3–6] has given new interest to the epidemiological data linking cervical cancer to a sexually transmitted agent. Consequently, careful vulvar examination and screening of the male partner have been proposed for women presenting with cervical intra-epithelial neoplasia (CIN) [6, 7]. This, together with the use of the colposcope and the application of acetic acid, has led to the better characterization of known HPV-associated lesions and to the description of new ones, mostly subclinical.

In fact, when male partners of women with HPV-associated genital disease are examined by colposcopy and the acetic acid test, about 40–50% of them show HPV-associated lesions, half of which are subclinical [8, 9]. About 20–30% of regular male partners of women with CIN have been showed to present histological features of penile intra-epithelial neoplasia (PIN) [8, 10]. Again, half of these lesions are subclinical. When a consecutive series of men with PIN is studied, 32% of the patients with PIN I are partners of women with high-grade cervical lesions (CIN II–III), whilst 72% of men with PIN II and 100% of men with PIN III lesions have female partners with CIN II–III [11].

The rate of PIN is much lower in studies from the USA [9]. This may be linked to the higher rates of circumcision, since, as we will see, circumcised male partners of women with CIN present a three-fold lower rate of PIN than uncircumcised men and, particularly, they do not present subclinical lesions, since those are mostly localized on the prepuce.

When the presence of HPV infection has been virologically studied by genital scrapings in both women with disease and their regular partners, one-third to one-half of couples have been shown to harbour the same HPV type, half of the males only presenting subclinical lesions [12–14]. Moreover, concomitant CIN and PIN mostly contain the same potentially oncogenic HPV types when PIN is subclinical [12]. These data underline the difficulties encountered in the study of sexual transmission, but also that subclinical lesions behave as the clinical ones as a reservoir of transmissible virus.

Cancer and Pre-cancer of the Cervix
Edited by D.M. Luesley and R. Barrasso. Published in 1998 by Chapman & Hall, London. ISBN 0 412 56600 1.

DIAGNOSTIC TOOLS

Clinical inspection is the basis of diagnosis of male HPV-associated lesions of external genitalia. The examination should be carried out carefully on the entire ano-genital region, with the help of a clear and powerful light and of a magnifying lens (at least 4×).

Virtually all abnormalities may be detected by careful clinical examination, except the changes appearing after application of acetic acid. This is why the latter are commonly defined as subclinical.

Careful inspection with a magnifying lens allows the observer to detect virtually all HPV-associated lesions and to make the differential diagnosis in most cases, as between warts and pearly papules [15]. Even after the acetic acid test, the white reactions will be mostly detected by the naked eye or a magnifying lens. The benefit of colposcopic viewing is the study of the angioarchitecture, which might be critical in diagnosing a few clinical and most subclinical HPV-associated lesions.

The application of 5% acetic acid, together with colposcopic magnification, has become the standard method for investigating discrete and subclinical HPV-associated lesions [8, 9].

The biological basis of acetowhitening is unknown. In lesions of the external genitalia, areas with surface nuclear activity (parakeratosis) and/or increased cellular density will react.

Acetic acid may be applied with a cotton ball mounted on a sponge-holding forceps or by self-application of a wet gauze. The application time and the time the operator waits before observation are critical. Acetic acid should be applied generously (this is why most clinicians prefer the self-application) and the operator has to wait 2–3 minutes before the observation. In fact, time before reaction is much longer than for the cervix. Paradoxically, the fastest reactions are mostly due to non-specific causes, such as inflammation, abrasion and infection.

Cytological examination of endourethral smears stained using Papanicolaou's method has been advocated, but studies performed have produced unsatisfactory results [13, 16]. Urethral scrapings for viral studies have also been proposed, in order to search for an urethral reservoir of latent HPV infection [17, 18].

Our data clearly show a high false-positive rate for urethral cytology, which is due to difficulties in identifying true koilocytes. The frequency of HPV infection as detected by Southern blot from urethral scrapes in men without colposcopically detected lesions is evaluated at 1–2% in partners of women with disease.

With the advent of the polymerase chain reaction (PCR) technique, the rate of HPV infection in males, as detected by penile and urethral scrapes, is around 15%, thus similar to that found in women (X. Bosch, personal communication).

In our experience [19], urethral scrapes are positive in men presenting with urethral lesions (more than 50% of HPV DNA prevalence, as detected by PCR) and in men without urethral lesions but presenting with penile lesions (about 15% of positive cases), but in less then 5% of men without colposcopically detected lesions. Thus, the urethral reservoir of infectious virus is infrequent. This is of clinical relevance, since patients who are lesion-free can be considered as mostly non-infectious.

As for the search of HPV in urine samples or in sperm, our data clearly show that this is an infrequent event. Semen from a cohort of male partners of women with histologically detected genital HPV infection was studied for HPV infection. PCR of semen showed the presence of HPV DNA in 48% of subjects with urethral lesions, including two cases of HPV 16. We also detected HPV DNA by means of PCR in 22% of semen samples from patients with penile lesions and no urethral lesions, and the HPV subtype was the same in the lesion as in the semen. HPV was detected in semen in only one patient without HPV-associated lesions. The study [19] does not confirm high HPV infection in semen from men without detectable lesions, suggesting that the mechanism for sperm contamination by HPV is the exfoliation of infected cells from urethral lesions during sperm transit and, probably, during masturbation in patients with penile lesions.

A biopsy may be needed in order to exclude the existence of cancer or intra-epithelial neoplasia (IN), mostly in older patients, or to confirm the diagnosis of HPV-associated lesion, for clinical or subclinical lesions.

Either a punch biopsy, such as Kevorkian or a 4 mm disposable Keyes punch, or a knife may be used.

MORPHOLOGY

CONDYLOMATA ACUMINATA

These warts are exophytic, white, greyish protuberances on keratinized skin, mostly papillary, with lobulated or irregular surface, pink-reddish through reddish-white on mucosal areas (Plate 31). Papillae may be prominent with finger-like projections. These projections mostly exhibit highly vascularized dermal cores appearing as typical punctated and/or loop-like patterns by magnifying equipment, unless they are heavily keratinized. Some warts will show large amounts of surface keratinization, thus appearing white.

Their number varies from a few to 50 or more lesions, and the sizes from 0.2 to 1.0 cm, but, if numerous, they become confluent, involving large areas of the genitalia.

Acuminate warts predominantly afflict areas submitted to trauma during intercourse. In the male, the lesions are prevalent on the inner aspect of the prepuce, at the frenulum and coronal sulcus; less frequently, they are located on the shaft, the glans, and within the urinary meatus. In circumcised men the shaft is often involved.

Acuminate warts afflicting either the meatal lips of the urinary meatus or the distal area of the urethra are far more common in the male (10–28%) than in the female (less than 5%). Condylomata extending to the surrounding area include the following: perineal, inguinal, anal and pubertal, which have a somewhat different gross appearance, forming a papillary growth of papilloma.

The incubation period may be as short as 3 weeks or as long as 8 months; about two-thirds of patient's sexual partners are afflicted after an average incubation period of 2.8 months [2].

Condylomata acuminata must be distinguished from other papillomatous lesions in this area. On the penis, a condition that may simulate warts has been described as pearly papules [15] or hirsutoid papillomas. Parallel rows of discrete acuminate structures distributed circumferentially around the coronal sulcus and on both sides of the frenumum (Plate 32) may resemble filiform warts. Histologically, these are hypertrophic papillae covered by a normal epithelium.

The colposcopic magnification allows distinction of pearly papules from condylomata acuminata since their surfaces are mostly smooth and dome-shaped, they do not show the typical vascular pattern presented by mucosal condylomata and they never coalesce.

Some warts may present as pigmented, mostly when they are located in skin areas. They mostly maintain their accuminate structure but in a few cases, differential diagnosis with seborrhoeic warts and with pigmented papules of IN may be needed.

The characteristic histologic feature of condylomata acuminata is epidermal proliferation with variously pronounced hyper- and parakeratosis.

In most cases, in the upper layers of the epidermis, there is a perinuclear vacuolization with usually slight koilocytotic atypia. In some cases koilocytosis is highly abundant, present throughout almost the whole epidermis. The normal mitotic figures are numerous and not infrequently scattered throughout the epidermis. In the corium there is usually abundant inflammatory infiltration and newly formed vessels, also seen within the elongated papillae, in spikes or papillary excrescences.

HPV 6 and 11 are detected in 70–95% of condylomata acuminata, independent of the location, extent and duration of lesions.

Potentially oncogenic HPVs are not detected in typical warts. When this is reported, careful morphologic analysis of lesions will invariably show papular lesions misclassified as warts.

GENITAL PAPILLOMAS

These proliferative lesions have a more hyperkeratotic and papillomatous surface and are larger. They lack the surface irregularities typical of condylomata acuminata. Usually pedunculated and darker than the surrounding skin, they may be brownish or even blackish. They are preferentially located on the skin. These lesions grow very slowly. This is why patients frequently, do not spontaneously refer themselves for therapy, convinced that they cannot be linked to an infectious process [12, 20].

The histological features include pronounced epidermal proliferation, elongation of rete pegs, hyperkeratosis and usually increased melanin accumulation in the basal cell layer. If hyperkeratosis is very extensive, they may be similar to seborrhoeic warts. Koilocytosis is discrete or absent. These papillomas mostly contain HPV type 6 or 11.

Differential diagnosis of genital papilloma includes true seborrhoeic warts, whose structure is similar. However, seborrhoeic warts mostly present with patchy black areas, due to focal melanocyte hyperproduction.

NON-PIGMENTED PAPULES

These lesions are clearly outlined, variably elevated but not pedunculated, with a round or dome-shaped, slightly hyperkeratotic or smooth surface [8]. The colour is that of normal skin, greyish, pinkish or, not infrequently, slightly brownish when located on genital skin sites, like the penile shaft, labia majora and perineum.

HPV-associated papules should be distinguished from psoriatic papules and from papules of lichen planus. The differential diagnosis is possible since HPV-associated papules mostly present with punctate vessels detectable by the colposcope on their surface, particularly on mucosal areas. Moreover, their relatively smooth surface may appear lobulated or micropapillary when viewed through magnifying equipment. HPV 6 and 11-associated papules are not pigmented, and they lack the surface exfoliation of psoriasis as well as surface striae of lichen planus. Finally, lichen planus and psoriasis also generally show associated extragenital lesions, which can be detected by careful examination.

HPV-associated papules are also easily distinguishable from the umbilicated pinkish–grey papules pathognomonic of molluscum contagiosum.

Sebaceous glands of the prepuce should not be mistaken for virally induced papules. Their colour is greyish-yellow, and the magnification systems allows their distinction.

When located on the mucosa, non-pigmented papules as well as small warts are translucent. They show a clear vascular punctation on the top, which allows their distinction from small pearly papules [8, 12, 20]. Often small, they may go undetected without the use of a magnifying lens or of a colposcope. The application of 5% acetic acid produces a strong, well-demarcated reaction, which usually preserves the punctate vessels. In most cases this will allow exact delineation of the lesions. The histology of papules differs from condylomata acuminata only by the absence of marked epidermal proliferation. In most lesions the dermo-epidermal border is flattened; in the upper layers of the epidermis there are variously abundant koilocytosis and, almost constantly, a discrete parakeratosis. There are no signs of IN. Mucosal acetowhite papules show the same histologic features, although koilocytosis is usually scarce or absent.

Non-pigmented papules of genital skin mostly contain HPV DNA types 6 or 11. Non-pigmented acetowhite mucosal papules may contain HPV DNA types 6 or 11, but also HPV 42.

MACULES

Macules are flat, well-demarcated areas of normal epithelium showing a white reaction after 5% acetic acid application (Plate 33) and containing easily detectable capillary loops [8].

They are usually seen on epithelial surfaces that had appeared perfectly normal before the test. However, at high magnification, it is not unusual to detect pale areas before the test, devoided of capillary structures. After the application of acetic acid, the neovascularization typical of most HPV-associated lesions appears, with the feature of capillary loops surrounded by a slowly appearing white, mostly opaque, reaction.

The surface of white areas is smooth, rarely micropapillary. The lack of detection of the normal branching vascular network within the white area together with the appearance of capillary loops after application of 5% acetic acid are the most important morphologic criteria to identify the lesions [12]. Their size may vary from 0.1 to 10 mm and their number from 1 to 10, rarely more.

They are usually found on mucosal surfaces (prepuce and glans in men), occasionally in the urethral meatus. Macules may be seen in association with clinically detectable lesions. They can be seen on the cutaneous part of external genitalia, although this is extremely rare.

The vast majority of typical, white-opaque macules are associated with HPV 42 infection [8, 20]. HPV 6 and 11 are rarely associated with subclinical lesions detected as a sole feature.

Psoriasis, lichen planus, folliculitis and contact dermatitis may react to the acetic acid test. The morphology of lesions before the test should allow for differential diagnosis.

In most cases, balanoposthitis will mimic subclinical lesions. In these cases, the white reaction is mostly diffuse, irregularly distributed and lacks the vascular pattern typical of macular HPV-associated lesions.

In some cases, frequently during yeast infection, multiple single white areas are detected (Plate 34). In these cases capillaries are evident all over the tissue, and not only within white areas, because of inflammation-associated hyperaemia.

Inflammatory and traumatic microabrasions will also react with acetic acid. Careful observation before the test allows the detection of the abraded border of those areas.

Histologically, most subclinical lesions show features of endophytic condyloma: parakeratosis, acanthosis and papillomatosis are usually present, koilocytosis is rare and, if present, focal [8]. In a few cases, koilocytosis will be abundant, realizing a typical endophytic wart (Plate 35). In some cases, features of IN may be detected [12].

If white areas that do not correspond to our description of subclinical lesions are biopsied, they will show non-specific inflammatory features and, sometimes, parakeratosis, but they will lack papillomatosis and, of course, koilocytosis.

Histological distinction between discrete HPV-associated features and the mild degrees of acanthosis and parakeratosis found in a non-specific acetowhite reaction is possible by careful morphological analysis, and is of major relevance for epidemiological studies and for clinical protocols [21].

These lesions are usually found in biopsies obtained from partners of women with disease: the pathologist may feel 'forced' to diagnose borderline condylomas and the clinician will consequently treat or follow men who will be told they have a sexually transmitted infection. In these cases, viral testing and, particularly, *in situ* hybridization, could be most helpful to avoid overtreatment and psychological problems.

All HPV types may be found in subclinical lesions. HPV 42, however, is the most frequent type in lesions showing discrete histologic features of endophytic condyloma, while HPV 6 or 11 are found in endophytic warts particularly rich in koilocytes [8, 12, 20].

In situ hybridization is positive in about 90% of exophytic condylomas, but also in 70% of subclinical lesions with histology of endophytic condyloma [8, 12, 20]. This suggests that HPV may replicate in 'true' subclinical lesions, and that those may consequently be infectious.

Thus, histology and *in situ* hybridization clearly show that from a biological point of view subclinical lesions and latent infection are different situations representing different virus–host interactions. The difference between subclinical lesions and latent infection has to be stressed, since it does not seem to be clear to most clinicians.

INTRA-EPITHELIAL NEOPLASIA OF YOUNG ADULTS

The term Bowenoid papulosis was coined in 1978 for clinically inconspicuous papular or papulo-macular lesions of the external genitalia of both sexes, showing histological features of squamous cell carcinoma *in situ* of Bowen's type [22]. These lesions are actually defined as vulvar and penile IN [23].

Pigmented papules are usually small, 0.2–3.5 cm in diameter (average 1.0 cm), often flat at the skin level in spite of the term papulosis, reddish or slightly brownish, with a smooth, glistening, sometimes verrucous surface. The pigmented features are produced by increased melanin synthesis and extrusion in the underlying papillary dermis (Plate 36).

Their number varies from single lesions to about 30. Lesions may coalesce to form plaques. Pigmented papules are mostly localized on the cutaneous side of external genitalia. On mucosal areas, red papules are also classified as Bowenoid papules, even if the amount of melanin is not increased. It is not unusual to observe a strong white reaction after the acetic acid test for mucosal red papules [12, 20].

Differential diagnosis has to be performed with pigmented warts, lentigo, naevi and eventually melanoma, seborrhoeic warts, angiokeratomas, lichen planus and localized lesions of psoriasis.

Pigmented warts may be indistinguishable from pigmented papules of Bowenoid papulosis. However, Bowenoid papules do not present the accuminate or spiked surface of most warts.

Malignant melanoma is rare on external genitalia. Its clinical presentation may mimic dark-pigmented papules. However, it occurs most often in non-hair-bearing areas. All localized pigmented lesions have to be biopsied, in order to rule out this malignancy.

Seborrhoeic warts are rare on the skin of external genitalia. If present, they are impossible to distinguish from verrucous pigmented lesions of IN and also from pigmented warts.

Lichen planus mostly affects young adults. The typical reddish–purplish, flat-topped papules of lichen planus are often followed by hyperpigmentation.

Leucoplastic lesions appear as white plaques, mostly raised, sometimes papular. Their surface is granular or micropapillary (Plate 37). Vessels are rarely detectable. Their histologic substratus consists of surface hyperkeratinization.

As a rule, lesions are numerous in young patients. They may also coalesce in large plaques. Leucoplastic lesions are mostly located on the cutaneous side of external genitalia. When detected on the mucosa, they maintain their vascular punctation and mostly present surface micropapillae.

Leucoplastic lesions should be distinguished from other situations rendering the epithelium white. Those may be associated with stromal hypovascularization, as in lichen sclerosus, or with depigmentation, as in vitiligo.

Differential diagnosis with vitiligo does not present problems. Lesions are not raised and the surface structure is the same as the surrounding skin. Moreover, they are often symmetrically distributed when located on the glans.

The differential diagnosis with lichen sclerosus as well as with leucoplastic invasive cancer is mostly determined by age and because lesions in young patients are multifocal.

Erythroplastic lesions are mostly flat. The observer detects a well-defined red or red-greyish area, showing evident and mostly atypical punctate vessels (Plate 38a). The colour depends on the increased vascularity of pre-cancers and on the existence of a decreased number of epithelial cell layers.

The area reacts strongly to the acetic acid test, usually preserving the punctation (Plate 38b). The reaction may turn opaque white or greyish.

Lesions are flat, non-erosive, well-demarcated, mostly multiple, invariably located on mucosal areas and mostly asymptomatic. Punctate vessels are obvious and mostly irregular in shape and distribution.

These criteria will mostly guide the differential diagnosis. Differential diagnosis has to be made with the erythema caused by most inflammations and infections (candidiasis, chemical or thermal burning, allergic responses, folliculitis). The existence of well-delimited lesions, the detection of the typical vascular punctation, the absence of symptoms, the persistence of lesions after anti-inflammatory treatment and, eventually, the biopsy will allow the correct diagnosis.

As we have seen, erythroplastic flat lesions are invariably located on genital mucosa, as well as subclinical lesions. Thus, it is interesting to observe that young circumcised men almost never show subclinical lesions or erythroplastic PIN [11]. PIN presents in these subjects only as pigmented or leucoplastic raised areas.

The implication of this is that a significant difference does exist between the frequency of IN in circumcised and uncircumcised patients [11], while the frequency of warts is not significantly different.

Circumcision does not prevent HPV from inducing lesions. However, local conditions could favour productive infection by potentially oncogenic HPVs, mostly on the prepuce.

The histology of IN in young adults [6] shows an exaggeration of the epithelial proliferation (acanthosis). Hyper- and parakeratosis are also invariably present. The cell nuclei are hyperchromatic, clumped and present a loss of organization, maturation and cohesion. Mitotic figures are generally numerous and abnormal forms are often present. Dyskeratosis is frequent. In most cases, koilocytosis appears in the upper third of the epithelium. In the corium, there are slight inflammatory infiltrates and dilated capillaries. Thus, the histologic picture is that of IN grade III.

Potentially oncogenic HPVs are usually associated with these lesions. The most frequently identified type is HPV 16 [3, 6, 20, 23].

The natural history of IN in young adults is not fully known. During the course of the disease, enlargement of lesions and inflammation are often seen. The course of the disease is usually chronic. On average, the duration of the disease is of about 2–3 years [1]. Lesions may spontaneously regress. However, in some cases they may last for as long as 8–10 years, and may progress to carcinoma. Although this disease is regarded as benign and self-limited, especially in younger men and women, the possibility exists that Bowenoid papulosis harbouring potentially oncogenic HPVs is also a risk factor with respect to the aetiology of carcinoma of the lower genital tract and of the anal canal [1]. Particularly, female patients with Bowenoid papulosis are at high risk for cervical carcinoma. In the majority of sexual partners of men with Bowenoid papulosis, cervical HPV infection is seen and high-grade IN can be detected [23]. Patients with multifocal or early vulvar carcinoma have a high incidence of cervical neoplasia. The same is true for vulvar IN in patients previously treated for cervical carcinoma. Thus, from a clinical point of view, all genital areas must be carefully examined in the presence of IN at one site; particularly, colposcopy is recommended in addition to cytology in women with IN of external genitalia.

INTRA-EPITHELIAL NEOPLASIA IN OLDER MEN

Intra-epithelial neoplasia in older patients mostly comprises the formerly defined Bowen's disease and Erythroplasia of Queyrat.

Clinically, IN in older patients mostly appears as a unique lesion, erythroplastic, leucoplastic or, usually, erythro-leucoplastic. The lesion is slightly elevated, plaque-like (Bowen) or totally flat (Queyrat). The erythroplastic form will appear as a brilliant, velvet-red area. The leucoplastic form will appear as a white plaque, with a smooth or verrucous surface. Localized at first, lesions may extend to large areas of external genitalia, including perineal and perianal areas.

The disease is mostly detected in men older than 50 and this might be due to decreased immunity. The intra-epithelial lesions may persist for a long time, but the risk of progression to cancer is high.

These lesions must be differentiated from a large number of other dermatological disorders.

Among white areas, lichen sclerosus is localized on the inner prepuce in men, as well as on the genito-crural folds and on the perianal area. Its early feature presents with flat, ivory papules. Papules quickly become confluent and form large plaques with satellite peripheral papules. Co-existing atrophy will obliterate the lesional contours. Pallor is striking, often with an added purpuric component. The main result in men will be phimosis.

Among red areas, lichen simplex may appear as a unique, usually very large red plaque, showing a smooth and brilliant surface, different from the one of HPV-associated lesions and lacking vascular punctation.

Psoriasis will present as a unique red or pink plaque, mostly localized on cutaneous tissue (preferentially perineum), showing a characteristic white exfoliation. Lesions are fissured when localized on skin folds. On mucosal areas psoriatic lesions will lack exfoliation, but their symmetrical distribution and the perfect-

ly delimited border, as well as co-existent lesions detected by a general clinical examination, will allow the differential diagnosis.

Zoon's plasmocytoma is a disease of unknown origin affecting mucosal areas of external genitalia. It presents as red, mostly unique large areas, which mimic IN. Those areas whiten after acetic acid application. Even though punctate vessels are mostly absent, only a biopsy will formally exclude the existence of IN.

It is not always possible to suspect early invasion in lesions of IN grade III. Erosive/ulcerated areas and atypical vessels within lesions of IN represent the most suspicious sign. In any case, treatment of IN in older patients will be excisional, in order to histologically analyse the entire lesion. Histology of IN of Bowen's type shows the same features as compared to IN in younger patients.

Intra-epithelial neoplasia in older patients is associated with the same HPV types found in IN in younger patients [12, 20]. It is still not clear if these are two different and unlinked diseases or if the persistence of potentially oncogenic HPVs may produce a long-term evolution of Bowenoid papulosis in Bowen's disease in the same patient.

CLINICAL APPROACH

Clinically detected HPV-associated lesions should be treated, mostly because of the risk of viral transmission. As for lesions of IN, a careful follow-up of the female partner has also to be performed, since those lesions contain potentially oncogenic viruses.

It is difficult to suggest treatment for subclinical lesions, since their biology is poorly known and the differential diagnosis with non-specific whitening does not seem to be easily reproducible. Thus, systematic treatment of subclinical lesions would cause overtreatment in numerous cases. However, it should be remembered that true subclinical lesions represent, warts with an endophytic growth and that *in situ* hybridization shows their infectivity.

Should the asymptomatic male partner of women with HPV-associated lesions be screened? A clear scientific answer to this question has not yet been offered and thus individual beliefs mostly guide clinicians. The author of this chapter strongly believes that screening the male partner is of value in the prevention of cervical cancer and that the demonstration will one day be furnished.

REFERENCES

1. Jablonska, S., Majewski, S. and Obalek, S. (1987) Age-related variability of ano-genital lesions induced by potentially oncogenic human papillomaviruses. *Giorn. Ital. Chir. Derm. Oncol.*, **2**, 347–52.
2. Oriel, J.D. (1971) Natural history of genital warts. *British Journal of Venereal Disease*, **47**, 1–13.
3. Ikenberg, H., Gissmann, L., Gross, G., Grussendorf-Conen, E.I. and zur Hausen, H. (1983) Human papillomavirus type 16-related DNA in genital Bowen's disease and in bowenoid papulosis. *International Journal of Cancer*, **32**, 563–5.
4. Iwasawa, A., Kumamoto, Y. and Fujinaga, K. (1993) Detection of human papillomavirus deoxyribonucleic acid in penile carcinoma by polymerase chain reaction and *in situ* hybridization. *Journal of Urology*, **149**, 59–63.
5. McCance, D.J., Kalashe, A. and Ashdown, K. (1986) Human papillomavirus types 16 and 18 in carcinomas of the penis from Brazil. *International Journal of Cancer*, **37**, 55–9.
6. Obalek, S., Jablonska, S., Beaudenon, S., Walkzak, S. and Orth, G. (1986) Bowenoid papulosis of the male and female genitalia: risk of cervical neoplasia. *Journal of the American Academy of Dermatology*, **14**, 433–44.
7. Campion, M.J., Singer, A., Clarkson, P.K. and McCance, D.J. (1985) Increased risk of cervical neoplasia in consorts of men with penile condylomata acuminata. *Lancet*, **I**, 943–6.
8. Barrasso, R., de Brux, J., Croissant, O. and Orth, G. (1987) High prevalence of papillomavirus-associated penile intraepithelial neoplasia in sexual partners of women with cervical intraepithelial neoplasia. *New England Journal of Medicine*, **317**, 916–23.
9. Levine, R.U., Crum, C.P., Herman, E., Silvers, D., Ferenczy, A. and Richart, R.M. (1984) Cervical papillomavirus infection and intra-epithelial neoplasia: a study of the male sexual partner. *Obstetrics and Gynecology*, **64**, 16–20.
10. Zanardi, C., Guerra, B., Martinelli, G., de Brux, J. and Barrasso, R. (1988) Papillomavirus-related genital lesions in male partners of women with genital condyloma or cervical intraepithelial neoplasia: diagnostic approach. *Cervix*, **6**, 127–34.
11. Aynaud, O., Ionesco, M. and Barrasso, R. (1994) Penile intraepithelial neoplasia: specific clinical features correlate with histologic and virologic findings. *Cancer*, **74**, 1762–7.
12. Barrasso, R. and Gross, G. (1997) Lesions of external genitalia: diagnosis. In *Human Papillomavirus Infection: A Clinical Atlas* (eds G. Gross and R. Barrasso). Ullstein Mosby, Berlin, pp. 290–361.
13. Schneider, A., Sawada, E., Gissmann, L. and Shah, K. (1987) Human papillomaviruses in women with a history of abnormal Papanicolau smears and in their male partners. *Obstetrics and Gynecology*, **69**, 554–62.
14. Wickenden, C., Hanna, N., Taylor-Robinson, D., *et al.* (1988) Sexual transmission of human papillomaviruses in heterosexual and male homosexual couples, studied by DNA hybridization. *Genitourinary Medicine*, **64**, 34–8.
15. Ackermann, A.B. and Kornberg, R. (1973) Pearly penile papules. Acral angiofibromas. *Archives of Dermatology*, **108**, 673–5.
16. Nahhas, W.A., Marshall, M.L. and Ponziani, J. (1986) Evaluation of urinary cytology of male sexual partners of women with cervical intraepithelial neoplasia and human papillomavirus infection. *Gynecologic Oncology*, **24**, 279–84.
17. Grussendorf-Conen, E.I., de Villiers, E.M. and Gissmann, L. (1986) Human papillomavirus genomes in penile smears of healthy men. *Lancet*, **I**, 1092 (letter).
18. Katakoa, A., Claesson, U., Hansson, B.G., Eriksson, M. and Lindh, E. (1991) Human papillomavirus infection of the male diagnosed by Southern-blot hybridization and polymerase chain reaction: comparison between urethra samples and penile biopsy samples. *Journal of Medical Virology*, **33**, 159–64.
19. Aynaud, O., Ionesco, M. and Barrasso, R. Lack of cytological and virological detection of papillomavirus infection in the urethra of healthy men. In press.
20. Barrasso, R. and Jablonska, S. (1989) Clinical, colposcopic and histologic spectrum of male human papillomavirus-associated genital lesions. *Clinical Practice in Gynecology*, **2**, 73–101.

21. Nuovo, G.J., Hochman, H.A., Eliezri, Y.D., Lastarria, D., Comite, S.L. and Silvers, D.N. (1990) Detection of human papillomavirus DNA in penile lesions histologically negative for condylomata. *American Journal of Surgical Pathology*, **14**, 829–36.

22. Wade, T.R., Kopf, A.W. and Ackermann, A.B. (1978) Bowenoid papulosis of the penis. *Cancer*, **42**, 1890–3.

23. Obalek, S., Jablonska, S. and Orth, G. (1985) HPV-associated intraepithelial neoplasia of external genitalia. *Clinics in Dermatology*, **3**, 104–13.

18

PROGNOSTIC FACTORS IN CERVICAL CANCER

Peter van Geene and John A. Tidy

INTRODUCTION

The evolution of our knowledge related to cervical cancer has resulted in the description of many factors which may impact on prognosis, and as we strive to provide more information to our patients, we may try to use these factors to individualize their risk. However, we should remain guarded in the use of this information as the data are not of uniform quality. In many studies the number of cases are small and some of the discrepancies between the reported results may reflect statistical problems rather than true biological differences. Because of the nature of the disease most of the factors described use retrospective case data and few have been subjected to prospective analysis.

Ideally, prognostic factors should be easily measured and reproducible; in other words, they should be the result of objective and not subjective observation and the information they provide should also be available at the outset of treatment to allow appropriate management planning. Tumour grade, for example, is easily assessed from a biopsy specimen, but may be subjective and inter-observer or even intra-observer variation is common. By comparison, measurement of tumour volume is objective but precise three-dimensional accuracy difficult to obtain and impossible if tumour excision is incomplete [1].

This chapter describes many of the prognostic factors related to cervical cancer. They can be divided into five broad groupings that relate to the patient, the tumour and its spread, the biology of the tumour and treatment-related factors (Table 18.1).

FACTORS RELATED TO THE PATIENT

AGE

Cervical cancer occurs with a bimodal distribution. Women under the age of 40 have been postulated to have a worse prognosis when compared with older women, although the published data are equivocal. Some studies have found a worse prognosis in this group [2–7], while others have found a better prognosis [8–12] and finally some have found no difference [13–16]. The variations between these data may be explained by the differences in trends in mortality rates experienced in each of the populations studied. While the overall age-adjusted mortality rates have declined in most countries the trends in age-specific rates have been variable [17]. An increase in the 30–39 year age-specific mortality rate has occurred since 1970 in England and Wales, Scotland, Ireland, Australia and New Zealand. A similar increase, but starting in the 1980s, is also apparent in some nordic and East European coun-

Cancer and Pre-cancer of the Cervix
Edited by D.M. Luesley and R. Barrasso. Published in 1998 by Chapman & Hall, London. ISBN 0 412 56600 1.

Table 18.1 Prognostic factors in cervical cancer

Factors related to the patient
- Age
- Race
- Social status
- Performance status and co-morbidity
- Parity
- Pregnancy

Factors related to the primary tumour
- FIGO stage
- Cell type
- Grade
- Tumour volume
- Lymphovascular space invasion
- Tumour angiogenesis
- Flow cytometry
- Host response

Factors related to the spread of the tumour
- Lymph node metastases

Factors related to the biology of the tumour
- Human papillomavirus
- Oncogenes and gene products
- Tumour markers
- Thrombocytosis

Factors related to treatment
- Inappropriate surgery
- Radiotherapy
- Chemotherapy

tries. However, in countries such as Canada, Japan and USA, the trend has been downward or static. Although there may be an increase in the age-specific mortality rate in young women it should be remembered that this rate is very low. In a multivariate analysis of 17,119 cases of cervical cancer obtained from the National Cancer Institutes database between 1973 and 1987, Kosary [12] found that increasing age was associated with a worsening 5-year relative survival for patients with equivalent stage, histological type, tumour grade and lymph node status. Women over the age of 70 at diagnosis were nearly 10-times more likely to die of their disease than women under 30 at diagnosis.

RACE

Racial factors have been implicated in the different incident rates for many cancers including gynaecological malignancies. However, the influence of race on survival is less well defined. It has been suggested that white races have a better prognosis than non-whites with cervical cancer [18, 19], although this finding may be due to differing population characteristics leading to bias in the groups studied. The incidence of cervical cancer in black women in the USA is higher compared with white women but the trend, over time, has been for a reduction in the incidence, which has been greatest amongst black women [17]. This may reflect improved health education and access to the health care system for black women. Social stigmata or taboos could lead to later presentation or decreased access to appropriate medical care in the poor prognosis group. When assessed by multivariate regression analysis and controlling for other variables such as histological type, grade and the International Federation of Gynaecology and Obstetrics (FIGO) stage, race ceases to be a significant prognostic factor [12].

SOCIAL STATUS

Women from lower socio-economic groups have an increased incidence of cervical cancer and are more likely to present with advanced stage disease; however, the outcome, stage for stage, is similar to that for women from higher socio-economic groups [20].

KARNOFSKY PERFORMANCE STATUS AND CO-MORBIDITY

The condition of a patient before treatment must always be taken into account if morbidity and mortality are to be minimized. Performance status probably gives additional prognostic information related to failure of radiotherapy and recurrence rates but data are equivocal [21]. In advanced-stage disease, co-morbidity from diabetes, hypertension and obesity may have little impact on survival; however, in early-stage disease they are associated with a reduction in survival rates [22, 23]

PARITY

Few data are available for the effect of parity on prognosis. One study of 167 Stage Ib and IIa patients from Sweden found the relative risk of recurrence in multiparous (three children or more) women to be 4.6. Interestingly, multiparity was more predictive of recurrence than residual tumour in the hysterectomy specimen (relative risk of recurrence 3.4), although this may be due to all patients receiving brachytherapy as well as radical hysterectomy and lymphadenectomy [24].

PREGNANCY

Carcinoma of the cervix may be diagnosed during pregnancy with an incidence of approximately 1 in 2200 pregnancies [25]. Pregnant women with cervical cancer generally present with early-stage disease and prognosis is similar to non-pregnant patients [25, 26]. When diagnosed in early pregnancy treatment options include immediate radical hysterectomy or to delay treatment until the foetus is viable, followed by caesarean section and radical hysterectomy. In Hacker's series there was no difference in survival between the two modes of treatment. The outlook, however, may be worse for patients with more advanced disease diagnosed in pregnancy.

FACTORS RELATED TO THE PRIMARY TUMOUR

FIGO STAGE

The clinical staging system adopted by FIGO and modified in 1994 [27] has a number of practical advantages over surgical staging but is not as accurate (Table 18.2). Prognostic infor-

mation is gathered from a physical examination under anaesthetic including cystoscopy and sigmoidoscopy, radiological studies including intravenous urogram, barium enema and chest X-ray and histological examination of a biopsy or conization specimen. Lymph node status is recorded in the FIGO stage but this information is often not available until laparotomy. Staging gives information principally relating to tumour extent. Early tumours, completely excised by conization, are measured to find their maximum depth and width of invasion. Where the maximum depth of invasion is 3 mm or less (Stage Ia$_1$), the risk of lymph node metastases is 0.3% and the risk of an invasive recurrence only 0.2% [28]. In the event of deeper invasion (Stage Ia$_2$) to 3.1–5 mm the risk of lymph node metastases is 7.4% and the local recurrence rate is 5.4%. The consensus of opinion is that Stage Ia$_1$ cancers of the uterine cervix behave in a similar way to cervical intra-epithelial neoplasia (CIN) and can therefore be managed in the same way with conservative treatment, usually either an adequate cone biopsy or simple hysterectomy [29–33]. Any tumours larger than this should be managed by radical hysterectomy because of the high risk of recurrence or spread from nodal disease.

Stage is the strongest predictor of prognosis in cervical cancer with survival times diminishing as stage at diagnosis increases (Table 18.3). Part of the reason for this is the increasing risk of lymph node metastases as tumour volume and vascular space involvement increase (Table 18.4). However, there is wide variation in survival times for individual patients within each stage due to additional prognostic factors that do not form part of the FIGO staging system. Hopkins and Morley [22], in a review of 175 cases of Stage IIIb and IV squamous cell carcinoma, found that the IVU status in Stage IIIb patients significantly predicted cumulative 5-year survival. Five-year survival in women with a normal IVU was 47%, ureteric obstruction without renal failure was associated with 29% survival and

Table 18.2 FIGO staging of cervical cancer from 1994

Stage	I	Carcinoma strictly confined to the cervix (extension to the corpus should be disregarded)
	Ia	Preclinical carcinomas of the cervix, that is, those diagnosed only by microscopy
	Ia_1	Less than 3 mm depth of invasion taken from the base of the epithelium, either surface or glandular, from which it originates, and a second dimension, the horizontal spread, must not exceed 7 mm
	Ia_2	Lesions with measured depth of invasion between 3 and 5 mm taken from the baser of the epithelium, either surface or glandular, from which it originates, and a second dimension, the horizontal spread, must not exceed 7 mm. Larger lesions should be staged as Ib
	Ib	Lesions of greater dimensions than stage Ia_2 whether seen clinically or not. Preformed space involvement should not alter the staging but should be specially recorded so as to determine whether it should affect treatment decisions in the future
	Ib_1	Lesions less than 4 cm in maximum diameter
	Ib_2	Lesions greater than 4 cm in maximum diameter
Stage	II	The carcinoma extends beyond the cervix but has not extended on to the side wall. The carcinoma involves the vagina, but not the lower third
	IIa	No obvious parametrial involvement
	IIb	Obvious parametrial involvement
Stage	III	The carcinoma has extended on to the pelvic side wall. On rectal examination there is no cancer-free space between the tumour and the pelvic wall. The tumour involves the lower third of the vagina. All cases with hydronephrosis or non-functioning kidney
	IIIa	No extension to the side wall
	IIIb	Extension on to the pelvic side wall and/or hydronephrosis or non-functioning kidney
Stage	IV	The carcinoma extends beyond the true pelvis or has clinically involved the mucosa of the bladder or rectum. A bulbous oedema as such does not permit a case to be alloted to stage IV
	IVa	Spread of the growth to adjacent organs
	IVb	Spread to distant organs

Table 18.3 Survival by stage and node status

Stage	n	5-Year survival (%)	
		All cases	Node-positive cases
Ib	6511[a]	80.1	67.3
IIa	2433[b]	62.6	48.4
IIb	2533[c]	55.5	42.7
III	2598[d]	37.9	30.0
IVa	1518[e]	10.5	7.0

Cumulative results from references: [a][12, 50, 128, 133–135]; [b][12, 128, 135–138]; [c][12, 128, 132, 134, 135, 138]; [d][12, 50, 128, 132, 138]; [e][12, 50, 128].

Table 18.4 Incidence of pelvic and para-aortic node metastases by stage

Stage	Pelvic nodes (%)	Para-aortic nodes (%)
Ia_1	0[a]	0[a]
Ia_2		
(1–3 mm)	0.6[a]	0[a]
(3–5 mm)	4.8[a]	<1.0[a]
Ib	20.8[b]	5.0[e]
IIa	26.5[c]	15.9[e,f]
IIb	41.5[c]	15.9[e,f]
III	44.8[d]	34.4[e]
IVa	55.0[d]	35.7[d]

Cumulative results from references: [a][139–142]; [b][143, 144]; [c][137, 143, 144]; [d][21]; [e][143, 145]. [f]Combined results for Stage II pelvic node involvement.

when renal failure occurred all patients were dead of disease by 16 months. The median survival of 18 patients who presented in renal failure and underwent nephrostomy was 8 months.

A large number of women who undergo surgery for early-stage cervical cancer will not have any evidence of metastatic nodal disease. The overall survival rate for this group is high but unfortunately, for some women, the disease will recur. Several studies have tried to identify women, within this group, who are at high risk of recurrence; however, the data are somewhat contradictory. Smiley *et al.* [34] found no correlation with age, depth of invasion and lymphovascular space involvement. Stockler *et al.* [35] found an increased risk of recurrence associated with these features. Both found poorly differentiated tumours to be at higher risk of recurrence. Stockler also reported an increased risk if the surgical clearance was less than 5 mm but adenocarcinomas were not linked with an increased risk of recurrence.

Histological examination of the biopsy or complete tumour specimen is used to provide important information used in treatment planning and every pathological report should include details of histological type, grade, tumour size in two dimensions, the presence or otherwise of vascular or lymphatic space involvement, the presence of CIN or VaIN (vaginal intra-epithelial neoplasia) and how close the edge of the tumour is to the resection margins.

HISTOLOGY

Most carcinomas of the cervix, about 85–90%, are of squamous type with adenocarcinomas accounting for 10–15%. Mixed adenosquamous carcinomas occur in about 7% of cases and other forms of carcinoma including adenoid basal cell carcinoma, glassy cell carcinoma and neuroendocrine carcinoma are rare. Other, non-epithelial, tumours of the cervix are also rare. These include lymphomas, sarcomas, germ cell tumours, malignant melanoma and trophoblastic tumours.

EPITHELIAL CANCERS

Squamous cell carcinoma (SCC)

Three histological groups of SCC are recognized by the World Health Organization (WHO): large cell keratinizing, large cell non-keratinizing and small cell carcinomas. Small cell carcinomas have been reported as having a worse prognosis [36, 37] but there is some disagreement in the more recent literature [38]. Large cell non-keratinizing tumours appear to respond better to radiotherapy than large cell keratinizing types but there is no difference in the prognosis when treatment is by surgery alone [39]. Overall, there is probably little difference in prognosis between the groups.

Adenocarcinoma

Some older reports using univariate analyses have suggested that adenocarcinoma of the cervix carries a worse prognosis than squamous carcinoma [40, 41] and if confirmed this would be a very significant finding in view of the increasing incidence of adenocarcinomas. However, more recent work, controlling for confounding factors, has not been able to confirm this finding [42, 43]. In a large population-based cohort study of the SEER data between 1973 and 1987, Platz and Benda [44] found that adenocarcinomas had similar survival patterns to squamous carcinomas in early-stage disease but worse prognosis in advanced disease. A smaller study by Kleine found the reverse, in that adenocarcinomas treated with radiotherapy alone had a worse prognosis than squamous carcinomas but only for patients with early stage (FIGO Stage I, II) disease. Hopkins and Morley [45] studied 959 patients with adenocarcinomas and squamous cell carcinomas and on multivariate analysis found that, stage for stage, adenocarcinomas had a poorer outcome. Other reports suggest that 5-year survival data do not truly reflect the poorer prognosis of adenocarcinomas due to a higher incidence of late metastases [46, 47].

Overall, the balance of opinion seems to be that adenocarcinomas have a worse prognosis than squamous cell carcinomas, and this view is supported by FIGO. Some reviewers have indicated that this may be due to lower sensitivity of glandular lesions to radiotherapy but there is no clear evidence to support this. Another explanation may be that the endophytic growth of adenocarcinomas results in fewer symptoms and delayed diagnosis, resulting in greater tumour volume when compared with similar stage squamous lesions. There is evidence that the incidence of adenocarcinoma of the cervix is increasing in some populations [17]. This may be reflected, to some degree, by a change in histological diagnosis but it may also be a consequence of a reduction in the incidence of squamous cancers secondary to effective screening.

A number of histological subtypes of adenocarcinoma have been described [48] and are listed in Table 18.5. Most reported studies have suggested that histological subtype has no influence on prognosis [42, 49, 50] but several studies have been published which contradict this evidence in some subtypes,

Table 18.5 Histological classification of cervical adenocarcinoma

- Endocervical type – usual
- Minimal deviation
- Endometrioid
- Clear cell
- Adenoid cystic
- Adenoid basal
- Mucinous
- Papillary
- Mesonephric
- Adenosquamous
- Squamous carcinoma with mucin production

mainly endometroid adenocarcinoma, minimal deviation adenocarcinoma and adenosquamous carcinoma.

Adenosquamous carcinomas

Both squamous cell carcinomas containing mucin but no histologically detectable glandular differentiation (immature adenosquamous carcinoma or with mucin secretion) and mature adenosquamous carcinomas where malignant glandular elements can be identified are said to have a worse prognosis than pure SCC. These tumours tend to occur at a younger age and metastasize to the lymph nodes earlier [51]. In a comparison, Bethwaite *et al.* [52] found that the 5-year survival of women with adenosquamous carcinomas was 52% whereas those women with pure SCC had a 75% 5-year survival. Colgan *et al.* [53] however, disagree with this opinion and found in a review no conclusive evidence to support the case for a worse prognosis in adenosquamous carcinomas.

Minimal deviation adenocarcinoma

These tumours, being very well differentiated, might be expected to demonstrate a favourable prognosis. However, the reverse seems to be the case and several studies using a univariate analysis report a poor or very poor prognosis [54, 55]. When stage is carefully controlled, some studies have shown no difference in survival between this tumour and the typical endocervical adenocarcinoma [56].

Endometrioid adenocarcinoma

Primary endometrioid adenocarcinomas of the cervix may be difficult to differentiate from direct invasion of an endometrial adenocarcinoma. Their microscopic appearance, clinical behaviour and prognosis are similar and two studies have been published confirming this impression [57, 58].

Veruccous carcinoma

These tumours, rare variants of SCC, are characterized by their slow, locally invasive growth, exophytic nature, papillary form, benign cytological appearance and lack of distant metastases. Prognosis is good provided a wide local excision is made, otherwise central recurrence is invariable. The lack of distant metastases makes lymph node dissection unnecessary. Radiotherapy, however, has no effect on local invasion and may be associated with conversion of the tumour to a highly aggressive anaplastic carcinoma [59] and is therefore contraindicated.

Adenoid basal cell carcinoma

Although rare and infrequently reviewed in the literature, it seems that this tumour behaves in a fairly benign manner with generally excellent prognosis. Most cases present as Stage I disease and mitotic activity is very low. Dinh and Woodruff [60] have suggested that hysterectomy without lymph node dissection should be adequate treatment.

Lymphoepithelioma-like carcinoma

Few reports exist in the literature of this rare variant of SCC. It is characterized by an intensive inflammatory infiltrate and this, it has been argued, may be the cause of its relatively good prognosis. Lymphatic metastases are uncommon, occurring in only 5.1% of patients in a small series in 1977 by Hasumi *et al.* [61].

RARE, NON-EPITHELIAL CANCERS OF THE CERVIX

A variety of rare malignant tumours may arise in the cervix or be metastatic from other sites. Undifferentiated small cell tumours were originally thought to be a subgroup of squamous cell carcinoma but the majority appear to demonstrate neuroendocrine differentiation. They are now generally referred to as small cell undifferentiated carcinomas (SCUC) or neuroendocrine tumours. Several types of sarcoma can arise in the cervix, they are all very rare and generally have a poor prognosis. Lymphoma, germ cell tumours, malignant melanoma and trophoblastic tumours can also arise in the cervix.

Neuroendocrine tumours (SCUC)

These tumours tend to metastasize early by both lymphatic and blood-borne routes, with widespread deposits occurring in regional and distant lymph nodes, bone marrow, lung, liver and brain. Few patients present with an abnormal cervical smear and most (75%) are diagnosed at Stage I or II, which is comparable to squamous cell carcinomas, but their growth is more aggressive and their prognosis worse. Overall 5-year survival is 14% [62], and the disease-free interval less than 2 years in most patients. Surgical staging is the best prognostic indicator with, in one study, eight out of 11 Stage I or II patients surviving 4 years and 22 out of 23 Stage III or IV patients dying of their disease [63]. In this same study, the presence of a squamous or glandular component was associated with a better prognosis, although the number of patients was small and other studies, also with small numbers, have not confirmed the difference [64, 65]. More recent studies using flow cytometry have suggested a DNA index of less than 1.5 to be associated with a better prognosis and longer survival [66].

Sarcomas

Endocervical stromal sarcoma

This very rare tumour with few reports in the literature exhibits poor prognosis. Abell and Ramirez [67] described a series of 13 patients, seven of whom died of their disease within 2 years of diagnosis, with the longest surviving patients being those with better differentiated tumours. Like other sarcomas, distant metastases to the lungs was common.

Embryonal rhabdomyosarcoma

This is the cervical equivalent of sarcoma botryoides, which occurs principally in the vagina of infants. It is rare, with only about 100 cases described, mainly in young women. The outlook for these patients is quite good, however, in comparison to sarcoma botryoides. Of patients with Stage I disease, 88% were alive and clinically disease-free after a mean follow-up period of 68 months [68] and for all stages the survival was 80%. The principal adverse prognostic factor in this tumour appears to be the depth of invasion as reported by Daya and Scully [69] and Perrone *et al.* [70]. It has also been suggested that as tumours containing an element of an alveolar rhabdomyosarcoma tend to be unusually aggressive these patients should be specifically highlighted in the pathology report [71].

Leiomyosarcoma, mixed Müllerian tumours

Primary tumours of this type are rare in the cervix but behave in a similar fashion to those arising in the uterus. Stage, tumour size and the presence of extracervical spread are probably the most important adverse prognostic factors, whilst mitotic rate and vascular invasion have not been found to be of prognostic value. No studies have been of sufficient size to reliably predict survival, but 5-year survival in the case of uterine primaries is 33–39%.

Malignant melanoma

Malignant melanoma metastatic to the cervix is probably much more common than a primary in this site but it is very difficult to distinguish between them. Data with regard to prognostic factors are sparse in the literature but the presence of metastatic lymph node involvement is associated with poor outcome. In 60% of cases there is evidence of metastatic spread at the time of diagnosis. Two-year survival based on reported cases is about 10%.

Trophoblastic tumours

Choriocarcinoma arising in the cervix is rare. Primary tumours of the uterus have a 5-year survival of greater than 90% but this falls once the tumour has spread to other sites. Those arising in the cervix are said to have a better prognosis than those arising in other sites.

GRADE

FIGO recognizes three histological grades of tumour based on the degree of cytological and histological differentiation. Grade 1 corresponds to well-differentiated tumours, grade 2 to moderate differentiation and grade 3 to poor differentiation. Grade appears to be an important prognostic indicator when all histological types of cervical carcinoma are considered together. However, although this prognostic significance persists in most studies when adenocarcinomas are considered alone [42, 72], several have suggested that grade carries no prognostic significance in squamous cell carcinomas [1, 34, 73]. Assessing grade in squamous cancers is a technical challenge with none of the three grading methods (Broder's, Ng or Reagan's) being reliably reproducible [1]. In a review by Griffin and Wells [74], 5-year survival in Stage I, grade 3 adenocarcinomas was found to be 41–62% compared to 80–90% for Stage I, grade 1 tumours.

TUMOUR VOLUME

Estimation of volume is most accurate in early-stage disease before tumour spread beyond the cervix has occurred. Technically volume can be difficult to measure and therefore depth and width of invasion are taken as markers for total volume. In advanced disease or disease treated non-surgically, tumour diameter is taken as a marker for tumour volume. Tumour volume, in all stages of disease, is intimately involved with increasing incidence of nodal spread and reduced disease-free survival (Tables 18.3 and 18.4). Ziano [1] examined 195 cases of Stage Ib squamous carcinomas and found a strong correlation between depth of invasion and the frequency of nodal disease and poor outcome. Eifel *et al.* [75, 76] and Fyles *et al.* [77] found that increasing tumour diameter, for both squamous and adenocarcinomas, was associated with reduced 5-year survival (88% for tumours less than 5 cm compared with 69% for tumours 5–8 cm).

LYMPHOVASCULAR SPACE INVASION

Invasion of the lymphatics and small vessels correlates strongly with the development of lymph node metastases, which is itself a poor prognostic indicator [34, 73, 78]. Whereas vascular invasion has been shown to be an adverse prognostic indicator, it is unclear from the literature whether it remains an independent prognosticator once tumour volume is controlled for [79]. Depth of stromal invasion is a stronger predictor of lymph node metastases and progression-free survival than lymphovascular space involvement [1]. The reliable interpretation of lymphovascular spaces remains a histopathological challenge and can be open to observer error since it can be mistaken for stromal reaction. However, Barber *et al.* [80], in a study of radical hysterectomy in 191 cases of Stage Ib carcinoma of the cervix, recorded a 5-year survival of 59.4% in cases with lymphovascular invasion and 90% where no lymphovascular invasion was present.

TUMOUR ANGIOGENESIS

As tumours grow, new blood vessels are induced to provide nutritional support for the enlarging mass of new cells. Abnormal surface capillaries can be seen by colposcopy and these constitute one of the criteria for the diagnosis of invasive carcinoma, but new capillaries can also be identified in stromal tissue by staining tumour sections with specific endothelial cell stains. In principle, the increased number of small vessels should correlate with an increased risk of metastasis formation, but as with many of the other prognostic factors the data are contradictory. Recent studies [81, 82] have failed to find an association between number of new vessels and lymph node status. Four of 29 patients [81] who developed recurrent disease within 1 year had high microvessel counts but interestingly none of these had positive nodes. Schlenger *et al.* [83] and Bremer *et al.* [84] also found a high tumour microvessel count to correlate significantly with both reduced disease-free interval and overall survival [83]. An alternative hypothesis, however, may be that decreased microvessel density results in areas of tissue hypoxia within a tumour, thus reducing sensitivity to radiotherapy or chemotherapy. Consistent with this view, Kainz found that a low microvessel density correlated with a poor recurrence-free interval [85]. Tumour angiogenesis may be an indicator of the rapidity of tumour growth and this could explain the correlation with recurrence-free survival time, or it could reflect tumour sensitivity to treatment. Either way, further studies are required to confirm microvessel counts as an independent prognostic factor.

FLOW CYTOMETRY

Recently, flow cytometry has been used to investigate several variables with a view to eliciting new prognostic factors. Two studies, including one large prospective study of 465 patients by Kristensen *et al.*, failed to find any prognostic significance of DNA ploidy analysis or of S-phase fraction [86, 87] on survival in cervical carcinoma. One study [88] reviewing the DNA index of 98 patients with Stage IIa carcinoma of the cervix found, using a multivariate analysis, that a DNA index of 1.7 was an independent prognostic factor for 5-year survival. The median 5-year survival for patients with a DNA index of 1.7 or greater was 36 months compared to 73.5 months for those with a DNA index of less than 1.7. Other studies, based on the assumption that rapidly dividing cells are more sensitive to radiotherapy and chemotherapy, have used flow cytometry to predict the response of cervical carcinomas to radiotherapy. Lutgens *et al.* [89] found that after stratification for stage, ploidy was significantly correlated with pelvic disease-free survival in Stage II patients although they admitted that stage was a better predictor overall. In a smaller study, Zolzer *et al.* [90] measured the S-phase fraction of carcinoma cells obtained from sequential biopsies taken during radiotherapy for cervical carcinoma. The pre-treatment S-phase fraction was not predictive of outcome but the combination of decrease in S-phase fraction and increase in micronucleus frequency during treatment reliably predicted survival, with 90% surviving 5-years compared to less than 30% for non-responders. A larger study, looking at 151 patients stratified for age and stage did, however, identify the S-phase fraction as the strongest predictor of survival, more so even than tumour grade, vascular invasion and other histopathological parameters [91].

HOST RESPONSE

The ability of the host to recognize and reject malignant cells is an important determinant of tumour invasion and may reflect the biological behaviour of the tumour independently of other prognostic factors. Various studies have reported a reduction in CD4 helper cells and an increase in CD8 suppresser/cytotoxic cells in cervical carcinoma, as well as impaired natural killer cell function and an increase in the monocyte numbers [92]. This host response can be determined by measuring lymphoid cell populations in blood or in the inflammatory stromal reaction (ISR) surrounding the tumour. Changes in the lymphocyte populations of peripheral blood films from patients with cervical cancer have been shown to be of no prognostic significance by some researchers [93]. Others have found a significant association between disease-free survival and

eosinophil counts [94], CD4 cell counts [95], circulating immune complex concentrations and interleukin 2 production following radiotherapy [96]. ISR also seems to correlate with the recurrence-free interval and overall survival time. Kainz and co-workers found that with reduced ISR the relative risk of dying from disease was 5.4 [85].

FACTORS RELATED TO THE SPREAD OF THE TUMOUR

LYMPH NODE METASTASES

The incidence of lymph node metastases increases with advancing stage, increasing tumour volume, the presence of lymphovascular space invasion and with certain histological types such as adenosquamous carcinomas. The first nodes to be involved tend to be the external iliac group, followed by the obturator and the hypogastric nodes. Further spread to the paraaortic or distant nodes is less common and generally carries a worse prognosis. Overall, positive pelvic lymph nodes are found in 16–55% of patents depending on stage and other factors (see Table 18.3). Survival data from studies of women, standardized for stage, with nodal metastases at the time of diagnosis are summarized in Table 18.4 and clearly show a significant reduction for patients with positive compared to negative nodes. It is useful during radical hysterectomy to record the site of the highest positive node as prognosis seems to depend more on the upper level of pelvic node involvement rather than the total number of nodes involved, although this is disputed by two recent studies which found the number of positive nodes to be of prognostic significance [22, 97]. Lymphangiography is no longer performed routinely in the majority of centres because of its relatively poor predictive value. Nevertheless, the information gained from lymphangiography may still be of use in prognosis as indicated by Roman *et al.* [98] and Fyles *et al.* [77] from a retrospective analysis of 965 patients with invasive cervical cancer. Women with para-aortic nodal involvement identified by lymphangiogram had a significant reduction in disease-free survival compared to women with pelvic lymph node involvement only.

FACTORS RELATED TO THE BIOLOGY OF THE TUMOUR

HUMAN PAPILLOMAVIRUS (HPV)

A large body of evidence exists showing a strong association between HPV and cervical cancer and current opinion certainly seems to implicate HPV with a role in the development of cervical carcinoma. It might be considered surprising, therefore, that HPV status has been shown to have a favourable effect on prognosis. Several studies have shown that patients where HPV DNA or RNA has been identified in tumours have a lower risk of recurrence and longer 5-year survival. Riou *et al.* [99], reported that in HPV DNA-negative tumours the recurrence rate was 2.6-times higher and the risk of distant metastases 4.6-times higher than in HPV DNA-positive tumours. These findings have been confirmed using HPV RNA by Higgins *et al.* [100] who also found an association between HPV RNA negativity and older patients with poor prognosis. Median survival in HPV-negative patients in one study was half that of HPV-positive patients [101]. Other reports, however, have found no significant association between HPV status and prognosis [102, 103]. The data relating to HPV type are just as conflicting as those relating to overall HPV status. HPV types 16 and 18 have been associated with a greater risk of progression of CIN to invasive carcinoma and many centres now routinely test for HPV type when treating patients with CIN. The presence of HPV 16 and 18 DNA sequences in cervical carcinoma has not, however, been shown unequivocally to be associated with an altered prognosis with reports both for [103, 104] and against

[105]. It seems unlikely that HPV typing will prove of clinical prognostic value in the near future but no doubt this will not prevent its continued evaluation in many centres.

ONCOGENES AND GENE PRODUCTS

Oncogenes play a central role in normal cell growth and proliferation by coding for proteins, cytokines and transmembrane receptors, which control cell growth by participating in the intracellular signal transduction network. The oncogenes mainly studied in cervical cancer include the RAS gene family, *c-myc* and *c-erbB2*. The tumour suppressor genes TP53 and retinoblastoma (RB1) are important regulators of the cell cycle and abrogation of their function by the E6 and E7 proteins of the oncogenic human papillomaviruses may be important in the development of cervical neoplasia.

C-erbB-2

C-erbB-2 is an oncogene encoding a 185 kDa transmembrane glycoprotein, $p185^{erbB2}$ with tyrosine kinase activity important for cell signalling. This protein shows extensive structural similarity to the epidermal growth factor receptor $p170^{EGFR}$ (EGFR) and both are thought to be growth factor receptors. Immunohistochemical staining has shown c-erbB2 to be present in 12.1% of early cervical carcinomas and overexpression of EGFR has been found in 25.8%. A study of 132 patients with Stage Ib cervical carcinoma found that overexpression of EGFR was an independent prognostic indicator associated with a shorter relapse-free survival time, and when considered in combination with immunostaining for cathepsin D was the strongest prognostic factor after tumour size [106].

TP53 and retinoblastoma gene

The tumour suppressor gene *TP53* has been implicated in the aetiology of many cancers. In cervical cancer the interaction with the E6 protein of the oncogenic HPV types probably plays an important function in the development of the tumour. This association leads to the inactivation of p53 protein so the incidence of *TP53* mutations in cervical cancer is low. Early publications demonstrated a mutually exclusive association where *TP53* mutations were only present in HPV negative cancers [107] whereas others have not been able to confirm this. The presence of aberrant p53 expression, detected by immunohistochemistry, has not always been confined to HPV-negative tumours and may occur in tumours with no evidence of *TP53* mutations; however, in some studies, the entire *TP53* gene has not been sequenced [108]. Studies looking at the expression of p53, however, found no correlation with prognosis [109].

RAS

The RAS gene family includes three structurally and functionally related genes: Ha-RAS, K-RAS and N-RAS, all of which code for proteins (p21) which bind to the inner surface of the cell membrane and transmit mitogenic signals from extracellular growth factors. Increased gene expression, as measured by immunohistochemical identification of p21, has been found in 50% of invasive cervical carcinomas [110]. Furthermore, this overexpression is associated with a greater risk of lymph node metastases and poorer prognosis [111].

c-myc

The *c-myc* oncogene codes for a protein that binds to the nucleus and regulates DNA replication and cell division. Normal expression is associated with passage of the cell from the resting phase of the cell cycle (G0) to the first phase of mitosis (G1). *C-myc* overexpression has been found in 0–33% of invasive cervical carcinomas depending on methodology and has been found to be predictive of survival in node-negative patients. Riou *et al.* reported 3-

year survival rates for node-negative patients to be 93% when *myc* was normal and 51% when *myc* was overexpressed [112].

CD44

The cell adhesion molecule CD44 is expressed on many cell types and abnormal expression of this molecule has been linked to other cancers. Expression of the spliced variant exon v-6 of CD44, in both early and advanced stage disease, is associated with a poor prognosis [113, 114]. However, the detection of CD44 variants in patients serum did not correlate with the presence of cervical cancer [115]. The expression of the variants may reflect a more malignant phenotype.

TUMOUR MARKERS

Squamous cell carcinoma antigen (SCC-ag)

Two groups, Kornafel *et al.* in Poland and Duk *et al.* from the Netherlands, have reported on the prognostic value of pre-treatment serum SCC-ag concentration [116, 117]. Both groups found that serum SCC-ag was an independent predictor of survival and strongly correlated with the presence of adverse clinicopathological features, especially the presence of lymph node metastases. The risk of recurrence in patients with elevated serum SCC-ag was three-times higher than in patients with normal concentrations of SCC-ag, even in node-negative patients. Bolger *et al.* [118] measured SCC-ag in 220 women with early stage cervical cancer. High levels were found in patients with larger tumours, nodal disease and Stage II disease but there was no correlation with tumour recurrence. Disappointingly, low levels of SCC-ag could not predict node-negative patients.

Human chorionic gonadotrophin

The detection of the beta core fragment of human chorionic gonadotrophin in pre-menopausal women is associated with a very poor outcome [119].

THROMBOCYTOSIS

Thrombocytosis is found frequently in association with malignant disease, although the mechanism for this and the clinicopathological implications apart from an increased risk of thrombosis are unclear. It has been suggested that a platelet count above $400 \times 10^9/l$ is a poor prognostic indicator in cervical carcinoma [120]. A large study by Lopes *et al.* [121] from Gateshead, UK, however, demonstrated that thrombocytosis was not related to the incidence of pelvic lymph node metastases and was not an independent prognostic factor.

FACTORS RELATED TO TREATMENT

INAPPROPRIATE SURGERY

Unfortunately some women may initially undergo inappropriate surgery, most frequently a simple hysterectomy has been performed without the surgeon being aware of the presence of cervical cancer. In most cases this could be avoided by the proper evaluation of the patient and their symptoms [122]. These women usually receive adjuvant radiotherapy but some may also undergo pelvic lymphadenectomy. Two published audits of the management of early stage cervical cancer from Scotland [123, 124] found that between 21 and 36% of women with cervical cancer, who were treated surgically, underwent inappropriate non-radical procedures. Although adjuvant radiotherapy was given to between 84 and 86% of these women, the reported 5-year survival for Stage Ib disease was reduced to 68% [124]. Hopkins *et al.* [125] reviewed the outcome for 92 women treated by simple hysterectomy and 78 received adjuvant radiotherapy. The 5-year survival for women with Stage I squamous cancers was 85% (88% for those who received adjuvant radiotherapy compared with 69% for those who were observed);

however the 5-year survival for Stage I adenocarcinoma was only 42%. In a series of 122 cases who had undergone simple hysterectomy followed by radiotherapy there was no difference in survival rates for squamous and adenocarcinomas. Survival was related to the presence of gross disease at the time of radiotherapy [98].

RADIOTHERAPY

As in other modes of therapy for cervical cancer, the response to radiotherapy will depend on the administration of optimal treatment as well as on aspects of tumour biology such as tissue oxygenation and tumour sensitivity to ionizing radiation. Tissue hypoxia may increase a tumour's malignant potential by increasing the number of non-cycling cells and therefore the resistance to ionizing radiation and chemotherapeutic agents. This hypothesis would seem to have little application as a clinically useful prognostic factor, as there is currently no reliable method for routinely measuring intra-tumoural pO_2. Hockel *et al.* [126], however, using a new technique now claim to be able to routinely measure the tissue pO_2 in tumours and have shown in a series of 35 patients followed up for 24 months that hypoxia is an adverse prognostic factor for both survival and recurrence-free interval. Unfortunately, to date no treatment has been found to alter tissue pO_2 in tumours although several have been tried and the authors have therefore recommended ultra-radical surgery and pelvic reconstruction as an alternative for these patients.

The Patterns of Care Study [127], utilizing data from the Surveillance, Epidemiology and End Results database (SEER) of the National Cancer Institute, identified radiotherapy techniques that constituted best practice and those that were associated with a poorer outcome. Patients receiving low energy teletherapy, insufficient radiation dose, a prolonged rest period between teletherapy and brachytherapy, a single brachytherapy insertion compared to two or three, or no brachytherapy at all were found to have a worse prognosis. The design of this study can be criticized, however, as the results were not controlled for patient case mix or high-/low-risk groups. Nevertheless, several studies have subsequently confirmed these results [77, 128, 129].

CHEMOTHERAPY

It seems logical to assume that patients receiving adequate treatment will have a better prognosis than those inadequately treated, particularly in early-stage disease when there is a good chance of eliminating all cancerous tissue. Certainly, patients with metastatic spread or with tumour present in resection margins have a worse prognosis. However, even when adequately treated, about 5–10% of Stage I node-negative patients survive less than 5 years. This may be due to undetected microscopic tumour spread at the time of initial treatment or it could also be due to inherent tumour biology and resistance to conventional therapy. Several studies have been performed where initial tumour response to treatment has been used as an assessment of inherent tumour biology and to provide prognostic information about overall outcome. In 1991, Panici *et al.* [130], published and subsequently updated [131, 132] a study showing that tumour response to neoadjuvant chemotherapy carried prognostic significance with regard to overall 5-year survival. Seventy-five patients with locally advanced cervical carcinoma were treated with neoadjuvant chemotherapy, including cisplatin, bleomycin and methotrexate, and their response assessed. A complete response was achieved by 15% of patients and 68% a partial response before proceeding to radical hysterectomy. Patients achieving a complete or partial response had a significantly improved 5-year survival, 77 and 14% respectively. Response to treatment also correlated with initial tumour size greater than 5 cm, parametrial involvement to the pelvic side wall and

pre-treatment serum squamous cell carcinoma antigen concentration. The presence of EGFR, oestrogen and progesterone receptors did not correlate with response to treatment.

SUMMARY

All cases of cervical cancer may demonstrate different facets of the disease process which may affect prognosis and outcome. The stage of disease at presentation is the single most important determinant of prognosis, while the data relating to other factors may be described as soft and at times contradictory. The influence of factors such as age may vary in different populations throughout the world. The current evidence suggests we must continue to remain guarded in our clinical use of prognostic factors, particularly when we are considering new or additional management strategies.

REFERENCES

1. Ziano, R.J., Ward, S., Delgado, G., *et al.* (1992) Histopathologic predictors of the behaviour of surgically treated Stage IB squamous cell carcinoma of the cervix. *Cancer*, **69**, 1750–8.
2. Stanhope, C.R., Smith, J.P. and Wharton, J.T. (1980) Carcinoma of the cervix; the effect of age on survival. *Gynecologic Oncology*, **10**, 188–93.
3. Prempree, T., Patanaphan, V. and Sewchand, W. (1983) The influence of patient's age and tumour grade on the prognosis of carcinoma of the cervix. *Cancer*, **51**, 1764–71.
4. Le Vecchia, C., Franceshi, S. and Decarli, A. (1984) Invasive cervical cancer in young women. *British Journal of Obstetrics and Gynaecology*, **91**, 1149–55.
5. Chapman, G.W., Abreo, F. and Thompson, H.E. (1988) Carcinoma of the cervix in young females (35 years and younger). *Gynecologic Oncology*, **31**, 430–4.
6. Fedorkow, D.M., Robertson, D.I. and Duggan, M.A. (1988) Invasive squamous cell carcinoma of the cervix in women less than 35 years old: recurrent versus non-recurrent disease. *American Journal of Obstetrics and Gynecology*, **158**, 307–11.
7. Kodama, S., Kanazawa, K. and Honma, A. (1991) Age as a prognostic factor in patients

with squamous cell carcinoma of the uterine cervix. *Cancer*, **68**, 2481–5.

8. Gynning, I., Johnsson, J.E. and Alm, P. (1983) Age and prognosis in stage Ib squamous cell carcinoma of the uterine cervix. *Gynecologic Oncology*, **15**, 18–26.

9. Russel, J.M., Blair, V. and Hunter, R.D. (1987) Cervical carcinoma: prognosis in younger patients. *British Medical Journal*, **295**, 300–3.

10. Meanwell, C.M., Kellyu, K.A. and Wilson, S. (1988) Young age as a prognostic factor in cervical cancer: analysis of population based data from 10,022 cases. *British Medical Journal*, **296**, 386–91.

11. Clarke, M.A., Naahas, W. and Markert, R.J. (1991) Cervical cancer: women aged 35 and younger compared to women aged 36 and older. *American Journal of Clinical Oncology*, **14**, 352–6.

12. Kosary, C.L. (1994) FIGO stage, histology, histologic grade, age and race as prognostic factors in determining survival for cancers of the female gynecological system: an analyses of 1973–87 SEER cases of cancers of the endometrium, cervix, ovary, vulva and vagina. *Seminars in Surgical Oncology*, **10**, 31–46.

13. Mann, W.J., Levy, D. and Hatch, K.D. (1980) Prognostic significance of age in stage I carcinoma of the cervix. *Southern Medical Journal*, **73**, 1186–8.

14. Hall, S.W. and Monaghan, J.M. (1983) Invasive carcinoma of the cervix in young women. *Lancet*, **2**, 731.

15. Mendenhall, W.M., Thar, T.L. and Bova, F.J. (1984) Prognostic and treatment factors affecting pelvic control of stage Ib and IIa–b carcinoma of the intact uterine cervix treated with radiation therapy alone. *Cancer*, **53**, 2649–54.

16. Carmichal, J.A., Clarke, D.H. and Mother, D. (1986) Cervical carcinoma in women aged 34 and younger. *American Journal of Obstetrics and Gynecology*, **154**, 263–9.

17. Beral, V., Hermon, C., Muñoz, N. and Devesa, S.S. (1994) Cervical cancer. *Cancer Surveys*, **19/20**, 265–85.

18. Myers, M.H. and Hankey, B.F. (1980) Cancer patient survival experience. Trends on survival 1960–63 to 1970–73. Comparison of survival for black and white patients. National Cancer Institute. NIH publication, 80–2148.

19. Howard, J., Hankey, B.F. and Greenberg, R.S. (1992) A collaborative study of differences in the survival rates of black patients and white patients with cancer. *Cancer*, **69**, 2349–60.

20. Lamont, D.W., Symonds, R.P., Brodie, M.M., Nwabineli, N.J. and Gillis, C.R. (1993) Age, socio-economic status and survival from cancer of cervix in the West of Scotland 1980–87. *British Journal of Cancer*, **67**, 351–7.

21. Marcial, V.A. and Marcial, L.V. (1993) Radiation therapy of cervical cancer. *Cancer*, **71**, 1438–45.

22. Hopkins, M.P. and Morley, G.W. (1993) Prognostic factors in advanced stage squamous cell cancer of the cervix. *Cancer*, **72**, 2389–93.

23. Peipert, J.F., Wells, C.K., Schwartz, P.E. and Feinstein, A.R. (1994) Prognostic value of clinical variables in invasive cervical cancer. *Obstetrics and Gynecology*, **84**, 746–51.

24. Gerdin, E., Cnattingius, S., Johnson, P. and Pettersson, B. (1994) Prognostic factors and relapse patterns in early stage cervical carcinoma after brachytherapy and radical hysterectomy. *Gynecologic Oncology*, **53**, 314–19.

25. Hacker, N.F., Berek, J.S., Lagasse, L.D., Charles, E.H., Savage, E.W. and Moore, J.G. (1982) Carcinoma of the cervix associated with pregnancy. *Obstetrics and Gynecology*, **59**, 735–46.

26. Zemlickis, D., Lishner, M., Degendorfer, P., Panzarella, T., Sutcliffe, S.B. and Koren, G. (1991) Maternal and fetal outcome after invasive cervical cancer in pregnancy. *Journal of Clinical Oncology*, **9**, 1956–61.

27. Creaseman, W.T. (1995) New gynecologic cancer staging. *Gynecologic Oncology*, **58**, 157–8.

28. Sevin, B.U., Nadji, M., Averette, H.E., Hilsenbeck, S., Smith, D. and Lampe, B. (1992) Microinvasive carcinoma of the cervix. *Cancer*, **70**, 2121–8.

29. Kolstad, P. (1989) Follow up study of 232 patients with stage Ia1 and 411 patients with stage 1a2 squamous cell carcinoma of the cervix (microinvasive carcinoma). *Gynecologic Oncology*, **33**, 265–72.

30. Tsukamoto, N., Kaku, T. and Matsukuma, K. (1989) The problem of stage Ia (FIGO, 1985) carcinoma of the uterine cervix. *Gynecologic Oncology*, **34**, 1–6.

31. Greer, B.E., Figge, D.C., Tamimi, H.K., Cain, J.M. and Lee, R.B. (1990) Stage Ia2 squamous carcinoma of the cervix: difficult diagnosis and therapeutic dilemma. *American Journal of Obstetrics and Gynecology*, **162**, 1406–9.

32. Burghardt, E., Girardi, F., Lahousen, M., Pickel, H. and Tamussino, K. (1991) Microinvasive carcinoma of the uterine cervix (International Federation of Gynecology and Obstetrics Stage Ia). *Cancer*, **67**, 1037–45.

33. Morgan, P.R., Anderson, M.C., Buckley, C.H., *et al.* (1993) The Royal College Of Obstetricians and Gynaecologists microinvasive carcinoma of the cervix study: preliminary results. *British Journal of Obstetrics and Gynaecology*, **100**, 664–8.

34. Smiley, L.M., Burke, T.M., Silva, E.G., Morris, M., Gershenson, D.M. and Wharton, J.T. (1991) Prognostic factors in stage Ib squamous cervical cancer patients with low risk for recurrence. *Obstetrics and Gynecology*, **77**, 271–5.

35. Stockler, M., Russell, P., McGahan, S., Elliot, P.M., Dalrymple, C. and Tattersall, M. (1996) Prognosis and prognostic factors in node-negative cervix cancer. *International Journal of Gynecological Cancer*, **6**, 477–82.

36. Wentz, W.B. and Louis, G.C.J. (1965) Correlation of histologic morphology and survival in cervical cancer following radiation therapy. *Obstetrics and Gynaecology*, **26**, 228–32.

37. Fidler, H.K., Boyes, D.A. and Worth, A.J. (1968) Cervical cancer detection in British Columbia. *Journal of Obstetrics and Gynaecology of the British Commonwealth*, **75**, 392–404.

38. Beacham, J.B., Halvorsen, T. and Kolbenstvedt, A. (1978) Histological classification, lymph node metastases, and patient survival in stage Ib cervical carcinoma. *Gynecologic Oncology*, **6**, 95–105.

39. Swan, D.S. and Roddick, J.W. (1973) A clinico-pathological correlation of cell type classification for cervical cancer. *American Journal of Obstetrics and Gynecology*, **116**, 666–70.

40. Milsom, I. and Friberg, L.G. (1983) Primary adenocarcinoma of the uterine cervix: a clinical study. *Cancer*, **52**, 942–7.

41. Moberg, P.J., Einhorn, N., Silfversward, C. and Soderberg, G. (1986) Adenocarcinoma of the uterine cervix. *Cancer*, **57**, 407–10.

42. Kilgore, L.C., Soon, S.-J., Gore, H., Shingleton, H.M., Hatch, K.D. and Partridge, E.E. (1988) Analysis of prognostic factors in adenocarcinoma of the cervix. *Gynecologic Oncology*, **31**, 137–48.

43. Vesterinen, E., Forss, M. and Nieminen, U. (1989) Increase in cervical adenocarcinoma: a report of 520 cases of cervical carcinoma including 112 tumours with glandular elements. *Gynecologic Oncology*, **33**, 49–53.

44. Platz, C.E. and Benda, J.A. (1995) Female genital tract: incidence and prognosis by histological type SEER population based data 1973–1987. *Cancer*, **75**, 270–9.

45. Hopkins, M.P. and Morley, G.W. (1991a) A comparison of adenocarcinoma and squamous cell carcinoma of the cervix. *Obstetrics and Gynecology*, **77**, 912–17.

46. Benda, J.A., Platz, C.E. and Buchsbaum, H. (1985) Mucin production in defining mixed carcinoma of the uterine cervix: a clinicopathological study. *International Journal of Gynecological Pathology*, **4**, 314–27.

47. Drescher, C.W., Hopkins, M.P. and Roberts, J.A. (1989) Comparison of the pattern of metastatic spread of squamous cell cancer and adenocarcinoma of the uterine cervix. *Gynecologic Oncology*, **33**, 340–3.

48. Fu, Y.S., Reagen, J.W., Fu, A.S. and Janiga, K.E. (1982) Adenocarcinoma and mixed carcinoma of the uterine cervix. *Cancer*, **49**, 2571–7.

49. Berek, J.S., Hacker, N.F., Fu, Y.S., Sokale, J.R., Leuchter, R.C. and Lagasse, L.D. (1985) Adenocarcinoma of the uterine cervix: histologic variables associated with lymph node metastasis and survival. *Obstetrics and Gynaecology*, **65**, 46–52.

50. Hopkins, M.P. and Morley, G.W. (1991b) Stage Ib squamous cell carcinoma of the cervix: clinicopathological features related to survival. *American Journal of Obstetrics and Gynecology*, **164**, 1520–9.

51. Buckley, C.H., Beards, C.S. and Fox, H. (1988) Pathological prognostic indicators in cervical cancer with particular reference to women under the age of 40 years. *British Journal of Obstetrics and Gynaecology*, **95**, 47–56.

52. Bethwaite, P., Yeong, M.L., Holloway, L., Robson, B., Duncan, G. and Lamb, D. (1992) The prognosis of adenosquamous carcinoma of the uterine cervix. *British Journal of Obstetrics and Gynaecology*, **99**, 745–50.

53. Colgan, T.J., Auger, M. and McLaughlin, J.R. (1993) Histopathologic classification of cervical carcinomas and recognition of mucin secreting squamous carcinomas. *International Journal of Gynaecological Pathology*, **12**, 64–9.

54. McKelvey, J.L. and Goodlin, R.R. (1963) Adenoma malignum of the cervix: a cancer of deceptively innocent histological pattern. *Cancer*, **16**, 549–57.

55. Gilks, C.B., Young, R.H., Aguirre, P., DeLellis, R.A. and Scully, R.E. (1989) Adenoma malignum (minimal deviation adenocarcinoma) of the uterine cervix: a clinicopathological and immunohistochemical analysis of 26 cases. *American Journal of Surgical Pathology*, **13**, 717–29.

56. Kaminski, P.F. and Norris, H.J. (1983) Minimal deviation adenocarcinoma (adenoma malignum) of the cervix. *International Journal of Gynecological Pathology*, **2**, 28–41.

57. Hurt, W.G., Silverberg, S.G., Frable, W.J., Belgrad, R. and Crooks, L.D. (1977) Adenocarcinoma of the cervix: histolopathologic and clinical features. *American Journal of Obstetrics and Gynecology*, **129**, 304–15.

58. Saigo, P.E., Cain, J.M., Kim, W.S., Gaynor, J.J., Johnson, K. and Lewis, J.L. (1986) Prognostic

factors in adenocarcnoma of the uterine cervix. *Cancer*, **57**, 1584–93.

59. Kraus, F.T. and Perez-Meza, C. (1966) Verrucous carcinoma: clinical and pathologic studies of 105 cases involving the oral cavity, larynx and genitalia. *Cancer*, **19**, 26–38.

60. Dinh, T.V. and Woodruff, J.D. (1985) Adenoid cystic and adenoid basal carcinomas of the cervix. *Obstetrics and Gynecology*, **65**, 705–9.

61. Hasumi, K., Sugano, H., Sakamoto, G., Masubachi, K. and Kubo, H. (1977) Circumscribed carcinoma of the uterine cervix, with marked lymphocytic infiltration. *Cancer*, **39**, 2503–7.

62. Abeler, V.M., Holm, R., Nesland, J.M. and Kjorstad, K.E. (1994) Small cell carcinoma of the cervix. *Cancer*, **73**, 672–7.

63. Silva, E.G., Gershenson, D., Sneige, N., Brock, W.A., Saul, P. and Copeland, L.J. (1989) Small cell carcinoma of the uterine cervix: pathology and prognostic factors. *Surgical Pathology*, **2**, 105–15.

64. Gersell, D.J., Mazoujian, G., Mutch, D.G. and Rudloff, M.A. (1988) Small cell undifferentiated carcinoma of the cervix: a clinicopathologic, ultrastructural and immunohistochemical study of 15 cases. *Surgical Pathology*, **12**, 684–98.

65. Walker, A.N., Mills, S. and Taylor, P.T. (1988) Cervical neuroendocrine carcinoma: a clinical and light microscopic study of 14 cases. *International Journal of Gynecologic Pathology*, **7**, 64–74.

66. Miller, B., Dockter, M., El Torky, M. and Photopulos, G. (1991) Small cell carcinoma of the cervix: a clinical and flow cytometric study. *Gynecologic Oncology*, **42**, 27–33.

67. Abell, M.R. and Ramirez, J.A. (1973) Sarcomas and carcinosarcomas of the uterine cervix. *Cancer*, **31**, 1176–92.

68. Brand, E., Berek, J.S., Nieberg, R.K. and Hacker, N.F. (1987) Rhabdomyosarcoma of the uterine cervix: sarcoma botyroides. *Cancer*, **60**, 1552–60.

69. Daya, D.A. and Scully, R.E. (1988) Sarcoma botyroides of the uterine cervix in young women: a clinicopathological study of 13 cases. *Gynecologic Oncology*, **29**, 290–304.

70. Perrone, T., Carson, L.F. and Dehner, L.P. (1990) Rhabdomyosarcoma with heterologous cartilage of the uterine cervix: a clinicopathalogical and immunohistochemical study of an aggressive neoplasm in a young female. *Medical and Paediatric Oncology*, **18**, 72–6.

71. Clement, P.B. (1995) Miscellaneous primary tumours and metastatic tumours of the uterine cervix. In *Haines and Taylor* (ed. Fox), 4th edn. Churchill Livingstone, Edinburgh, pp. 345–64.

72. Goodman, H.M., Buttlar, C.A. and Niloff, J.M. (1989) Adenocarcinoma of the uterine cervix: prognostic factors and patterns of recurrence. *Gynecologic Oncology*, **33**, 241–7.

73. Hale, R.J., Wilcox, F.L., Buckley, C.H., Tindall, V.R., Ryder, W.D.J. and Logue, J.P. (1991) Prognostic factors in uterine cervical carcinoma: a clinicopathological analysis. *International Journal of Gynecological Cancer*, **1**, 19–23.

74. Griffin, N.R. and Wells, M. (1995) Premalignant and malignant glandular lesions of the cervix. In *Haines and Taylor* (ed. Fox), 4th edn. Churchill Livingstone, Edinburgh, pp. 323–44.

75. Eifel, P.J., Morris, M., Wharton, J.T. and Oswald, M.J. (1994) The influence of tumor size and morphology on the outcome of patients with FIGO stage IB squamous carcinoma of the uterine cervix. *International Journal of Radiation Oncology, Biology and Physiology*, **17**, 9–16.

76. Eifel, P.J., Burke, T.W., Morris, M. and Smith, T.L. (1995) Adenocarcinoma as an independent risk factor for disease recurrence in patients with stage IB cervical carcinoma. *Gynecologic Oncology*, **59**, 38–44.

77. Fyles, A.W., Pintilie, M., Kirkbride, P., Levin, W., Manchul, L.A. and Rawlings, G.A. (1995) Prognostic factors in patients with cervix cancer treated by radiation therapy: results of a multiple regression analysis. *Radiotherapy Oncology*, **35**, 107–17.

78. Van Nagell, J.R., Donaldson, E.S., Wood, E.G. and Parker, J.C. (1978) The significance of vascular invasion and lymphocytic infiltration in invasive cervical cancer. *Cancer*, **41**, 228–34.

79. Coppleson, M. (1992) Early invasive squamous and adenocarcinoma of cervix: (FIGO stage IA) clinical features and management. In *Gynecologic Oncology* (ed. M. Coppleson), 2nd edn. Churchill Livingstone, Edinburgh, pp. 631–48.

80. Barber, H.R.K., Sommers, S.C., Rotterdam, H. and Kwan, T. (1978) Vascular invasion as a prognostic factor in stage Ib cancer of the cervix. *Obstetrics and Gynecology*, **52**, 343–8.

81. Wiggins, D.L., Granai, C.O., Steinhoff, M.M. and Calabresi, P. (1995) Tumour angiogenesis as a prognostic factor for cervical carcinoma. *Gynecologic Oncology*, **56**, 353–6.

82. Kainz, C., Speiser, P., Wanner, C., Obermair, A., Tempfer, C.S.G., Reinthaller, A. and Breitenecker, G. (1995) Prognostic value of tumour microvessel density in cancer of the uterine cervix stage Ib to IIb. *Anticancer Research*, **15**, 1549–51.

83. Schlenger, K., Hockel, M., Mitze, M., Schaffer, U., Weikel, W., Knapstein, P.G. and Lambert, A. (1995) Tumour vascularity – a novel prognostic factor in advanced cervical cancer. *Gynecologic Oncology*, **59**, 57–66.

84. Bremer, G.L., Tiebosch, A.T., van-der-Putten, H.W., Schouten, H.J., de-Haan, J. and Arends, J.W. (1996) Tumor angiogenesis: an independent prognostic parameter in cervical cancer. *American Journal of Obstetrics and Gynecology*, **174**, 126–31.

85. Kainz, C., Gitch, G., Tempfer, C., Heinzl, H., Koelbl, H., Breitecker, G. and Reinthaller, A. (1994) Vascular space invasion and inflammatory stromal reaction as prognostic factors in patients with surgically treated cervical cancer stage Ib to IIb. *Anticancer Research*, **14**, 2245–8.

86. Jelen, I., Valente, P.T., Gautreaux, L. and Clark, G.M. (1994) Deoxyribonucleic acid ploidy and S-phase fraction are not significant prognostic factors for patients with cervical cancer. *American Journal of Obstetrics and Gynecology*, **171**, 1511–16.

87. Kristensen, G.B., Kaern, J., Abeler, V.M., Hagmar, B., Trope, C.G. and Pettersen, E.O. (1995) No prognostic impact of flow cytometric measured DNA ploidy and S-phase fraction in cancer of the uterine cervix: a prospective study of 465 patients. *Gynecologic Oncology*, **57**, 79–85.

88. Nguyen, H.N., Sevin, B.U., Averette, H.E., Ganjei, P., Perras, J., Ramos, R., Angioli, R., Donato, D. and Penalver, M. (1993) The role of DNA index as a prognostic factor in early cervical carcinoma. *Gynecologic Oncology*, **50**, 54–9.

89. Lutgens, L.C., Schutte, B., de Jong, J.M. and Thunnissen, F.B. (1994) DNA content as prognostic factor in cervix carcinoma stage Ib/III treated with radiotherapy. *Gynecologic Oncology*, **54**, 275–81.

90. Zolzer, F., Alberti, W., Pelzer, T., Lamberti, G., Hulskamp, F.H. and Streffer, C. (1995) Changes in S phase fraction and micronucleus frequency as prognostic factors in radiotherapy of cervical carcinoma. *Radiotherapy Oncology*, **36**, 128–32.

91. Strang, P., Bergstrom, R. and Stendahl, U. (1992) Prognostic factors in cervical carcinoma (Meeting Abstract). *Anticancer Research*, **12**, 1919–20.

92. Raddhakrishna-Pillai, M., Balaram, P. and Nair, M.K. (1992) Role of immune response in the prognosis of carcinoma of the uterine cervix: can *in vitro* analysis provide a better framework for more effective management. *Tumori*, **78**, 87–93.

93. Onsrud, M., Grahm, I., Gaudernack, G. and Trope, C. (1992) Lymphoid cell distribution as a prognostic factor in carcinoma of the uterine cervix. *Acta Obstetrica et Gynecologica Scandinavica*, **71**, 135–9.

94. Raddhakrishna-Pillai, M., Balaram, P., Bindu, S., Hareendran, N.K., Padmanabhan, T.K. and Nair, M.K. (1990) Radiation associated eosinophilia and monocytosis in carcinoma of the uterine cervix: a simple reliable prognostic and clinical indicator. *Neoplasma*, **37**, 91–6.

95. Raddhakrishna-Pillai, M., Balaram, P., Chidambaram, S., Padmanabhan, T.K. and Nair, M.K. (1990) Development of an immunological staging to prognosticate disease course in malignant cervical neoplasia. *Gynecologic Oncology*, **37**, 299–305.

96. Raddhakrishna-Pillai, M., Balaram, P., Bindu, S., Hareendran, N.K., Padmanabhan, T.K. and Nair, M.K. (1989) Interleukin 2 production in lymphocyte cultures: a rapid test for cancer associated immunodeficiency in malignant cervical neoplasia. *Cancer Letters*, **47**, 205–10.

97. Lai, C.H., Chang, H.C., Chang, T.C., Hsueh, S. and Tang, S.G. (1993) Prognostic factors and impacts of adjuvant therapy in early stage cervical carcinoma with pelvic node metastases. *Gynecologic Oncology*, **51**, 390–6.

98. Roman, L.D., Morris, M., Mitchell, M.F., Eifel, P.J., Burke, T.W. and Atkinson, E.N. (1993) Prognostic factors for patients undergoing simple hysterectomy in the presence of invasive cancer of the cervix. *Gynecologic Oncology*, **50**, 179–84.

99. Riou, G., Favre, M., Jeannel, D., Bourhis, J., Le Doussal, V. and Orth, G. (1990) Association between poor prognosis in early stage invasive cervical carcinoma and non-detection of HPV DNA. *Lancet*, **335**, 1171–4.

100. Higgins, G.D., Davy, M., Roder, D., Uzelin, D.M., Phillips, G.E. and Burrell, C.J. (1991) Increased age and mortality associated with cervical carcinomas negative for human papilloma virus (HPV) RNA. *Lancet*, **338**, 910–13.

101. De Britton, R.C., Hildersheim, A., De Lao, S.L., Brinton, L.A., Sathys, P. and Reeves, W.C. (1993) Human papillomavirus and other influences on survival from cervical cancer in Panama. *Obstetrics and Gynecology*, **81**, 19–24.

102. Jarrell, M.A., Heintz, N., Howard, P., Collins, C., Badger, G. and Belinson, J. (1992) Squamous cell carcinoma of the cervix HPV 16 and DNA policy as predictors of survival. *Gynecologic Oncology*, **46**, 361–6.

103. Rose, B.R., Thompson, C.H., Simpson, J.M., Jarrett, C.S., Elliott, P. and Tattersall, M.H.N. (1995) HPV DNA as a prognostic indicator in early stage cervical cancer: a possible role for HPV 18. *American Journal of Obstetrics and Gynecology*, **173**, 1416–68.

104. Konya, J., Veress, G., Hernadi, Z., Soos, G.C.J. and Gergely, L. (1995) Correlation of human papillomavirus 16 and 18 with prognostic factors in invasive cervical neoplasms. *Journal of Medical Virology*, **46**, 1–6.

105. King, L.A., Tase, T., Twiggs, L.B., Okagaki, T., Savage, J.E. and Adcock, L.L. (1989) Prognostic significance of the presence of human papillomavirus DNA in patients with invasive carcinoma of the cervix. *Cancer*, **63**, 897–900.

106. Kristensen, G.B., Holm, R., Abeler, V.M. and Trope, C.G. (1996) Evaluation of the prognostic significance of cathepsin D, epidermal growth factor receptor, and c-erbB2 in early cervical squamous cell carcinoma. An immunohistochemical study. *Cancer*, **78**, 433–40.

107. Crook, T., Wrede, D., Tidy, J.A., Mason, W.P., Evans, D.J. and Vousden, K.H. (1992) Clonal p53 mutation in primary cervical cancer: association with human-papillomavirus-negative tumours. *Lancet*, **339**, 1070–3.

108. Schneider, J., Rubio, M.P., Rodriguez-Escudero, F.J., Seizinger, B.R. and Castresana, J.S. (1994) Identification of p53 mutations by means of single strand conformation polymorphism analysis in gynaecological tumours: comparison with the results of immunohistochemistry. *European Journal of Cancer*, **30A**, 504–8.

109. Kainz, C., Kohlberger, P., Gitsch, G., Sliutz, G., Breitenecker, G. and Reinthaller, A. (1995) Mutant p53 in patients with invasive cervical cancer stages IB to IIB. *Gynecologic Oncology*, **57**, 212–14.

110. Sagae, S., Kudo, R., Kuzumaki, N., Hisada, T., Mugikura, Y., Nihei, T., Takeda, T. and Hashimoto, M. (1990) Ras oncogene expression and progression in intraepithelial neoplasia of the uterine cervix. *Cancer*, **66**, 295–301.

111. Hayashi, Y., Hachisuga, T. and Iwasaka, T. (1991) Expression of RAS oncogene product and EGF receptor in cervical squamous cell carcinomas and its relationship to lymph node metastases. *Gynecologic Oncology*, **40**, 147–51.

112. Riou, G., Sherg, Z.M. and Zhou, D. (1990) C-myc and c-Ha-ras proto-oncogenes in cervical cancer: prognostic value. *Bulletin of Cancer*, **77**, 341–7.

113. Kainz, C., Kohlberger, P., Sliutz, G., Tempfer, C., Heinzl, H., Reinthaller, A., Breitenecker, G. and Koelbl, H. (1995) Splice variants of CD44 in human cervical cancer stage IB to IIB. *Gynecologic Oncology*, **57**, 383–7.

114. Kainz, C., Kohlberger, P., Tempfer, C., Sliutz, G., Gitsch, G., Reinthaller, A. and Breitenecker, G. (1995) Prognostic value of CD44 splice variants in human stage III cervical cancer. *European Journal of Cancer*, **31A**, 1706–9.

115. Kainz, C., Tempfer, C., Winkler, S., Sliutz, G., Koelbl, I.H. and Reinthaller, A. (1995) Serum CD44 splice variants in cervical cancer patients. *Cancer Letters*, **90**, 231–4.

116. Kornafel, J., Wawrzkiewicz, M. and Blaszczyk, J. (1993) Is the serum SCC concentration a prognostic factor in uterine cervix cancer patients? Meeting Abstract. *Gynaecological Oncology, 8th International Meeting of the European Society of Gynaecological Oncology*, 9–12 June, 1993, Barcelona, Spain.

117. Duk, J.M., Grounier, K.H., de Bruijn, H.W., Hollema, H., ten Hoor, K.A., van der Zee, A.G. and Aalders, J.G. (1996) Pretreatment serum squamous cell carcinoma antigen: a newly identified prognostic factor in early stage cervical carcinoma. *Journal of Clinical Oncology*, **14**, 111–18.

118. Bolger, B., Dabbas, M., Lopes, A. and Monaghan, J. (1996) Prognostic value of preoperative squamous cell carcinoma antigen level in patients surgically treated for cervical carcinoma. *British Journal of Obstetrics and Gynaecology*, **103**, 1263.

119. Carter, P.G., Iles, R., Neven, P., Ind, T.E., Shepherd, J.H. and Chard, T. (1994) The prognostic significance of urinary beta core fragment in premenopausal women with carcinoma of the cervix. *Gynecologic Oncology*, **55**, 271–6.

120. Rodriguez, G.C., Clarke-Pearson, D.L., Soper, J.T., Berchuck, A., Synan, I. and Dodge, R.K. (1994) The negative prognostic implications of thrombocytosis in women with stage Ib cervical carcinoma. *Obstetrics and Gynecology*, **83**, 445–8.

121. Lopes, A., Daras, V., Cross, P.A., Robertson, G., Beynon, G. and Monaghan, J.M. (1994) Thrombocytosis as a prognostic factor in women with cervical cancer. *Cancer*, **74**, 90–2.

122. Roman, L.D., Morris, M., Eifel, P.J., Burke, T.W., Gershenson, D.M. and Wharton, J.T. (1992) Reasons for inappropriate simple hysterectomy in the presence of invasive cancer of the cervix. *Obstetrics and Gynecology*, **79**, 485–9.

123. Gaze, M.N., Kelly, C.G., Dunlop, P.R.C., Redpath, A.T., Kerr, G.R. and Cowie, V.J. (1992)

Stage IB cervical carcinoma: a clinical audit. *British Journal of Radiology*, **65**, 1018–24.

124. Bissett, D., Lamont, D.W., Nwabinelli, N.J., Brodie, M.M. and Symonds, R.P. (1994) The treatment of stage I carcinoma of the cervix in the west of Scotland 1980–1987. *British Journal of Obstetrics and Gynaecology*, **101**, 615–20.

125. Hopkins, M.P., Peters, W.A., Andersen, W. and Morley, G.W. (1990) Invasive cervical cancer treated initially by standard hysterectomy. *Gynecologic Oncology*, **36**, 7–12.

126. Hockel, M., Knoop, C., Vorndran, B., Baussmann, E., Schlenger, K. and Knapstein, P.G. (1993) Tumour oxygenation: a new prognostic factor influencing survival in advanced cancer of the uterine cervix (Meeting Abstract). *24th Annual Meeting of the Society of Gynecologic Oncologists*, 7–10 February, Palm Desert, CA, A25.

127. Hanks, G.E., Herring, D.F. and Kramer, S. (1983) Patterns of care outcome studies: results of the National Practice Survey in cancer of the cervix. *Cancer*, **51**, 959–67.

128. Busch, M., Duhmke, E., Kuhn, W. and Teichmann, A. (1991) Definitive radiation therapy in the treatment of carcinoma of the uterine cervix. Treatment results and prognostic factors. *Strahlenther Onkol*, **167**, 628–37.

129. Arthur, D., Kaufman, N., Schmidt, U.R., Kavenagh, B., Simpson, P., Hill, M. and Ali, M. (1995) Heuristically derived tumour burden score as a prognostic factor for stage IIIb carcinoma of the cervix. *International Journal of Radiation Oncology, Biology and Physiology*, **31**, 743–51.

130. Panici, P.B., Scambia, G., Baiocchi, G., Greggi, S., Ragusa, G., Gallo, A., Conte, M., Battaglia, F.L.G. and Rabitti, C. (1991) Neoadjuvant chemotherapy and radical surgery in locally advanced cervical cancer. Prognostic factors for response and survival. *Cancer*, **67**, 372–9.

131. Panici, P.B., Greggi, S., Scambia, G., Salerno, M.G., Battaglia, F., Amoroso, M. and Mancuso, S. (1993) Responsiveness to neoadjuvant chemotherapy (NAC) as a major prognostic factor in locally advanced cervical cancer (LACC). Meeting abstract. *8th International Meeting of the European Society of Gynaecological Oncology*.

132. Panici, P.B., Greggi, S., Scambia, G., Battaglia, F., Amoroso, M., Salerno, M.G. and Mancuso, S. (1995) Response to neoadjuvant chemotherapy (NAC) as a major prognostic factor in locally advanced cervical cancer (LACC). Meeting abstract. *Annual Meeting of the American Society of Clinical Oncologists*, **14**, A792.

133. Monaghan, J.M., Ireland, D., Mor-Yosef, S.,
Pearson, S., Lopes, A. and Sinha, D. (1990) Role of centralisation of surgery in stage Ib carcinoma of the cervix: a review of 498 cases. *Gynecologic Oncology*, **37**, 206–9.

134. Burghardt, E. (1993) Cervical cancer: results. In *Surgical Gynecologic Oncology* (eds E. Burghardt, M.J. Webb, J.M. Monaghan and G. Kindermann). Thieme, Stuttgart, pp. 302–15.

135. Werner-Wasik, M., Schmid, C.H., Bornstein, L., Ball, H.G., Smith, D.M. and Madoc-Jones, H. (1995) Prognostic factors for local and distant recurrence in stage I and II cervical carcinoma. *International Journal of Radiation Oncology, Biology and Physiology*, **32**, 1309–17.

136. Fuller, A.F., Elliott, N. and Kosloff, C. (1989) Determinants of increased risk of recurrence in patients undergoing radical hysterectomy for a stage Ib and IIa carcinoma of the cervix. *Gynecologic Oncology*, **33**, 34–9.

137. Lee, Y.N., Wang, K.L. and Lin, M.H. (1989) Radical hysterectomy with pelvic lymph node dissection for treatment of cervical carcinoma: a clinical review of 954 cases. *Gynecologic Oncology*, **31**, 135–42.

138. Averette, H.E., Nguyen, H.N., Donayo, D.M., Penalver, M.A., Sevin, B.U., Estape, R. and Little, W.A. (1992) Radical hysterectomy for invasive cervical cancer. *Cancer*, **71**, 1422–37.

139. Lohe, K.J. (1978) Early squamous cell carcinoma of the uterine cervix. I. Definition and histology. *Gynecologic Oncology*, **6**, 10.

140. Morrow, P. (1980) Panel Report: Is pelvic irradiation beneficial in the postoperative management of stage Ib squamous cell carcinoma of the cervix with pelvic node metastases treated by radical hysterectomy and pelvic lymphadenectomy? *Gynecologic Oncology*, **54**, 3035.

141. Boyce, J., Fruchter, R. and Nicastri, A. (1981) Prognostic factors in stage I carcinoma of the cervix. *Gynecologic Oncology*, **12**, 154.

142. Inoue, T. (1984) Prognostic significance of the depth of invasion relating to nodal metastases, parametrial extension and cell types. *Cancer*, **54**, 3035.

143. Di Re, F., Fontanelli, R., Rasplagiesi, F. and Di Re, E. (1990) Surgery in the treatment of stage Ib–II cervical cancer. *Cervix*, **8**, 89–98.

144. Girardi, F. and Haas, J. (1993) The importance of the histologic processing of pelvic lymph node in the treatment of cervical cancer. *International Journal of Gynecological Cancer*, **3**, 12–17.

145. Sevin, B.U. and Averette, H.E. (1988) Staging laparotomy and hysterectomy for cancer of the cervix. *Balliére's Clinical Obstetrics and Gynecology*, **2**, 761–8.

G. Kenter

INTRODUCTION

Early carcinoma of the uterine cervix is defined by the International Federation of Gynaecology and Obstetrics (FIGO) as FIGO Stage I and II (see Table 19.1). Intensive screening strategies and improvements in the conservative management of pre-invasive disease have resulted in a reduction in the incidence of cervical cancer. Unfortunately, this has only occurred in certain countries where such health measures have been put in place. The results of the treatment for early carcinoma of the uterine cervix have improved tremendously over the past century. Mortality rates of surgical treatment have dropped from 50% to almost zero and the morbidity figures are also acceptably low. However, far less progress has been made in the assessment and management of patients with poor prognostic factors and in those who eventually develop recurrent disease. These women still pose the greatest problem in the management of invasive cervical cancer.

PRESENTATION

Abnormal vaginal bleeding is the most common method of presentation although an increasing number of incident cases are now detected as a result of cervical screening. By definition these women are asymptomatic.

Bleeding following trauma (usually post-coital) and bleeding after the menopause are classically associated with cervical malignancy and any such symptom should always be fully investigated. Vaginal discharge may also occur and if due to tumour necrosis and secondary infection can be quite offensive. Pain and disturbance of bowel or bladder function are usually associated with late disease. Finally, some cases may be discovered because of a suspicious appearance of the cervix. This could occur at any time when a vaginal and speculum examination are performed and, as with a high-grade smear, is an indication in itself for a colposcopic assessment. Once suspected, the next phase is confirmation or diagnosis.

DIAGNOSIS AND STAGING

Diagnosis requires a representative biopsy. If the cervix is macroscopically abnormal a large diagnostic biopsy can be removed from the cervix at the time of formal staging (see below). If only Stage Ia_1 or Ia_2 disease is suspected, then a large cone specimen should be removed. It is important to include all of the atypical epithelium in such biopsies for if the lesion is deemed incompletely excised, it is not possible to accurately allocate either Stage Ia substages.

In large macroscopic lesions several punch biopsies may be taken and will nearly always confirm the diagnosis.

Cancer and Pre-cancer of the Cervix
Edited by D.M. Luesley and R. Barrasso. Published in 1998 by Chapman & Hall, London. ISBN 0 412 56600 1.

Table 19.1   FIGO classification in cervical carcinoma from 1995

Stage 0	Carcinoma *in situ*, intra-epithelial carcinoma	
Stage I	Carcinoma strictly confined to the cervix (extension to the corpus should be disregarded)	
	Stage Ia	Invasive cancer identified only microscopically. All gross lesions even with superficial invasion are Stage Ib cancers Invasion limited to measured stromal invasion with maximam depth of 5.0 mm and no wider than 7.0 mm[a]
		Stage Ia$_1$ — Measured invasion of stroma no greater than 3.0 mm in depth and no wider than 7 mm
		Stage Ia$_2$ — Measured invasion of stroma greater than 3 mm and no greater than 5 mm and no wider than 7 mm
	Stage Ib	Clinical lesions confined to the cervix or preclincal lesions greater than Stage Ia
		Stage Ib$_1$ — Clinical lesions no greater than 4 cm in size
		Stage 1b$_2$ — Clinical lesions greater than 4 cm in size
Stage II	The carcinoma extends beyond the cervix but has not extended onto the pelvic wall. The carcinoma involves the vagina, but not the lower third.	
	Stage IIa	No obvious parametrial involvement
	Stage IIb	Obvious parametrial involvement
Stage III	The carcinoma has extended onto the pelvic wall. On rectal examination, there is no cancer-free space between the tumour and the pelvic wall. The tumour involves the lower third of the vagina. All cases with hydronephrosis or non-functioning kidney should be included unless they are due to other causes	
	Stage IIIa	No extension onto the pelvic wall but involvement of the lower third of the vagina
	Stage IIIb	Extension onto the pelvic wall and/or hydronephrosis or non-functioning kidney
Stage IV	The carcinoma has extended beyond the true pelvis or has clinically involved the mucosa of the bladder or the rectum. A bulbous oedema as such does not permit a case to be allotted to Stage IV	
	Stage IVa	Spread of the growth to adjacent organs
	Stage IVb	Spread to distant organs

[a]The depth of invasion should not be more than 5 mm taken from the base of the epithelium, either the surface or glandular, from which it originates. Vascular space involvement, either venous or lymphatic, should not alter the staging.

STAGING

FIGO staging

The extent of local disease can be assessed by clinical staging, usually carried out according to the FIGO definitions. FIGO staging includes a bimanual vagino-rectal examination, some- times in combination with cystoscopy and proctoscopy, performed under general anaes- thesia, plus chest X-ray and intravenous urogram (IVU). As a rule, the clinical staging is not changed on the grounds of subsequent surgical or histological findings, and if there is any doubt at all, a lower stage is chosen. In many centres the use of general anaesthesia is no longer a routine part of the staging procedure, and ultrasonography of the kidneys is used as an alternative for the IVU.

The FIGO definitions have been changed several times since they were first formulated in 1958. Table 19.1 shows the latest definitions dating from 1995 [1].

One of the more contentious areas has been the definition of Stage Ia, i.e. microinvasive carcinoma. Although the FIGO stage is not necessarily regarded as a guideline to management, the former FIGO definition was confusing regarding the size of the lesion in Stage Ia because it included all lesions up to 5 mm in depth of penetration and up to 7 mm in horizontal extent. Furthermore, FIGO does not take any other possible prognostic factors into account, such as lymph-vascular space invasion. This situation has resulted in differing treatment modalities in early/microinvasive carcinoma. Burghardt advocated radical hysterectomy only in cases with tumours larger then 500 mm^3 [2]. However, most centres do not use volume measurement as a routine as this is very labour-intensive and time-consuming. In 1974, the Society of Gynecologic Oncologists (SGO) in the USA defined microinvasive carcinoma as a lesion in which neoplastic epithelial cells invade the epithelium to a depth of 3 mm or less, and in which no lymphatic or blood vessel involvement can be demonstrated. This definition has been successfully used in most centres in the USA, Japan and in several European cancer centres to distinguish between simple and radical treatment [3–5]. In the latest FIGO definition a distinction is made between lesions up to 3 mm and between 3 and 5 mm in depth of invasion [3]. Microinvasive disease is covered in further detail in Chapter 15.

Clinical staging

Clinical staging is extremely inaccurate. A report from the University of Southern California indicates that the correlation between the clinical stage and the subsequent surgical findings was only 52% [6]. The majority of patients were found to have occult metastases in pelvic or para-aortic lymph nodes and had therefore been understaged pre-operatively; a smaller proportion with benign pelvic pathology had been overstaged [7]. However, neither ultrasound nor computed tomography scanning have proven to be superior to clinical staging [8–10]. The latest reports on the results of magnetic resonance imaging (MRI) have shown improved results in the staging of cervical carcinoma with a sensitivity for stage of 75–90%, for parametrial involvement of 87–94% and 86–88% for lymph nodes [11–13]. The current literature suggests that MRI offers more reliable results in predicting pathological outcome than clinical staging [14].

Surgical staging

The lack of reliable diagnostic methods for detecting small lymph node metastases has led to the introduction of the pre-treatment staging laparotomy or laparoscopy in order to identify patients with metastases in the pelvic and/or para-aortic lymph nodes. Surgical staging can be performed via the transperitoneal or the extraperitoneal route, with similar results in terms of detecting lymph nodes [15]. The extraperitoneal approach has the advantage of less enteric side-effects in those patients who subsequently require radiotherapy. Although the morbidity of the surgical staging procedure is acceptable, it has failed to realize its intended goal of improving survival [16]. This is probably due to the lack of effective systemic treatment modalities for distant metastases. Other, non-invasive methods are needed to predict the outcome of the disease and to indicate adjustments to the treatment. Surgical staging by laparoscopy remains unproven in terms of outcome benefit and should only be performed within the context of randomized controlled trials.

Once diagnosis and disease extent have been confirmed, management planning can commence.

MANAGEMENT STRATEGIES IN CERVICAL CARCINOMA STAGES I AND II

The three basic modalities in the treatment of early invasive carcinoma of the cervix are surgery, radiation or a combination of surgery and radiotherapy. In some centres patients with large tumours or high-risk patients are given chemotherapy in combination with either surgery or radiotherapy. The results of radical surgery and radiotherapy are essentially the same in low stage (Stages Ib and IIa) cervical carcinoma. With the current treatment modalities for Stage Ib and IIa cervical carcinoma 5-year survival figures of about 85% are generally reported [2, 17–20].

HISTORICAL BACKGROUND

At the end of the 19th century patients with carcinoma of the cervix were treated by surgery. Mortality rates exceeded 50% due to bleeding and/or post-operative infections. Patients who survived often died some time later because of local recurrence or distant metastases. As no alternatives to surgical excision existed, patients with advanced or large tumours were also treated surgically, although we recognize now that this was suboptimal.

The first radical abdominal hysterectomy was performed in 1889 by Wertheim [21]. To increase the radicality of the operation he combined it with a pelvic lymphadenectomy. Schauta, who did not believe lymphadenectomy was a method that enhanced radicality, improved the vaginal radical hysterectomy, which was introduced by Schuchardt in 1893. By using paravaginal incisions he tried to remove the parametrium more radically [22]. Schauta believed that the lymphadenectomy was of little value because he considered it could never be complete, i.e. not radical enough; furthermore, the next lymph node group in the chain could also be involved. Finally, and with particular regard to small tumours confined to the cervix, up to 80% of the cases would have negative nodes. Given the morbidity of surgery at that time, this was a major factor in attempts to minimize such morbidity. In the meantime, Wertheim, who originally started working as an assistant of Schauta and later became head of the department in Vienna, published the results of 500 abdominal operations in 1911. The debate over the most appropriate route to radically remove the cervix has continued since that time.

The introduction of radiotherapy in 1903 resulted in a significant reappraisal of management largely as a result of the morbidity advantage associated with radiotherapy. Only in the late 1930s was surgery reintroduced by some European surgeons like Bouwdijk Bastiaanse in Amsterdam, Bonney in London and by several others in the German-speaking countries [23–25]. Taussig and Bonney improved the lymphadenectomy, which was not routinely performed by the Germans. Cure rates of 40%, combined with mortality rates of 14–25%, were recorded by these workers. The gradual improvements in results over the years were the result of better patient selection. Meigs [26] from Boston described a technique which had greater radicality. First he introduced the routine lymphadenectomy in all radical operations, which had only been done occasionally before. Second, he took up the so-called Latzko technique, which allowed for better dissection of the different tissues of the parametrium: the vesico-utero-vaginal or Mackenrodt's ligament, the recto-vaginal and uterosacral ligaments, by opening the paravesical and pararectal spaces. By performing the lymphadenectomy at the beginning of the operation, these spaces and ligaments can be found, and the Mackenrodt's ligament (or 'web') is clamped and cut on the lateral side of the ureter. The next essential part of Meigs' technique is an extensive release of the ureter from its surrounding tissue. The 5-year survival rate of his patients was 75% for Stage I and 54% for Stage II. Fistulas were seen in 9% of patients. After the introduction of sulphonamides he was the first to report a series of 100 patients without operative mortality.

In Japan, Okabayashi modified the Wertheim technique in a different way [27]. He incised the pouch of Douglas and developed the prerectal space between the rectum and posterior vaginal wall before opening the parametrium, thus facilitating the mobilization of the uterus for removal. Many authors have published the results of a variety of surgical techniques since then. In 1954 Mitra [28] described the combination of a Schauta radical vaginal hysterectomy with extraperitoneal lymphadenectomy: the Mitra–Schauta procedure, with a 5-year survival rate of 61%. In 1957 Sindram [29] developed a technique that combined the Wertheim and the Schauta procedure together with the abdominal lymphadenectomy according to Taussig in one single operation called AVRUEL (abdominal vaginal radical uterus extirpation with transperitoneal lymphadenectomy). The advantage of this combined procedure was based on the principle that in the abdominal phase attention could be concentrated on the local extent of the tumour and any metastatic growth but that lymphadenectomy could also be performed. The vaginal phase makes the radical removal of the paracolpium easier than in the Wertheim procedure. However, the radicality of the operation was not associated with improvements in morbidity and survival [20].

SURGERY

Micro-invasive cervical carcinoma with a lesion less than 3 mm in depth (Stage Ia$_1$) and without signs of vascular invasion can be treated safely either by conization or by simple hysterectomy [3, 30]. Whether a cone biopsy or a hysterectomy is chosen will depend on the age of the patient, the desire to retain fertility or indeed the patient's personal preference. When a cone biopsy is performed, the lesion must be totally confined to the cone with surgical margins that are sufficiently free of atypical epithelium.

In Stage Ia$_2$ the chance of finding positive lymph nodes is between 5 and 10%, which may justify the need for pelvic lymphadenectomy

[4]. This procedure can be combined with a type 1 radical hysterectomy. Currently, experimental treatments aimed at uterine conservation such as radical trachelectomy are being evaluated. This is the removal of the cervix and parametria through a vaginal route in combination with removal of the lymph nodes. Results from larger series in terms of survival and subsequent fertility are still awaited.

Stage Ib and IIa are treated by radical hysterectomy with pelvic lymphadenectomy. The operation consists of extirpation of the uterus, the paracervical and paravaginal tissue, plus a portion of the upper vagina and the perivascular fatty and connective lymph-bearing tissues on the lateral pelvic wall. In young women the surgical approach has the advantage over radiotherapy of preserving the ovaries and retaining a functional vagina. Both are important factors for the general feeling of well-being and sexual functioning following treatment in patients already at risk of negative body image and self-esteem. Another advantage is retaining radiation treatment in case of recurrent disease after surgical treatment. Furthermore, the surgical approach enables further determination of the disease status. This is important in view of the poor reliability of clinical staging. For these reasons radical surgery is considered the treatment modality of choice in early cervical carcinoma. Even in very obese patients the results of radical surgery are not compromised in terms of survival and the incidence of severe complications is not increased, although the operative technique is more difficult, the procedure lasts longer and the surgery is associated with greater blood loss [31].

Age in itself is not a contraindication to surgery. In geriatric patients radical hysterectomy can be tolerated well and is associated with good results as long as the patients are well selected [32] and appropriately prepared for radical surgery.

The most frequently employed procedure for Stage Ib and IIa carcinoma of the uterine cervix is a modification of the Wertheim operation with influences of Meigs and/or Okabayashi.

Variations exist in the extent of resection of the upper vagina, the lateral extent of parametrial excision, the amount of mobilization of the ureters from their bed and the sacrifice of the vasculature to ureter and bladder. Piver *et al.* [33] and Rutledge [34] have tried to classify different procedures in an attempt to simplify recording and still be able to individualize radical surgery for cervical carcinomas. Five types of operations are described in increasing radicality:

- Type I, which is essentially the extrafascial hysterectomy described by Te Linde *et al.* [35], ensures complete removal of the cervix and 1–2 cm of vaginal cuff but leaves the ureters lateral. This operation can be used in microinvasive carcinomas (without removal of the vaginal cuff), in Stages Ib and IIa after pre-treatment radiotherapy or in cases with bulky tumours where adjuvant radiotherapy will be given.
- Type II preserves the blood supply to the lower ureters and the vagina by leaving the superior vesicle artery intact. The ureters are freed from the paracervical position but are not dissected out of the pubovesicle ligament. The medial two-thirds of the parametria and the upper third of the vagina are removed.
- In Type III, resection of the parametria follows the pelvic wall; dissection of the ureter from the pubovesicle ligament is complete to entry into the bladder (except a small lateral portion between the lower end of the ureter and the superior vesicle artery) and the uterosacral ligaments are excised at their sacral attachments.
- Type IV differs from the former in three aspects: (a) the ureter is completely dissected from the pubovesicle ligament, (b) the superior vesicle artery is sacrificed and (c) three-quarters of the vagina is removed.
- Type V is more radical still and includes removal of parts of the urinary tract.

From the description of several techniques it might be clear that surgical treatment for carcinoma of the cervix can only be performed by skilled gynaecologists in centres where post-operative care – including psychological after-care – for these patients is a matter of routine [36].

RADIOTHERAPY

Radiotherapy, when used as a primary treatment, consists of external beam therapy to the whole pelvis combined with intracavity radiation to deliver a total dose of 75–80 Gy. Indications for radiotherapy rather than surgery will vary from centre to centre. Most, however, would consider radiotherapy as the primary option in women considered unfit for surgery (although some might argue they will be equally unfit for radical pelvic radiation), where the nodes are known to be positive (previous staging laparotomy) or when the tumour is a Stage Ib$_2$, i.e. the diameter is greater than 4 cm.

The logic in choosing radiotherapy as the primary treatment is to attempt to avoid surgery followed by radiotherapy. The morbidity of combined approaches usually exceeds that associated with the sum of the two separate approaches. Most gynaecologists favour post-operative radiation in cases shown to be node-positive, and many also do for bulky tumours, particularly if the excision margins are compromised. For more advanced disease that has spread beyond the cervix (regional spread) radiotherapy offers greater advantages than surgery. The 5-year survival rate described by several authors is 83–90% in Stage Ib, 53–61% in bulky Stage Ib, 75% in Stage IIa, 72% in non-bulky and 50% in bulky Stage IIb [37, 38]. Although in the latter group some centres prefer radical hysterectomy instead of radiotherapy, in about a quarter of the patients the operation cannot be completed and 30–40% have lymph node metastases [39].

SPECIAL ISSUES

LYMPHADENECTOMY

The role of pelvic lymphadenectomy is to attempt to define a group of patients at high risk of relapse (positive nodes), and thus justify adjuvant therapy. In addition, it has also been proposed that the procedure itself may have some therapeutic value. Some have suggested that by adding pelvic lymphadenectomy prior to definitive pelvic irradiation, improvements in survival of 10–15% might be expected [16, 25, 40]. This could suggest that radiotherapy itself is not capable of sterilizing all pathologic nodes. Inoue and Morita [41] reported a significantly better 5-year survival rate in a group of 61 patients with resectable macroscopic nodal metastases compared to ten patients with non-resectable nodes (82 vs 20%). Hirabayashi [42] compared patients in whom a lymphangiography was performed before and after operation. The more radical the lymph node dissection, the more favourable the 5-year survival rate. Furthermore, the more extensively the lymphadenectomy is performed, the higher the amount of positive nodes [43]. Girardi *et al.* even found 55% positive nodes in Stage Ib patients treated between 1980 and 1989 compared to 33% in those treated before 1980. The median number of nodes removed per patient differed from 24 before 1980 to 35 after 1980 [43]. In a study by Benedetti and co-workers, a mean of 48 pelvic and 22 aortic nodes were removed [44].

Randomized studies to prove the effect of lymphadenectomy on survival rates have not been performed but whilst surgery alone is unlikely to eradicate disease in these circumstances the current data suggest a possible beneficial additive effect. Although it is not likely that it will ever be possible to cure vascular spread of the disease with surgical methods alone, it must be concluded from these data that the lymphadenectomy might have a favourable effect on survival rates.

Para-aortic lymphadenectomy should not be performed on a routine basis. However, there may be some justification in performing a para-aortic lymphadenectomy if there is a high suspicion that metastases may be involved. The presence of clinically suspicious pelvic or para-aortic nodes or extracervical spread of tumour are significant predictors of para-aortic metastases [45].

PARAMETRIA

The aim of radical surgery in malignant disease is the removal of the primary tumour with an adequate, tumour-free resection margin. However, in carcinoma of the uterine cervix the removal of the parametria can be problematic, leading to excess blood loss at the time of surgery and occasionally post-operative problems with micturition. Although the parametrial tissue is considered the primary site for lateral extension, not only by direct spread but also through embolization of tumour cells via lymph vessels, in most studies a higher proportion of pelvic node metastases than parametrial nodes is found [43]. Tumour spread into the parametria can occur in all four directions and is found more frequently with increasing depth of stromal invasion, and when the margin of unaffected cervical stroma is narrower [46]. Microscopic infiltration of the parametria has an unfavourable effect on prognosis. The number of positive parametrial nodes is clearly related to the size of the tumour; 4% in small tumours vs >35% in bulky tumours. Also, tumour spread into the parametria is associated with pelvic lymph node metastasis (Table 19.2).

BULKY TUMOURS

Bulky Stage Ib tumours form a difficult treatment group due to their higher risk of local failure and poorer survival. Some authors consider a visible tumour of >3 or 4 cm a bulky tumour, while others define the barrel-shaped cervix with a size of >5 cm as such. Some use the tumour/cervix quotient (TCQ) of more than 40% as a definition of 'bulky' [2]. Different treatment modalities for this group of patients have been proposed. Radiation therapy as a sole treatment is associated with high failure rates when administered in standard doses. A combination of external irradiation

Table 19.2 Incidence of pelvic and para-aortic nodal metastases by stage in early cervical carcinoma

Stage	Pelvic (%)	CIPAN (%)	References
Ia$_1$	0	0	152–154
Ia$_2$	5	0	45, 152–155
Ib$_1$	15–20	5–7	2, 18, 29, 45, 56, 58, 62, 121, 156, 157
Ib$_2$	25–35	5–?	45, 49
IIa	16–25	2.4–12	45, 62, 135
IIb	43	16	2, 62, 157

CIPAN = common iliac and/or para-aortic nodes.

and intracavitary insertion with higher doses of brachytherapy produced better results, with a 5-year survival rate of 76% [37]. A decrease in failure has been documented if extrafascial hysterectomy or radical hysterectomy is performed after radiotherapy [47, 48]; however, the results seem contradictory and high complication rates have been reported [37, 49].

Although pre-treatment radiotherapy is often proposed because of fear of per-operative complications, reasonable results are obtained by the treatment of bulky tumours with radical surgery followed by radiation therapy [50, 51]. More recently, encouraging results have been reported using pre-operative chemotherapy [52, 53].

VAGINAL CUFF

The rationale for the removal of the vaginal cuff has never been a subject of much debate. If the vagina is unaffected, extirpating one-quarter or one-third of it does not improve the cure rate. The percentage of women who can no longer have intercourse because of shortening of the vagina due to a radical hysterectomy can amount to 25%, whereas tumour growth in the vagina in Stage Ib is generally found in only 5–7% [2, 20]. Furthermore, the mean incidence of vaginal recurrence following radical hysterectomy is 2.5% [54]. This is a strong argument in favour of performing colposcopy of the vagina pre-operatively in all women undergoing radical hysterectomy for cervical cancer and only removing a mar-

gin of normal tissue to allow for adequate radical clearance.

NODE-POSITIVE PATIENTS

Pelvic lymph node metastases are found in 15–20% of patients with low-stage cervical carcinoma and are considered to be the single most important prognostic factor. Five-year survival rates for cervical carcinoma Stage Ib falls from 90% when nodes are negative to 50% when nodes are positive [3, 55–58] (Table 19.3). In addition, the number and the site of the positive nodes has been found to be important [56, 59]. When the common iliac nodes are positive, the 5-year survival rate is 25% vs 65% when the pelvic nodes only are involved. Inoue and Morita found 5-year survival rates of 89, 81, 63, 41 and 23% in patients with, respectively, 0, 1, 2 or 3, 4, and non-resectable positive nodes [60]. The regional lymph nodes most often affected are the parametrial, the external iliac, the obturator, the common iliac, and the para-aortic lymph nodes [43, 44, 61] (Figure 19.1). The percentage of positive nodes increases with increasing number of removed nodes. As the number of positive pelvic nodes increases, the percentage of positive common iliac and para-aortic nodes increases [62] (Table 19.2).

As already discussed, the issue is whether surgery can be of help to patients with positive nodes. Burghardt *et al.* [2] believe that surgical removal of involved nodes can be curative. However, randomized studies have

Table 19.3 Percentage of positive pelvic lymph nodes and 5-year DFS by different prognostic factors with current treatment modalities

	Percentage positive nodes	*5-year DFS*	*References*
Pelvic nodes			
negative		85–90	
positive		50–65	58, 121, 158
Tumour size			
<4 cm	15	87	
>4 cm	23	74	5, 123
Depth of infiltration			
<15 mm	15	86	
>15 mm	38	71	121
Vaso-invasion			
absent	8–15	94	
present	25–58	71	5, 58
Parametrial infiltration			
absent	13–15	90	
present	43–68	75	5, 55, 58, 127
SCC–ag			
<1.9	15	87	
>1.9	42	74	123

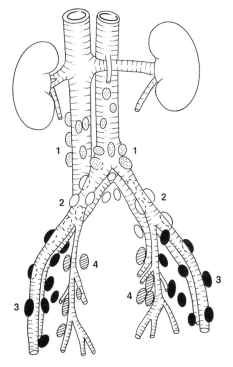

Figure 19.1 Lymph nodes draining the cervix presented schematically. 1: Para-aortic nodes (inconsistent number); 2: common iliac nodes ($n = 6$); 3: external iliac nodes ($n = 7$); 4: obturator ($n = 5$) [60, 61].

not been performed and adjuvant and subsequent therapies differ worldwide.

When macroscopically positive nodes are found during radical surgery there may be justification in attempting their removal alone and not complete the hysterectomy. However, if the tumour is also bulky, surgical removal of the tumour load might also be performed to enhance the effect of subsequent radiotherapy. In such cases a Rutledge I or extrafascial hysterectomy might be sufficient, provided that free surgical margins can be obtained.

CERVICAL STUMP CARCINOMA

Carcinoma in a cervical stump can be treated either by radiotherapy or by surgery. Both seem to provide equal results for loco-regional control compared to cervical carcinoma in an intact uterus. However, in surgically treated patients more severe complications are seen compared to patients treated with radiotherapy alone [63].

CARCINOMA FOUND IN SIMPLE HYSTERECTOMY SPECIMENS

When a previously unsuspected carcinoma is found in a simple hysterectomy specimen the patient can be treated by radiotherapy or by re-operation. Re-operation consists of pelvic lymphadenectomy, radical parametrectomy and removal of the upper part of the vagina. In young patients re-operation has the same advantage of preserving ovarian and vaginal function compared to radiation as the primary radical surgery. Results in terms of survival are similar to that for radical hysterectomy. Complications were found to be similar to those associated with primary radical surgery in a group studied by Orr *et al.* However, 73% of the patients required adjuvant radiotherapy [64]. In cases with larger tumours, tumour-positive margins or in elderly patients, radiation treatment is preferred, with comparable results to radical surgery followed by radiation; 78% disease-free 5-year survival in Stage Ib and 67% in Stage IIb [65].

PREGNANCY

Between 1–3% of cervical carcinomas are diagnosed during pregnancy. Controlled studies show equal results of treatment between pregnant women and non-pregnant controls [66, 67]. Once suspected, a large representative biopsy is taken to confirm the diagnosis despite the known increased risk of haemorrhage. Small directed biopsies in this situation can be misleading. When diagnosed in the first or second trimester, if the foetus is to be sacrificed (only after full and in-depth counselling with the couple concerned), surgery is usually performed with the foetus in the uterus. In the third trimester a delay can be optional with the purpose of achieving foetal viability. Large or controlled studies on the effect of the delay on prognosis are not available. Authors have reported delays varying from 53 to 212 days, which did not seem to effect prognosis [68, 69].

Radical hysterectomy with lymphadenectomy during pregnancy has proven acceptable morbidity and survival rates in Stages Ib and IIa [70].

NEOADJUVANT AND ADJUVANT THERAPY

ADJUVANT RADIOTHERAPY

Indications for adjuvant radiotherapy are spread of the disease into the parametria, incomplete excision margins or pelvic node metastases. The poorer survival in patients who have large (>4 cm) tumours, and or those with evidence of vascular invasion, is also cited as some as justification for adjuvant radiotherapy to improve local control [59, 71].

Postoperative radiotherapy generally consists of a combination of external therapy to the pelvis to treat the nodes and internal radiation to the vaginal vault. The principle of adjuvant radiotherapy, which as yet has not been shown to influence survival, is local and regional control [72–75]. However, the complications following this form of combined treatment can be severe [37, 76]. Apart from the effects on the ovaries and the vagina, all pelvic organs are at risk of radiation injury, especially the bladder, rectosigmoid and the small intestine, particularly when adherent to the pelvic floor structures. The small bowel is the most radiosensitive organ and is therefore the limiting factor in delivering pelvic irradiation [77, 78]. In an attempt to reduce the radiation related side-effects the successful use of an absorbable mesh has been reported by Snijders and Trimbos [79]. The mesh can be sutured in the abdominal cavity and can temporarily lift the small bowel loops out of the pelvis and thus out of the radiation field. Using this method it has been possible to demonstrate a decrease in severe morphological changes and maintenance of function [80].

Para-aortic or extended field radiotherapy carries a considerable morbidity [81]. Long-term complications can be especially severe [37, 41]. There is no consensus on its value with respect to survival in cases at risk of or known to have positive para-aortic nodes. In a large randomized EORTC study there was no significant difference in terms of local control, distant metastases and survival between the patients treated with pelvic irradiation only and those treated with pelvic and para-aortic irradiation. However, the incidence of severe gastrointestinal tract complications was significantly higher in the group treated with para-aortic irradiation [83, 84]. Although Rotman *et al.* found a significant improvement in 10-year overall survival in a group of 367 patients randomized to treatment with either extended field irradiation or pelvic irradiation only, the disease-free survival was similar in both groups. Higher complication rates were found in those patients who had undergone prior abdominal surgery [83].

Although extended-field radiation therapy might benefit a small group of patients, its routine use in high-risk patients is of limited value.

NEOADJUVANT AND ADJUVANT CHEMOTHERAPY

Traditionally, chemotherapy has been limited to a palliative role in patients with distant metastatic disease at presentation or recurrent disease following primary local treatment. There have been numerous reports of both single agent and combination chemotherapy in recurrent cervical carcinoma [84–87]. In general, response rates with both single agent and combination regimens rarely exceed 25%, with a median duration of 2–5 months. Most patients

with recurrent disease present with and die of pelvic disease. Pelvic sites of disease will, in the majority of cases, have been previously subjected to local radiotherapy. As a consequence of pelvic radiotherapy and resultant fibrosis and vasculitis the blood supply is often impaired, compromising delivery of drugs to the tumour site. This may explain the poor results of chemotherapy in recurrent cervical carcinoma (see Chapters 23 and 24).

The role of chemotherapy as a part of the primary treatment for patients with early cervical carcinoma might be more successful in two situations: in a neoadjuvant role prior to surgery in bulky disease (NACT) and as an adjuvant in patients with positive lymph nodes in combination with post-operative radiotherapy in an effort to reduce distant recurrences. Data on NACT before surgery seem promising and show better results than the data on NACT before radiotherapy (Table 19.4) [88–93]. Overall response rates with cisplatin-based combination chemotherapy are high: 78–94% (of which 11–44% are complete responses) with a median follow-up of 16–36 months. The most interesting observation is the reduced incidence of positive pelvic lymph nodes in these patients [91]. Benedetti *et al.* described a 5-year survival rate of 71% in a group of 42 patients with adenocarcinoma pre-treated with chemotherapy followed by radical surgery [93]. Namkoong *et al.* found a significantly better survival in a randomized study of 92 patients with locally advanced Stage I and II cervical carcinoma pre-treated with chemotherapy before radical hysterectomy compared to surgery alone [91]. More randomized studies are warranted to

Table 19.4 Results of trials utilizing NACT followed by surgery or radiotherapy

Reference	Stage	N	NACT/FB	Response (%)	Follow-up (month)	Survival (%)
89	IIb	21	MVC/RS	90	–	–
90	Ib–IIb	111	P5FU/RT	86	60	78
91	Ib–IIb	54	PVB/RS	94	36	
92	Ib–IIb	92	VBP/RS	87	48	82.5
93	Ib–IIb	27	PB/RS	78		
94	Ib–IIIb	42	PBM/RS	79	54	88

RS = radical surgery; RT = radiotherapy.

confirm these results and to establish the morbidity of this treatment strategy.

At present the few data available on the value of chemotherapy as an adjuvant in women with early-stage cervical cancer who are at high risk of relapse after primary treatment show no difference in survival between radiotherapy alone or in combination with chemotherapy [94–96]. More prospective randomized trials focused on this area of management are currently in progress.

ADVERSE EFFECTS

Despite the progress that has been made in the surgical and radiotherapeutic treatment of low-stage cervical cancer, urologic and sexual complications remain a problem.

Complications of radical surgery with lymphadenectomy are mostly urologic in origin: ureteric obstruction (4%), dysfunction of the bladder (3.5%), fistulas (0.5–5%) and incontinence (12%) [2, 17–20, 97]. Lymphocysts can occur in up to 20%, but are usually asymptomatic and rarely need treatment [98]. Severe constipation (<5%) and lymphoedema can also occur [20].

After radiotherapy, complications such as bowel obstruction (1–5%), proctitis (1%), bladder dysfunction (5–10%), fistulas (2.5%), vaginal vault necrosis (2–6%) and vaginal stenosis (2–25%) can occur [37, 38, 99–104]. Most complications are grade 1–2, which means that they do not warrant hospital admission and surgical treatment [37, 99]. Complications of radiotherapy occur more often in patients with a history of pelvic surgery and can occur up to 10 years following treatment [37, 38, 105].

UROLOGIC PROBLEMS

The deeper the surgical dissection extends into the pelvic cavity, the greater the damage to the pelvic neural plexus. The bladder is supplied by nerve fibres from the pelvic plexus, the major part of which is located in the paracervical and paravaginal tissues; topographically the pelvic plexus is closely related to the pelvic vessels. After a radical hysterectomy the rest-tone and the filling pressure of the bladder increase, whereas pressure in and along the length of the urethra decreases. Loss of compliance is caused by direct surgical injury to the bladder wall, lymphostasis, interruption of the blood supply, neural denervation of the bladder and the urethra and fibrosis of the urethra [106–108]. This may lead to urinary retention, and straining or inability to void, and to a lesser extent to urge and stress incontinence. Both the nerve and vascular supply to the lower part of the bladder and the urethra is important in maintaining urethral pressure and for appropriate detrusor function [109]. It is for this reason that many surgeons advocate sparing of the superior vesical artery. By using a modified surgical technique designed to spare as much vaginal and paravaginal tissue as possible, the urological complications can be minimized [110].

The main causes for fistula development [18, 20, 111, 112] after radical hysterectomy are trauma during the procedure, ischaemia resulting from damage to the ureteral vessels, infection and stasis of urine in the urinary tract. To reduce morbidity the vascular supply to the ureter should be kept intact and bladder drainage instituted for at least 7 days in uncomplicated cases [113]. Post-operative transient ureteral dilatation occurs quite often in asymptomatic patients [114].

SEXUAL PROBLEMS

Sexual dysfunction is more common after radiotherapy, although women who have undergone radical surgery also suffer sexual impairment [115, 116]. In 1974 Abitbol and Davenport [104] carried out a retrospective study to compare the sexual functioning of 28 patients who had received radiotherapy with that of 32 patients who had undergone a radical hysterectomy and 15 after combined treatment for cervical carcinoma. Sexual activity was markedly reduced or absent in 79% of the radiotherapy group, 6% of the surgery group and in

33% of the group that received a combined treatment. Narrowing of the vagina occurs more often after radiotherapy or after combined surgery and radiotherapy than after surgery alone, and is correlated strongly with the patient's age at the time of the treatment: 1% for women < 40 years at the time of treatment vs 5% for women >50 years [99]. In a detailed prospective study, Schover *et al.* [117] describe sexual function, sexual behaviour and marital happiness in 26 women treated with surgery, and 37 treated with radiotherapy with or without surgery for low-stage cervical carcinoma. In the first 6 months after treatment no differences were found between the two groups: sexual activity with a partner decreased significantly, but the women's sexual satisfaction and frequency of masturbation remained stable. By 1 year, however, the radiotherapy group had developed dyspareunia and had significantly more problems related to sexual desire and arousal. Weymar Schultz *et al.* [118] reported a controlled prospective study on sexual functioning before and after treatment for cervical carcinoma in 26 couples. At 1 year, sexual functioning of these couples was similar to that of couples where the women had undergone simple hysterectomy for a benign disease: in both populations the sexual response was significantly disturbed whereas current sexual behaviour and motivation for sexual interaction were within the normal range. At 2 years, the negative genital sensations during sexual arousal and orgasm were found to have increased in the cervical cancer group. These findings indicate that although women experience many changes after cervical cancer treatment they manage to maintain some form of sex-life despite poor sexual response and annoying physical complaints.

BOWEL PROBLEMS

Few studies mention the problem of severe constipation after radical hysterectomy. As with bladder dysfunction, it seems that the frequency of this complication depends on the radicality of the procedure, especially the extent to which the uterosacral ligaments are dissected [20]. Bowel perforation during the operation can occur, but should be recognized and corrected during the procedure.

After irradiation, small bowel obstruction is seen in about 4% of the cases and can be severe. This complication can also occur many years after treatment [99]. Diarrhoea forms one of the main complaints during treatment but is mostly a grade 1 complication and amenable to simple medical management.

Complications of surgery and radiotherapy are summarized in Table 19.5.

RISK FACTORS FOR RELAPSE

Recurrence affects 15% of patients after primary treatment for early-stage cervical carcinoma [119, 120]. The majority of recurrences occur during the first 2 years after treatment. Symptoms include pain, vaginal bleeding, discharge, disturbance of bowel or bladder function or they may be asymptomatic [105]. To date, a large number of different factors have been suggested as having an important prognostic role in surgically managed cervical carcinoma. It is surprising that few of these studies have incorporated a thorough statistical multivariate analysis. Using the Cox proportional hazards regression model, the following independent prognostic factors have been identified:

- the presence of lymph node metastasis;
- vascular invasion;
- the size and depth of the tumour;
- pre-treatment SCC-ag levels [58, 121–123].

Other previously identified significant factors in univariate analyses are likely to be related to one or other of these.

The presence of lymph node metastases is generally considered to be the most important prognostic factor in carcinoma of the cervix [2, 17, 56–58]. However, different studies have also identified a range of other morphological factors as playing a role. It is possible that the

Table 19.5 Percentage of complications by different treatment modalities for cervical carcinoma Stage I and II [2, 17–20, 37, 38, 97–104, 158, 159]

	Surgery	*Radiotherapy*	*Combination*
Wound infection	2.5–4		
Bladder			
dysfunction	3.5–12	5–11	
incontinence	12		
ureteric obstruction	4	2	2.5
Bowel			
ileus	2–4		
diarrhoea	4		
proctitis	0	2	
small bowel obstruction	0	1–5	4
constipation	5		
Fistulas	1.5–2	2–4	3–12
Vaginal stenosis	3–6	5–25	1
Vaginal necrosis	0	2–6	1
Lymphoedema	8	0.5–2	2

importance of these other factors is related indirectly to their association with lymph node metastases (Table 19.3).

The presence of tumour cells in the lymph vessels has a considerable adverse effect on the 5-year survival rate [2, 58, 124] and recurrence rate [109]. There have been several reports of a direct correlation between the presence of vascular invasion and lymph node metastasis [58, 125–127].

Cell type and tumour grading have not been identified as having an important prognostic value [126–131]. Some authors, however, mention a worse prognosis associated with adeno-carcinoma or adenosquamous cell carcinoma and grade of differentiation appears to be more important in these tumour types [119, 120, 132, 133].

Generally, the depth of invasion is a good indicator of tumour load [121, 130, 134, 135], and methods have been developed to measure tumour volume in an attempt to assess its influence on prognosis [2, 135]. Burghardt *et al.* reported the prognostic significance of the

tumour size/cervix size ratio, but this also appeared to be related to an increase in involved nodes [2].

An increase in the number of positive lymph nodes has been recognized in cases where there is evidence of parametrial involvement [2, 136]. Comparisons between various published series is made difficult by a lack of uniformity in stages included by different authors. For example, Burghardt includes Stage IIb, whereas others confine their selection of patients to Stage I–IIa. Another explanation for the contradictory results concerning the effect of parametrial invasion is that the extent of pre-operative staging may not be uniform. Some surgeons may not operate when involvement of the parametria is suspected, whereas others perform laparotomies in those cases and report successful radical hysterectomies in 77% [39].

Although the existence of an aggressive form of cervical cancer affecting young women has been the subject of several studies, none of the authors could prove that age had an independent effect on survival [137–140].

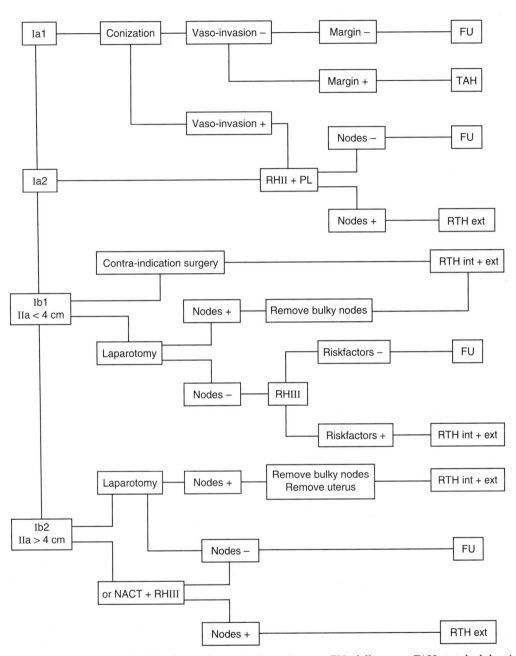

Figure 19.2 Treatment schedule for early cervical carcinoma. FU, follow-up; TAH, total abdominal hysterectomy; RHII, radical hysterectomy class II; RHIII, radical hysterectomy class III; PL, pelvic lymphadenectomy; RTH, radiotherapy; NACT, neoadjuvant chemotherapy.

In a large Gynecologic Oncology Group study by Delgado *et al.*, the authors proposed a model to predict the relative risk of recurrence, which might be used to select those patients at risk [121]. For example, the relative risk of recurrence for a patient with an occult tumour, <3 mm without capillary space and lymph node involvement is <7.5, whereas for a patient with 14 mm invasion, a tumour size of 4 cm and positive capillary space involvement the relative risk of recurrence is >120. Using this model patients with low, intermediate and high risk can be identified.

More recent techniques seek to exploit various biological factors felt to be related to tumour behaviour. These include DNA-ploidy measurement, methods for virus detection, immunochemistry and oncogene expression. The development of DNA aneuploidy is a frequently observed phenomenon in human cancer and its prognostic value has been reported for several malignancies [17, 141]. A clear relationship has been found between the percentage of aneuploid cells in pretreatment biopsies and the recurrence rate in patients with advanced squamous cervical carcinoma treated with radiotherapy [142]. However, none of the studies on patients with low-stage carcinoma of the cervix treated with radical surgery have shown any prognostic value of tumour ploidy [143, 144]. An elevated S-phase fraction has probably more prognostic value but at this moment the series are too small to be clinically useful [144, 145].

A number of tumour markers have been investigated in cervical cancer. Carcino-embryonic Antigen (CEA) and Cancer Antigen 125 (CA 125) may have some value in cervical adenocarcinoma [146, 147]. In 1977 Kato *et al.* published results on the development of a radioimmunoassay based on a polyclonal antiserum directed against a purified homogenate of cervical cancer tissue called Ta-4 [148]. They found that the presence of squamous carcinoma of the cervix corresponded to elevated serum levels of Ta-4 (SCC-ag). In later reports, elevated levels of this marker was shown to be associated with advanced tumour stage, and rising levels after initial treatment were associated with recurrence or metastases [149, 150]. Duk *et al.* found an independent relationship of the pretreatment level of SCC-ag with survival in a large group of patients with Stage Ib and IIa cervical carcinoma [123]. Conversely, in a prospective study by Gaarenstroom *et al.* pretreatment serum levels of neither SCC-ag, Cyfra-21 nor TPA were sufficiently reliable to identify patients at risk of lymph node metastases or parametrial involvement [151]. SCC-ag levels obtained during follow-up appeared to be the best parameter in predicting tumour recurrence. However, up to now the clinical use of these markers has been restricted by their sensitivity and specificity and the lack of effective intervention in the majority of cases of recurrence.

CONCLUSIONS

Changes and improvements in the treatment of cervical carcinoma in the last few decades have shown a gradual movement towards individualization of treatment. Although the results of treatment by radical surgery and radiotherapy in early cervical carcinoma are similar, surgery has many advantages as long as the procedures are performed by skilled gynaecological oncologists.

Figure 19.2 provides an overview of proposed treatment schedules for Stage I and II cervical carcinoma including the recommended treatment modalities that are currently available.

REFERENCES

1. Creasman, W.T. (1995) New gynaecologic cancer staging (editorial). *Gynecologic Oncology*, **58**, 157.
2. Burghardt, E., Pickel, H., Haas, J. and Lahousen, M. (1987) Prognostic factors and operative treatment of stages IB to IIB cervical cancer. *American Journal of Obstetrics and Gynecology*, **156**, 988–96.
3. Trelford, J.D., Tesluk, H., Franti, C.E., Bradfield, G., Ordorica, E. and Deer, D.

(1992) 20 Year follow-up on micro-invasive squamous carcinoma of the cervix. *European Journal of Gynaecologic Oncology*, **37**, 155–9.

4. Kolstad, P. (1989) Follow-up study of 232 patients with stage Ia1 and 411 patients with stage Ia2 squamous cell carcinoma of the cervix (microinvasive carcinoma). *Gynecologic Oncology*, **34**, 265–72.

5. Delgado, G., Bundy, B.N., Fowler, W.C., Steliman, F.B., Sevin, B., Creasman, W.T., Major, F., DiSaia, P. and Zaino, R. (1989) A prospective surgical pathological study of stage 1 squamous carcinoma of the cervix: a gynaecologic oncology group study. *Gynecologic Oncology*, **35**, 314–20.

6. Potter, M.E., Spencer, S., Soong, S.I. and Hatch, K.D. (1993) The influence of staging laparotomy for cervical cancer on patterns of recurrence and survival. *International Journal of Gynecological Cancer*, **3**, 169–74.

7. La Polla, J.P., Schlaerth, J.B., Gaddis, O. and Morrow, C.P. (1986) The influence of surgical staging on the evaluation and treatment of patients with cervical carcinoma. *Gynecologic Oncology*, **24**, 194–206.

8. Moore, D.H., Dotters, D.I. and Fowler, W.C., Jr (1992) Computed tomography: does it really improve treatment of cervical carcinoma? *American Journal of Obstetrics and Gynecology*, **167**, 768–71.

9. Brodman, M., Friedman, F., Dottino, P., Janus, C., Plaxe, S. and Cohen, C. (1990) A comparative study of computerized tomography, magnetic resonance imaging, and clinical staging for the detection of early cervix cancer. *Gynecologic Oncology*, **36**, 409–12.

10. Mamsen, A., Ledertoug, S., Hørlyck, A., Knudsen, H.J., Rasmussen, K.L., Nyland, M.H. and Jakobsen, A. (1995) The possible role of ultrasound in early cervical cancer. *Gynecologic Oncology*, **56**, 187–90.

11. Kim, S.H., Choi, B.I., Han, I.K., Kim, H.D., Le, H.P., Kang, S.B., Lee, J.Y. and Han, M.C. (1993) Preoperative staging of uterine cervical carcinoma: comparison of CT and MRI in 99 patients. *Journal of Computer Assisted Tomography*, **17**, 633–40.

12. Subak, L.L., Hricak, H., Powell, B., Azizi, L. and Stern, J. (1995) Cervical carcinoma: computed tomography and magnetic resonance imaging for preoperative staging. *Obstetrics and Gynecology*, **86**, 43–50.

13. Hricak, H., Quivey, J.M., Campos, Z., *et al.* (1993) Carcinoma of the cervix: predictive value of clinical and magnetic resonance imaging assessment of prognostic factors. *International Journal of Radiation Oncology, Biology and Physiology*, **27**, 791–801.

14. Winter, R., Lahousen, M. and Haas, J. (1996) Magnetic resonance imaging beim cervixkarzinom. *Gynacologe*, **29**, 207–12.

15. Weiser, E.B., Bundy, B.N., Hoskin, W.J., *et al.* (1989) Extraperitoneal versus transperitoneal selective paraaortic lymphadenectomy in the pretreatment surgical staging of advanced cervical carcinoma (A Gynecologic Oncology Group Study). *Gynecologic Oncology*, **33**, 283–9.

16. Potish, R.A., Downey, G.O. and Adcock, L.L. (1989) The role of surgical debulking in cancer of the uterine cervix. *International Journal of Radiation Oncology, Biology and Physiology*, **17**, 979–84.

17. Shephard, J.H. (1985) Surgical management of early invasive cervical cancer. *Clinics in Obstetrics and Gynecology*, **12**, 183–202.

18. Webb, M.J. and Symmonds, R.E. (1979) Wertheim hysterectomy – A reappraisal. *Obstetrics and Gynecology*, **542**, 140–5.

19. Hoskins, W.J., Ford, J.H., Lutz, M.H. and Averette, H.E. (1976) Radical hysterectomy and pelvic lymphadenectomy for the management of early invasive cancer of the cervix. *Gynecologic Oncology*, **4**, 278–90.

20. Kenter, G.G., Ansink, A.C., Heintz, A.P.M., Aartsen, E.J., Delemarre, J. and Hart, A.A.M. (1989) Carcinoma of the uterine cervix stage I and IIA: results of surgical treatment: complications, recurrence and survival. *European Journal of Surgical Oncology*, **15**, 55–60.

21. Wertheim, E. (1911) *Die erweiterte abdominale operation beim carcinoma colli uteri*. Urban & Schwarzenberg, Berlin.

22. Schauta, F. (1908) *Die erweiterte vaginale Totalextirpation des uterus bei Kollum Karzinom*. Verlag Safar, Berlin.

23. Bouwdijk Bastiaanse v. M.A. (1956) Treatment of cancer of the cervix uteri. *American Journal of Obstetrics and Gynecology*, **72**, 100.

24. Bonney, V. (1941) The results of 500 cases of Wertheims operation for carcinoma of the cervix. *Journal of Obstetrics and Gynaecology of the Emipre*, **48**, 421–35.

25. Taussig, F.J. (1943) Iliac lymphadenectomy for group II cancer of the uterine cervix. *American Journal of Obstetrics and Gynecology*, **45**, 733–48.

26. Meigs, J.V. (1945) The Wertheim operation for carcinoma of the cervix. *American Journal of Obstetrics and Gynecology*, **49**, 542–53.

27. Okabayashi, H. (1921) Radical abdominal hysterectomy for cancer of the cervix uteri. Modifications of the Takayama operation. *Surgical Gynecology and Obstetrics*, **33**, 335.

28. Mitra, S. (1954) Radical vaginal hysterectomy. In *Surgical Treatment of Cancer of the Cervix*, (ed. J.V. Meigs). Grune & Stratton, New York, pp. 267–80.

29. Sindram, I.S. (1959) A new combined approach in the treatment of cancer of the uterine cervix. *Acta Union International Contre le Cancer*, **15**, 403.

30. Morgan, P.R., Anderson, M.C., Buckley, C.H., Murdoch, J.B., Lopes, A., Duncan, I.D. and Monaghan, J.M. (1993) The Royal College of Obstetricians and Gynaecologists micro-invasive carcinoma of the cervix study: preliminary results. *British Journal of Obstetrics and Gynaecology*, **100**, 664–8.

31. Soisson, A.P., Soper, J.T., Berchuck, A., Dodge, R. and Clarke-Pearson, D. (1992) Radical hysterectomy in obese women. *Obstetrics and Gynecology*, **80**, 940–3.

32. Kinney, W.K., Egorshin, E.V. and Podratz, K.C. (1988) Wertheim hysterectomy in the geriatric population. *Gynecologic Oncology*, **31**, 227–32.

33. Piver, M.S., Rutledge, F.N. and Smith, P.J. (1974) Five classes of extended hysterectomy for carcinoma of the cervix. *Obstetrics and Gynecology*, **44**, 265–72.

34. Rutledge, J.H. (1974) Radical hysterectomy. In *Gynecologic Surgery* (ed. J.H. Ridley). Williams & Wilkins, Baltimore, MD, pp. 261–77.

35. Te Linde, R.W. (1954) Hysterectomy for carcinoma *in situ*. In *Surgical Treatment of Cancer of the Cervix* (ed. J.V. Meigs). Grune & Statton, New York.

36. Covens, A., Rosen, B., Gibbons, A., *et al.* (1993) Differences in the morbidity of radical hysterectomy between gynecological oncologists. *Gynecologic Oncology*, **51**, 39–45.

37. Perez, C.A., Grigsby, P.W., Camel, H.M., Galakatos, A.E., Mutch, D. and Locket, M.A. (1995) Irradiation alone or combined with surgery in stage Ib, IIa, and IIb carcinoma of the uterine cervix: update of a non-randomized comparison. *International Journal of Radiation Oncology, Biology and Physiology*, **31**, 703–16.

38. Werner-Wasik, M., Schmid, C.H., Bornstein, L., Ball, H.G., Smith, D.M. and Madoc-Jones, H. (1995) Prognostic factors for local and distant recurrence in stage I and II cervical carcinoma. *International Journal of Radiation Oncology, Biology and Physiology*, **32**, 1309–17.

39. Kamura, T., Tsukamoto, N., Tsuruchi, N., Kaku, T., Saito, T., To, N., Akazawa, K. and Nakano, H. (1993) Histopathologic prognostic factors in stage IIB cervical carcinoma treated with radical hysterectomy and pelvic node dissection – an analysis with mathematical statistics. *International Journal of Gynecologic Cancer*, **3**, 219–25.

40. Kjorstad, K.E., Kolbenstvedt, A. and Strickert, T. (1984) The value of complete lymphadenectomy in radical treatment of cancer of the cervix, stage IB. *Cancer*, **54**, 2215–19.

41. Inoue, T. and Morita, K. (1995) Long term observations of patients treated by postoperative extended field irradiation for nodal metastases from cervical carcinoma stages Ib, IIa and IIb. *Gynecologic Oncology*, **58**, 4–10.

42. Hirabayashi, K. (1988) Individualisation of post. operative irradiation in the treatment of stage I and II cervical cancer. *Ballière's Clinical Obstetrics and Gynecology*, **2**, 1013–22.

43. Girardi, F., Pickel, H. and Winter, R. (1993) Pelvic and parametrial lymph nodes in the quality control of the surgical treatment. of cervical cancer. *Gynecologic Oncology*, **50**, 330–3.

44. Benedetti-Panici, P., Manesch, F., Scambia, G., *et al.* (1996) Lymphatic spread of cervical cancer: an anatomical and pathological study based on 225 radical hysterectomies with systemetic pelvic and aortic lymphadenectomy. *Gynecologic Oncology*, **62**, 19–24.

45. Hackett, Th.E., Olt, G.O., Sorosky, J.I., Podczaski, E., Harrison, T.A. and Mortel, R. (1995) Surgical predictors of para-aortic metastases in early-stage cervical carcinoma. *Gynecologic Oncology*, **59**, 15–19.

46. Landoni, F., Bocciolone, L., Perego, P., Maneo, A., Bratina, G. and Mangioni, C. (1995) Cancer of the cervix, FIGO stage IB and IIA: patterns of local growth and paracervical extension. *International Journal of Gynecologic Cancer*, **5**, 329–34.

47. Rutledge, F.N., Wharton, S.T. and Fletcher, G.H. (1976) Clinical studies with adjunctive surgery and radiation therapy in treatment of carcinoma of the cervix. *Cancer*, **38**, 596–602.

48. Rampone, J.F., Klem, V. and Kolstad, P. (1973) Combined treatment in stage IB carcinoma of the cervix. *Obstetrics and Gynecology*, **41**, 163–7.

49. Coleman, D.L., Gallup, D.G., Wolcott, H.D., Otken, L.B. and Stocck, R.J. (1992) Patterns of failure of bulkybarrel carcinomas of the cervix. *American Journal of Obstetrics and Gynecology*, **166**, 916–20.

50. Rettenmaier, M.A., Casanova, D.M., Mischa, J.P., Moran, M.F., Ramsanghani, N.S., Syed, N.A., Puthawala, A. and Disaia, P.J. (1989) Radical hysterectomy and tailored postoperative radiation therapy in the management of bulky stage IB cervical cancer. *Cancer*, **63**, 2220–3.

51. Bloss, J.D., Berman, M.L., Mukhererjee, J., Manetta, A., Emma, D., Ramsanghani, N.S.

and Disaia, P.J. (1992) Bulky stage IB cervical carcinoma managed by primary radical hysterectomy followed by tailored radiotherapy. *Gynecologic Oncology*, **47**, 21–7.

52. Sardi, J.E., *et al.* (1986) A possible new trend in the management of carcinoma of the cervix uteri. *Gynecologic Oncology*, **25**, 139–49.

53. Benedetti-Panici, P., *et al.* (1991) High dose Cisplatin and Bleomycin chemotherapy plus radical surgery in locally advanced cervical carcinoma: a preliminary report. *Gynecologic Oncology*, **41**, 212–16.

54. Chen, N.J., Okuda, H. and Sekiba, K. (1985) Recurrent carcinoma of the vagina following Okabayashi's radical hysterectomy for cervical carcinoma. *Gynecologic Oncology*, **20**, 10–16.

55. Bleker, O.P., Ketting, B.W., Van Wayjen-Eecen, B. and Kloosterman, G.J. (1983) The significance of microscopic involvement of the parametrium and/or pelvic lymph nodes in cervical cancer stages IB and IIA. *Gynecologic Oncology*, **16**, 56–62.

56. Martimbeau, P., Kjorstad, K. and Iversen, T. (1982) Stage IB carcinoma of the cervix; the Norwegian Radium Hospital II. Results when pelvic nodes are involved. *Obstetrics and Gynecology*, **60**, 215–18.

57. Baltzer, J., Lohe, K.J., Kopcke, W. and Zander, J. (1982) Histologic criteria for prognosis in patients with operated squamous cell carcinoma of the cervix. *Gynecologic Oncology*, **13**, 184–94.

58. Kenter, G.G., Ansink, A.C., Heintz, A.P.M., Delemarre, J., Aartsen, E.J. and Hart, A.A.M. (1988) Low stage invasive carcinoma of the uterine cervix: morphological prognostic factors. *European Journal of Surgical Oncology*, **14**, 187–92.

59. Monk, B.J., Cha, D.S., Walker, J.L., *et al.* (1994) Extent of disease as an indication for pelvic radiation following radical hysterectomy and bilateral lymph node dissection in the treatment. of stage IB and IIA cervical carcinoma. *Gynecologic Oncology*, **54**, 4–9.

60. Inoue, T. and Morita, K. (1990) The prognostic significance of number of positive nodes in cervical carcinoma stages IB, IIA and IIB. *Cancer*, **65**, 1923–7.

61. Reiffenstuhl, G. (1957) *Das lymphsystem des weiblichen Genitale*. Urban & Schwarzenberg, Wien.

62. Inoue, T., Chihara, T. and Morita, K. (1986) Postoperative extended field irradiation in patients with pelvic and/or common iliac node metastasis from cervical carcinoma stages IB to IIB. *Gynecologic Oncology*, **25**, 234–43.

63. Barillot, I., Cuisenier, J., Pigneux, J., *et al.* (1993) Carcinoma of the cervical stump: a review of 213 cases. *European Journal of Cancer A General Topics*, **29**, 1231–6.

64. Orr, J.W., Ball, G.C. and Soong, S.J. (1986) Surgical treatment of women found to have invasive cervical cancer at the time of total hysterectomy. *Obstetrics and Gynecology*, **68**, 353–6.

65. Heller, P.B., Barnhill, D.R. and Mayer, A.R. (1986) Cervical carcinoma found identically in a uterus removed for benign indications. *Obstetrics and Gynecology*, **67**, 187–90.

66. Van der Vange, N., Weverling, G.J., Ketting, B.W., Ankum, W.M., Samlal, R. and Lammes, F.B. (1995) The prognosis of cervical cancer associated with pregnancy: a matched cohort study. *Obstetrics and Gynecology*, **85**, 1022–6.

67. Zemlickis, D., Lishner, M.N., Degendorfer, P., Panzarella, T., Sutcliffe, S.B. and Koren, G. (1991) Maternal and foetal outcome after invasive cervical carcinoma. *Journal of Clinical Oncology*, **9**, 1956–61.

68. Sorosky, J.I., Squatrito, R., Ndubisi, B.U., Anderson, B., Podczaski, E.S., Mayr, N. and Buller, R.E. (1995) Stage I squamous cell cervical carcinoma in pregnancy: planned delay in therapy awaiting fetal maturity. *Gynecologic Oncology*, **59**, 207–10.

69. Duggan, B., Muderspach, L., Roman, L., Curtin, J., d'Ablaing, G. and Morrow, C.P. (1993) Cervical cancer in pregnancy: reporting on planned delay in therapy. *Obstetrics and Gynecology*, **82**, 598–602.

70. Monk, B.F. and Montz, F.J. (1992) Invasive cervical cancer complicating intrauterine pregnancy: treatment with radical hysterectomy. *Obstetrics and Gynecology*, **80**, 199–203.

71. Francke, P., Maruyama, Y., Nagell van, J. and DePriest, P. (1996) Lymphovascular invasion in stage IB cervical carcinoma: prognostic significance and role of adjuvant radiotherapy. *International Journal of Gynecological Cancer*, **6**, 208–12.

72. Morrow, C.P., Shingleton, H.M. and Austin, J.M. (1980) Is pelvic radiation beneficial in the post operative management of stage IB squamous cell carcinoma of the cervix with pelvic node metastasis treated by radical hysterectomy and pelvic lymphadenectomy. *Gynecologic Oncology*, **10**, 105–10.

73. Kinney, W.K., Alvarez, R.D., Reid, G.C., *et al.* (1989) Value of adjuvant whole-pelvis irradiation after Wertheim hysterectomy for early stage carcinoma of the cervix with pelvic nodal metastasis: a matched-control study. *Gynecologic Oncology*, **34**, 258–62.

74. Remy, J.C., Di Maio, Th., Fruchter, R.G., Sedlis, A., Boyce, J.G., Sohn, C.K. and Rotman, M. (1990) Adjuvant radiation after radical hysterectomy in stage 113 squamous cell carcinoma of the cervix. *Gynecologic Oncology*, **38**, 161–5.

75. Larson, D.M., Stringer, C.A., Copeland, L.J., Gershenson, D.M., Malone, J.M. and Rutledge, F.N. (1987) Stage IB cervical carcinoma treated with radical hysterectomy and pelvic lymphadenectomy: role of adjuvant radiotherapy. *Obstetrics and Gynecology*, **69**, 378–81.

76. Barter, I.F., Soong, S.J., Shingleton, H.M., Hatch, K.D. and Orr, W. (1989) Complications of combined radical hysterectomy–postoperative radiation therapy in women with early stage cancer. *Gynecologic Oncology*, **32**, 292–6.

77. Strockbine, M.F., Hancock, J.E. and Fletcher, G.H. (1970) Complications of 813 patients with squamous cell carcinoma of the intact uterine cervix treated with 3000 rads or more whole pelvis irradiation. *American Journal of Roentgenology*, **108**, 293–304.

78. Warren, S. and Froiedman, N.B. (1942) Pathology and pathologic diagnosis of radiation lesions in the gastrointestinal tract. *American Journal of Pathology*, **18**, 499–507.

79. Snijders-Keilholz, A. and Trimbos, J.B. (1991) A preliminary report on new efforts to decrease radiotherapy related small bowel toxicity. *Radiotherapy Oncology*, **22**, 206–8.

80. Devereux, D.F., Thompson, D., Sandhaus, L., Sweeney, W. and Haas, A. (1987) Protection from radiation enteritis by an absorbable polyglycolic-acid mesh sling. *Surgery*, **101**, 123–9.

81. Cunningham, M.J., Dunton, C.J., Corn, B., *et al.* (1991) Extended field radiation in early-stage cervical carcinoma: survival and complications. *Gynecologic Oncology*, **43**, 51–4.

82. Haie, C., Pejovic, M.H., Gerbaulet, A., *et al.* (1988) Is prophylactic para-aortic irradiation worthwhile in the treatment of advanced cervical carcinoma? Results of a controlled clinical trial of the EORTC radiotherapy group. *Radiotherapy Oncology*, **11**, 101–12.

83. Rotman, M., Pajak, F., Choi, K., Clery, M., Marcial, V., Grigsby, P.W., Cooper, J. and John, M. (1995) Prophylactic extended-field irradiation of the para-aortic lymph nodes in stage IB and Iia cervical carcinomas. Ten year treatment results of RTOG 79–20. *Journal of the American Medical Association*, **274**, 387–93.

84. Bonomi, Ph., Blessing, J., Ball, H., Hanjani, P. and Disaia, Ph.J. (1989) A Phase II evaluation of Cis-platin and 5-Fluoro-uracil in patients with advanced squamous cell carcinoma of the cervix: a Gynecologic Oncology Group study. *Gynecologic Oncology*, **34**, 357–61.

85. Bonomi, P., Blessing, J., Stehman, F., Disaia, P., Walton, L. and Major, F. (1985) Randomized trial of three Cis-platin dose schedules in squamous-cell carcinoma of the cervix: a Gynecologic Oncology Group study. *Journal of Clinical Oncology*, **8**, 1079–85.

86. Vermorken, J.B., Mangioni, C., Van der Burg, M.E.L., *et al.* (1987) Mitomycin-C/Cisplatin combination chemotherapy in recurrent and/or metastatic squamous cell carcinoma of the uterine cervix (SCCUC). The EORTC Gynaecological Cancer Cooperative Group (GCCG) experience. *Proceedings 1st Meeting International Gynecologic Cancer Society*, Amsterdam, p. 31.

87. Alberts, D.S., Martimbeau, P.W., Surwit, E.A. and Oishi, N. (1981) Mitomycin-C, bleomycin, vincristine and cis-platinum in the treatment of advanced, recurrent squamous cell carcinoma of the cervix. *Cancer Clinical Trials*, **4**, 313–16.

88. Park, S.Y., Kim, B.G., Kim, J.H., *et al.* (1995) Phase I/II study of neo-adjuvant intraarterial chemotherapy with mitomycin-C, vincristine, and cisplatin in patients with stage IIB bulky cervical carcinoma. *Cancer*, **76**, 814–23.

89. Park, T.K., Choi, D.H., Kim, S.N., *et al.* (1991) Role of induction chemotherapy in invasive cervical cancer. *Gynecologic Oncology*, **41**, 107–12.

90. Kim, D.S., Moon, H., Kim, K.T., Hwang, Y.Y., Cho, S.H. and Kim, S.R. (1989) Two-year survival: preoperative adjuvant chemotherapy in the treatment of cervical cancer stage IIB and II with bulky tumor. *Gynecologic Oncology*, **33**, 225–30.

91. Namkoong, S.E., Park, J.S., Kim, J.W., *et al.* (1995) Comparative study of the patients with advanced stages I and II cervical cancer treated by radical surgery with and without preoperative adjuvant chemotherapy. *Gynecologic Oncology*, **59**, 136–42.

92. Fontanelli, R., Spatti, G., Raspagliesi, F. *et al.* (1992) A pre-operative single course of high-dose cisplatin and bleomycin with gluthathione protection in bulky stage IB/II carcinoma of the cervix. *Annals of Oncology*, **3**, 117–21.

93. Benedetti-Panici, P., Greggi, S., Scambia, G., *et al.* (1996) Locally advanced cervical adenocarcinoma: is there a place for chemo-surgical treatment? *Gynecologic Oncology*, **61**, 44–9.

94. Killackey, M.A., *et al.* (1993) Adjuvant chemotherapy and radiation in patients with poor prognostic stage IB/IIA cervical cancer. *Gynecologic Oncology*, **49**, 377–9.

95. Tattersall, M.H.N., Rainirez, C. and Coppleson, M. (1992) A randomized trial of adjuvant chemotherapy after radical hysterectomy in stage IB–IIA cervical cancer patients with pelvic lymph node metastases. *Gynecologic Oncology*, **46**, 176–81.

96. Curtin, I.F., Hoskins, W.J., Venkatraman, E.S., *et al.* (1996) Adjuvant chemotherapy versus chemotherapy plus pelvic irradiation for high-risk cervical cancer patients after radical hysterectomy and pelvic lymphadenectomy. *Gynecologic Oncology*, **61**, 3–10.

97. Kadar, M., Saliva, N. and Nelson, J.H. (1983) The frequency, causes and prevention of severe urinary dysfiinction after radical hysterectomy. *British Journal of Obstetrics and Gynecology*, **90**, 858–63.

98. Petru, E., Tamussino, K., Lahousen, M., Winter, R., Pickel, H. and Haas, J. (1989) Pelvic and para-aortic lymphcysts after radical surgery because of cervical and ovarian cancer. *Obstetrics and Gynecology*, **161**, 937–41.

99. Eifel, P.J., Levenback, C., Wharton, J.T. and Oswald, M.J. (1995) Time course and incidence of late complications in patients treated with radiation therapy for FIGO stage Ib carcinoma of the uterine cervix. *International Journal of Radiation Oncology, Biology and Physiology*, **32**, 1289–300.

100. Coia, L., Won, M., Lanciano, R., Marcial, V.A., Martz, K. and Hanks, G. (1990) The patterns of care outcome study for cancer of the uterine cervix; results of the second National Practice Survey. *Cancer*, **66**, 2451–6.

101. Horiot, J.C., Pigneux, J., Pourquier, H., *et al.* (1988) Radiotherapy alone in carcinoma of the intact uterine cervix according to G.H. Fletcher guidelines: a French cooperative study of 1383 cases. *International Journal of Radiation Oncology, Biology and Physiology*, **14**, 605–11.

102. Perez, C.A., Fox, S., Locket, M., Grigsby, P.W., *et al.* (1991) Impact of dose in outcome of irradiation alone in carcinoma of the uterine cervix. *International Journal of Radiation Oncology, Biology and Physiology*, **21**, 885–98.

103. Choy, D., Wong, L.C., Sham, J., Ngan, H.Y.S. and Ma, H.K. (1993) Dose–tumor response for carcinoma of the cervix: an analysis of 594 patients treated by radiotherapy. *Gynecologic Oncology*, **49**, 311–17.

104. Abitbol, M.M. and Davenport, J.H. (1974) Sexual dystunction after therapy for cervical carcinoma. *American Journal of Obstetrics and Gynecology*, **119**, 181–9.

105. Gerdin, E., Cnattingius, S., Johnson, P. and Petterson, B. (1994) Prognostic factors and relapse patterns in early-stage cervical carcinoma after brachytherapy and radical hysterectomy. *Gynecologic Oncology*, **53**, 314–19.

106. Ralph, G., Tamussino, K. and Lichtenegger, W. (1988) Urodynamics following radical abdominal hysterectomy for cervical cancer. *Archives of Gynecologic Obstetrics*, **243**, 215–20.

107. Lowe, J.A., Manger, G.H. and Carmichael, J.H. (1981) The effect of the Wertheim hysterectomy on bladder and urethral function. *Obstetrics and Gynecology*, **139**, 826–34.

108. Westby, M. and Asmussen, M. (1985) Anatomical and tunctional changes in the lower urinary tract after radical hysterectomy with LND as studied by dynamic urethrocystography and simultaneous urethrocystometry. *Gynecologic Oncology*, **21**, 261–76.

109. Asmussen, M. and Muller, A. (1983) *Clinical Gynaecological Urology*. Blackwell Scientific Publications, Oxford, pp. 119–51.

110. Vervest, H.A.M., Barents, J.W., Haspels, A.A. and Debruyne, F.M.J. (1989) Radical hysterectomy and the function of the lower urinary tract. Urodynamic quantification of changes and storage and evacuation function. *Acta Obstetrica et Gynecologica Scandinavica*, **68**, 331–40.

111. Riss, P., Koebl, H., Neinteufel, W. and Janish, H. (1988) Wertheim radical hysterectomy 1921–1986: changes in urologic complications. *Archives in Gynecological Obstetrics*, **241**, 249–53.

112. Sivaneratnam, V., Sen, D.K., Jayalaksmi, P. and Ong, G. (1993) Radical hysterectomy and pelvic lymphadenectomy for early invasive cancer of the cervix – 14-year experience. *International Journal of Gynecological Cancer*, **3**, 231–8.

113. Nwabineli, N.J., Walsh, D.J. and Davis, J.A. (1993) Urinary drainage following radical hysterectomy for cervical carcinoma – a pilot comparison of urethral and suprapubic routes. *International Journal of Gynecological Cancer*, **3**, 208–10.

114. Larson, D.M., Malone, J.M., Copeland, L.J., Gershenson, D.M., Kline, R.C. and Stringer, C.A. (1987) Ureteral assessment after radical hysterectomy. *Obstetrics and Gynecology*, **67**, 612–16.

115. Decker, W.H. and Schwartzman, E. (1962) Sexual functioning following treatment for carcinoma of the cervix. *American Journal of Obstetrics and Gynecology*, **83**, 401–5.

116. Andersen, B.L. and Hacker, N.F. (1983) Treatment for gynecological cancer: a review of the effects on female sexuality. *Health Psychology*, **2**, 203–21.

117. Schover, L.R., Fife, M. and Gershenson, D.M. (1989) Sexual dysfunction and treatment. for early stage cervical cancer. *Cancer*, **63**, 204–12.

118. Weijmar Schultz, W.C.M., Wiel, v. d. H.B.M. and Bouma, J. (1991) Psychosexual functioning after treatment. for cancer of the cervix, a comparative and longitudinal study. *International Journal of Gynecologic Cancer*, **1**, 37–46.

119. Burke, T.W., Hoskins, W.J., Heller, P.B., Shen, M.C., Weiser, E.W. and Park, R.C. (1987) Clinical patterns of tumour recurrence after radical hysterectomy in stage IB cervical carcinoma. *Obstetrics and Gynecology*, **69**, 382–5.

120. Larson, D.M., Copeland, L.J., Stringer, C.A., Gershenson, D.M., Malone, J.M. and Creighton, L.E. (1988) Recurrent cervical carcinoma after radical hysterectomy. *Gynecologic Oncology*, **30**, 381–7.

121. Delgado, G., *et al.* (1990) Prospective surgical–pathological study of disease-free interval in patients with Stage IB squamous cell carcinoma of the cervix: a Gynecologic Oncology Group study. *Gynecologic Oncology*, **38**, 352–7.

122. Sevin, B.U., Nadji, M., Lampe, B., Lu, Y., Hilsenbeck, S., Koechli, O.R. and Averette, H.E. (1995) Prognostic factors of early cervical cancer treated by radical hysterectomy. *Cancer*, **76**, 1978–86.

123. Duk, J.M., Groenier, K., Bruijn De, H., *et al.* (1996) Pretreatment serum squamous cell carcinoma antigen: a newly identified prognostic factor in early-stage cervical carcinoma. *Journal of Clinical Oncology*, **14**, 111–18.

124. White, C.D., Morley, G.W. and Kumar N. (1984) The prognostic significance of tumour emboli in lymphatic or vascular spaces of the cervical stroma in stage IB squamous cell carcinoma of the cervix. *American Journal of Obstetrics and Gynecology*, **149**, 342–9.

125. Boyce, J.G., Fruchter, R.G., Nicastri, A.D., *et al.* (1984) Vascular invasion in stage I carcinoma of the cervix. *Cancer*, **53**, 1175–80.

126. Shingleton, H.M., Gore, H., Soong, H.J., Orr, J.W., Hatch, K.D., Austin., J.M. and Partridge, E.E. (1983) Tumour recurrence and survival in stage IB cancer of the cervix. *American Journal of Clinical Oncology*, **6**, 265–72.

127. Inoue, T. and Okura, M. (1984) The prognostic significance of parametrial extension in patients with cervical carcinoma stage IB, IIA and IIB. *Cancer*, **54**, 1714–19.

128. Nagell, v. J.R., Donaldson, E.S., Parker, J.C., Van Dijke, A.H. and Wood, E.G. (1977) The prognostic significance of cell type and lesion size in patients with cervical cancer treated by radical surgery. *Gynecologic Oncology*, **5**, 142–51.

129. Shingleton, H.M., Bell, M.C., Fremgen, A., *et al.* (1995) Is there really a difference in survival of women with squamous cell carcinoma, adenocarcinoma and adenosquamous cell carcinoma of the cervix. *Cancer*, **76**, 1948–55.

130. Crissmann, J.D., Malluch, R. and Budhraja, M. (1985) Histopathologic grading of squamous cell carcinoma of the uterine cervix. *Cancer*, **55**, 1590–6.

131. Anton-Culver, H., Bloss, J.D., Bringinan, D., Lee-Feldstein, A., Disia, P. and Manetta, A. (1992) Comparison of adenocarcinoma and squamous cell carcinoma of the uterine cervix: a population based epidemiologic study. *American Journal of Obstetrics and Gynecology*, **166**, 1507–14.

132. Tinga, D.J., Bouma, J. and Aalders, J.G. (1992) Patients with squamous cell versus (adeno) squamous carcinoma of the cervix, what factors determine the prognosis? *International Journal of Gynecologic Cancer*, **2**, 83–91.

133. Eifel, P.J., Burke, T.W., Morris, M. and Smith, T.J. (1995) Adenocarcinoma as an independent risk factor for disease recurrence in patients with stage Ib cervical carcinoma. *Gynecologic Oncology*, **59**, 38–44.

134. Zander, J., Baltzer, J., Lohe, K.J., Ober, K.G. and Kaufmann, C. (1981) Carcinoma of the cervix: an attempt to individualize treatment. *American Journal of Obstetrics and Gynecology*, **139**, 752–9.

135. Inoue, T. (1984) Prognostic significance of the depth of invasion relating to nodal metastases, parametrial extension and cell type. *Cancer*, **54**, 3035–42.

136. Matsyama, T., Inoue, I., Tsukamoto, N., Kashimura, M., Kashimura, T., Saito, T. and Uchino, H. (1984) Stage Ib, IIa, and IIb cervix cancer, postsurgical staging, and prognosis. *Cancer*, **54**, 3072–7.

137. Poka, R., Juhash, B. and Lampe, L. (1994) Cervical cancer in young women: a poorer prognosis? *International Journal of Gynecological Obstetrics*, **46**, 33–7.

138. Austin, J.P., Degefu, S., Torres, J., Bush, D.J., O'Quinn, A.G., Ozmen, N. and Rice, J. (1994) Cervical carcinoma in women less than 35 years of age. *Southern Medical Journal*, **87**, 375–9.

139. Mariani, L., Iacovelli, A., Vincenzoni, C., Diotallevi, F.F., Atlanta, M. and Lombardi, A. (1993) Cervical carcinoma in young patients: clinical and pathological variables. *International Journal of Gynecological Obstetrics*, **41**, 61–6.

140. Free, K., Roberts, S., Bourne, R., Dickie, G., Ward, B., Wright, G. and Hill, B. (1991) Cancer of the cervix – old and young, now and then. *Gynecologic Oncology*, **43**, 129–36.

141. Koss, L.G., Czerniak, B., Herz, F. and Wersto, R.P. (1989) Flow cytometric measurements of DNA and other cell components in human tumours: a critical appraisal. *Human Pathology*, **20**, 528–48.

142. Jacobsen, A., Bichel, P. and Vaeth, M. (1985) New prognostic factors in squamous cell carcinoma of the cervix uteri. *American Journal of Clinical Oncology*, **8**, 39–43.

143. Kenter, G.G., Cornelisse, C.J., Aartsen, E.J., Mooi, W., Hermans, J., Heintz, A.P.M. and Fleuren, G.J. (1990) DNA ploidy level in low stage carcinoma of the uterine cervix. *Gynecologic Oncology*, **39**, 181–5.

144. Naus, G.J. and Zimmerman, R.L. (1991) Prognostic value of cytometric DNA content analysis in single treatment stage IB–IIA squamous cell carcinoma of the cervix. *Gynecologic Oncology*, **43**, 149–53.

145. Strang, P., Stendahl, U., Bergstrom, R., Frankendahl, B. and Tribukait, B. (1991) Prognostic flow cytometric information in cervical squamous cell carcinoma: a multivariate analysis of 307 patients. *Gynecologic Oncology*, **43**, 3–8.

146. Nagell, v. J.R., Donaldson, E.S., Gay, E.C., Rayburn, P., Powell, D.F. and Goldenberg, D.M. (1978) Carcinoembryonic antigen in carcinoma of the uterine cervix. *Cancer*, **42**, 2428–34.

147. Duk, J.M., Aalders, J.G., Fleuren, G.J., Kians, M. and Bruijn, de H.W.A. (1989) Tumor markers CA-125, squamous cell carcinoma antigen, and carcinoembryonic antigen in patients with adenocarcinoma of the uterine cervix. *Obstetrics and Gynecology*, **73**, 661–7.

148. Kato, H. and Torigoe, T. (1977) Radioimmunoassay for tumour antigen of human cervical squamous cell carcinoma. *Cancer*, **40**, 1621–8.

149. Kato, H., Morioka, H., Nagai, M., Nagaya, T. and Torigoe, T. (1984) Tumour antigen Ta-4 in the detection of recurrence in cervical squamous cell carcinoma. *Cancer*, **54**, 1544–6.

150. Gaarenstroom, K.N., Bonfrer, J.M.G., Kenter, G.O., Korse, C.M., Hart, A.A.M., Trimbos, J.B. and Helmerhorst, Th.J.M. (1995) Clinical value of pretreatment serum Cyfra 21-1, tissue polypeptide antigen, and squamous cell carcinoma antigen levels in patients with cervical cancer. *Cancer*, **76**, 807–13.

151. Bonfrer, J.M.G., Gaarenstroom, K.N, Korse, C.M., Bunningen, van B.N.F.M. and Kenemans, P. (1997) Cyfra 21-1 in monitoring cervical cancer: a comparison with tissue polypeptide antigen and squamous cell carcinoma antigen. *Anticancer Research*, **17**, 2329–34.

152. Leman, M.H., Benson, W.L., Kurman, R.J. and Park, R.C. (1976) Microinvasive carcinoma of the cervix. *Obstetrics and Gynecology*, **48**, 571–8.

153. Hasumi, K., Sakamoto, A. and Sugano, H. (1980) Microinvasive carcinoma of the uterine cervix. *Cancer*, **45**, 928–31.

154. Nagell, van J.R., Greenwell, N., Powell, D.F., Donaldson, E.S., Hanson, M.B. and Gay, E.C. (1983) Microinvasive carcinoma of the cervix. *American Journal of Obstetrics and Gynecology*, **145**, 981–91.

155. Creasman, W.T., Fetter, B.F., Clarke-Pearson, D.L., Kaufmann, L. and Parker, R.T. (1985) Management of stage IA carcinoma of the cervix. *American Journal of Obstetrics and Gynecology*, **153**, 164–72.

156. Terada, K.Y., Morley, G.W. and Roberts, J.A. (1988) Stage B carcinoma of the cervix with lymph node metastases. *Gynecologic Oncology*, **31**, 389–95.

157. Berman, M., Keys, N., Creasman, W. and DiSaia, P. (1984) Survival and patterns of recurrence in cervical cancer metastatic to paraaortic lymph nodes. *Gynecologic Oncology*, **19**, 8–16.

158. Monaghan, J.M., Ireland, D., Mor, Y., Pearson, S.E., Lopes, A. and Sinha, D.P. (1990) Role of centralisation of surgery in stage IB carcinoma of the cervix: a review of 498 cases. *Gynecologic Oncology*, **37**, 206–9.

159. Magrina, I.F., Goodrich, M.A., Weaver, A.L. and Podratz, K.C. (1995) Modified radical hysterectomy: morbidity and mortality. *Gynecologic Oncology*, **59**, 277–82.

Robin Crawford and John H. Shepherd

INTRODUCTION

In this chapter, radical surgery for treatment of cervical cancer is presented. The selection of the patient is then considered with detailed comment on the techniques as performed in our department. The role of lymphadenectomy and available methods are reviewed. The laparoscopically assisted radical vaginal hysterectomy and variations on radical vaginal surgery are briefly discussed. The role of the radical surgery as part of salvage therapy is also considered when a 'cut-through' operation has occurred as well as its position relating to treatment during pregnancy and following recurrence. Throughout this chapter, parametrium is used as the collective term to include all the tissue between the lower part of the uterus, cervix and upper part of the vagina and the pelvic side wall.

HISTORY

In 1898 in Vienna, Wertheim recorded the first radical procedure for treatment of cervical cancer. At this operation, he removed the cervical tumour with the parametrium and sampled pelvic lymph nodes. Five years later, he reported his first 270 cases [1]. The operation did not receive the appropriate recognition due to the high surgical mortality and morbidity from urinary and bowel fistulae. Also in Vienna, Schauta [2] described the radical vaginal hysterectomy.

As radium was introduced for the treatment of cervical cancer, radical surgery (with its high attendant morbidity and mortality) was less frequently used.

At the Middlesex Hospital in London in the 1920s and 1930s, Bonney had an acceptable cure rate in his series of 483 radical hysterectomies using the Wertheim method, especially when one considers that he often dealt with more advanced stage disease than seen today.

The next major step in the development of the radical hysterectomy came from Boston where Meigs [3] advised the removal of all pelvic lymph nodes and thus described a more extensive operation than the original Wertheim's procedure. Meigs reported good survival rates, which were comparable to radiotherapy (75% 5-year survival rate for Stage I) but this was at the cost of a high postoperative morbidity – the fistula rate was 9%. The extended surgery was introduced before the scientific assessment of techniques and therefore, to date, we do not know the full value of lymphadenectomy in the treatment of cervical cancer.

Appropriate anaesthesia, fluid management, blood transfusion, antibiotic and thrombosis prophylaxis and bladder drainage have reduced the operative morbidity and mortality of radical surgery significantly. The ureteric complication rate, which was more than 10%, has fallen to approximately 1–2% [4–9].

In the USA and the UK, there have been few proponents of the Schauta radical vaginal

Cancer and Pre-cancer of the Cervix
Edited by D.M. Luesley and R. Barrasso. Published in 1998 by Chapman & Hall, London. ISBN 0 412 56600 1.

hysterectomy because of the need to carry out a lymphadenectomy through separate incisions. However, the reduced morbidity and the successful local treatment with comparable survival rates have contributed to the renewed interest in this technique [10].

NEW ADVANCES WHICH MAY CHANGE THE ROLE OF SURGERY IN CARCINOMA OF THE CERVIX

The review of less extensive radical procedures and the use of laparoscopy may be important. Operative laparoscopy may assist radical vaginal surgery both by performing the lymphadenectomy and preparing the vascular pedicles and upper part of the parametrium. The place of neo-adjuvant chemotherapy is being evaluated prior to surgery and may be helpful to down-stage patients [11, 12] – although other reports show no benefit [13, 14]. Molecular biology and other pathological studies may identify groups who would benefit from surgery [15]. These advances remain to be assessed in good clinical trials.

STAGING

The woman should undergo an examination under anaesthetic (EUA) to establish the correct International Federation of Gynecology and Obstetrics (FIGO) stage, which allows the surgeon to make a decision whether the lesion is suitable for operative treatment. If there is any doubt about the likely operability of the lesion, a radiotherapist (clinical oncologist) should be involved in the woman's management. FIGO staging is a clinical method rather than a surgical–pathological technique. There is an overstaging of up to 40% [7]. At present, the authors feel that Stages Ib, IIa and some early IIb tumours are appropriate for surgical management if the woman is fit. Age and obesity are not contraindications. There is no difference in outcome in women over 65 years of age compared to the younger age group if they are fit (American Society of

Anaesthiologists Physical Score I to III) [16]. Obesity is associated with an increased blood loss and longer operating time but otherwise the outcome is similar to operations on slimmer women [17, 18]. At the EUA, cystoscopy is usually performed. If the cystoscopy is normal, then urine cytology is not required. Examination of the cervix will reveal the tumour and a biopsy is taken if previous histology is not available. It is important to have an histological diagnosis before proceeding with radical treatment. Hysteroscopy is not normally performed through a frank carcinoma. If it is considered that perforation of the tumour or the uterus may occur, then it is best to abandon the curettage. Assessment of the uterine cavity and endocervical canal is necessary if a conservative procedure is considered. Should perforation occur during the staging procedure, urgent treatment with either radical surgery or radiotherapy should be given. As there are anecdotal reports of these women having a worse outcome, the definitive treatment should be started within 48 hours.

Sigmoidoscopy is favoured by the senior author, although a digital examination of the rectum is sufficient if the tumour is mobile and felt to be operable. Bimanual examination with the index finger vaginally and middle finger rectally is used to assess the uterosacral and lateral ligaments and the mobility of the cervix.

PREPARATION FOR RADICAL SURGERY

Staging requires an intravenous urogram and chest X-ray in assessing women for carcinoma of the cervix. This is the minimum data set of radiological imaging required by FIGO. These investigations should be available in all departments, even those without access to more expensive imaging. In our department, we now perform magnetic resonance imaging of the pelvis including the cervix because this gives the most information about the tumour, parametrial spread and lymphadenopathy

[19]. Until recently a CT scan of the abdomen and pelvis was performed. If the scan showed obvious malignant lymph nodes, a fine needle aspiration under radiological control would be performed, because if distant metastasis is confirmed, radical surgery is unsuitable. Prior to surgery, some form of bowel preparation is useful as an empty colon facilitates operating in the pelvis and, in the rare event of a bowel complication, allows primary closure. In addition, the woman is more comfortable in the immediate postoperative period without gross faecal loading. Our present regime is to use two sachets of sodium picosulphate (Picolax [Nordic]) 1 day apart with clear fluids on the 2 days prior to surgery. At both the Royal Marsden Hospital and St Bartholomew's Hospital, a trained nurse counsellor is an integral member of the gynaecological oncology team and will meet the woman in outpatients as well as on the ward, at her staging procedure and prior to her admission for radical surgery. This counsellor is able to provide a valuable supportive role prior to surgery and subsequent in her care and management [20].

Blood is cross-matched as a routine for radical hysterectomy. Other tests are as per the woman's medical condition. However, it is usual that a woman undergoing radical surgery would be generally fit.

As all these women are at high risk for a thrombo-embolic complication, prophylaxis is given. On admission, they have knee-length anti-thrombotic stockings fitted, which are then worn until their discharge. During surgery, mechanical compression of the calf or foot is used. Subcutaneous heparin is used twice daily. This is continued until they are fully mobile. Prophylactic antibiotics are given with induction. Cefuroxime and metronidazole are used at present.

Following general anaesthesia with or without regional block, the woman is placed supine on the operating table. A midline incision is used as this allows easy access to the upper abdomen for the para-aortic dissection. On occasions, a transverse incision can be used, especially if no para-aortic dissection is intended. Following a thorough laparotomy, a large ring retractor is then used. The wound edges are protected with two large packs. The bowel is packed away in the upper abdomen. This is done using three large moist packs, with the aim to keep the abdominal contents out of the pelvis in a bag arrangement. The first is inserted unfolded over the abdominal contents and the remaining two are folded into quarters and placed in the paracolic gutters.

LYMPHADENECTOMY

Meigs added the systematic and complete bilateral pelvic lymphadenectomy to the Wertheim's radical hysterectomy as one of his modifications. It is not clear whether the value of the lymphadenectomy is both diagnostic and therapeutic, but the pelvic lymphadenectomy is now performed as part of the treatment of early carcinoma of the cervix. The range of positive lymph node metastases to the pelvic lymph nodes in Stage Ib disease ranges from 9% [21] to 31% [22]. A collection of published reports covering more than 1000 patients showed that 17% of women with Stage Ib had positive lymph nodes [23].

A para-aortic lymphadenectomy is part of the procedure in large tumours as approximately 7.5% of Stage Ib carcinomas have metastatic disease in the para-aortic nodes [24]. It is felt that there is continuous spread from the cervical lesion out to the pelvic side walls and then via pelvic lymphatic chains onto the para-aortic region. There have been isolated case reports of positive para-aortic nodes with negative pelvic nodes [25, 26]. We perform a frozen section on the para-aortic node sampling and would consider abandoning the radical hysterectomy if it is positive. In our view, the woman who has positive para-aortic nodes will require postoperative therapy and, at present, we would offer chemo-radiotherapy. Intra-operative samples from the pelvic lymphadenectomy are not sent for frozen section because it will not alter our management of the radical case at that stage. Women who have significant positive nodes after radical surgery would be offered radiotherapy. The urinary fistula and bowel obstruction rate in those women with a large tumour removed by radical surgery followed by treatment with adjuvant radiotherapy is 14.2% [27]. Postoperative radiotherapy may control pelvic recurrence [28] but does not improve survival [29].

The technique employed in this department for para-aortic lymphadenectomy is similar to that described by Oram and Bridges [30]. The abdomen is opened in an extended midline incision. After a thorough laparotomy, the small bowel is packed away in the upper abdomen. A broad bladed retractor is placed with its tip under the third part of the duodenum to aid exposure. The para-aortic region is palpated and two stay sutures are placed in the retroperitoneum, overlying the aorta about 5 cm above the bifurcation. In cervical cancer, the para-aortic lymphadenectomy is performed in the area below the inferior mesenteric artery and exposure of the renal vessels is not necessary. A midline incision is made between the stay sutures and the loose areolar tissue is divided. The landmarks are clearly identified (the aorta, the inferior vena cava and the ureters). The fatty nodal tissue is picked up with tissue forceps and removed using a mixture of sharp and blunt dissection with diathermy and titanium clips (Liga) for haemostasis. Care is taken to avoid damaging the sympathetic chain and the inferior mesenteric artery. Some authors [24] favour a para-aortic lymphadenectomy up to the renal vessels and they suggest ligating and dividing the inferior mesenteric artery to avoid accidental damage to this vessel. The open retroperitoneal dissection often leads to a significant ileus, which is not seen using operative laparoscopy.

The pelvic lymphadenectomy is then performed in a standard way. The round ligament is divided between the stay suture on the lateral side and the clamp on the medial side. This is approximately one-third of the distance from the pelvic side wall. The areolar tissue is separated and the great vessels are identified. The paravesical space is defined anteriorly and the pararectal space posteriorly. At this time, the ureter is clearly seen. If an oophorectomy is to be carried out, then the infundibulo-pelvic ligament would be divided and double ligated. Care must be taken that the ureter is not included in this bundle. The ureter is dissected from the medial aspect of the peritoneum and can be supported with a sling of soft rubber. The pelvic lymphadenectomy can be aided by tilting the table approximately 15° away from the operator. The procedure is started by removing the fatty lymphatic tissue from the adventitia of the external iliac vessels, having divided the chain at its distal margin, the superficial circumflex iliac vein. The dissection is performed using both sharp and blunt techniques. Both diathermy and haemostatic clips are used for haemo- and lymphostasis. Care is taken to dissect behind the external iliac artery and between the external iliac artery and vein. During the dissection of the external iliac group of lymph nodes, the genito-femoral nerve is clearly identified. It would only be sacrificed if it was involved in obviously malignant nodes. The retroperitoneal space is opened further cranially and the lymphatic tissue overlying the common iliac artery is then also removed. Care must be taken that no damage occurs to the ureter at this point. The external iliac vein is carefully retracted using a vein retractor and the obturator fossa is viewed. Using a mixture of palpation and direct vision, the obturator nerve is identified. The lymphatic tissue is separated from this important structure and then ligated at the distal end using haemostatic clips. Using Babcock tissue forceps, this cord of tissue is then peeled off the side wall of the pelvis, away from the obturator nerve and removed after further haemostatic clips have been applied. This comprises both the deep and superficial obturator group of nodes and is perhaps the most important lymph node group to be sampled on the side wall. Occasionally, an aberrant obturator vessel can give rise to substantial haemorrhage. The nodal tissue between the vessels (interiliac) is cleared. The internal iliac nodes are rarely removed separately and are included in the lymphatic tissue in the parametrium. Haemostasis is obtained using diathermy. The use of a hot pack can paradoxically reduce venous ooze. Damage to the great vessels is repaired immediately using fine (5–0) prolene sutures. In 1967, FIGO stated a radical lymphadenectomy must yield at least 20 nodes [26]. The use of lymph node count as a marker of completeness of surgery is fraught with problems. Obviously, the more tissue that is removed the more lymph nodes will be recovered, but the total number will also depend on the woman's anatomy and the pathological processing. The use of intra-operative nuclear scanning to ensure completeness of the lymphadenectomy is not asscociated with a better survival outcome.

The use of frozen section in assessing lymph node status was reviewed by Bjornsson *et al.* [31]. The sensitivity of frozen section as compared to paraffin section was 68% with a specificity of 100%. The only metastases missed by frozen section were less than 2 mm in size, although all these nodes were regarded as suspicious on palpation by a pathologist. The use of lymphangiography to indicate pelvic lymph node involvement is not entirely reliable [32]. In small tumours (Stage Ib1), we do not perform a para-aortic lymphadenectomy unless there is a specific indication such as poor prognosis histology. This practice is supported by the evidence from Patsner *et al.* [33], where in 125 cases of small tumours (less than 3 cm) there were only two positive para-aortic nodes. Both of these cases had grossly positive pelvic nodes.

The use of peritoneal cytology is well known in the staging of ovarian cancer and has recently been introduced into the staging of endometrial cancer. In the surgical management of early cervical cancer, peritoneal washings are of no value [34].

Following pelvic lymphadenectomy at the completion of the surgery, it is the practice of the senior author to use a closed suction drain which is left *in situ* for 7–10 days or until there have been two consecutive days with less than 20 ml drainage. Although the aim of these drains is to reduce lymphocyst formation, prospective trials have shown that drainage of the pelvic side walls has no effect on pelvic morbidity and lymphocyst formation [35, 36].

It is a matter of personal preference as to whether the pelvic lymphadenectomy is performed before or after the radical hysterectomy.

OOPHORECTOMY OR OVARIAN CONSERVATION?

One of the advantages of radical surgery for treatment of carcinoma of the cervix is that the ovaries may be retained. There is no such option when using radical radiotherapy. Ovarian metastases have been reported in about 0.5% of cases of squamous carcinoma of the cervix [37]. There is a bias in reporting, as studies include those woman with ovarian metastases discovered at autopsy (i.e. who died of the cervical cancer), women who had their ovaries removed at the time of radical surgery and then the group of women presenting with symptomatic ovarian metastases at a date after their initial treatment. Young *et al.* [38] suggest that ovarian metastases from cervical primaries of adenocarcinoma and histological types other than squamous cell are more common than from squamous cell carcinoma. Ovarian conservation should be attempted in a young woman with a good prognostic tumour. We would electively perform bilateral oophorectomy or consider oophoropexy in a woman undergoing surgery who had a poor prognosis tumour where we anticipated a probability of pelvic radiotherapy. Over the age of 45, we would suggest that the woman had bilateral oophorectomy at the same time as the hysterectomy.

Lateral ovarian transposition to move the ovaries out of a possible radiation field is fraught with its own problems. Oophoropexy can be performed both at open surgery and laparoscopically. The ovary is marked with metal clips (two parallel clips applied adjacent to the lower pole is an easy method) to identify its position for both the radiotherapist and in case of subsequent problems. Chambers *et al.* [39] reported that there was significant cyst formation in 24% of women with ovarian transposition. Anderson *et al.* [40] showed that in a group that had ovarian transposition and radiation only a third had residual functional ovarian tissue. This was compared with 65% who had transposition and no radiation and 90% whose ovaries were left unmoved. In the group with transposition of the ovaries who received no radiation, 17.6% had subsequent oophorectomy for management of painful ovarian cysts. Although we suggest and encourage the use of hormone replacement therapy for women who have a bilateral oophorectomy, the problems of compliance can be considerable. Parker *et al.* [41] report only a 71% compliance with HRT following bilateral salpingo-oophorectomy, with compliance higher (93%) in the younger age group (<40 years). Not surprisingly, there is a difference in psychosexual problems in those women who have had oophorectomy. In the women who retained their ovaries, 89% reported either improved or no change in their sexual relations prior to surgery, compared to 68% of the women who had both ovaries removed.

THE RADICAL ABDOMINAL HYSTERECTOMY

Following the lymphadenectomy, the remaining part of the operation is now to remove the cervix with the appropriate parametrium and cuff of vagina. It is this radical dissection that is the curative management of the early stage cancer of the cervix.

The uterovesical fold of peritoneum is incised and the bladder is reflected down-wards with blunt and sharp dissection. The ureteric dissection is performed as follows. Using a Lahey right angle clamp, the ureteric tunnel is opened and the uterine vessels are divided laterally at their origin from the internal iliac vessels. As the ureteric tunnel is opened further, care must be taken not to damage the bladder. At this point, the bladder pillar is developed and divided, leaving the bladder wall intact. The ureter is then seen in its deroofed tunnel all the way down to its insertion in the trigone. It may then be rolled laterally with a mixture of blunt and sharp dissection to enable its clearance from the parametrium. Once the ureter has been clearly defined, the peritoneum in the Pouch of Douglas is incised. This allows access to the rectovaginal septum. This plane is developed with a sweeping movement of the hand. This leaves the uterosacral ligaments prominent between the pre-rectal space and the para-rectal space. These are divided usually in two parts (the superficial and deep uterosacral ligaments) and ligated. Care must be taken that the rectum is not included in these ligatures. With the ureter retracted laterally, the lateral ligament is then divided close to the pelvic side wall. We favour a technique using clamps such as Roberts or Moynihan type and then stitching the pedicle. Other authors [42] favour a cut and clip process to this area as they feel this increases the radicality. However, we find that the use of haemostatic clips on the pelvic side wall can be dislodged easily or snag the operator's hand.

The final part of the radical hysterectomy relies on ensuring an adequate cuff of vagina distal to the tumour. Posteriorly, this adequacy has been achieved by dividing the uterosacral ligaments flush with the rectum and dividing the cardinal and laterally ligaments lateral on the side wall down to the pelvic floor. Anteriorly, the bladder is reflected further with a mixture of sharp and blunt dissection. The tube of vagina may then be transected at the appropriate level. To ensure a sufficient length of this cuff is removed, a curved clamp can be placed each side on the vagina just below the cervical tumour. A further pair of curved clamps can then be placed 3 cm distal so that an adequate cuff is taken. Heavy curved scissors are used to transect the vagina. Various sets of custom-fitting instruments are available where the designs incorporate long handles for pelvic work and curved heads with an adequate clamp for the tissue of the parametrium and vagina. The vaginal angles are then secured both at the heel and the toe of the clamp and tied. The intervening space between the two angles is closed with 2–3 sutures depending on the width of the vagina. The specimen is then sent for histology. The central suture on the vaginal vault is left long to aid with traction during the assessment of the pelvis for haemostasis. The area in the pelvis is not reperitonealized and, therefore, care is taken to ensure that the sigmoid colon lies deep in the pelvis covering the ureters with the small bowel above it. The authors use polyglactin sutures and ties throughout the surgery. The radical abdominal hysterectomy outlined is between a Rutledge class 2 and 3 radical hysterectomy [43]. The surgery achieves the balance between radicality to achieve cure and an acceptable level of morbidity.

CLOSURE AND POST-OPERATIVE CARE

A mass closure technique using looped Polydioxonone (PDS) is performed. This suture is monofilament and is absorbable. In the long midline incision, it is important to ensure that the appropriate suture is used and that the suture is placed 1 cm from the edge of the rectus sheath and the sutures are 1 cm apart. The suture material should be three times the wound length to ensure a safe closure. Attention should be paid particularly at the lower end, where skimping on suture material and poor attention to technique can lead to incisional hernias. Skin closure is of cosmetic significance. The use of disposable staples can reduce time at the end of a long operation.

Ideally the woman should be nursed on a high dependency unit for 24 hours following the surgery. During this time, it will become apparent whether an ileus develops due to the para-aortic node and other retroperitoneal dissection. Intensive physiotherapy is given to prevent chest infection and early mobilization is encouraged.

Post-operative care is routine, as for all major surgery. Pain relief may be given as an opiate infusion either via an epidural catheter or as a patient-controlled analgesia system (PCAS). Rectal administration of non-steroidal anti-inflammatory analgesia is very useful as it provides effective, long-lasting analgesia without drowsiness. The bladder is drained for 5–10 days. Prior to the removal of the catheter, the woman is seen again by the physiotherapist and together with written information, the importance of pelvic floor exercises are explained. These help with regaining control of continence as well as ensuring complete voiding. All women are asked to measure their urine output following catheter removal and a residual volume ideally measured by ultrasound is performed 6–8 hours after removal of the catheter. If this residual volume is less then 150 ml, then it is considered that the woman is voiding sufficiently well to avoid a further catheter. If an ultrasound facility is not available to assess pre- and post-voiding, then a catheter is inserted for 20 minutes. If the residual is greater than 150 ml, the catheter is left in for a further 48 hours. The skin clips are removed at 8 days. The woman is seen by the nurse counsellor postoperatively and is then reviewed on a regular basis in out-patients. Psychological support is especially important around the time of the pathology results, as this will determine whether further therapy is required.

COMPLICATIONS OF RADICAL SURGERY

The notable complications of radical surgery for cervical cancer are haemorrhage and sepsis. Haemorrhage may occur during the pelvic lymphadenectomy, during mobilization of the bladder and division of the lateral ligaments. Damage to the major vessels is repaired directly. In a woman with previous pelvic irradiation, we do not routinely perform a lymphadenectomy unless there are palpable nodes, which are then biopsied. This selective lymph node sampling in the post-radiation case reduces the risk of haemorrhage. Prolonged operating time in a difficult case leads to increased postoperative complications. These may range from atelectasis from the prolonged anaesthetic to pneumonia or even adult respiratory distress syndrome. This syndrome may occur especially with pre-existing chest disease and major intra-operative haemorrhage. The likelihood of this syndrome occurring can be reduced by maintaining the patient's core temperature by using warmed intravenous fluids, a warming mattress and accurate blood replacement. The use of pneumatic calf compression as well as anti-embolus stockings reduces the effect of vasculostasis in the lower limbs. This, coupled with the use of prophylactic heparin, will reduce the incidence of deep vein thrombosis, a serious complication seen in women who have prolonged pelvic surgery and malignancy. Nerve damage during the surgery may take several forms. The neurapraxia caused by dissection or traction may lead to a temporary weakness, which improves with time and appropriate physiotherapy. The nerves particularly at risk are the obturator nerve during the dissection in the obturator fossa and the femoral nerve compressed by a self-retaining retractor. Areas of anaesthesia and hyperaesthesia may occur from damage to the genito-femoral nerve and the lateral cutaneous nerve of the thigh. Although these latter two are a minor inconvenience for the woman, they can be a major source of irritation.

The bladder and ureters are considered the structures most at risk in the Wertheim's hysterectomy. The more radical the surgery, the more bladder dysfunction occurs with impaired bladder sensation, bacteruria, increased residual volumes, decreased flow rates and abnormal bladder compliance [44]. The instance of ureteric fistula is just over 1% [8, 45]. It is our usual practice to drain the bladder for 5–10 days with a urethral catheter. There seems to be little advantage in choosing a suprapubic catheter although there may be less associated bacteruria with this method. The use of bladder antiseptics depends on the individual's choice. Frequent cultures of the urine may be appropriate as the bladder is often a source for pyrexia. An attempt to reduce the damage to the parasympathetic nerve bundle in the lateral ligaments is described by Sakamoto and Takizawa [46]. This is not always so simple in practice at open surgery.

A pyrexia following radical surgery occurs in between 10–20% of cases. The aetiology of the pyrexia is similar to that following most major abdominal surgery. In the initial stages, pulmonary atelectasis with mucus plugging is common. Wound infection and haematoma are frequent causes of a low-grade temperature. Urinary tract infections are notable and should be considered due to the long-term bladder drainage. Deep vein thrombosis should always be remembered as a cause for a later rise in temperature. The presence of pelvic cellulitis or pelvic abscess will be noted by a swinging fever.

Longer term problems include lymphocysts. These occur in up to a third of women [47–49]. The development of a lymphocyst may be the first indication of recurrent disease. If the lymphocyst causes symptoms, then drainage may be required and if persistent, then marsupialization is indicated. After ovarian conservation, there is a risk of cyst formation and this may necessitate further surgery. Although hormone replacement therapy is prescribed after oophorectomy, compliance can be a problem. As mentioned earlier, psychosexual problems relating to the fears of cancer and the effects of the treatment are best dealt with by supportive counselling starting pre-operatively and being followed in the short and long term postoperatively. Reliance on expectant measures to solve psychosexual dysfunction is inappropriate and intervention with counsellors and/or psychological help is important.

SUPERFICIAL INVASIVE CARCINOMA OF THE CERVIX

It is now well accepted that conservative treatment of superficial invasive carcinoma of the cervix [50] is appropriate [51]. This could be done with a cone biopsy, which would remove the complete focus of a squamous carcinoma. If the woman is over 45 or does not wish for further children, then the authors feel that an extended hysterectomy (Rutledge type 1) [43] can be considered. Lymphadenectomy would only be performed in that group where there was a significant risk of nodal involvement (Stage IA2, lymphovascular space involvement, poor differentiation, adverse histology). In the USA the Society of Gynecologic Oncologists (SGO) criteria of microinvasion (3 mm depth of invasion with no lymphovascular space involvement) determines their level of conservatism. A recent technique for the treatment of invasive carcinoma of the cervix which requires radical therapy but maintains fertility is the radical trachelectomy with laparoscopic pelvic lymph node sampling [52]. However, this management remains to be verified as safe with long-term follow-up in other groups.

CARCINOMA OF THE CERVIX IN PREGNANCY

The incidence of invasive carcinoma complicating pregnancy is between 1 and 13 per 10,000 pregnancies [53, 54]. Surgical treatment of early invasive carcinoma is indicated. In the first and second trimesters, the Wertheim's hysterectomy can be performed with the foetus *in utero*. In the late second and early third trimester, the use of steroids will allow maturation of the foetal lungs. This delay is probably not detrimental and a classical Caesarean section is performed. The foetus is delivered prematurely and will require appropriate neonatal support. A lower segment Caesarean section is not suitable due to increased vascularity and close proximity of the incision to the tumour. An emergency Caesarean section is required if the woman goes into labour as a vaginal delivery can disseminate the carcinoma. The Wertheim's hysterectomy with pelvic lymphadenectomy is as previously described. The results of surgical treatment in pregnancy are comparable to the non-pregnant group [55]. The women who were treated in the puerperal period fared worse than the antenatal group and this was related to poor prognostic factors.

CARCINOMA OF THE CERVICAL STUMP

Carcinoma of the cervical stump is a rare entity and occurs late after the initial hysterectomy (average time 26 years [56]). This type of cervical cancer should be decreasing because of the reduced number of sub-total procedures and the use of cervical screening. Although the morbidity is high following radiotherapy [57] the morbidity following radical surgery is higher [58]. The surgical technique is similar to that required for completion of radical surgery following simple hysterectomy. The anatomy is often considerably distorted by the previous surgery.

FOLLOWING A 'CUT-THROUGH' HYSTERECTOMY

The unexpected finding of cervical carcinoma in a hysterectomy specimen can be easily avoided if the pre-operative assessment is appropriate. In 69% of cases reported by Roman [59] there was inadequate assessment of an abnormal smear and in the remaining cases with normal or inflammatory smears, 93% had abnormal bleeding which had not been investigated. The 5-year survival rates from surgical treatment with completion of radical surgery are better (89% [60], 82% [61]) than compared to radiotherapy (no residual disease: 81% [62];

75% [63]; with residual disease: 56%, 39%, respectively). The operation consists of a pelvic lymphadenectomy and radical parametrectomy with upper colpectomy. Damage to the bladder and rectum is more common. The fistula rate is raised to 7% [61].

RADICAL HYSTERECTOMY IN RECURRENT DISEASE

The place of the radical hysterectomy in recurrent disease is not entirely clear. Women who have a recurrence of carcinoma following radiotherapy need to be carefully assessed. If there is no sign of distal spread, then it may be appropriate to perform an examination under anaesthetic to assess the feasibility of surgical removal of the central recurrence. In the majority of these women, the procedure of choice will be some form of exenteration. In our series of exenterations [64] the number of women undergoing a salvage radical hysterectomy rather than exenteration was too small for statistical comparison with the whole group. Terada and Morley [65] reported a 5-year survival rate of only 27% in their group treated with radical hysterectomy, although the survival rate increased to 54% after women with micrometastatic disease were excluded. This would suggest that occasionally the radical hysterectomy, a less morbid procedure than exenteration, may achieve a similar survival rate to exenteration, although this has not been supported in the literature to date.

In those women who undergo exploratory laparotomy to consider salvage treatment for a solitary central recurrence, a thorough examination of the abdominal cavity is performed. Any suspicious nodules or lymph nodes are sent for frozen section and these results are awaited before proceeding with the definitive surgery. The role of intra-operative radiotherapy or the placement of guide tubes for after-loading techniques to an area of fixed malignancy or positive resection margin at surgery may have a role in the future. However, there are major technical difficulties associated with these procedures and only a few centres have the equipment and the expertise to offer this treatment [66]. For women in whom exenteration or a salvage procedure is not suitable, good palliative care and support is required for both the patient and her family.

LAPAROSCOPY AND RADICAL SURGERY

With the advent of advanced laparoscopic equipment that allows accurate manipulation and dissection of tissue and good haemostasis, the prospect of a staging lymphadenectomy performed using these techniques has to be reviewed. The laparoscopic technique can mirror the open operation and so lead to the use of the radical vaginal approach. Laparoscopic pelvic lymphadenectomy was described in 1991 [67]. The technique involves using a camera mounted on a 10 mm laparoscope, which is ideally placed through a sub-umbilical incision. Three or four ports are used for manipulation, diathermy and specimen retrieval. Two 5 mm portals are used in the iliac fossae and a 10–12 mm portals is used in the midline suprapubic area. This larger port allows the use of a ligaclip applicator or linear stapler/cutter and also allows retrieval of the tissue. As usual for operative laparoscopy, a thorough inspection of the abdomen is performed. If suitable, then the round ligament is elevated and divided with diathermy to enable access to the pelvic side wall. Once the major vessels on the pelvic side wall are exposed, the sheath of nodal tissue and fat can be stripped cleanly. Diathermy can be used to maintain haemostasis. Due to the magnifying effect of the laparoscope, a good view is obtained. The paravesical and pararectal spaces are developed with blunt dissection as in the open lymphadenectomy and the obturator fossa is cleared of nodal tissue. Attention is paid to preserving the obturator nerve and either liga-clips or diathermy is used if an aberrant obturator vessel is damaged. The samples of tissue can be removed through the 10 mm port.

Alternatively, a 'ceolio-extractor' (Lépine, Lyon, France) can be used to retrieve the tissue. Para-aortic lymph node sampling has also been described [68]. Laparoscopic lymphadenectomy allows sampling of the pelvic and para-aortic nodes. Fowler *et al.* [69] compared the yield from the laparoscopic approach to an open lymphadenectomy. In this study, between 62 and 97% of the total lymph nodes removed were sampled laparoscopically. The range of lymph nodes removed in total was from 11 to 46 at open operation, with between 7 and 33 being removed laparoscopically. However, the authors point out that there is a significant learning curve with this procedure and the percentage recovered laparoscopically in the second half of their series was significantly greater than in the first half (85% vs 63%, $p < 0.005$). Interestingly, no patient with negative nodes at laparoscopy had positive pelvic nodes found at laparotomy. Another group [70] reported a 91% yield in lymph nodes removed laparoscopically, compared to that at laparotomy. Our view is that a lymphadenectomy will demonstrate the absence of lymph node metastases and, therefore, allow a less morbid but equally radical vaginal procedure in the management of early cervical cancer. Downey *et al.* [71] suggested that the surgical clearance of macroscopically involved pelvic nodes in the 41% of their patients with Stage Ib–IIa cervical cancer with positive nodes had a survival benefit. This point is disputed by Potter *et al.* [72], as there was no survival benefit in their study when completion of the radical hysterectomy in the presence of metastatic lymph nodes with adjuvant radiotherapy was compared to abandoning the surgery leaving the uterus *in situ* for easier application of the postoperative radiotherapy. As the data using laparoscopic techniques are small case series, it is not known what effect the laparoscopic management has on the morbidity with subsequent surgery or other modalities of treatment. There is a learning curve for the laparoscopic procedure and the operating time can be protracted. The

length of procedure and the presence of insufflated gas may also have an effect on implantation of tumour at distant sites.

The laparoscopic approach has also been used to aid radical removal of the uterus and cervix vaginally. This has been described by Querleu [73] and Dargent and Mathevet [74]. Following the laparoscopic lymphadenectomy, Dargent and Mathevet divide the uterine artery at its origin using the linear stapler cutter. The cardinal ligament is then divided with a second application of the instrument. Following this procedure, the radical vaginal hysterectomy with an appropriate cuff of vagina is performed. The division of the uterine artery from above makes the mobilization of the ureter during this procedure considerably easier. However, as Dargent comments in his overview of laparoscopic surgery [75], the position of the laparoscopic assisted radical vaginal hysterectomy remains to be evaluated against either the Wertheim's hysterectomy or the radical Schauta–Amreich vaginal hysterectomy on its own. Laparoscopic techniques are discussed in detail in Chapter 21.

RADICAL VAGINAL HYSTERECTOMY

The vaginal route for operating has not had major support in the English-speaking world. However, with the apparent ease of laparoscopic assessment and removal of lymph nodes, the more radical local treatment of the cervix via the vaginal route may undergo a renaissance. The Schauta–Amreich radical vaginal hysterectomy has less severe complications than the Wertheim's Meigs hysterectomy, with the same survival rate in the node negative group [10].

The operation is started using a Schuchardt incision. This is a mediolateral episiotomy which cuts into the pubococcygeous and ileococcygeous muscle. Cutting diathermy reduces blood loss. The rectum should be protected during this incision and this can be done with the left index finger. The upper limit of the Schuchardt incision is taken to the level where the vaginal cuff will begin. A circumferential incision is made in the vagina and a flap of epithelium is raised distal to the cervix. Injection of the subcutaneous tissue with saline with or without adrenaline can help separate the tissue planes. With a mixture of blunt and sharp dissection, the vaginal cuff is developed. This may be stitched together covering the cervical tumour. The sutures can then be used for traction. The dissection is developed further with the bladder and rectum being carefully reflected in the midline. The cervicovesical ligament anteriorly is put under tension and divided. By retracting the vaginal cuff to one side and retracting the bladder anteriorly and to the same side, the paravesical fossa can be developed with a mixture of blunt and sharp dissection. A similar space is created for the pararectal fossa and the pelvic fascia between the two spaces is then divided after ligation. In the bladder pillar, which has now been developed, lies the genu (knee) or curve of the ureter as it loops under the uterine vessels. The ureter can usually be palpated easily in this pillar. Careful sharp dissection enables the ureter to be identified and exposed. Once the ureter has been clearly identified, the lateral portion of the bladder pillar can be ligated with a pedicle needle. The use of a pedicle needle facilitates the placement of sutures. The medial part of the bladder pillar contains the uterine vessels. The ureter is reflected upwards, allowing the uterine vessels to be ligated. This dissection is much easier when the uterine pedicle has already been divided laparoscopically. The cervix with its vaginal cuff is then elevated and the Pouch of Douglas is opened. The uterosacral ligaments are then divided laterally. The adnexa can be removed at this stage. Laparoscopy can aid adenexectomy by dividing the infundibulo-pelvic ligament from above. The cardinal ligaments are now the sole attachments of the uterus and can be divided laterally. The pedicles are inspected and care should be taken to ensure that neither vessels of the ureters are not damaged at this stage. The vaginal vault is closed and Schuchardt incision is then repaired in layers. The bladder is drained either urethrally or suprapubically. The discomfort from this operation is mainly due to the Schuchardt incision and so most women will mobilize well following this surgery. The bladder should be drained for between 5 and 10 days.

The complications of the Schauta radical vaginal hysterectomy are three times less frequent and less serious than those following a Wertheim's hysterectomy [75]. The fistulae seen are related to direct surgical injury and should be repaired at the time. The incidence of fistulae following vaginal surgery is much lower than after abdominal surgery.

REFERENCES

1. Wertheim, E. (1905) Discussion on the diagnosis and treatment of carcinoma of the uterus. *British Medical Journal*, **2**, 689.
2. Schauta, F. (1902) Die Operation des Gebarmutterkrebses Mittles des Schuchardt'schen Paravaginatschnittes. *Monatsschrift Geburtshilfe und Gynäkologie*, **15**, 133–52.
3. Meigs, J.V. (1944) Radical hysterectomy with bilateral pelvic node dissection – report on 100 cases operated on 5 years or more. *American Journal of Obstetrics and Gynecology*, **62**, 854–66.
4. Novac, F. (1963) Procedure for the reduction of the number of uterovaginal fistulae after Wertheim's operations. *Proceedings of the Royal Society of Medicine*, **56**, 881–4.
5. Webb, M.J. and Symmonds, R.E. (1979) (a) Wertheim's hysterectomy: a reappraisal. *Obstetrics and Gynecology*, **54**, 140–5. (b) Radical hysterectomy: influence of recent conisation on morbidity and complications. *Obstetrics and Gynecology*, **53**, 290–2.
6. Bennedet, J.L., Turko, N., Boyes, D.A., et al. (1980) Radical hysterectomy in the treatment of cervical cancer. *American Journal of Obstetrics and Gynecology*, **137**, 254–62.
7. Zander, J., Baltzer, J., Lohe, K.J., et al. (1981) Carcinoma of the cervix – an attempt to individualise treatment: results of a 20 year cooperative study. *American Journal of Obstetrics and Gynecology*, **139**, 752–9.
8. Monaghan, J.M., Ireland, D., Mor-Yosef, S., Pearson, S.E., Lopes, A. and Sinha, D.P. (1990) Role of centralisation of surgery in stage 1b carcinoma of the cervix: a review of 498 cases. *Gynecologic Oncology*, **37**, 206–9.

9. Averette, H.E., Nguyen, H.N., Donato, D.M., Penalver, M.A., Sevin, B.U., Estape, R. and Little, W.A. (1993) Radical hysterectomy for cervical cancer. A 25 year prospective experience with the Miami technique. *Cancer*, **71**, 142–437.

10. Massi, G., Savino, L. and Sozinni, T. (1993) Schauta–Amreich vaginal hysterectomy and Wertheim–Meigs abdominal hysterectomy in the treatment of cervical cancer: a retrospective study. *American Journal of Obstetrics and Gynecology*, **168**, 930–4.

11. Dottino, P.R., Plaxe, S.C., Beddoe, A.M., Johnston, C. and Cohen, C.J. (1991) Induction chemotherapy followed by radical surgery in cervical cancer. *Gynecologic Oncology*, **40**, 7–11.

12. Sardi, J., Giaroli, A., Sananes, C., Rueda, N.G., Vighi, S., Ferreira, M., Bastardas, M., Paniceres, G. and di Paola, G. (1993) Randomized trial with neoadjuvant chemotherapy in stage IIIB squamous carcinoma cervix uteri: an unexpected therapeutic management. *International Journal of Gynecological Cancer*, **6**, 85–93.

13. Zanetta, G., Landani, F., Colombo, A., Pellegrino, A., Maneo, A. and Leventis, C. (1993) Three year results after neoadjuvant chemotherapy, radical surgery and radiotherapy in locally advanced cervical carcinoma. *Obstetrics and Gynecology*, **82**, 447–50.

14. Tattersall, M.H., Lorvidhaya, V., Vootiprux, V., Cheirsilpa, A., Wong, F., Azhar, T., Lee, H.P., Kang, S.B., Manalo, A., Yen, M.S., *et al.* (1995) Randomized trial of epimbian and cisplatin chemotherapy followed by pelvic radiation in locally advanced cervical cancer. Cervical Cancer Study Group of the Asian Oceanian Clinical Oncology Association. *Journal of Clinical Oncology*, **13**, 444–51.

15. Jakobsen, A., Bichel, P., Ahrous, S., Nyland, M. and Knudsen, J. (1990) Is radical hysterectomy always necessary in early cervical cancer? *Gynecologic Oncology*, **39**, 80–1.

16. Fuchtner, C., Manetta, A., Walker, J.L., Emma, D., Berman, M. and DiSaia, P.J. (1992) Radical hysterectomy in the elderly patient: analysis of morbidity. *American Journal of Obstetrics and Gynecology*, **166**, 593–7.

17. Soisson, A.P., Soper, J.T., Berchuck, A., Dodge, R. and Clarke-Pearson, D. (1992) Radical hysterectomy in obese women. *Obstetrics and Gynaecology*, **80**, 940–3.

18. Levrant, S.G., Fruchter, R.G. and Maiman, M. (1992) Radical hysterectomy for cervical cancer: morbidity and survival in relation to weight and age. *Gynecologic Oncology*, **45**, 317–22.

19. Kim, S.H., Choi, B.I., Han, J.K., Kim, H.D., Lee, H.P., Kang, S.B., Lee, J.Y. and Han, M.C. (1993) Pre-operative staging of uterine cervical carcinomas: comparison of CTT and MRI in 99 patients. *Journal of Computerized Assisted Tomography*, **17**, 633–640.

20. Crowther, M.E., Corney, R.H. and Shepherd, J.H. (1994) Psychosexual implications of gynaecological cancer: talk about it. *British Medical Journal*, **308**, 869–70.

21. Artman, R.E., Hoskins, W.J., Bybrowe, M.C., *et al.* (1987) Radical lymphadenectomy and pelvic lymphadenectomy for stage 1b carcinoma of the cervix: 21 years experience. *Gynecologic Oncology*, **28**, 8–13.

22. Burghardt, E., Pickle, H., Haas, J., *et al.* (1987) Prognostic factors and operative treatment of stages 1b to 2b cervical cancer. *American Journal of Obstetrics and Gynecology*, **156**, 988–96.

23. Hoskins, W.J. (1988) Prognostic factors with a risk of recurrence in stages 1b and 2a cervical cancer. *Baillère's Clinical Obstetrics and Gynaecology*, **2**, 817–28.

24. Winter, R., Petru, E. and Haas, J. (1988) Pelvic and para-aortic lymphadenectomy in cervical carcinoma. *Ballière's Clinical Obstetrics and Gynaecology*, **2**, 857–66.

25. Chung, C.K., Nahahs, W.A., Stryker, J.A., *et al.* (1980) Analysis of factors contributing to treatment failures in stage 1b and 2a carcinoma of the cervix. *American Journal of Obstetrics and Gynecology*, **138**, 550–6.

26. Ferraris, G., Lanza, A., D'Addato, F., *et al.* (1988) Techniques of pelvic and para-aortic lymphadenectomy in the surgical treatment of cervix carcinoma. *European Journal of Gynaecological Oncology*, **9**, 83–6.

27. Bloss, J.D., Berman, M.L., Mukhererjee, J., Manetta, A., Emma, D., Ramsanghani, N.S. and Disaia, P.S. (1992) Bulky stage Ib cervical carcinoma managed by primary radical hysterectomy followed by tailored radiotherapy. *Gynecologic Oncology*, **47**, 21–7.

28. Remy, J.C., di Maio, T., Fruchter, R.G., Seolis, A., Boyce, J.G., Sohn, C.K. and Rothman, M. (1990) Adjunctive radiation after radical hysterectomy in stage Ib squamous cell carcinoma of the cervix. *Gynecologic Oncology*, **38**, 161–5.

29. Morrow, C.P. (1980) Panel report: Is pelvic radiation beneficial in the post-operative management of stage 1b squamous cell carcinoma of the cervix with pelvic node metastases treated by radical hysterectomy and pelvic lymphadenectomy? *Gynecologic Oncology*, **10**, 105–10.

30. Oram, D. and Bridges, J. (1987) Para-aortic lymphadenectomy. *Ballière's Clinical Obstetrics and Gynaecology*, **1**, 369–81.

31. Bjornsson, B.L., Nelson, B.E., Reale, F.R. and Rose, P.G. (1993) Accuracy of frozen section for lymph node metastasis in patients undergoing radical hysterectomy for carcinoma of the cervix. *Gynecologic Oncology*, **51**, 50–3.

32. Stellato, G., Tikkala, L., Makela, P. and Kajanoja, P. (1992) Pelvic lymph node metastases in cervical cancer: comparison of lymphangiography, inspection radiography and histologic examination of lymph nodes. *European Journal of Gynecological Oncology*, **13**, 161–6.

33. Patsner, B., Seolacek, T.V. and Lovecchio, J.L. (1992) Para-aortic node sampling in small (3 cm or less) stage Ib invasive cervical cancer. *Gynecologic Oncology*, **44**, 53–4.

34. Morris, P.C., Haugen, J., Anderson, B. and Buller, R. (1992) The significance of peritoneal cytology in stage Ib cervical cancer. *Obstetrics and Gynecology*, **80**, 196–8.

35. Lopes, A. de B., Hall, J.R. and Monaghan, J.M. (1995) Drainage following radical hysterectomy and pelvic lymphadenectomy: dogma or need? *Obstetrics and Gynecology*, **86**, 960–3.

36. Barton, D.P., Cavanagh, D., Roberts, W.S., Hoffman, M.S., Fiorica, J.V. and Finan, M.A. (1992) Radical hysterectomy for treatment of cervical cancer: a prospective study of two methods of closed-suction drainage. *American Journal of Obstetrics and Gynecology*, **166**, 533–7.

37. Toki, N., Tsukamoto, N., Kahn, T., Toh, N., Saito, T., Kaimura, T., Matsukuma, K. and Nakano, H. (1991) Microscopic ovarian metastasis of the uterine cervical cancer. *Gynaecologic Oncology*, **41**, 46–51.

38. Young, R.H., Gersell, D.J., Roth, L.M. and Scully, R.E. (1993) Ovarian metastases from cervical carcinomas other than pure adenocarcinomas. *Cancer*, **71**, 407–18.

39. Chambers, S.K., Chambers, J.T., Holm, C., Peschel, R.E. and Schwartz, P.E. (1990) Sequelae of lateral ovarian transposition in unirradiated cervical cancer patients. *Gynecological Oncology*, **39**, 155–9.

40. Anderson, B., La Polla, J., Turner, D., Chapman, G. and Buller, R. (1993) Ovarian transposition in cervical cancer. *Gynecologic Oncology*, **49**, 206–14.

41. Parker, M., Bosscher, J., Baumhill, D. and Park, R. (1993) Ovarian management during radical hysterectomy in the premenopausal patient. *Obstetrics and Gynecology*, **82**, 187–90.

42. Lichtenegger, W., Anderhuber, F. and Ralph, G. (1988) Operative anatomy and technique of radical parametrial resection in the surgical treatment of cervical cancer. *Ballière's Clinical Obstetrics and Gynaecology*, **2**, 841–56.

43. Piver, M.S., Rutledge, F. and Smith, J.P. (1974) Five classes of extended hysterectomy for women with cervical cancer. *Obstetrics and Gynaecology*, **44**, 265–72.

44. Ralph, G., Winter, R., Michelitsch, L. and Tamussino, K. (1991) Radicality of parametrial resection and dysfunction of the lower urinary tract after radical hysterectomy. *European Journal of Gynaecological Oncology*, **12**, 27–30.

45. Shepherd, J.H. and Crowther, M.E. (1986) Complications of gynaecological cancer surgery: a review. *Journal of the Royal Society of Medicine*, **79**, 89–93.

46. Sakamoto, S. and Takizawa, K. (1988) An improved radical hysterectomy with fewer urological complications, no loss of therapeutic results for invasive cervical cancer. *Ballière's Clinical Obstetrics and Gynaecology*, **2**, 953–63.

47. Todd, G.D., Rutledge, R. and Wallace, S. (1970) Post-operative pelvic lymphocysts. *American Journal of Roentgenology*, **108**, 312–23.

48. Kaser, O., Ikle, F.A. and Hiersch, H.A. (1973) Atlas der Gynakologachen Operationen Unter Berucksichitigung. *Gynakologisch-She-Urologische*, EINGRIFFE 3, Thieme, Stuttgart.

49. Fung, Y.M. and Wong, W.S. (1992) Malignant lymphocyst after Wertheim's operation. *Gynecologic Oncology*, **44**, 288–90.

50. Shepherd, J.H. (1996) Cervical and vulva cancer: changes in FIGO definitions of staging. *British Journal of Obstetrics and Gynaecologic*, **103**, 405–6.

51. Jones, W.B., Mercer, G.O., Lewis, J.L., Jr, Rubin, S.C. and Hoskins, W.J. (1993) Early invasive carcinoma of the cervix. *Gynecologic Oncology*, **51**, 26–32.

52. Dargent, D., Brun, J.L., Roy, M., Mathevet, P. and Remy, I. (1994) La Trachélectomie Élargie (TE). Une alternative à l'hystérectomie radicale dans le traitement des cancers infiltrants développés sur la face externe du col utérin. *JOBGYN* **2**, 285–92.

53. Kistner, R.W., Gorbach, A.L. and Smith, G.V. (1957) Cervical cancer in pregnancy: review of the literature and presentation of 30 additional cases. *Obstetrics and Gynecology*, **9**, 554–60.

54. Shingleton, H.M. and Ore, J.W. (1983) Cancer complicating pregnancy. In *Cancer of the Cervix; Diagnosis and Treatment* (eds H.M. Shingleton and J.W. Ore). Churchill Livingstone, Edinburgh pp. 105–25.

55. Monk, B.J. and Moritz, F.J. (1992) Invasive cervical cancer complicating intrauterine pregnancy: treatment with radical hysterectomy. *Obstetrics and Gynecology*, **80**, 199–203.

56. Kovalic, J.J., Grigsby, P.W., Perez, C.A. and Locken, M.A. (1991) Cervical stump carcinoma. *International Journal of Radiation Oncology, Biology and Physiology*, **20**, 933–8.

57. Petersen, L.K., Mamsen, A. and Jakobsen, A. (1992) Carcinoma of the cervical stump. *Gynecologic Oncology*, **46**, 199–202.

58. Barillot, I., Horiot, J.C., Cuisenier, J., *et al.* (1993) Carcinoma of the cervical stump: a review of 213 cases. *European Journal of Cancer*, **29(A)**, 1231–6.

59. Roman, L.D., Morris, M., Eifel, P.J., Burke, T.W., Gershenson, D.M. and Wharton, J.T. (1992) Reasons for inappropriate simple hysterectomy in the presence of invasive cancer of the cervix. *Obstetrics and Gynecology*, **79**, 485–9.

60. Chapman, J.A., Manuel, R.S., DiSaia, P.J., Walker, J.L. and Berman, M.C. (1992) Surgical treatment of unexpected invasive cervical cancer found at total hysterectomy. *Obstetrics and Gynecology*, **80**, 931–4.

61. Kinney, W.K., Egorshin, W.V., Ballard, D.J. and Podratz, K. (1992) Long term survival and sequelae after surgical management of invasive cervical carinoma diagnosed at the time of simple hysterectomy. *Gynecologic Oncology*, **44**, 24–7.

62. Fang, F.M., Yeh, C.Y., Lai, Y.L., Chion, J.F. and Chang, K.H. (1993) Radiotherapy following simple hysterectomy in patients with invasive carcinoma of the uterine cervix. *Journal of the Formosan Medical Association*, **92**, 420–5.

63. Roman, L.D., Morris, M., Mitchell, M.F., Eifel, P.J., Burke, T.W. and Atkinson, E.N. (1993) Prognostic factors for patients undergoing simple hysterectomy in the presence of invasive cancer of the cervix. *Gynecologic Oncology*, **50**, 179–84.

64. Shepherd, J.H., Ngan, H.Y.S., Neven, P., Fryatt, I., Woodhouse, C.R.J. and Hendry, W.F. (1994) Multivariate analysis of factors affecting survival in pelvic exenteration. *International Journal of Gynaecological Oncology*, **4**, 361–70.

65. Terada, K. and Morley, G.W. (1987) Radical hysterectomy as surgical salvage therapy for gynaecologic malignancy. *Obstetrics and Gynaecology*, **70**, 913–15.

66. Garton, G.R., Gunderson, L.L., Webb, M.J., Wilson, T.O., Martenson, J.A., Cha, S.S. and Podratz, K.C. (1993) Intraoperative radiation therapy in gynecologic cancer: the Mayo Clinic experience. *Gynecologic Oncology*, **48**, 328–32.

67. Querleu, D., LaBlanc, E. and Castelain, B. (1991) Laparoscopic pelvic lymphadenectomy in the staging of early carcinoma of the cervix. *American Journal of Obstetrics and Gynecology*, **164**, 579–81.

68. Querleu, D. (1993) Laparoscopic para-aortic node sampling Gynaecologic Oncology: preliminary experience. *Gynecologic Oncology*, **49**, 24–9.

69. Fowler, J.M., Carter, J.R., Carlson, J.W., Maslonkowski, R., Byers, L.J., Carson, L.F. and Twiggs, L.B. (1993) Lymph node yield from laparoscopic lymphadenectomy and cervical cancer: a comparative study. *Gynecologic Oncology*, **51**, 187–92.

70. Childers, J.M., Hatch, K. and Surwit, E.A. (1992) The role of laparoscopic lymphadenectomy in the management of cervical carcinoma. *Gynecologic Oncology*, **47**, 38–43.

71. Downey, G.O., Pottish, R.A., Adcock, L.L., Prem, K.A. and Twiggs, L.B. (1989) Pre-treatment surgical staging and cervical carcinoma: therapeutic efficacy of pelvic lymph node resections. *American Journal of Obstetrics and Gynecology*, **160**, 1055–61.

72. Potter, M.E., Abrovez, R.D., Shingleton, H.M., Soong, S.J. and Hatch, K.D. (1990) Early invasive cervical cancer with pelvic lymph node involvement: to complete or not to complete radical hysterectomy? *Gynecology Oncology*, **37**, 78–81.

73. Querleu, D. (1991) Hysterectomie de Schauta Amreich Schauta Stoeckel Assistees par Coelioscopie. *Journal de Gynaecologique Obstetrique et Biologie de Reproduction*, **20**, 747–8.

74. Dargent, D.E. and Mathevet, P. (1992) Hysterectomie Élarge Laparoscopico-Vaginale. *Journal de Gynaecologique Obstetrique et Biologie de Reproduction*, **21**, 709–10.

75. Dargent, D.E. (1993) Laparoscopic surgery in gynaecologic cancer. *Current Opinion in Obstetrics and Gynaecology*, **5**, 294–300.

D. Dargent

Minimal access surgery (MAS) has become quite fashionable in modern medicine and there are few areas where it has not had an effect. Regarding cervical cancer, MAS was initially considered some years ago when laparoscopic lymphadenectomy was introduced as an option for selected patients who may have been thought suitable for vaginal radical hysterectomy [1]. Since then, further uses for the laparoscope have been introduced, although not necessarily without controversy. The object of this chapter is to discuss such controversies, look at the techniques and try to arrive at a balance as to where the role of MAS lies in this common cancer.

THE PLACE OF SURGERY IN THE MANAGEMENT OF CERVICAL CANCER

This topic has been covered elsewhere in this text, both in relation to early cervical disease and also the more palliative roles that surgery may have in patients presenting with advanced or recurrent disease. In summary, the role of the surgeon has been to try and improve the radicality and therefore improve the cure, whilst at the same time minimizing morbidity. The development of surgery was somewhat affected by the introduction of effective radiation techniques and even now the advantages and disadvantages of treating this disease with either radiotherapy or surgery continue to be debated.

With little to choose between in terms of efficacy or cure rate, there has been an increasing focus on the morbidity associated with treatment. It is often forgotten that radical hysterectomy, particularly those more radical types, are associated with not insignificant morbidities, particularly those affecting the bladder but also, and to a less well-known extent, sexual function. The current consensus is that radical surgery is best reserved for the smaller cervical lesions, particularly Stage Ib_1 of the most recent FIGO classification.

THE CONCEPT OF STAGING SURGERY

According to International Federation of Gynaecology and Obstetrics (FIGO) guidelines cervical cancer is still staged clinically, although it is well accepted that lymph node status is one of the most important prognostic factors determining not only outcome, but also further therapy after surgery has been performed. It is not surprising, therefore, that there has been a need and a desire to assess the status of lymph nodes prior to undertaking definitive therapy. The insensitivity and unpredictability of imaging techniques over the last 20 years led to the concept of staging surgery being introduced by Nelson *et al.* in

Cancer and Pre-cancer of the Cervix
Edited by D.M. Luesley and R. Barrasso. Published in 1998 by Chapman & Hall, London. ISBN 0 412 56600 1.

the USA [2]. Initially the intent was to identify those in whom extended field radiotherapy should be scheduled as part of primary radiation treatment and not to identify those who might be suitable for a surgical approach. However, it was only a short conceptual step towards developing staging surgery as it is now known, i.e. selecting those women with cervical disease who have no lymph node involvement.

It might be expected that this procedure would be associated with improved results, and indeed for patients selected for radical hysterectomy the rate of survival was higher than usual, being 89% for Stage I disease and 59% for Stage IIa disease in the series published by Averette *et al.* [3]. However, if the whole population among whom the true Stage I and IIa cases were selected is examined, the overall rate of survival did not improve, being 70% for Stage I and 56% for Stage II, thus the apparent improvement was a pure selection bias. When the advanced cases were examined no improvement in survival was observed at all, the rate being 64.5% and 57.1% for surgically staged IIb and III vs 62.8% and 60% for the non-surgically staged patients in Nelson's series [4]. One might conclude that staging surgery does not really improve survival in early or advanced cases.

If there are few or any advantages associated with staging surgery, could there be disadvantages? One might assume that previous surgery should increase the complications associated with subsequent treatment because of adhesion formation. In Nelson's series [4] the rate of recto-vaginal fistula was 4.5% in the surgically staged patients versus 2.1% in the non-surgically staged patients. The use of a retroperitoneal approach for surgical staging might significantly reduce the risk of complications and this was reported as being 3.9% versus 11.5% in the Gynecological Oncology Group prospective and randomized study [5]. Even this, however, is a fairly high rate and it was because of this and its lack of impact on survival that surgical staging became gradual-

ly less fashionable and was virtually discontinued by the end of the 1980s.

LAPAROSCOPIC SURGERY IN STAGING CERVICAL CANCER

It is perfectly feasible to perform an adequate pelvic lymphadenectomy using the laparoscope. In addition to the laparoscope itself, the additional instruments required include grasping and dissecting forceps, washing and aspirating cannulae and extraction forceps. All of these instruments are introduced using either 5 mm trocars or 10–12 mm trocars with the appropriate reduction devices. The placement of the instruments varies according to the targeted area, whether this be pelvic or aortic, and the chosen approach, whether this is going to be transperitoneal or retroperitoneal (anterior retroperitoneal approach).

THE RETROPERITONEAL APPROACH

The retroperitoneal approach was the one we initially used and still electively use 10 years after its introduction. There are some differences depending upon the targeted areas but the general rules are the same regardless and these are summarized as follows.

The approach is made through a microcutaneous incision and the successive layers of the abdominal wall are crossed up to the level of the fascia parietalis, which must be preserved as much as possible. The space between this and the abdominal wall is identified and developed in two stages. The first is by blunt and blind dissection, the second is by sharp dissection under direct vision.

PELVIC DISSECTION

If a pelvic dissection is being performed, this may be done through a suprapubic or subumbilical micro-incision. The suprapubic approach was the first one we experimented with and this was introduced early in 1996. We continue to use this in specific, although uncommon, circumstances, such as cervical carcinoma detected during pregnancy. The subumbilical route, however, seems to suit our practice better and in our opinion is more appropriate for the pelvic dissection. The development of the retroperitoneal space can be achieved directly using the cutting and transparent trocar (Optiview; Visipor). The separate layers of the abdominal wall are crossed successively and the blind dissection ceases as soon as the fascia parietalis is reached. At this point the cutting trocar is removed. A carbon dioxide tube is attached to the sheath of the trocar and the laparoscope are introduced. The retroperitoneal space can then be developed by direct mechanical pressure of the endoscope. The direction of the dissection proceeds obliquely until Cooper's ligament is reached (suprapubic incision) or proceeds vertically to achieve contact with the symphisis pubis (infra-umbilical incision). On slight withdrawal of the laparoscope one then obtains a view that appears very much like a tunnel filled by cobwebs. A further trocar is then introduced to the distal extremity of this tunnel and initially a dissecting instrument is introduced (monopolar scissors). By the use of the dissecting scissors the posterior aspect of the abdominal wall can then be prepared under direct vision, this will then allow the introduction of the other ancillary instruments.

When a suprapubic approach is used the first ancillary port is introduced laterally at the level of the pubic spine. The dissecting instrument is introduced through this port allowing the posterior aspect of the abdominal wall to be prepared and a second port is opened approximately 6 cm cranially to the first one, both of these ports are placed medial to the epigastric vessels. Having introduced a second dissecting instrument into the second port, the dissection can proceed. After the identification of Cooper's ligament one approaches the external iliac vein and this is followed in a dorsal direction. The round ligament may cause some resistance during the mobilization of the created peritoneal bag, but this resistance can usually be overcome. After identification of the round ligament, the superior vesical artery is identified and by following this one reaches the level of the common iliac bifurcation. When an infra-umbilical approach is used, the first ancillary port is placed in the midline in the suprapubic area. The dissecting instrument introduced through this port enables one to dissect clear to Cooper's ligament and then to identify the external iliac vessels and their epigastric collaterals. By passing behind the vessels one can proceed along the posterior aspect of the abdominal wall up to the level of McBurney's point. A second ancillary port is introduced at this level. The procedure is repeated on the opposite side. In order to allow wide mobilization of the created bag of peritoneum, we recommend cutting the round ligament in its pre-peritoneal course. By doing this, one can detach the peritoneum up to the level of the bifurcation of the common iliac vessels. Should one wish to extend the dissection dorsally, it is necessary to cut the insertion of the fascia parietalis at the level of the iliac spine (Douglas line). This can be achieved by using the contra-lateral iliac port for introducing the laparoscope and the vacated port can then be used to introduce a working instrument. The view is similar to that obtained in the supra-pubic approach but one can extend the dissection further along the iliac vessels.

AORTIC DISSECTION

Dissection of the para-aortic lymph nodes can be scheduled as a continuation of the pelvic dissection or as an independent procedure. In the former, one proceeds as described above and in order to extend the dissection in a cranial direction one opens further ancillary ports as the peritoneum is continuously detached.

When an aortic dissection is being performed as an independent procedure the main port has to be placed in McBurney's area. It can be opened directly using the cutting and transparent trocar but the progression through the successive layers of the abdominal wall can be difficult, particularly in the appropriate identification of the retroperitoneal space at the outset. The major risk in this part of the operation is accidental peritoneal laceration. We would normally prefer to start with a transumbilical, transperitoneal laparoscopy. This enables us to make an assessment of the peritoneal cavity, a step of the staging procedure which is an important part of the indication for aortic dissection (see later). This transperitoneal approach also allows the progression of the retroperitoneal dissection to be directly visualized from within the peritoneum and digital preparation of the retroperitoneal space can be carried out again under direct vision, thus minimizing the risk of trauma and lacerating the peritoneum accidentally. Once the retroperitoneal space has been developed it is progressed in a dorso-lateral direction. After preparation of the space a self-retaining pneumostatic trocar is introduced (Blunt-Tip Origin). The retroperitoneal space is filled with carbon dioxide through the trocar, whilst at the same time the peritoneal cavity is deflated and the carbon dioxide removed. Two ancillary ports are introduced into the previously created retroperitoneal space in the same vertical line as the primary trocar. The dissecting instruments introduced through these ports enable continued detachment of the peritoneum in a medial direction and identification of the psoas muscle. The ureter and ovarian vessels are left attached to the peritoneum and mobilized with it in a cranial direction up to the level of the renal vein.

Two mini-laparotomies (one on each side) are essential for dissecting the lateral aspects of the aorta. Experience has taught us that it is perfectly feasible to assess the right aspect of the aorta whilst approaching it from the left-hand side. One does, however, have to cut and ligate the two fourth lumbar arteries in order to open the space between the aorta and vena cava and to clear enough tissue from the right aspect of the aorta. It is also possible, whilst ligating the fourth left lumbar vein, to proceed along the dorsal aspect of the vena cava and enter the space on the right of the vena cava. These procedures should allow an adequate aortic lymphadenectomy.

THE TRANSPERITONEAL APPROACH

Most gynaecologists will be more familiar with the transumbilical, transperitoneal approach to laparoscopy. The transperitoneal approach has another advantage in that it enables one to assess at the same time the intra- and extra-peritoneal aspects of the disease in a much more panoramic way, and whilst various ancillary ports have to be placed, the main port remains the same. Whilst this might seem to be the approach favoured by the majority of endoscopic gynaecologists, the postoperative peritoneal adhesions, although less dense and extended than after laparotomy, are certainly felt to be more so than after extra-peritoneal laparoscopy. This has been demonstrated elegantly by Denis Querleu on a porcine model where 3 weeks after aortic dissection the adhesion score was 2 versus 0.5. As a result of this observation we would favour the retroperitoneal approach, except in cases where this approach may pose additional hazards, such as previous laparotomy.

PELVIC LYMPH NODE DISSECTION TECHNIQUE

Pelvic node dissection following the transumbilical laparoscopic approach is commenced once the ancillary ports have been opened, one in the midline in the suprapubic area and two laterally in McBurney's area. The peritoneum is opened along the pelvic brim and the dorsal sheet of the broad ligament pushed medially, whilst preserving the infundibulopelvic ligament and allowing the ureter to remain attached to the peritoneum in its lateral aspect. The paracervical space is opened, pushing the inferior hypogastric artery and the lateral aspect of the bladder medially. This then allows the pararectal space to be developed, and whilst mobilizing the peritoneum medially the point at which the ureter crosses the uterine artery can be identified. The lymph node dissection itself is performed in the same way with a transperitoneal approach as in the retroperitoneal approach, the technique we favour being the most simple one, which consists of using grasping forceps to take hold of the lymph nodes and dissecting forceps to separate the connecting fibre and lymphatic channels attaching the nodes to the surrounding structures. We do not use laser or ultrasonic dissecting tools and rarely use monopolar scissors or bipolar cauterization. The various lymphatic sites on the pelvic side wall are dissected one after the other, usually starting in the area located between the external iliac vein and the obturator nerve, taking care to avoid damage to the inferior obturator vein. The space is then located between the external iliac vein and the artery, and this is opened. The lymph nodes here are usually few, but by opening this space access is gained to the area between the lateral aspect of the iliac veins and the pelvic wall. The important lymph nodes here can then be removed. The last area to dissect is that located medially to the common iliac artery's bifurcation. By mobilizing the tissues in a caudal direction this dissection naturally arrives at the level of the origin of the collaterals of the internal iliac artery. By dissecting these collaterals for the first 2 cm or so of their route, one performs a lateral paracervical lymphadenectomy.

AORTIC DISSECTION

Two possible approaches are available for para-aortic node dissection. In the first, described first by Childers, the surgeon stands on the left-hand side of the patient, the patient herself being in a steep Trendelenburg position, allowing intestinal loops to be mobilized and pushed up into the left diaphragmatic dome. The peritoneum is opened along the axis of the aorta and the distal peritoneal flap is elevated. The ureter and right ovarian artery are identified and also elevated. The dissection then commences on the right-hand side in a similar manner to that adopted for pelvic lymphadenectomy and is repeated for the nodes on the left-hand side of the aorta. In the second technique described by Querleu, the surgeon stands between the patient's legs and observes the video monitor placed at the head of the operating table. The peritoneum is opened beneath the caudal brim of the ileocaecal junction and after the upper flap has been elevated the dissection proceeds along the vena cava and aorta in a cranial direction. In this technique, as in Childer's, three ancillary ports are usually used, one being used to introduce a retracting instrument.

WHAT ROLE CAN BE PLAYED BY LAPAROSCOPY AND RADICAL SURGERY?

The first role is more conceptual than technical and it is in the role of selecting patients. Selection is usually based upon the balance between risk to the patient and benefit that may accrue. In Stage Ib$_1$ most patients will benefit from radical surgery, however, there is a sub-population within this group where the risks of surgery may outweigh the benefits. This group is more readily identified as those who have three or more lymph nodes involved and it is obvious that laparoscopy may have a potential role in allowing selection of these patients before proceeding to radical surgery. A second role of laparoscopy is that in assisting the radical surgery, where two different options again exist, the laparoscopic radical hysterectomy (LRH) or the laparoscopically assisted vaginal radical hysterectomy (LAVRH). It is not the place of this chapter to describe in detail the techniques of laparoscopic-assisted vaginal radical hysterectomy or laparoscopic radical hysterectomy, and those interested should refer to more detailed surgical texts. It is, however, worthy to note at this point some of the developments in a more general way. Perhaps the most exciting development has been the true laparoscopic radical hysterectomy. This innovation is exciting and may well provide a standard by which we will set all procedures in the future. At present, the purely laparoscopic technique, however, has certain disadvantages. The first and most obvious is the additional time taken in the operating theatre, which in general is three or more times longer than the time needed for a standard radical hysterectomy. The second disadvantage is related to the radicality, which remains uncertain, either for the vaginal cuff or for the paracervical and paravaginal tissues. The third and final disadvantage is more theoretical than actual but is strongly linked with the first two. In vaginal surgery we have an excellent tool to enable time-saving and obtain a radical outcome. Some authors have, however, pointed out that most gynaecological oncologists are not trained to this level in vaginal surgery. This is currently true but rather than remain as a criticism, it should become an objective for training in the future.

LAPAROSCOPY IN ITS ROLE OF INCREASING THE RADICALITY OF VAGINAL HYSTERECTOMY

The vaginal approach to radical hysterectomy combines the advantages of speed and safety. The major disadvantage is that of not being able to assess the lymph nodes. Laparoscopic surgery, on the other hand, makes this assessment possible. Moreover, laparoscopic surgery can also assist in making the vaginal approach more radical and at the same time less problematic.

When we started using laparoscopy for the management of cervical cancer, this was one of our primary objectives, i.e. to make the vaginal operation more radical. We described the new operation in 1992 [6] and

later called it a coelio-Schauta. In this operation the paracervical ligaments are transected under laparoscopic guidance, approaching the area through a lateral port, and using an endostapler we are able to cut the ipsilateral ligament at the level of its origin. Because of the funnel shape of the pelvic cavity it is much easier to divide the paracervical ligaments close to the pelvic wall operating from above rather than from below. As a consequence of this, the radicality is improved. There are, however, some problems associated with this approach in that the increased radicality may induce more problems with bladder function subsequently.

Post-radical hysterectomy bladder dysfunction results from the interruption of the parasympathetic innervation of the bladder and this results in spastic contracture of the ureter and bladder. The relevant nerve fibres cross the lateral part of the inferior brim of the paracervical ligament. It is theoretically possible to spare these, however, the anatomical disposition of the fibres is highly variable and it is impossible to be certain of preserving them when these ligaments are divided in their most lateral part. The main objective during this part of the operation, that is whilst dividing the ligament close to its pelvic insertion, is to completely remove the lymphocellular tissue which is contained within the vascular and nervous network of the ligament. This network is much more complex in its medial part and is relatively simple in its lateral part, and removing the lymphovascular component of it is possible whilst not damaging the vessels and nerves. This procedure is made much simpler by laparoscopic magnification and the atraumatic laparoscopic instrumentation. Surgical principles such as these have allowed us, along with Denis Querleu in 1995, to establish the rationale for this operation termed laparoscopic vaginal radical hysterectomy (LVRH).

This operation is more than a laparoscopically assisted vaginal hysterectomy, in effect being a true combined operation, in which the two approaches are used to the best of their potential. The LVRH starts with a pelvic lymphadenectomy performed either through the retroperitoneal or transperitoneal approach. Once the pelvic lymphadenectomy has been completed the dissection is extended to the lateral part of the cardinal ligament, as described above. The development of the paravesical and pararectal spaces is the first step of this so-called paracervical lymphadenectomy. The uterine artery is transected at the level of its origin and whilst the remaining stump of the artery is pulled medially and detached from the lateral aspect of the ureter, the actual deroofing of the ureteric canal is not carried out at this phase but is done during the vaginal step of the operation.

The vaginal step of the operation is very similar to that described by Schauta, the only difference being the increased ease with which the vesico-vaginal spaces can be entered because of the previous laparoscopic preparation.

Whilst the operative specimen looks very much like the one achieved following a modified radical hysterectomy (Piver 2), the LVRH is a truly radical operation. The parts of the paracervical ligament attached to the specimen are only the medial parts but the lateral parts of them have also been removed during the laparoscopic preparation. This distribution of resecting roles facilitates maximum radicality without impairing bladder function.

THE PLACE OF LAPAROSCOPY IN THE MANAGEMENT OF EARLY CERVICAL CANCER

Laparoscopy as described above has two major roles in the management of early cases, firstly in selecting those who would be suitable for radical surgery and secondly as an adjuvant to surgical management itself.

With regards to selection, the first discriminant is the tumour diameter. Magnetic resonance imaging is valuable in this phase of selection. The second discriminant is lymph node status and laparoscopy is the tool of choice in determining whether or not the lymph nodes are involved. Both of these selection phases are built into the clinical algorithm we use for case management (Figure 21.1). Stage Ib_1 cases with negative nodes we would describe as the true early cases. These are our most numerous, representing 85% of all Stage Ib. These are treated by surgery and the laparoscopic vaginal approach is the one we would advocate.

In cases where one or two lymph nodes only are involved, again Stage Ib_1, it is possible to obtain up to 60% cure rate using surgery alone, including systematic lymphadenectomy. Whilst this is possible using the laparoscopic vaginal approach many of those who have been practising surgery for many years would still believe that this degree of radicality can only realistically be achieved by an open abdominal procedure.

Whether there are three or more lymph nodes involved, the chances of cure are poor (no more than 15%) and in this situation it is difficult to justify the use of radical surgery, and radiotherapy would probably be the treatment of choice.

Our own experience of laparoscopy can really be defined in two phases. In the first, we

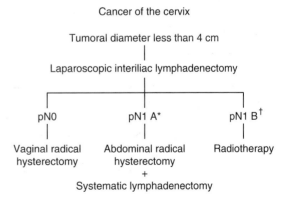

Cancer of the cervix

Tumoral diameter less than 4 cm

Laparoscopic interiliac lymphadenectomy

| pN0 | pN1 A* | pN1 B[†] |
| Vaginal radical hysterectomy | Abdominal radical hysterectomy + Systematic lymphadenectomy | Radiotherapy |

* pN1 A = one or two positive lymphnodes
[†] pN1 B = three or more positive lymphnodes

Figure 21.1 Algorithm for management of uterine cervical cancer Stage Ib_1.

largely use the laparoscope as a selection tool and only perform laparoscopic lymphadenectomies proceeding to radical hysterectomy using the Schauta technique in those selected to proceed. The first 6 years of this experience has previously been published [7].

Among the 98 scheduled patients, laparoscopy had to be abandoned in favour of a mini-laparotomy for performing the lymphadenectomy in three instances. There were seven instances of urinary tract injuries occurring during the vaginal part of the surgery, four of these being cystotomies and three ureteric transections. There were, however, no subsequent fistulae and ureteric stenosis was only recorded on one occasion. There were no other major complications and the mean hospital stay was 11 days. Within this group there were 81 patients whose tumour diameter was less than 4 cm and the 3-year disease-free survival in this group was 89%.

The second option that was introduced in July 1992 was initially the laparoscopically assisted vaginal radical hysterectomy, the so-called coelio-Schauta operation described previously. This operation has a high degree of radicality but we also noted a high degree of morbidity. Amongst the first 20 cases, two had to be converted to laparotomies. There was one transection of the ureter and two emergency postoperative laparotomies, one because of bleeding and one because of intestinal obstruction. Urinary retention was noted in seven cases at the time of catheter removal. It was largely because of this high morbidity that we started to develop the technique of laparoscopic radical hysterectomy also described previously. This we have used electively since July 1995. Although we have not accumulated enough data as yet to be sure of its outcome, our initial impressions are that it is less morbid but gives similar results.

There is now a third option and this has been mentioned in Chapter 20. This option is that of radical trachelectomy, which would appear to be suitable for highly selected

young patients affected by cervical cancer where that cancer is largely confined to the surface of the cervix. Between April 1987 and July 1995, we have attempted radical trachelectomy in 33 patients. In three instances the attempt had to be abandoned because of obvious involvement at the upper end of the cervical resection margin. The complications in this series were limited to two pelvic haematomata, one treated by suprapubic mini-laparotomy and one by mini-colpotomy. Two recurrences have been noted, one after 18 months (this patient had a 22 mm adenocarcinoma and at the time of recurrence was noted to have pulmonary metastases). A further recurrence was noted after 8 years and this patient had a 12 mm squamous carcinoma and recurred on the lateral pelvic side wall. The recurrence was resected and she had intraoperative and postoperative radiotherapy and remains well with no evidence of recurrent disease 2 years later. Fourteen of the 33 selected patients actually attempted to become pregnant and 10 have succeeded, with a total of 17 pregnancies in all. Eleven of these 17 pregnancies have ended with the birth of a normal living baby.

THE PLACE OF LAPAROSCOPY IN THE MANAGEMENT OF ADVANCED DISEASE

Apart from being able to identify those whose disease appears early on clinical grounds, yet who have lymphatic disease that would categorize them as advanced, laparoscopy may play another role and that is in selecting the therapeutic modalities which best suit the specific situation of each individual. Laparoscopic aortic dissection plays an important role here. Selecting the therapy schedule on the basis of the outcome of aortic dissection has only one aim and that is to avoid overtreatment for the patient who does not need it. The two situations where decisions really have to be made are those candidates who may best be managed by radiotherapy and those candidates who might be able to go forward to pelvic exenteration as part of salvage. Candidates for radiotherapy are those considered to be in Stage Ib_2 and more, but also include Ib_1 cases who have positive lymph nodes requiring adjuvant treatment. Nelson demonstrated in the 1970s that aortic lymph nodes may often be involved in this category of patient and in this group pelvic radiotherapy is obviously insufficient to provide any chance of cure. The only hope of curing these patients lies in extended field irradiation. It is this that underlines the simple algorithm in Figure 21.2.

We use a simple step-by-step technique to assess the lymph nodes in those patients who have Stage Ib_1 disease. This is based on our belief that it is highly unlikely to have involved para-aortic lymph nodes in the presence of normal pelvic lymph nodes and therefore we commence with a pelvic lymph node dissection performed laparoscopically and proceed in a step-wise fashion only if we find positive pelvic lymph nodes. This allows us first to decide which patients will benefit from radiotherapy and also the extent of the radiation field. Obviously in patients who will require irradiation because of the characteristics of the primary tumour, this step-by-step technique has no utility. In these we go to a direct aortic assessment in order to determine whether in addition to pelvic irradiation the patient will require extended fields.

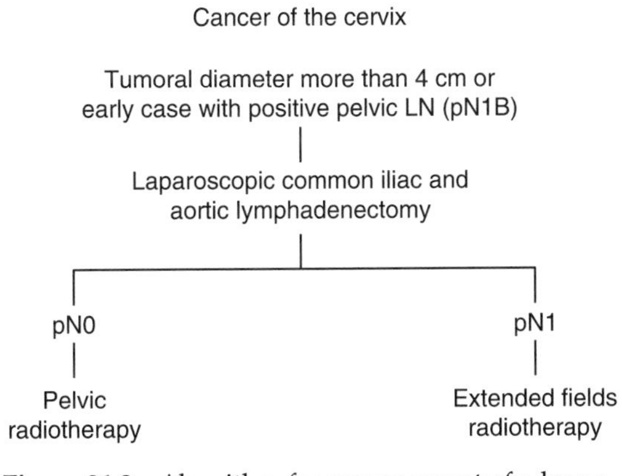

Cancer of the cervix

Tumoral diameter more than 4 cm or
early case with positive pelvic LN (pN1B)

|

Laparoscopic common iliac and
aortic lymphadenectomy

|

pN0   pN1

| |

Pelvic   Extended fields
radiotherapy   radiotherapy

Figure 21.2 Algorithm for management of advanced uterine cervical cancer.

With regard to selecting patients for attempted pelvic exenteration, laparoscopy also has an obvious role. These patients are generally those affected by central pelvic recurrence of a cancer of the uterus, cervix or vagina. The chances of cure after pelvic exenteration directly relate to the state of the lymph nodes. Massive extension to the pelvic lymph nodes is so highly significant as to be a contraindication to surgery. Extension to the aortic lymph nodes, even if microscopic, is also considered to be a contraindication. Based on these concepts, laparoscopic aortic dissection appears to be very important in the management of such cases. The procedure can be carried out without directly assessing the pelvic lymph nodes that we would normally assess by using imaging such as MRI and CT scan and only consider gross involvement of the pelvic lymph nodes, a contraindication to proceeding with an exenterative procedure.

CONCLUDING REMARKS

Over the last century we have reached a point where we can perform a radical hysterectomy for cervical cancer without opening the abdomen. The disease, particularly with its propensity to involve local and regional lymph nodes, still poses significant problems for the contemporary clinician. We believe that here lies the true place of laparoscopy, in the assessment of the pelvic and para-aortic lymph nodes. The accuracy of laparoscopic assessment of these nodal groups is outstanding and certainly equals that of open-laparotomy assessment with significantly less morbidity. That such procedures can be performed with very short stays in hospital has obvious health economic advantages but also has psychological advantages for the patient. It also allows the decision to proceed with radical surgery to be made on the basis of solid histological evidence, bearing in mind that frozen sections of lymph nodes are associated with varying false-negative rates as high as 30%. Whilst one cannot underestimate the advantages that laparoscopic nodal assessment has provided for the cervical cancer surgeon, one must be realistic, for in the future as imaging techniques improve, particularly in combination with fine needle stereotactic puncture, it is likely that the role of laparoscopic nodal assessment may decline. Nevertheless, we believe that it certainly has a role to play now, both in lymphatic staging and in assisting more radical but less morbid pelvic procedures.

REFERENCES

1. Dargent, D. (1987) A new future for Schauta's operation through pre-surgical retroperitoneal pelviscopy. *European Journal of Gynaecology and Oncology*, **8**, 292–6.
2. Nelson, J.H., Macasaet, M.A., *et al.* (1974) The incidence and significance of paraortic lymph nodes metastases in late invasive carcinoma of the cervix. *American Journal of Obstetrics and Gynecology*, **118**, 749–56.
3. Averette, H.E., Sevin, B.U., Bell, J.G., *et al.* (1987) Surgical staging of cervical cancer. *Cancer*, **60**, 2010–20.
4. Nelson, J.H., Boyce, J., Macasaet, M., *et al.* (1977) Incidence, significance and follow up of paraortic lymph nodes metastases in late invasive carcinoma of the cervix. *American Journal of Obstetrics and Gynecology*, **128**, 336–40.
5. Weiser, F.B., Bundy, B.N., Hoskins, W.J., *et al.* (1989) Extraperitoneal versus transperitoneal selective paraortic lymphadenectomy in the treatment surgical staging of advanced cervical cancer (a GOG study). *Gynecologic Oncology*, **33**, 283–9.
6. Dargent, D. and Mathevet, P. (1992) Hysterectomie élargie laparoscopico-vaginale. *Journal of Gynecology, Obstetrics and Biology of Reproduction*, **21**, 709–10.
7. Dargent, D. and Mathevet, P. (1996) Radical vaginal hysterectomy in the primary treatment of invasive cervical cancer. In *Cervical Cancer and Preinvasive Neoplasia* (eds S.C. Rubin and W.J. Hoskins). Lippincott Raven, Philadelphia, PA, pp. 207–17.

BIBLIOGRAPHY

Canis, M., Mage, G., Wattiez, A., *et al.* (1990) La chirurgie endoscopique a-t-elle une place dans la chirurgie radicale du cancer du col utérin? (lettre) *Journal of Gynecology, Obstetrics and Biology of Reproduction*, **19**, 921.

Childers, J.M., Hatch, K.D., Tran, A.N. and Surwitt, E.A. (1993) Laparoscopic paraaortic lymphadenectomy in gynecologic malignancy. *Obstetrics and Gynecology*, **82**, 741–7.

Clark, J.G. (1985) A more radical method for performing hysterectomy for cancer of the uterus. *Bulletin of Johns Hopkins Hospital*, **6**, 120.

Meigs, J.V. (1945) Wertheim operation for carcinoma of the cervix. *American Journal of Obstetrics and Gynecology*, **49**, 542.

Piver, M.S., Rutledge, F. and Smith, J.P. (1974) Five classes of extended hysterectomy for women with cervical cancer. *Obstetrics and Gynecology*, **44**, 265–7.

Querleu, D. (1993) Laparoscopic paraaortic lymphadenectomy: a preliminary experience. *Obstetrics and Gynecology*, **68**, 90–3.

Schauta, F. (1908) *Die Erweiterte Vaginale Totalextirpation des Uterus bei Kollumkarzinome*. Journal Safar Pub Vienna.

Spirtos, N.M., Schlaerth, J.B., Kimball, R.E., *et al.* (1996) Laparoscopic radical hysterectomy (Type III) with aortic and pelvic lymphadenectomy. *American Journal of Obstetrics and Gynecology*, **6**, 1763–8.

Wertheim, E. (1911) *Die erweiterte abdominale Operation bei Carcino colli uteri*. Urban und Schwarzenberger, Berlin.

Kathryn M. Greven and Rachelle Lanciano

INTRODUCTION

Since the 1920s, radiation has been used to treat and cure cervical cancer effectively. Several treatment systems have been developed and form the basis of our knowledge about the radiotherapeutic management of cervical cancer. Thousands of patients have been treated and analysed so that benchmarks for survival, local control, and complication rates by stage are available. Various patient-, tumour-, and treatment-related variables are known to influence the outcome of radiotherapeutic management, and these will be discussed here.

Research continues in order to define the optimum radiation regimen. The need for adequate dose through the use of brachytherapy and reduction of overall treatment time has already been established. The use of high dose rate afterloading techniques are more generally accepted world-wide and research continues to determine the best optimal fractionation schemes. Three-dimensional treatment planning and improvement in dose specification are research agenda for the twenty-first century. Strategies utilizing combinations of radiation, chemotherapy and surgery continue to be investigated.

TECHNICAL FEATURES

INTRACAVITARY TREATMENT

The treatment of cervical cancer with radiation therapy generally requires the use of external beam irradiation combined with intracavitary brachytherapy. External beam irradiation is used to treat the pelvic nodes and parametria, while the central disease is treated primarily by the intracavitary implant. The use of intracavitary radiation is associated with improved pelvic control and survival over external beam radiation alone. Data from the Patterns of Care Study in the USA showed that patients with Stage I–III cervical cancer had 4-year survival and pelvic control rates of 36% and 46%, respectively, without the use of brachytherapy compared with 67% and 78% with brachytherapy [1].

Since the first use of intracavitary brachytherapy in the early part of the twentieth century, many types of applicators have been developed for the treatment of cervical carcinoma. All of these applicators utilize radioactive isotopes, such as caesium-137, which are inserted into the uterine cavity and vaginal fornices. The most commonly used applicator in the USA is the Fletcher-suit type because it provides a fairly stable relationship to the anatomy and its afterloading, which decreases radiation exposure to medical

Cancer and Pre-cancer of the Cervix
Edited by D.M. Luesley and R. Barrasso. Published in 1998 by Chapman & Hall, London. ISBN 0 412 56600 1.

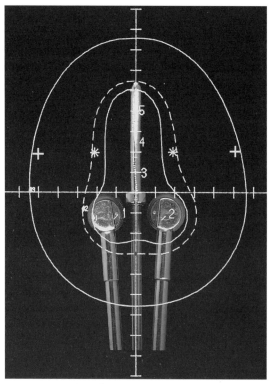

Figure 22.1 Fletcher tandem and ovoids with isodose curves to demonstrate radiation distribution. — 60 cGy per hour line; * point A; + point B.

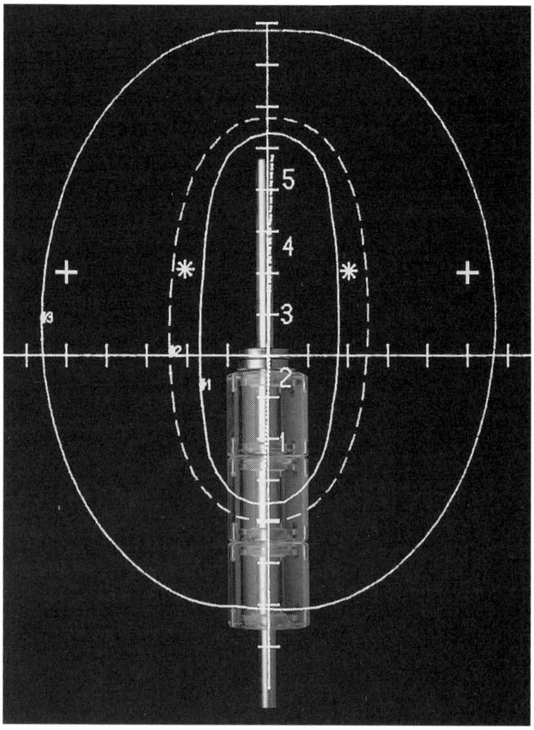

Figure 22.2 Fletcher tandem and cylinder with isodose curves to demonstrate radiation distribution. — 60 cGy per hour line; * point A; + point B.

personnel. The use of a Fletcher-suit application with an intrauterine tandem and an ovoid (colpostat) in each fornix provides a pear-shaped isodose distribution for the treatment of the cervix (Figure 22.1). If the vaginal vault is narrowed or a bulky tumour prevents the placement of colpostats, vaginal rings may be substituted. This linear source distribution may not provide an optimum dose distribution for treatment (Figure 22.2). Technically accurate intracavitary insertions with proper geometric relationship have been demonstrated to result in improved local control when compared to patients treated with unsatisfactory placements [2]. In the situation of a poor placement or in the event of a bulky cervical cancer with extensive parametrial involvement that may not be covered with the iso-

dose distributions of an intracavitary placement, some authors have advocated the use of a perineal interstitial implant [3, 4]. This interstitial placement is frequently performed with a template and involves inserting radioactive sources through the perineum into the parametrium. It may be performed with surgical guidance from above via the peritoneal cavity. Despite the encouraging results from individual centres, there have been no randomized studies to document the superiority of either approach for advanced stage tumours.

Prior to the advent of modern computerized dosimetry, dose calculations were expressed in terms of 'milligram hours'. It is now preferable to obtain radiographs of each brachytherapy placement and calculate doses to critical structures – vagina, bladder, and rectum [5]. Doses to pelvic points – A, B, and P –

are calculated to ensure adequate delivery of dose to eradicate the tumour. Adjustments in the strength or positioning of the sources can be made to optimize the insertion. Point A is located 2 cm cephalad from the cervical os and 2 cm lateral to the uterine canal. It is termed the paracervical point of reference. Anatomically, it represents the medial parametrium/lateral cervix, approximately where the ureter and uterine artery cross. Point B is 5 cm lateral to the centre of the pelvis at the same level as point A and approximates the region of the obturator node or lateral parametrium. Point P is located along the bony pelvic side wall at its most lateral point and represents the minimum dose received by the external iliac lymph node.

Standard dose rates at point A are typically 40–60 cGy per hour. Total doses per insertion may vary from 20 to 40 Gy depending on the stage of disease and adequacy of the placement. The insertion is performed under anaesthesia and the patient is hospitalized at bedrest for a period of 2–3 days during the implant. One or two implants are usually performed. Despite the fact that two insertions may allow time for regression of disease between placements, there are no data to support improved pelvic control or survival rates when compared to one insertion [6].

The use of high dose rate brachytherapy has been increasing over the last decade. Dose rates are typically 200–300 cGy per minute. Smaller doses per insertion with three to ten insertions per patient are performed on an out-patient basis. Several institutions have reported comparable disease control rates and complication rates to low dose irradiation [7, 8]. Fractionation schemes vary but complication rates appear lower in those patients with lower total dose per insertion and more frequent insertions.

EXTERNAL FIELDS

Pelvic radiation fields extend superiorly from the top of the fifth lumbar vertebral body, infe-riorly to cover the tumour in the cervix or vaginal disease with at least a 3 cm margin, and laterally to encompass the pelvic brim with a minimum 2 cm margin lateral to the widest diameter of the pelvic outlet. If the para-aortic region is to be included the superior extent of the field is at the top of the twelfth vertebral body.

When anterior–posterior portals are used, high energy photons (18 MV) usually provide the best dose distribution with sparing of the anterior bowel. Multiple field arrangements with typically a four-field distribution using lateral and anterior–posterior opposed fields are used with lower energy photons. Lateral fields have the same superior and inferior extent. Anteriorly, the field should take into account the external iliac node chain as well as the extent of the uterus on a pre-treatment computerized tomography (CT) scan. Posteriorly, the field must cover the pre-sacral nodes and sparing of the posterior rectum may not be possible without sparing some of the uterosacral ligaments, which may harbour disease [9].

Depending on the bulkiness of the cervical cancer, the midline structures of the pelvis may be shielded during a portion of the whole-pelvis irradiation. This shielding helps to reduce the dose to the bladder and rectum, which may allow better tolerance to the intracavitary radiation. The midline shield can be performed in a variety of ways but is typically 4–5 cm wide and extends from the inferior border of the pelvic field to the bottom of the sacrum.

External pelvic irradiation is given daily to all treatment fields per day with fraction sizes of 1.8–2.0 Gy per day. Total doses range from 40 to 50 Gy. Table 22.1 demonstrates a treatment schema integrating external beam radiation and intracavitary irradiation by stage.

DOSE RESPONSE AND OVERALL TREATMENT TIME

Radiation factors including overall treatment time and total dose to point A have been

Table 22.1 Radiation treatment schema for cervical cancer

		External beam	Intracavitary	Total point A doses
Ia		None	60–75 Gy in two insertions	60–75 Gy
Ib/IIa	Non-bulky	40 Gy[a]	40 Gy in two insertions	80–85 Gy
Ib–IIb	Bulky	40–45 Gy[b]	40–50 Gy in two insertions	85–95 Gy
IIIb	Bulky	45–50 Gy[b]	35–40 Gy in one or two insertions	85–95 Gy

[a]If mid-line block is inserted, intracavitary doses may be increased. Preferred doses to normal tissue reference points: vaginal surface dose: 110–130 Gy; bladder: 75 Gy; rectum: 70 Gy.
[b]Residual parametrial/pelvic wall disease or positive nodes are boosted to 60–65 Gy.

investigated as important prognostic factors for local control in cervical cancer. Lanciano *et al.* demonstrated a highly significant decrease in survival and pelvic control as the total treatment time was increased from less than 6 weeks to greater than 10 weeks. Stage III accounted for the majority of the adverse effects from the prolongation of treatment time. Multivariate analysis suggested total treatment time was an independent prognostic factor for cervical cancer [6]. Fyles *et al.* found a very similar effect of treatment duration on pelvic control in 830 patients with cervical cancer treated by radiation at the Princess Margaret Hospital [10]. In multivariate analysis, overall treatment time was independently correlated with pelvic control [10]. The magnitude of the time effect in cervical cancer was approximately 1% loss of local control per day of delay. At present, five large retrospective experiences of radiotherapy for cervical cancer confirm the importance of overall treatment time on outcome and suggest overall treatment time should be limited to 8 weeks, particularly for locally advanced disease [6, 10–13].

A radiation dose response has been demonstrated for Stage IIIb cervical cancer in the Patterns of Care database for 4-year survival and pelvic control [14]. When doses above 85 Gy were delivered to point A, the infield failure rate was reduced to 36% and the survival improved to 49%. Choy *et al.* [15] reviewed 594 cervical cancer patients treated with a combination of external and intracavitary brachytherapy and demonstrated a dose response for Stages IIb and III with doses greater than 85 Gy associated with significantly improved central control. There was a statistically significant improvement in survival for Stage IIb disease but not Stage III disease. These analyses, as well as institutional experience reported by centres of excellence for locally advanced cervical cancer, suggest that an optimal dose for point A should be in the range of 85–90 Gy [16].

TREATMENT OF EARLY DISEASE

STAGE IA

For those women with microinvasive cervical cancer who are not candidates for surgery, irradiation with intracavitary treatment provides excellent survival and pelvic control rates. Grigsby *et al.* [17] reported a progression-free survival rate of 100% at 10 years for 34 patients with Stage Ia cervical cancer. Estimates of nodal involvement in these patients varies from 4% to 10% with the size of the lesion and the presence of lymph vascular involvement. Treatment of lymph nodes with the addition of pelvic radiation in these patients should be considered on an individual basis. Patients with lesions less than 3 mm of invasion deep to the basement membrane and without lymphovascular invasion do not require treatment of lymph nodes.

STAGE IB/IIA

Patients with early-stage disease may be definitively treated with either radical hysterectomy or irradiation. Preliminary results from the only randomized prospective trial comparing the two treatments demonstrate no difference in survival or relapse rate with an increase in morbidity for initial hysterectomy [18]. Retrospective reports of survival rates for patients with early-stage cervical cancer treated with radiation therapy are comparable to those treated with radical surgery and range from 80% to 90%. A consensus conference held by the National Institute of Health in 1996 concluded that patients with Stages Ib and IIa cervical cancer are appropriately treated with either radical hysterectomy with pelvic lymphadenectomy or radiation therapy, with equivalent results [19]. Choice of treatment may vary based on the patient's general condition and preference, physician preference and characteristics of the lesion.

Various prognostic factors that were not included in the prior staging system have been reported to influence the outcome of patients with cervical cancer. This probably accounts for the variation in reported outcomes between different investigators. The most important factor includes tumour volume, which was adopted into the current staging system in 1994 [20]. Other factors include lymph node status, patient age, histology and pre-treatment haematocrit.

TUMOUR VOLUME

Tumour volume is the most important prognostic factor for determining outcome for patients with carcinoma of the cervix. Larger tumour volumes are associated with a higher incidence of pelvic nodes, pelvic failure and decreased survival. Table 22.2 demonstrates the survival rates and pelvic recurrence rates for patients based on the size of the tumour. Because pelvic tumour recurrences may be

Table 22.2   Results after radiation alone for Stage Ib and/or IIa cervical cancer

	No.	5-year survival Ib/IIa (%)	5-year DFS Ib/IIa (%)	5-year pelvic recurrence Ib/IIa (%)	Serious complications (%)[d]
Horoit *et al.* [53]	533	89/85	–	7	5
Lanciano *et al.* [14]	618				10
Non-bulky	82[a]	–	6	–
Bulky	75[a]	–	18	–
Lowrey *et al.* [25]	194	–			5
0–3 cm	–	93/75	3/19	–
4–5 cm	–	77/79	15/12	–
≥6 cm	–	65/60	20/40	–
Eifel *et al.* [27][b]	1526	–			
≤4 cm	–	86–94[c]	≤5	–
5.0–5.9 cm	–	69[c]	15	–
6.0–6.9 cm	–	69[c]	21	–
7.0–7.9 cm	–	58[c]	19	–
≥8 cm	–	40[c]	43	–
Perez *et al.* [73]	552				7
Non-bulky	90/75	–	12/17	–
Bulky	61/69	–	31/12	–

[a]4-year survival.
[b]Ib only.
[c]Disease specific survival.
[d]Crude rates.

higher in 'bulky' Stage I tumours, some groups have advocated the addition of an extra-fascial hysterectomy to improve the outcome of such lesions.

An early report from the M.D. Anderson Cancer Center reported a lower pelvic recurrence rate for patients with bulky endocervical tumours treated with external beam and intracavitary irradiation followed by extra-fascial hysterectomy than for those treated with radiation alone [21]. The authors recommended combined treatment for patients with tumours greater than or equal to 6 cm in diameter, and this approach was adopted by many groups as a standard approach to bulky Stage Ib or IIa disease. Subsequent reports suggested that adjuvant hysterectomy after radiotherapy reduced the likelihood of central recurrence and resulted in survival rates that were comparable to those observed in similarly staged 'non-bulky' lesions [22, 23].

Several retrospective reports fail to support the value of adjuvant hysterectomy for bulky lesions. An update of the M.D. Anderson Cancer Center experience demonstrated a number of biases that had influenced the selection of treatment and suggested that the differences observed in earlier reports may have resulted from a tendency to choose patients with very massive tumours (>8 cm) or clinically positive nodes for treatment with radiation alone [24]. When such patients were excluded, pelvic control and survival rates between the two approaches were similar.

The University of Florida published an analysis comparing patients treated with radiation to those treated with radiation and adjuvant hysterectomy. There was no difference in pelvic control, distant metastasis, or cause-specific survival between the two groups [25]. There was an increased incidence of severe complications in the group of patients receiving surgery and radiation compared with radiation alone, 18% versus 7%, respectively.

Perez and Kao reported a series of 128 patients with Stage Ib and IIa–b cervical cancers greater than 5 cm in diameter treated with irradiation alone in 75% of the cases or with combinations of irradiation and surgery for the remainder [26]. They found that central recurrences were rare in patients with Stage Ib tumours that measured greater than 5 cm in diameter if adequate doses of irradiation (>80 Gy to point A) were delivered. Likewise, an analysis of patients from the M.D. Anderson Hospital with bulky (>6 cm) Stage Ib–IIb tumours reported a pelvic recurrence rate of 33% and 16% for patients with lower doses to point A compared with higher doses, respectively [27].

A prospective randomized trial has been performed by the Gynecologic Oncology Group (GOG). Between 1984 and 1991, 269 patients with Stage Ib tumours greater than or equal to 4 cm were randomized to receive radiation alone or radiation with adjuvant hysterectomy. The GOG is expected to analyse and report these results soon.

OTHER PRE-TREATMENT FACTORS

A low pre-treatment haemoglobin has been reported to affect the outcome of patients in several clinical studies of patients with cervical cancer. At the Institute Gustave-Roussy, patients with advanced cervical carcinoma were noted to have an increased risk of local regional failure if haemoglobin concentration during treatment was below 10 g/dl [28]. Bush *et al.* [29] also noted a 10% lower overall survival in patients with Stage IIb/III cervical carcinoma and haemoglobin levels <12 g/dl because of failure to control pelvic disease. Lowrey *et al.* [25] demonstrated that patients with Stage Ib and IIa–b disease with low pretreatment haemoglobin had inferior relapse-free survival and increased distant metastasis after multivariate analysis. It is not clear whether these observations are consistent with the fact that radiation is less effective at lower oxygen tensions or that hypoxia may induce enhancement of the metastatic potential and create more aggressive tumours [30].

Several authors have also documented young age as a factor associated with inferior outcomes. Lanciano noted a significant decrease in pelvic control for younger Stage I and II patients in multivariate analysis [14]. Kapp *et al.* reported a significant increase in local failure for patients age <55 years in multivariate analysis for all stages of cervical cancer [31]. Lowrey *et al.* found that young age was an independent predictor of pelvic control and relapse-free survival after multivariate analysis [25].

The majority of cervical cancers are squamous cell carcinomas. Some have suggested that the major variant histologies – adenocarcinoma, adenosquamous carcinoma and neuroendocrine small cell carcinoma – have an inferior prognosis. Eifel *et al.* reported a 5-year relapse-free survival rate of 88% for 91 patients with Stage Ib adenocarcinomas of less than 3 cm treated with definitive radiation [32]. This compares favourably with patients treated with radical hysterectomy at the same institution. For adenocarcinoma greater than or equal to 3 cm, radical hysterectomy had inferior results compared with radiation. Patients with Stage Ib cervical cancers treated primarily by radiation and stratified by size had a lower survival rate for adenocarcinoma compared with squamous cell carcinoma, primarily due to a higher rate of distant metastasis [33].

Patients with carcinoma of the cervical stump appear to have similar cure rates to those patients with an intact uterus if adequate treatment can be delivered. In one series locoregional control was 81.5% in patients receiving external radiation with intracavitary versus 38.5% in patients receiving external radiation alone [34].

POST-OPERATIVE RADIATION FOLLOWING RADICAL HYSTERECTOMY

Several pathological factors have been demonstrated to influence disease control following radical hysterectomy. Involvement of the pelvic lymphatics strongly influences the outcome of these patients. One report documented a 5-year survival rate of 91% for Stage Ib cervical cancer patients when the lymphatics were not involved, compared to 55% when one or more lymph nodes contained disease [35]. Hsu *et al.* reported that patients with one to four positive nodes had a 48% 5-year survival rate compared with 19% 5-year survival rate for women with disease involving more than four nodes [36]. In addition, large tumour size, deep invasion in the cervix, capillary lymphatic space involvement, occult parametrial involvement and positive or close surgical margins have all been associated with recurrence after radical hysterectomy. Thomas and Dembo concluded that 72% of recurrences after radical hysterectomy are associated with pelvic relapse [37]. For this reason, numerous investigators have suggested that the addition of post-operative irradiation may be beneficial.

Most reports of post-operative pelvic radiotherapy suggest a reduction in pelvic relapses with no affect on survival. It may be that patients with positive nodes have an increased risk of distant metastasis as well as pelvic recurrence so that a negligible benefit in terms of survival would be expected. In 1980, a panel report conducted by the Society of Gynecologic Oncologists asked the question 'is pelvic radiation beneficial in the post-operative management of Stage Ib squamous cell carcinoma of the cervix with pelvic node metastasis treated by radical hysterectomy and pelvic lymphadenectomy?' [38]. No conclusions could be drawn from this report but there was a suggestion of improved pelvic control in the group of patients receiving radiation compared to those who did not. A recent multicentre randomized phase III trial reported no difference in recurrence patterns or survival in 'high-risk' patients treated with either chemotherapy alone or chemotherapy with pelvic radiation [39]. High-risk patients had either positive nodes, bulky tumours, parametrial involvement or non-squamous histology.

Criticisms of this trial included a low radiation dose (40 Gy), delayed onset to time of radiation (as much as 12 weeks) and insufficient numbers of patients necessary to demonstrate a difference between the two arms. A recently completed Gynecologic Oncology Group trial (109) randomized node-positive patients or patients with positive surgical margins to pelvic external beam radiation versus pelvic external beam radiation plus concurrent 5-FU and cisplatin following radical hysterectomy–pelvic lymphadenectomy. This study may help to determine whether concurrent chemoradiation cannot only reduce the rate of distant metastases but also improve local control.

Patients most likely to benefit from improved pelvic control by the addition of pelvic radiation may be those with high risk features but negative nodes. The Gynecologic Oncology Group has recently completed a trial randomizing patients to pelvic external beam radiation or no further therapy following radical hysterectomy and pelvic lymphadenectomy. Eligible patients had various combinations of risk factors that included: primary tumour >4 cm, deep cervical stromal invasion and capillary–lymphatic space invasion. Analysis of this trial is pending longer follow-up of patients.

At the present time, the use of adjuvant pelvic radiotherapy needs to be individualized for patients with negative nodes who are at risk for pelvic failure, and remains standard post-operative treatment for patients with positive lymph nodes. Treatment consists of external pelvic radiation (45–50 Gy) with specific sites boosted with further external beam, intracavitary or interstitial radiation as needed.

POST-OPERATIVE RADIATION FOLLOWING INADVERTENT HYSTERECTOMY

Patients who undergo simple hysterectomy for presumed benign disease and are found to have invasive cervical cancer are considered candidates for post-operative radiation therapy. Hopkins *et al.* reported that radiation therapy and observation alone following inadvertent hysterectomy produced a survival rate of 88 and 69%, respectively [40]. A report of 122 patients stated that patients with and without gross disease at the start of post-hysterectomy treatment had 5-year survival rates of 39 and 75%, respectively [41]. The most common site of treatment failure was the pelvis. Radiation usually consists of external pelvic radiation (45–50 Gy) with an intracavitary vaginal cuff boost designed to deliver an additional 40 Gy to the mucosal surface.

TREATMENT OF ADVANCED DISEASE

The standard treatment approach for Stages IIb–IVa cervical cancer is radiation alone with combined external pelvic radiation and brachytherapy. Neoadjuvant chemotherapy prior to radiation has not improved survival in locally advanced cervical cancer and is therefore considered investigational [42–44]. Concurrent chemotherapy has been studied extensively by the Gynecologic Oncology Group (GOG) and the most active agents include hydroxyurea (HU), 5-fluorouracil (5-FU) and cisplatin. The earliest study conducted by the GOG compared standard radiation with placebo to radiation plus HU [45]. Median survival was better for patients treated with HU compared with placebo, 19.5 months vs 10.7 months. However, criticisms of this study include a large number of inevaluable patients, low radiation doses and lower than expected survival rates for each arm. A later GOG protocol compared the hypoxic cell sensitizer, misonidazole, and radiation to HU and radiation [46]. A significant improvement in progression-free interval for Stages IIb–IVa and survival for Stages III–IVa was demonstrated for HU and radiation. However, other studies have shown a trend to better survival with placebo over misonidazole, suggesting that misonidazole may not have been an appropriate control [47]. A recent GOG protocol compared 5-FU and cisplatin to HU and radiation and early results suggest a survival benefit with concurrent 5-FU and cisplatin. The current GOG trial is a three arm comparison of weekly cisplatin vs 5-FU/cisplatin/HU vs HU alone as an adjunct to radiotherapy.

While other studies of radiation with 5-FU or 5-FU and cisplatin are ongoing, no results are yet available. The Radiation Therapy Oncology Group (RTOG) has recently completed a study of concurrent 5-FU/cisplatin and pelvic radiation therapy (RT) in a phase III trial compared to pelvic plus para-aortic RT for advanced stage cervical cancer. The next GOG trial that is planned will test 5-FU/cisplatin with concurrent radiation vs radiation alone. Presently there is no evidence that hydroxyurea or any other concomitant chemotherapy agent should be incorporated into standard practice [19].

PROGNOSTIC FACTORS

Detection of para-aortic and pelvic lymph nodes has considerable impact on outcome of patients with advanced cervical cancer. In the GOG experience with selective para-aortic lymphadenectomy, the incidence of involved para-aortic lymph nodes was 24% for advanced cervical cancer (21% for Stage IIb and 31.5% for Stage III). CT scan and ultrasound as diagnostic tools for evaluating para-aortic lymph node metastases were associated with false-negative rates of 18% and 22%, respectively, and therefore were not clinically useful. However, a negative lymphangiogram was associated with only an 8% risk of missing extrapelvic lymph node metastases and is the best non-surgical method for assessing the peri-aortic lymph nodes [48]. Unfortunately, the use of lymphangiography is declining due to lack of technical expertise in performing and interpreting the study. Surgical staging of the para-aortic region, considered the 'gold standard' for evaluating the para-aortic region, is still considered investigational since there is no proven survival advantage [49].

Involvement of the para-aortic lymph nodes is the most important adverse prognostic factor for advanced cervical cancer and reduces survival by half. Twiggs *et al.* reported a 90% 3-year survival rate if pelvic and peri-aortic lymph nodes were uninvolved compared to 45% if pelvic and peri-aortic lymph nodes were involved [50]. The GOG reported peri-aortic lymph nodes metastases, followed by pelvic lymph nodes metastases and tumour bulk to be the most important negative prognostic factors for Stage IIb–IVa cervical cancer [51]. Tumour bulk has been defined in various ways for advanced cervical cancer. Reports from four large retrospective studies found tumour bulk within each stage to have prognostic significance [14, 52–54]. For Stage IIb, bulky disease was defined as lateral or bilateral parametrium, or barrel-shaped endocervical lesion greater than 6 cm in size, and resulted in a 5-year survival rate of 78% compared with a rate of 86% for medial or unilateral parametrium [14]. For Stage III, bulky disease was defined as bilateral side wall fixation or lower third vaginal involvement and resulted in a 5-year survival rate of 45% compared with a rate of 68% for unilateral side wall fixation [14, 53]. The results of radiotherapy for advanced cervical cancer by stage is shown in Table 22.3.

TREATMENT OF THE PARA-AORTIC REGION

Patients with documented para-aortic lymph node metastases are usually treated with extended fields to encompass the retroperitoneum. In order to keep bowel complications to a minimum, the dose of radiation to the peri-aortic region should not exceed 45–50 Gy and surgical staging, if performed, should be limited to the extraperitoneal approach. Five-year survival rates range from 20% to 40% in recent series for extended field radiation, with most patients having low volume pelvic disease which is more likely to be controlled [55]. Likewise, survival rates are higher for patients with microscopic or small para-aortic nodes

Table 22.3 Results after radiation alone for Stage II/III cervical cancer

Author	No. of patients	Stage II		Stage III	
		5-year survival IIa/IIb (%)	5-year pelvic recurrence IIa/IIb (%)	5-year (%) survival	5-year pelvic recurrence (%)
Horoit *et al.* [53]	1111	85/76	12/20	50	43
Lanciano *et al.* [14]	901	67/65	17/20	40[a]	47[a]
Thomas and Dembo [74]	455	82/73	16/21	46[b]	40[b]
Perez *et al.* [75]	774	70/72	19/23	52[c]	41[c]
Eifel *et al.* [76]	903	–	–	35	–

[a]4-year survival.
[b]Disease specific survival.
[c]Disease free survival.

compared with massive para-aortic involvement [56]. The GOG recently completed a phase II study of combined infusional 5-FU, cisplatin and concurrent radiation with extended fields and results are pending. The RTOG recently opened a phase II trial of therapeutic lymph node dissection followed by extended field radiation. Since distant failure is a predominant pattern of failure for these patients, improvements in systemic therapy are necessary before an increase in survival can be appreciated.

Prophylactic para-aortic irradiation is an alternative to surgical staging for advanced cervical cancer. Patients felt to be at high risk for para-aortic lymph node metastases may be treated with extended field radiation for potential microscopic disease. Two randomized trials have studied the relative benefits of prophylactic radiation to the region. An EORTC study found no survival advantage to prophylactic para-aortic radiation [57]. The majority of patients in this study had bulky Stage II or III disease with a high local regional failure rate, which may have overwhelmed any beneficial effect of prophylactic para-aortic radiation [57]. The RTOG published results of a phase III trial demonstrating a statistically increased survival rate at 10 years of 55% vs

44% for prophylactic para-aortic radiation compared to pelvic radiation alone. This improvement was primarily due to a reduced rate of distant metastases in patients achieving pelvic control in the para-aortic radiation arm [58]. This study did not include Stage III cervical cancer, which may be the reason for the positive results. The RTOG has recently completed a randomized trial comparing pelvic and para-aortic radiation to pelvic radiation and 5-FU/cisplatin and further follow-up is needed before the data will be mature.

TREATMENT OF LOCAL–REGIONAL RECURRENCE

Following radical hysterectomy for Stage I–IIa cervical carcinoma, 10–15% of women will have a recurrence. Of these recurrences, 60% will be located in the pelvis alone. The treatment of a localized pelvic recurrence depends on a number of factors, but the most important are a history of previous radiation to the pelvis and the location of the recurrence. For patients treated with radical hysterectomy, radiation with or without chemotherapy can lead to disease-free survival rates of 20–50% and local control rates of 20–60%, and should be considered as first line treatment [59]. For patients with a history of radiation who have central recurrence without associated pelvic or para-aortic lymphadenopathy, exenterative procedures have resulted in survival rates of 30–60% [60]. For patients with a history of previous radiation who have side wall or nodal recurrences, a combination of surgery and brachytherapy or intra-operative radiotherapy may improve survival for these women who were previously considered incurable. However, this requires special expertise and technology not available in most centres [59].

COMPLICATIONS

Sequelae following definitive radiation for cervical cancer can be divided into acute or chronic effects. Acute reactions occur during the treatment and generally resolve following treatment. These may include fatigue, bladder irritation, diarrhoea or bowel cramping and rectal or anal irritation. Diminished white blood cell counts may also be seen with radiation to the pelvis.

Chronic complications may occur months to years following radiation and are frequently reported as minor or major complications. Major complications necessitate hospitalization or surgery, or cause death. The major complication rate in most series of patients treated with definitive radiation ranges from 5% to 15% [61]. Multivariate analysis demonstrated the following factors to increase the incidence of complications: young age, prior surgery, dose to point A and dose per fraction [62]. Several reports have suggested that the external beam dose is the most important factor in determining the incidence and severity of complications [63–65]. An increased complication rate has been reported for patients treated with a tandem and vaginal line source when compared with a standard Fletcher-suit applicator, 29% vs 5% [66].

The most frequently observed complications are gastrointestinal with small bowel obstruction, chronic diarrhoea and rectal proctitis, ulceration or fistula. Although poorly documented, anorectal dysfunction can occur because of weakening of the sphincter and stiffening of the rectal wall [67]. Bladder complications with cystitis, haematuria or fistula occur less frequently than bowel complications. In a report of 1784 patients, less than 2% were identified who had severe haemorrhagic cystitis [68]. The time to occurrence of bowel complications is usually shorter compared with that for bladder complications. Kottmeier reported that the usual time for bowel complications is 6–18 months, whereas bladder complications were seen at 18–48 months [69]. Eifel *et al.* reported that the majority of serious complications occurs by 5 years. However, there is a small but continuous risk of complications that can be documented as far as 20 years after treatment [61]. If radiation sequelae are suspected, biopsy of bladder or bowel ulceration should be avoided since it may precipitate necrosis and fistula formation.

Other more uncommon or less well documented complications include lumbosacral plexopathy, pelvic insufficiency fractures, ureteral stenosis and vaginal stenosis. Georgiou *et al.* reported four cases of lumbosacral plexopathy among 2410 patients treated to the pelvis [70]. Ureteral stenosis following radiation is similarly rare and recurrent tumour should always be suspected [68]. Pelvic insufficiency fractures probably occur more frequently than are recognized but usually heal with symptomatic treatment [71].

Vaginal stenosis and sexual function have not been well documented. However, one study has demonstrated a decrease in vaginal length, sexual frequency and sexual satisfaction following treatment for cervical cancer with radiation [72]. Prophylactic treatment with vaginal dilators and oestrogen may help to alleviate this side-effect.

CONCLUSIONS

Radiation is the most active curative agent for cervical cancer. Brachytherapy is the most important component of the radiation treatment plan. With aggressive radiotherapeutic management, tumours with significant bulk are effectively treated and cured. The benefit of concurrent chemotherapy will be better defined following maturation of a generation of GOG and RTOG trials recently closed or analysed. The challenge of the twenty-first century is better understanding of the three-dimensional dose distribution of both brachytherapy and external beam in order to reduce late complications while improving dose delivery and ultimately increasing the cure rate.

REFERENCES

1. Coia, L., Won, L., Lanciano, R., *et al.* (1990) The patterns of care outcome study for cancer of the uterine cervix: results of the 2nd National Practice Survey. *Cancer*, **66**, 2451.
2. Corn, B., Hanlon, A., Pajak, T., *et al.* (1994) Technically accurate intracavitary insertions improve pelvic control and survival among patients with locally advanced carcinoma of the uterine cervix. *Gynecologic Oncology*, **53**, 294–300.
3. Prempree, T. (1983) Parametrial impact in stage IIIB cancer of the cervix. III. A five-year study. *Cancer*, **52**, 748.
4. Martinez, A., Edmundson, G., Cox, R., *et al.* (1985) Combination of external beam irradiation and multiple-site perineal applicator (Mupit) for treatment of locally advanced or recurrent prostatic, anorectal, and gynecologic malignancies. *International Journal of Radiation Oncology, Biology and Physiology*, **11**, 391.
5. ICRU Report No. 38 (1985) Dose and volume specification for reporting intracavitary therapy in gynecology. International Commission of Radiation Units and Measurements, Bethesda, MD, pp. 1–16.
6. Lanciano, R., Pajak, T., Martz, K., *et al.* (1993) The influence of treatment time on outcome for squamous cell cancer of the uterine cervix treated with radiation: a patterns of care study. *International Journal of Radiation Oncology, Biology and Physiology*, **25**, 391–7.
7. Patel, F., Sharma, S., Negi, P. *et al.* (1993) Low dose rate vs high dose rate brachytherapy in the treatment of carcinoma of the uterine cervix: a clinical trial. *International Journal of Radiation, Oncology, Biology and Physicology*, **28**, 335–41.
8. Teshima, T., Inoue, T., Ikeda, H. *et al.* (1983) High-dose rate and low-dose rate introeavitery therapy for carcinoma of the uterine cervix. *Cancer*, **72**, 2408–14.
9. Russell, A., Walter, J., Anderson, M., *et al.* (1992) Sagittal magnetic resonance imaging in the design of lateral radiation treatment portals for patients with locally advanced squamous cancer of the cervix. *International Journal of Radiation Oncology, Biology and Physiology*, **23**, 449.
10. Fyles, A., Keane, T., Barton, M., *et al.* (1992) The effect of treatment duration in the local control of cervix cancer. *Radiotherapy Oncology*, **25**, 273–9.
11. Girinsky, T., Rey, A., Roche, B., *et al.* (1993) Overall treatment time in advanced cervical carcinomas: a critical parameter in treatment outcome. *International Journal of Radiation Oncology, Biology and Physiology*, **27**, 1051–6.
12. Perez, C., Grigsby, P., Castro-Vita, H., *et al.* (1995) Carcinoma of the uterine cervix. I. Impact of prolongation of overall treatment time and timing of brachytherapy on outcome of radiation therapy. *International Journal of Radiation Oncology, Biology and Physiology*, **32**, 1275–88.
13. Petereit, D., Sarkaria, J., Chappell, R., *et al.* (1995) The adverse effect of treatment prolongation in cervical carcinoma. *International Journal of Radiation Oncology, Biology and Physiology*, **32**, 1301–7.
14. Lanciano, R., Won, M., Coia, L., *et al.* (1991) Pretreatment and treatment factors associated with improved outcome in squamous cell carcinoma of the uterine cervix: a final report of the 1973 and 1978 patterns of care studies. *International Journal of Radiation Oncology, Biology and Physiology*, **20**, 667–76.
15. Choy, D., Wong, L., Sham, J., *et al.* (1993) Dose–tumor response for carcinoma of cervix: an analysis of 594 patients treated by radiotherapy. *Gynecologic Oncology*, **49**, 311–17.
16. Eifel, P., Morris, M., Wharton, J., *et al.* (1994) The influence of tumor size and morphology on the outcome of patients with FIGO stage IB squamous cell carcinoma of the uterine cervix. *International Journal of Radiation Oncology, Biology and Physiology*, **29**, 9–16.

17. Grigsby, P.W. and Perez, C. (1991) Radiotherapy alone for medically inoperable carcinoma of the cervix: stage IA and carcinoma *in situ*. *International Journal of Radiation Oncology, Biology and Physiology*, **21**, 375–8.

18. Landoni, F., Maneo, A., Colombo, A., *et al.* (1997) Radical surgery or radiotherapy for cervical carcinoma stage IB–IIA. A randomized study. *Lancet*, **350**, 535–40.

19. NIH Consensus Statement Online (1996) 1–3 April **14**(1), 1–38.

20. Shephard, J. (1995) Staging announcement, FIGO staging of gynecologic cancers; cervix and vulva. *International Journal of Gynecological Cancer*, **5**, 319.

21. Nelson, A., Fletcher, G. and Wharton, T. (1975) Indications for adjunctive conservative extrafascial hysterectomy in selected cases of carcinoma of the uterine cervix. *American Journal of Roentgenology, Radium Therapy and Nuclear Medicine*, **123**, 91.

22. Einhorn, N., Patek, E. and Sjoberg, B. (1985) Outcome of different treatment modifications in cervix carcinoma, stage IB and IIA: observation in a well-defined Swedish population. *Cancer*, **55**, 949.

23. Gallion, H., Van Nagell, J., Donaldson, G., *et al.* (1985) Combined radiation therapy and extrafascial hysterectomy in the treatment of stage IB barrel-shaped cervical cancer. *Cancer*, **56**, 262.

24. Thoms, W., Jr, Eifel, P., Smith, T., *et al.* (1992) Bulky endocervical carcinoma: a 23-year experience. *International Journal of Radiation Oncology, Biology and Physiology*, **23**, 491–9.

25. Lowrey, G., Mendelhall, W. and Million, R. (1992) Stage IB or IIA–B carcinoma of the intact uterine cervix treated with irradiation: a multivariate analysis. *International Journal of Radiation Oncology, Biology and Physiology*, **24**, 205–10.

26. Perez, C. and Kao, M. (1985) Radiation therapy alone or combined with surgery in the treatment of barrel-shaped carcinoma of the uterine cervix (stages IB, IIA, IIB). *International Journal of Radiation Oncology, Biology and Physiology*, **11**, 1903–9.

27. Eifel, P., Thoms, W., Smith, T., *et al.* (1994) The relationship between brachytherapy dose and outcome in patients with bulky endocervical tumors treated with radiation alone. *International Journal of Radiation Oncology, Biology and Physiology*, **28**, 113–18.

28. Girinsky, T., Pejovic-Lenfant, M., Bourhis, J., *et al.* (1989) Prognostic value of hemoglobin concentrations and blood transfusions in advanced carcinoma of the cervix treated by radiation therapy: results of a retrospective study of 386 patients. *International Journal of Radiation, Biology and Physiology*, **16**, 37–42.

29. Bush, R., Jenkin, R., Ailt, W., *et al.* (1978) Definitive evidence for hypoxic cells influencing cure in cancer therapy. *British Journal of Cancer*, **37**, 302–6.

30. Hockel, M., Schlenger, K., Mitze, M., *et al.* (1996) Hypoxia and radiation response in human tumors. *Seminars in Radiation Oncology*, **6**, 3–9.

31. Kapp, D., Fischer, D., Gutierrez, E., *et al.* (1983) Pretreatment prognostic factors in carcinoma of the uterine cervix: a multivariate analysis of the effects of age, stage, histology and blood counts on survival. *International Journal of Radiation Oncology, Biology and Physiology*, **9**, 445–55.

32. Eifel, P.J., Morris, M., Oswald M.J., *et al.* (1990) Adenocarcinoma of the uterine cervix. Prognosis and patterns of failure in 367 cases. *Cancer*, **65**, 2507–14.

33. Eifel, P., Burke, T., Morris, M., *et al.* (1995) Adenocarcinoma as an independent risk factor for disease recurrence in patients with stage IB cervical carcinoma. *Gynecologic Oncology*, **59**, 38–44.

34. Barillot, I., Horiot, J., Cuisenier, J., *et al.* (1993) Carcinoma of the cervical stump: a review of 213 cases. *European Journal of Cancer*, **29A**, 1231–6.

35. Piver, M. and Chung, W. (1975) Prognostic significance of cervical lesion size and pelvic node metastasis in cervical carcinoma. *Obstetrics and Gynecology*, **46**, 507–10.

36. Hsu, C., Cheng, Y. and Su, S. (1972) Prognosis of uterine cervical cancer with extensive lymph node metastasis: special emphasis on the value of pelvic lymphadenotomy in the surgical treatment of uterine cervical cancer. *American Journal of Obstetrics and Gynecology*, **114**, 954–62.

37. Thomas, G. and Dembo, A. (1991) Is there a role for adjuvant pelvic radiotherapy after radical hysterectomy in early stage cervical cancer? *International Journal of Gynecological Cancer*, **1**, 1–8.

38. Morrow, C. (1980) Is pelvic radiation beneficial in the postoperative management of stage Ib squamous cell carcinoma of the cervix with pelvic node metastasis treated by radical hysterectomy and pelvic lymphadenectomy? *Gynecologie Oncology*, **10**, 105–10.

39. Curtin, J., Hoskins, W., Venkatraman, E., *et al.* (1996) Adjuvant chemotherapy versus chemotherapy plus pelvic irradiation for high-risk cervical cancer patients after radical hysterec-

tomy and pelvic lymphadenectomy (RH-PLND): a randomized phase III trial. *Gynecologic Oncology*, **61**, 3–10.

40. Hopkins, M., Peters, W.A., Anderson, W., *et al.* (1990) Invasive cervical cancer treated initially by standard hysterectomy. *Gynecologic Oncology*, **36**, 7–12.

41. Roman, L.D., Morris, M., Mitchell, M.F. *et al.* (1993) Prognostic factors for patients undergoing simple hysterectomy in the presence of invasive cancer of the cervix. *Gynecologic Oncology*, **50**, 179–84.

42. Souhami, L., Gil, R., Allan, S., *et al.* (1991) A randomized trial of chemotherapy followed by pelvic radiation therapy in stage IIIB carcinoma of the cervix. *Journal of Clinical Oncology*, **9**, 970–7.

43. Kumar, L., Kaushal, M., Biswal, S., *et al.* (1994) Chemotherapy followed by radiotherapy versus radiotherapy alone in locally advanced cervical cancer: a randomized study. *Gynecologic Oncology*, **54**, 307–15.

44. Tattersall, M., Lorvidhaya, V., Vootiprux, V., *et al.* (1995) Randomized trial of epirubicin and cisplatin chemotherapy followed by pelvic radiation in locally advanced cervical cancer. *Journal of Clinical Oncology*, **13**, 444–51.

45. Hreshchyshyn, M., Aron, B., Boronow, R., *et al.* (1979) Hydroxyurea or placebo combined with radiation to treat stages IIB and IV cervical cancer confined to the pelvis. *International Journal of Radiation Oncology, Biology and Physiology*, **5**, 317–22.

46. Stehman, F., Bundy, B., Thomas, G., *et al.* (1993) Hydroxyurea versus misonidazole with radiation in cervical carcinoma: long-term followup of a Gynecologic Oncology Group trial. *Journal of Clinical Oncology*, **11**, 1523–8.

47. Overgaard, J., Bentzen, S., Kolstad, P., *et al.* (1989) Misonidazole combined with radiotherapy in the treatment of carcinoma of the uterine cervix. *International Journal of Radiation Oncology, Biology and Physiology*, **16**, 1069–72.

48. Heller, P., Malfetano, J., Bundy, B., *et al.* (1990) Clinical-pathologic study of stage IIB, III, and IVA carcinoma of the cervix: extended diagnostic evaluation for paraaortic node metastasis – a Gynecologic Oncology Group study. *Gynecologic Oncology*, **38**, 425–30.

49. Lanciano, R. and Corn, B. (1994) The role of surgical staging for cervical cancer. *Seminars in Radiation Oncology*, **4**, 46–51.

50. Twiggs, L., Potish, R., George, R., *et al.* (1984) Pretreatment extraperitoneal surgical staging in primary carcinoma of the cervix uteri. *Surgical Gynecologic Obstetrics*, **158**, 243–50.

51. Stehman, F., Bundy, B., DiSaia, P., *et al.* (1991) Carcinoma of the cervix treated with radiation therapy. A multivariate analysis of prognostic variables in the Gynecologic Oncology Group. *Cancer*, **67**, 2776–85.

52. Fletcher, G. (1980) *Textbook of Radiotherapy*, 3rd edn, Lea & Febiger, Philadelphia, PA, pp. 720–89.

53. Horiot, J., Pigneux, J., Pourquier, H., *et al.* (1988) Radiotherapy alone in carcinoma of the intact uterine cervix according to G. H. Fletcher guidelines: a French cooperative study of 1383 cases. *International Journal of Radiation Oncology, Biology and Physiology*, **14**, 605–11.

54. Perez, C., Grigsby, P., Nene, S., *et al.* (1992) Effect of tumor size on the prognosis of carcinoma of the uterine cervix treated with irradiation alone. *Cancer*, **69**, 2796–806.

55. Rotman, M., Aziz, H. and Eifel, P. (1994) Irradiation of pelvic and para-aortic nodes in carcinoma of the cervix. *Seminars in Radiation Oncology*, **4**, 23–9.

56. Vigliotti, A., Wen, B., Hussey, D., *et al.* (1992) Extended field irradiation for carcinoma of the uterine cervix with positive periaortic nodes. *International Journal of Radiation Oncology, Biology and Physiology*, **23**, 501–9.

57. Haie, L., Pesovic, M., Gerbaulet, A., *et al.* (1988) Is prophylactic para-aortic irradiation clinical trial of the EORTC radiotherapy group. *Radiotherapy Oncology*, **11**, 101–12.

58. Rotman, M., Pajak, T., Choi, T., *et al.* (1995) Prophylactic extended-field irradiation of para-aortic lymph nodes in stages IIB and bulky IIA cervical carcinomas. Ten year treatment results of RTOG 79-20. *Journal of the American Medical Association*, **274**, 387–93.

59. Lanciano, R. (1996) Radiotherapy for the treatment of locally recurrent cervical cancer. *Journal of the NCI Monographs*, **21**, 113–15.

60. Penalver, M., Barreau, G., Sevin, B., *et al.* (1996) Surgery for the treatment of locally recurrent disease. *Journal of the National Cancer Institute Monographs*, **21**, 117–22.

61. Eifel, P.J., Levenback, C., Wharton, J.T., *et al.* (1995) Time course and incidence of late complications in patients treated with radiation therapy for FIGO stage IB carcinoma of the uterine cervix. *International Journal of Radiation Oncology, Biology and Physiology*, **32**, 1289–300.

62. Lanciano, R., Martz, K., Montana, G., *et al.* (1992) Influence of age, prior abdominal surgery, fraction size, and dose on complications after radiation therapy for squamous cell cancer of the uterine cervix. A patterns of care study. *Cancer*, **69**, 2124–30.

63. Hamberger, A., Unal, A., Gershenson, D., *et al.* (1983) Analysis of the severe complications of irradiation of carcinoma of the cervix: whole pelvis irradiation and intracavitary radium. *International Journal of Radiation Oncology, Biology and Physiology*, **9**, 367–71.

64. Stryker, J., Bartholomew, M., Velkley, D., *et al.* (1988) Bladder and rectal complications following radiotherapy for cervix cancer. *Gynecologic Oncology*, **29**, 1–11.

65. Perez, C., Camel, H., Kuske, R., *et al.* (1986) Radiation therapy alone in the treatment of carcinoma of the uterine cervix: a 20-year experience. *Gynecologic Oncology*, **23**, 127–40.

66. Crook, J., Esche, B., Chaplain, G., *et al.* (1987) Dose–volume analysis and the prevention of radiation sequelae in cervical cancer. *Radiotherapy Oncology*, **8**, 321–32.

67. Yeoh, E., Sun, W., Russo, A., *et al.* (1996) A retrospective study of the effects of pelvic irradiation for gynecological cancer on anorectal function. *International Journal of Radiation Oncology, Biology and Physiology*, **35**, 1003–10.

68. Levenback, C., Eifel, P., Burke, T., *et al.* (1994) Hemorrhagic cystitis following radiotherapy for stage Ib cancer of the cervix. *Gynecologic Oncology*, **55**, 206–10.

69. Kottmeier, H. (1964) Complications following radiation therapy in carcinoma of the cervix and their treatment. *American Journal of Obstetrics and Gynecology*, **88**, 854–6.

70. Georgiou, A., Grigsby, P. and Perez, C. (1993) Radiation induced lumbosacral plexopathy in gynecologic tumors: clinical findings and dosimetric analysis. *International Journal of Radiation Oncology, Biology and Physiology*, **26**, 479–82.

71. Mumber, M., Greven, K. and Haygood, T. (1997) Pelvic insufficiency fractures associated with radiation atrophy: clinical recognition and diagnostic evaluation. *Skeletal Radiology*, **26**, 94–9.

72. Bruner, D., Lanciano, R., Keegan, M., *et al.* (1993) Vaginal stenosis and sexual function following intracavitary radiation for the treatment of cervical and endometrial carcinoma. *International Journal of Radiation Oncology*, **27**, 825–30.

73. Perez, C., Grigsby, P., Camel, H., *et al.* (1995) Irradiation alone or combined with surgery in stage IB, IIA, and IIB carcinoma of uterine cervix: update of a nonrandomized comparison. *International Journal of Radiation Oncology, Biology and Physiology*, **31**, 703–16.

74. Thomas, G. and Dembo, A. (1991) Carcinoma of the cervix. In *Combined Radiotherapy and Chemotherapy in Clinical Oncology* (ed. A. Horwich). Edward Arnold, London, pp. 150–64.

75. Perez, C., Fox, S., Lockett, M., *et al.* (1991) Impact of dose in outcome of irradiation alone in carcinoma of the uterine cervix: analysis of two different methods. *International Journal of Radiation Oncology, Biology and Physiology*, **21**, 885–98.

76. Eifel, P. and Logsdon, M. (1996) FIGO stage IIIB squamous cell carcinoma of the uterine cervix: natural history, treatment results, and prognostic factors. *International Journal of Radiation Oncology, Biology and Physiology*, **36**, 217.

77. Fagundes, H., Perez, C., Grigsby, P., *et al.* (1992) Distant metastases after irradiation alone in carcinoma of the uterine cervix. *International Journal of Radiation Oncology, Biology and Physiology*, **24**, 197–204.

23

CHEMOTHERAPY FOR CERVICAL CARCINOMA

C.J. Poole

INTRODUCTION

Despite its first evaluation in the 1960s (reviewed in Malkasian *et al.* [1]), and the demonstration of response rates approaching 70% with combination regimens in the late 1980s (Buxton [2]), the ability of chemotherapy to alter the natural history of cervical carcinoma remains unproven, and its role in the management of this disease remains limited. As well as summarizing the state of the art, this chapter will endeavour to identify the reasons for this lack of progress and speculate about its further development.

The usual rationale for the use of chemotherapy in malignant disease is predicated on the limitations of local treatments in the face of a metastatic process and the axiomatic desire to address a systemic disease with a systemic therapy. For some tumour types, adjuvant chemotherapy, used after surgery in patients presenting with ostensibly localized disease, can prolong both relapse-free and overall survival through its effects on occult micrometastases. In other circumstances, primary pre-operative or neoadjuvant chemotherapy may improve the outcome of treatment for locally advanced tumours, by reducing the extent of the surgery otherwise required, or facilitating surgery for otherwise inoperable disease. Neoadjuvant chemothera-

py can thus preserve limb function in osteosarcoma, for example, or enhance surgical cosmesis in breast cancer – whilst evidently providing the same degree of control over micrometastases as adjuvant post-operative therapy.

For cervical carcinoma in developed countries, preoccupation with such concerns has been tempered by its increasing presentation as an anatomically localized disease, in which frank or occult metastases have proven diminishingly frequent: in contrast with most other common solid tumours, the great majority of patients with cervical neoplasia now present with non-invasive or early stage disease, and are successfully treated with surgery (see Chapters 19–21) or radical radiotherapy (see Chapter 22). This good fortune owes much to the practical efficacy of population screening with Papanicolaou smears in reducing the number of patients presenting with advanced disease, and in the USA, deaths have fallen from 60,000 per year in 1950 to fewer than 5000 per year in 1995 [3].

However, there remains an unfortunate minority of patients with cervical carcinoma who are failed by screening programmes and latter-day therapeutics, and for whom more effective local and systemic treatments are necessary. They include women presenting with bulky pelvic disease or tumours with

Cancer and Pre-cancer of the Cervix
Edited by D.M. Luesley and R. Barrasso. Published in 1998 by Chapman & Hall, London. ISBN 0 412 56600 1.

lymphovascular invasion, which are poorly controlled by radiotherapy or radical surgery [4–8], women presenting with distant metastases, and those relapsing after local radical therapy. Ironically, their decreasing numbers have made the systematic development of new treatments through large randomized clinical trials increasingly difficult, and have led to the premature closure of several regional studies in the UK, North America and Australasia through poor accrual [9, 10].

This 'problem' has had self-evident implications for the design and execution of future trials, and there is therefore an emerging consensus that definition of the role of chemotherapy in the management of cervical carcinoma requires a coordinated international effort to examine its use in several different clinical contexts and alongside other treatment modalities. These include: as adjuvant therapy after surgery when histological features or pathological stage indicate an increased risk of systemic relapse; as primary neoadjuvant treatment in those patients presenting with bulky disease, for whom radiotherapy alone often fails and in whom primary surgery is impracticable; as concurrent chemo-radiotherapy (radiation sensitization), an alternative approach in locally advanced disease, and lastly, as palliative therapy in the management of recurrent or metastatic tumours. The conventions of evidence-based medicine demand that each new approach be compared with 'standard treatment' within randomized prospective phase III clinical trials. Integrated multimodality adjuvant and neoadjuvant approaches are discussed in Chapter 19. This chapter will next address the use of chemotherapy in advanced disease and review its development through clinical trials.

CHEMOTHERAPY FOR RECURRENT OR METASTATIC CERVICAL CARCINOMA

Chemotherapy was first evaluated in cervical carcinoma in patients with recurrent disease, or those presenting with metastases at the out-

set, and to date this remains the context for trials of new agents. Unfortunately, comparative phase III trials in advanced disease are few, the majority of studies being uncontrolled. Review of the available literature can therefore yield only clues rather than conclusions about the relative potency and toxicity of different regimens. In the view of most authors, single agent cisplatin represents the most rational choice in these circumstances: certainly, cisplatin's activity is best documented, with several hundred patients' experiences reported, and its use as a single agent obviously avoids the increased toxicity intrinsic to many combination regimens. However, the poverty of the response rates generally reported for cisplatin, typically just 20–30%, has the implication that few clinicians would hesitate to offer a patient with recurrent disease treatment with an investigational phase II drug as an alternative, retaining an option for platinum second-line.

Guidelines for the conduct of phase II clinical trials in previously untreated patients with chemoresponsive (though incurable) tumours have been suggested by Cullen, following the demonstration that exposure to inactive experimental agents may prejudice subsequent salvage therapy with standard drugs [11, 12]. An arguably better approach might be to apply prognostic models for platinum sensitivity, to identify poor risk patients who might better be offered investigational agents first-line. This would avoid pointless quality-of-life-degrading toxicity in patients who were unlikely to respond to platinum, and eventually discriminate drugs with activity complementary to platinum. Although well documented, such prognostic factors may be generally less well appreciated than those applying to ovarian cancer or breast cancer, and have certainly had rather less impact on day-to-day clinical practice and trial design.

TRIAL DESIGN

Historically, the activity of new chemotherapeutic drugs in particular tumour types has been explored in phase II clinical trials, undertaken in small groups of patients with measurable disease whose other therapeutic options are limited. A drug's 'activity' in this context is generally defined by an overall response rate of just 20%, a figure whose modesty reflects the relatively refractory nature of such tumours, and whose conventional significance lies not in its determining clinical utility, but in providing the rationale for the drug's further evaluation, perhaps at an earlier point in the natural history of a disease, and in combination with other drugs. Phase II trials typically employ a two-stage design to provide a ≤5% probability of falsely dismissing a drug as inactive, but accommodate a 50% chance of incorrectly identifying a drug as active [13–15]. This inequity reflects the paucity of active agents for solid tumours in general and the more serious implications of a false-negative result. An alternative approach is the Flemming one-sample multiple testing procedure. This may be particularly relevant to cervical cancer. It recognizes that the potential value of a new compound depends on its activity in relation to other drugs in the same disease (e.g. cisplatin), and defines a response probability which, if attained, might imply the necessity of formal comparison with standard treatment [16].

For an endpoint such as a response rate of 20% to have any reference value, it has become increasingly recognized that care must be taken to standardize patient selection criteria to limit the impact of non-treatment-related prognostic variables on response [17]. For several other solid tumours, these have been defined in great detail: typically, they may include performance status, serum albumin, tumour stage at presentation, treatment-free interval or time to relapse, extent of previous treatment, and number and location of metastatic sites and histological grade [18–24]. In ovarian carcinoma, for example, most phase II trials now specify exposure to one prior platinum course only, and a treatment-free interval of up to 12 months. This

approach recognizes that patients with a longer disease-free interval are generally better served by platinum re-exposure, with the expectation of a response rate in excess of 20%, and focuses efforts on identifying drugs with platinum non-cross resistance: furthermore, it avoids the inclusion of good prognostic patients in a study of a new and active drug, which might well provide a misleadingly high response rate.

Unfortunately, there is no tradition of standardizing patient selection criteria for known prognostic influences on treatment outcome in phase II trials in cervical carcinoma, and most studies have been undertaken in biologically heterogeneous populations, in respect of both tumour-specific and host-specific factors. Whilst this heterogeneity may reflect the relatively small numbers of candidate patients, it is now nearly 10 years since multivariate techniques were first applied to the analysis of factors determining response, and the frequently inadequate definition of patient characteristics in more recent studies represents a general failure of the peer review system. Without relevant explicit information about the characteristics of the population studied, false negative or false positive results remain a considerable risk. A randomized phase II design may also be useful in highlighting false positive results, in which case better than expected activity is easily identified in the reference (standard treatment) arm. This approach seems particularly applicable to platinum combination studies in cervical cancer, but it seems sadly neglected [25].

PROGNOSTIC FACTORS FOR CHEMOTHERAPY RESPONSE AND SURVIVAL

Much of the debate about prognostic factors for response to chemotherapy concerns the influence of prior radiotherapy and the anatomical distribution of measurable lesions in relation to the field applied. There is now considerable evidence to suggest that response rates tend to be lower when the indicator lesion, or measurable disease, falls within a previously irradiated field [26–29]. Ironically, standard entry criteria for phase II new drug trials in most other tumour types now specifically exclude such patients, in recognition that the information gained is of limited significance in determining a drug's eventual potential in the treatment of early-stage disease.

Whilst cross-resistance between radiotherapy and chemotherapy was once held to reflect poor tumour vascularization and consequently impaired tumour tissue access by cytotoxic drugs, we now recognize that radio-resistant tumours may also manifest increased DNA repair capability, enhanced free radical quenching or raised threshold for apoptosis, phenotypic qualities which may all impair chemosensitivity [30–34]. Furthermore, pelvic radiotherapy may also deplete haematopoetic stem cells, exacerbate chemotherapy-induced myelosuppression, necessitate dose reduction and prejudice response. Although there is no universal consensus about prior irradiation as a poor prognostic factor, it seems likely that inconsistencies in this regard reflect other variables, such as the influence of time elapsed since irradiation. Late recurrences tend to be more sensitive.

Unfortunately, the overall picture is further confused by semantic inconsistencies in analytical methodology. For example, some studies suggest that the presence of a 'pelvic mass' may be a worse prognostic factor for response than 'prior pelvic radiotherapy'. Multivariate analysis could yield such a counterintuitive finding if the majority of patients had received previous 'pelvic radiotherapy', in which case the category 'pelvic mass' might be a surrogate for an irradiated indicator lesion. The category of 'prior pelvic radiotherapy' might then not discriminate non-responders if a substantial proportion of those previously irradiated had no pelvic mass but responsive measurable extra pelvic-metastatic disease. Irradiated pelvic masses are frequently associated with fistulae and hydronephrosis complications which increase the risks of

chemotherapy: both may predispose to infection, and hydronephrosis may impair the clearance of drugs such as cisplatin, methotrexate and bleomycin and the metabolites of others such as ifosfamide, all with potentially adverse implications for serious toxicity and treatment withdrawal [35–37].

The dose of radiotherapy previously delivered might also exert an effect on drug sensitivity at relapse: for example, in some of the studies where the radiotherapy doses are documented, it is clear they often fail to match the 75–90 Gy usually specified in North American radiotherapy protocols [38]. Plausibly, this might be relevant to the discrepancies evident between various reports of ifosfamide activity [39, 40].

Another interesting prognostic factor is the presence of measurable metastases outside the irradiated field. Although these are often held to be intrinsically more chemo-responsive, there are two important caveats to this generalization. First, the presence of distant metastases may itself have adverse implications for tumour natural history and imply early disease progression, perhaps confounding the potential impact of a response on survival. This may dissolve the paradox of why some combination regimens have been shown to have high response rates but little or no survival benefit. Second, distant metastases may themselves contribute little to the bulk of measurable disease in patients with concurrent pelvic disease, as distant lesions are typically much smaller. The application of UICC response criteria in these circumstances (a reduction of 50% in the sum of the products of the perpendicular diameters of all measurable lesions) undoubtedly weights the response towards that of the irradiated site, where a large pelvic mass all too often fails to respond. The literature addressing prognosis following chemotherapy for advanced or recurrent disease is now usefully reviewed in detail.

Bonomi *et al.* analysed the following characteristics of 444 patients treated in a randomized trial comparing three different platinum regimens for their influence on response and survival: International Federation of Gynaecology and Obstetrics (FIGO) stage, histological grade, age, sites of disease, sites of measurable lesions, performance status, previous radiation, body surface area, weight and disease-free interval [41]. Univariate analysis showed that better performance status, ($p = 0.0001$), increasing age ($p = 0.0002$); longer disease-free interval ($p = 0.0005$) and more poorly differentiated tumours ($p = 0.07$) were associated with a greater likelihood of achieving a complete or partial response. In contrast, larger tumours were associated with a lower response rate. Performance status, age, disease-free interval and tumour grade were also significantly related to response in multivariate analysis, but the influence of tumour size failed to reach significance in this sample. Better performance status, longer disease-free survival, increasing age and smaller tumour masses were associated with longer survival.

Potter *et al.* undertook a retrospective analysis of the outcome of treatment in 74 patients treated with cisplatin 50–100 mg/m^2 3-weekly in Birmingham, Alabama between 1977 and 1987 [26]. Of these, 68 patients were evaluable for response using UICC criteria. Patient charts were reviewed for age, FIGO stage, date of diagnosis and recurrence, location of disease and type of previous therapy; 53 patients had been treated with radiotherapy and 15 with both radiotherapy and surgery. The overall response rate was 40%, including 16% with complete responses.

In this series, univariate analysis revealed a significant relationship between tumour site and probability of response. Overall response rates for isolated chest metastases were 73%, compared with 22% in patients with pelvic recurrence ($p = 0.0007$). For disease in the chest and elsewhere, the response was 33% ($p < 0.05$); for pelvis and other sites, 29%; for chest and pelvis, just 17%. The complete response rate for the chest was 53% (eight patients); there were no complete responses in the pelvis. A further three patients achieved a complete response in the lungs but were classed differently because of non-response elsewhere. Despite these findings, there was no significant difference between the survival achieved by patients with pelvic disease (14 months) and those with isolated chest disease (23 months, $p = 0.25$), but the survival of patients with both pelvic and chest disease was significantly shorter (8 months vs 23 months, $p = 0.04$). There was a non-significant trend towards better survival in early stage patients, and higher response rate in pelvic tumours ≤ 6 cm in diameter. However, there was no evidence of a relationship between response or survival and interval from diagnosis to relapse, or age.

Buxton analysed the results of five phase II trials of chemotherapy involving a total of 172 patients with recurrent disease, undertaken between 1982 and 1988 by two UK cooperative groups, the West Midlands Gyneoncology Group and the London Gyne Group. He used univariate and multivariate techniques to determine the factors predicting response and survival, and to elucidate whether the striking results previously reported for bleomycin, ifosfamide and cisplatin (BIP) [2] might better be explained by differences in patient selection or a treatment-specific effect [42]. As well as the BIP study, the analysis included two separate trials addressing ifosfamide [36, 43], ifosfamide with cisplatin [44] and ifosfamide with bleomycin. More detailed patient characteristics were presented than had previously been made available and were uniformly applied to all studies.

Buxton's analysis revealed significant differences between the BIP study population and the others: these including histological type (squamous vs non-squamous; $\chi^2 = 8.01$, $p = 0.0047$), FIGO stage at presentation (Stage I vs Stages II–IV; $\chi^2 = 4.97$, $p = 0.026$), sites of relapse (pelvic alone vs other $\chi^2 = 13.22$, $p = 0.0013$), WHO performance status at entry (0 vs 1–3; $\chi^2 = 19.67$, $p = 0.00001$), as well as response to chemotherapy ($\chi^2 = 22.55$, $p = 0.00001$). There were no significant differences between studies in terms of patient age, time to relapse after primary therapy, tumour differentiation, type of primary treatment and irradiation status of relapse site. Univariate analysis of patient characteristics showed type of chemotherapy regimen (BIP vs others; $\chi^2 = 25.7$, $p < 0.0001$), and WHO performance status (0 vs 1–3; $\chi^2 = 13.3$, $p = 0.0004$) to be significantly associated with likelihood of response. Multivariate analysis revealed that performance status remained an independent determinant of response after controlling for chemotherapy regimen, but only just retained significance. All other variables were non-significant. Only WHO performance status and complete response to chemotherapy proved predictive of survival.

These provide extremely interesting data about the BIP trials, but prompt more questions still: it seems very likely that most patients were initially selected using a nomogram previously developed to minimize the risk of ifosfamide-related encephalopathy [39]. The nomogram utilizes serum albumin, a well-known powerful prognostic indicator for a range of other conditions (including ovarian cancer [21]), serum creatinine and the presence or absence of a pelvic mass to calculate the probability of treatment remaining uncomplicated. The nomogram was not used in the other studies, and this raises the possibility of its role in the occult selection of better prognosis patients. It is ostensibly surprising that time to relapse following 'primary therapy' is insignificant in this analysis. However, although we are given no specific data about time elapsed since 'radiotherapy', a characteristic whose significance is agreed by several other prognostic studies, the proportion of patients whose primary therapy included radiotherapy varied between 84% and 87% across the four studies supplying 95% of the patients analysed and it seems admittedly unlikely that this biologically more relevant characteristic might supply a different answer in this group. Futhermore, it seems improbable that the use of median and range to describe the distribution of time to progression values might mask biologically relevant interstudy differences.

Another retrospective study using multivariate analytical techniques was undertaken by Zanetta *et al.* in Milan, in 140 patients treated with various platinum-containing regimens for advanced, recurrent or persistent squamous cell carcinoma no longer amenable to radical surgery or radiotherapy [45]. One-hundred-and-fifteen patients had received prior radiotherapy, 61 with surgery in addition. Four patients had surgery alone; 12 had treatment that included chemotherapy and nine had no prior treatment. Age, performance status, histological type and grade, previous irradiation, interval from start of primary treatment and from irradiation, site of tumour and therapeutic regimen were all considered as possible predictors of response and survival. Stage was not analysed.

The six different platinum regimens used and their respective response rates are shown in Table 23.1. Of 140 patients, 137 were followed until death. Four died of treatment complications. Two patients were alive in complete remission 74 and 84 months from the start of chemotherapy, after partial responses consolidated with radiotherapy and tamoxifen, respectively. A further two partial responses to tamoxifen were seen in patients with progressive disease, out of a total of 117 treated with 40 mg/day on completion of chemotherapy. The median duration of response was comparable in all treatment regimens, between 5 and 9 weeks.

Response by tumour site is shown in Table 23.2: lung, liver and soft tissue metastases were more likely to respond than disease elsewhere. Predictors of response are shown in Table 23.3: they include site of disease, interval from irradiation, and performance status. Tumours relapsing within 12 months of radiotherapy were less likely to respond than those recurring later (21% vs 41%). Time from diagnosis had no significance. However, in multivariate analysis, only performance status and interval from radiotherapy retained significant predictive power.

Table 23.1 Platinum regimens employed by Zanetta *et al.* [45]: factors predicting response to chemotherapy and survival in patients with metastatic or recurrent squamous cell cervical carcinoma: a multivariate analysis

| Regimen | Number of patients | Response | | | |
		CR	PR	Total	%
BEMP	48	4	13	17	35
BIP	18	–	6	6	33
P	28	–	6	6	21
MP	21	1	6	7	33
VBMP	25	1	4	5	20

BEMP, bleomycin 15 mg days 2, 3 and 4; vindesine 3 mg/m^2 days 1, 8; mitomycin-C, 8 mg/m^2 day 5 (course 1, 3); cisplatin 50 mg/m^2 day 1 (repeated 3-weekly).
BIP, bleomycin 30 mg day 1, ifosfamide 5 g/m^2 day 2, mesna 5 + 3 g/m^2 days 2 and 3; cisplatin 50 mg/m^2, day 2 (repeated 3-weekly).
P, cisplatin 1 mg/kg, day (repeated weekly).
Cisplatin 50 mg/m^2, day 1 (repeated 3-weekly).
MP, mitomycin-C 6 mg/m^2 day 1; cisplatin 50 mg/m^2 day 1 (repeated 4-weekly).
VBMP, vincristine 2 mg, day 1; bleomycin 15 mg, days 2–3; mitomycin-C, 8 mg/m^2, day 5 cycles 1 and 3; cisplatin 50 mg/m^2, day 1 (repeated 3-weekly).

Table 23.2 Metastatic or recurrent squamous cell cervical carcinoma: chemotherapy response rate by tumour site [45]

Site	Locations	Objective responses	Percentage
Pelvis	97	16	16.5
Distant nodes	66	23	34
Lung	27	13	48
Liver	10	4	40
Soft tissues	8	7	87.5
Bone	8	0	0
Brain	2	0	0

Table 23.3 Factors predicting response to chemotherapy in patients with metastatic or recurrent squamous cell cervical carcinoma [45]

Factors predicting response	Univariate	Multivariate
Previous irradiation	**$p = 0.168$**	**$p = 0.799$**
Performance status < 1	$p = 0.025$	$p = 0.048$
Interval from irradiation, >1	$p = 0.006$	$p = 0.030$
Extra-pelvic tumour	$p = 0.003$	

Note: non-significant characteristics and *p* values are in bold-face type.

In univariate analysis, interval from diagnosis, objective response, interval from irradiation, tumour site and performance status were all significantly related to survival (see Table 23.4). However, in multivariate analysis, only objective response to chemotherapy, interval from diagnosis (<12 months) and tumour site (pelvic or not) retained significance.

Table 23.4 Factors predicting survival after chemotherapy for metastatic or recurrent squamous cell cervical carcinoma [45]

Factors predicting survival	Univariate	Multivariate
Previous irradiation	**$p = 0.0883$**	**0.9167**
Performance status < 1	$p = 0.0160$	**0.2622**
Pelvic tumour	$p = 0.0027$	0.0031
Interval from irradiation, < 1 year	$p = 0.0005$	**0.7456**
Objective response	$p = 0.0001$	0.0001
Interval from diagnosis, < 1 year	$p = 0.0001$	0.0002

Note: non-significant characteristics and *p* values are in bold-face type.

Conversely, where phase II trial populations also include patients presenting *de novo* with advanced tumours, untreated primary tumours may often prove rather more sensitive than distant metastases, so the proportion of each in a mix of patients' measurable lesions may be a determinant of overall response rate within a particular treatment population. For example, initial reports of cisplatin activity in squamous cell carcinoma of the cervix documented response rates of 44%, with no apparent difference in response by disease location [46]. However, 50% of pelvic responders in this study had received no previous therapy and of those categorized as having 'extra-pelvic' disease, many had concurrent pelvic tumour, a group shown by the Potter study to have a bad prognosis [26]. Thus, it is not surprising perhaps that subsequent trials with single agent cisplatin have shown lower overall response rates [47, 48].

DETERMINANTS OF CHEMOTHERAPY RESPONSE IN PREVIOUSLY UNTREATED PATIENTS

Women presenting with bulky cervical tumours are rarely considered for primary surgery but are instead usually offered radiotherapy, and constitute a poor risk group given their propensity for relapse [5, 49]. Neoadjuvant chemotherapy ahead of radiotherapy has failed to fulfil its initial promise in this context in randomized trials ([10, 50–52]; see Chapter 19 for full discussion) and interest now focuses on alternative approaches using simultaneous chemo-irradiation, or neoadjuvant chemotherapy ahead of surgery; (radiation later ± [53]). In the latter situation, disease progression on chemotherapy augurs ill, as such tumours are often cross-resistant to salvage radiotherapy [54–56]. It would therefore be extremely desirable to identify patients less likely to respond to standard chemotherapy at the outset, and offer them either primary radiotherapy (despite its limitations) or treatment with investigational agents.

Several groups have recently addressed the question of what constitute the dominant prognostic factors for response to primary chemotherapy. Their analyses have yielded different, though not perhaps mutually exclusive, conclusions. These findings are now discussed.

Benedetti Panici *et al.* analysed the results of neoadjuvant chemotherapy in 77 patients presenting with Stage Ib–III tumours, using a regimen comprising cisplatin 100 mg/m^2 day 1, bleomycin 15 mg day 1 and day 8, and methotrexate 300 mg/m^2 day 8 (with folinic acid rescue), repeated every 21 days for three cycles [56]. Tumour size was assessed by EUA and pelvic ultrasonography. No attempt was made to gauge pelvic lymph node involvement pre-operatively. Data were analysed separately for relationship to response and recurrence.

A tumour size of >5 cm, and parametrial involvement were associated with poor responsiveness to chemotherapy. The overall response rate for patients with tumours smaller or equal to 5 cm was 93%, but the percentage dropped to 68% when the tumour was larger than 5 cm. Of the six patients with tumours larger than 6 cm in diameter, three showed no change and one progressed on chemotherapy. Furthermore, only one of seven patients with bilateral pelvic side wall involvement responded, compared with 22 of 24 with less extensive disease. Neither age, histology or tumour grade were significantly associated with response. The trend for the earlier stage to be associated with a higher response did not achieve significance, perhaps reflecting that only nine out of 77 patients had Stage Ib–IIa disease (see Table 23.5). Of the 62 responding patients who underwent surgery, 24% were found to have pathological evidence of lymph node involvement.

At a median follow-up of 27 months (range 15–48), 50 of 62 (81%) patients who had undergone surgery showed no evidence of recurrence. Of 12 patients with recurrence, nine had isolated pelvic recurrences, six were central

Table 23.5   Factors determining response to primary chemotherapy for previously untreated cervical carcinoma (after Benedetti *et al.* [56])

Patient characteristic	Number	Response CR	(%)	PR	(%)	NC/P	(%)	p value
FIGO stage								
Ib–IIa	9	3	(33)	5	(55)	1	(11)	
IIb	35	5	(14)	26	(74)	4	(11)	
III	31	3	(3)	20	(64)	8	(26)	0.18
Histology								
Squamous	60	10	(17)	41	(68)	9	(15)	
Adenocarcinoma	15	1	(7)	10	(67)	4	(27)	0.49
Grade								
Well differentiated	6	1	(17)	5	(83)	0	(–)	
Moderately well diff.	43	5	(12)	31	(72)	7	(16)	
Poorly differentiated	26	5	(19)	15	(58)	6	(23)	0.50
Tumour size (cm)								
≤5 cm	43	9	(21)	31	(72)	3	(7)	
>5 cm	32	2	(6)	20	(62)	10	(32)	0.02
Parametrial involvement (Stage III patients only), clinically defined								
Complete bilateral	7	0	(–)	1	(14)	6	(86)	
< Complete bilateral	24	3	(12)	19	(79)	2	(9)	0.005

CR, complete response; PR, partial response; NC, no change (stable); P, progression.
Complete bilateral: complete bilateral involvement to pelvic side wall.
< Complete bilateral: less than complete bilateral involvement to pelvic side wall.

and three side wall. Two had both pelvic and distant failure, and one had isolated distant metastases. All relapses occurred within 24 months of surgery. Univariate analysis suggested pathologically defined parametrial involvement was the strongest predictor of relapse; 44% of such patients relapsed, compared with 9% of those without parametrial invasion. Relapse also occurred in one out of 20 (5%) patients with cervical infiltration of less than 5 mm, compared with 11 out of 42 (26%) of those with deeper tumour infiltration, and in five of 31 (16%) patients with Stage IIb tumours, compared with seven out of 23 (30%) presenting Stage III. However, Cox multivariate analysis confirmed only parametrial involvement as predictive of relapse; cervical invasion, stage and pathological response added nothing significant to the power of the model. These data are shown in Table 23.6.

These results were updated in 1997, and extended in a larger series of 130 patients with >4 cm Stage Ib–III disease [57]. Factors determining response and 10-year survival were analysed. One-hundred-and-twenty-eight patients were evaluable, with an overall response rate of 83% and a complete response rate of 15%. Cox regression analysis showed FIGO stage, tumour size, parametrial involvement and histology were highly predictive of response and survival. Ten-year survival was 91% for patients presenting with Stage Ib–IIa tumours, 80% for Stage IIb, and 34.5% for Stage III.

Bolis *et al.* [58, 59] by contrast, treated 79 women with histologically confirmed invasive bulky (>4 cm) squamous cell cervical carcinoma, Stages Ib–IIb, with a weekly cisplatin (40 mg/m²) for seven cycles, and ifosfamide 3.5 g/m² (and mesna) with cycles 1, 4 and 7.

Table 23.6 Disease recurrence after chemotherapy and surgery in relation to clinical stage and pathological variables (after Benedetti *et al.* [56])

Characteristic	Number of patients	Number of relapses	Percentage	p value
FIGO stage				
Ib–IIa	8	0	–	
IIb	31	5	16	
III	23	7	30	0.08
Lymph node status				
Negative	47	8	17	
Positive	15	4	26	0.31
Parametrial involvement				
Absent	44	4	9	
Present	18	8	44	0.003
Cervical infiltration				
<5 mm	20	1	5	
≥5 mm	42	11	26	0.04

Pretreatment CT scans were used to gauge pelvic lymph node involvement, but response to chemotherapy was determined using transvaginal ultrasound and clinical assessment. Sixty-six of 79 women underwent subsequent surgery, which included routine pelvic lymphadenectomy. The overall response rate was 70% with 5% complete responses. The response distribution by various patient characteristics is shown in Table 23.7.

Table 23.7 Determinants of response to a cisplatin-based regimen as neoadjuvant chemotherapy in Stage Ib–IIb invasive cervical cancer after Bolis *et al.* [59]

Characteristic	No. of patients	CR	PR	Odds ratio (95% CI)[a]
Total	79	4	51	–
Age (years)				
≤45	26	0	19	1[b]
≥45	53	4	32	0.8 (0.2–2.7)
Stage				
Ib	39	1	27	1[b]
IIa	15	1	9	0.8
IIb	25	2	15	0.8
χ^2 trend				p = not significant
Grade				
2	39	3	26	1[b]
3	40	1	25	0.6 (0.2–1.6)
Lymph node status				
Negative	42[c]	2	33	1[b]
Positive	24	0	13	0.2 (0.1–0.4)

[a]CI, confidence interval. [b]Reference category. [c]In 13 cases lymph node status was not available.

In this series, the major determinant of response to chemotherapy was the pre-treatment lymph node status, information available for 66 out of 79 women analysed. In women with positive nodes as gauged on computerized tomography (CT) scan, the response rate was 50%; in women with no apparent lymphadenopathy, the response rate was 85%. Unfortunately, CT definitions of lymph node involvement are not explicitly defined, although the sensitivity and specificity of the means of assessment is referenced at 80% [60]. However, of 24 women with pre-treatment-assessed positive nodes, 15 had tumour involvement confirmed at surgery and nine were negative. Of 42 judged as having negative nodes initially, only two were subsequently shown to have pathological nodes at surgery, and 40 were negative. Several other studies have related pelvic lymph node status after chemotherapy to survival [61, 62].

Other authors' findings are broadly similar, if less detailed: they indicate tumour size to be all important, with reduced responsiveness above a 5 cm threshold [63, 64]. Some have suggested that the number of chemotherapy cycles may impact on response, but that stage may have more influence in determining complete response rate [63, 65]. Others claim depth of tumour invasion is critical in determining response to neoadjuvant chemotherapy [66].

INFLUENCE OF HISTOLOGY ON OUTCOME OF CHEMOTHERAPY

ADENOCARCINOMA AND SMALL CELL CARCINOMA OF THE CERVIX

Little is known for certain about the impact of cervical tumour histology on chemotherapy outcome. Contrary to tradition, the available data suggest little if any difference between adenocarcinoma and squamous cell carcinoma [56, 67–69].

Small cell carcinoma of the cervix is a very different disease and by consensus merits exclusion from studies involving squamous cell or adenocarcinoma. Clinically, it behaves very aggressively with a marked propensity for early recurrence, reminiscent of small cell carcinoma of the lung [70, 71]. Similarly, neuroendocrine differentiation is common. The majority of neuroendocrine cases have been found to express HPV 18 mRNA, rather than the HPV 16 commoner in other histologies [72]. Unfortunately, its rarity results in the literature addressing its treatment being largely anecdotal [73].

Surgery or radiotherapy alone for early-stage disease has been shown to be inadequate. In one series, 12 out of 14 patients who received surgery or radiotherapy for Stages Ib or IIa tumours were dead of disease between eight and 31 months, and the other two patients were both alive in relapse. Surgery revealed 57% had pelvic lymph node metastases. Ironically, one patient excluded from the series because of disseminated disease, but treated instead with cyclophosphamide, doxorubicin and vincristine, was alive 3 years later [74]. Modern management recognizes this tendency for haematogenous spread, and exhibits early chemotherapy [75].

Data from two series utilizing early chemotherapy in small cell carcinoma of the cervix are presented in Table 23.8: the larger, consisting of ten patients prospectively treated at the M.D. Anderson center, employed cisplatin 50 mg/m^2 and doxorubicin 50/m^2 on day 1, with etoposide 75 mg/m^2 days 1–3 [76]; the second, from Ohio State University, reported four patients treated with a similar regimen comprising cisplatin 100 mg/m^2, doxorubicin 50–60 mg/m^2 and etoposide 100 mg/m^2 days 1–3 [77].

In summary, these data show seven out of 13 patients treated with chemotherapy for Stage I or II disease in the adjuvant or neoadjuvant context surviving disease-free 7–60 months, median 23 months. They offer no clues as to whether surgery of radiotherapy should be the preferred local treatment, or whether an adjuvant or neoadjuvant strategy is better.

Table 23.8 Patient characteristics after combination chemotherapy with cisplatin, doxorubicin and etoposide for small cell carcinoma of the cervix: after Morris *et al.* [100], and Lewandowski and Copeland [77]

Age	FIGO stage	Tumour size (cm)	Histol.	Prior Rx.	Adjuv. or meas. dis.	Cycles chemo	CR	Further treatment	Current status	Follow-up (months)
MD Anderson, University of Texas									
50	Ib	4	Pure	Surg	Adj.	6	NE	None	NED	60
31	Ib	1	Pure	Surg	Adj.	6	NE	None	NED	54
32	Ib	2	Mixed	Surg	Adj.	6	NE	XRT	DOD	14
46	Ib	2	Mixed	None	Meas.	2	Stable	XRT	NED	21
28	Ib	4	Mixed	None	Meas.	3	CR	XRT	DOD	14
27	Ib	5	Pure	None	Meas.	3	CR	XRT	NED	7
34	Ib	4	Pure	XRT	Meas.	2	Prog.	None	DOD	11
40	IIa	4	Pure	None	Meas.	2	Stable	XRT	AWD	18
61	IIb	6	Pure	None	Meas.	4	CR	XRT	DOD	12
56	IIb	5	Pure	None	Meas	4	PR	XRT	DOD	14
Ohio State University									
31	Ib	5	Pure	Surg.	Adj.	4	NE	None	NED	51
31	IVb	4	Pure	None	Meas.	6	CR	XRT/chemo	DOD	23
44	IIa	7	Pure	None	Meas.	3	CR	Surg/chemo XRT	NED	14
57	IIb	10	Pure	None	Meas.	2	CR	Surg/chemo	NED	12

PHASE II TRIALS OF SINGLE AGENT CHEMOTHERAPY IN CERVICAL CARCINOMA

For all the reasons mentioned in the above discussions of prognosis, any cursory tabulation of reported phase II response rates in advanced cervical cancer is of limited value, and interpretation of such data merits considerable caution, above and beyond that more usually engendered by the wide confidence intervals intrinsic to small studies, or cynical concerns about 'publication-bias'. Furthermore, many of the earlier studies predate the universal application of sophisticated cross-sectional imaging to clinical trials and the acceptance of standardized response criteria. Some are disarmingly frank about the idiosyncratic approaches to assessment employed [78]; others may have been less than candid. However, most would agree about the very real difficulties in assessing clinical response in women who have received prior radiothera-

py: pelvic fibrosis may be indistinguishable from tumour, and in neoadjuvant chemotherapy-then-radiotherapy, for example, some authors have commented on the similar progression-free survival of partial and complete responders [50].

Trials that report response rates at odds with those seen in other studies with the same drug merit special scrutiny. Some such paradoxes may be unravelled by examining the patient characteristics in each trial – where they are stated. Ironically though, on the occasions when it seems such data might yield the most telling perspective, the information is often missing.

Only the most tentative conclusions can be advanced from these uncontrolled data derived from single arm studies in patients with advanced disease. First, in general, it appears that response rates are worse in those trials accruing patients with a high proportion of irradiated indicator lesions; second, it is evident that prior chemotherapy reduces the probability of response; third, very few complete responses are seen and there is little suggestion that chemotherapy might prolong survival; and fourth, chemotherapy often seems to provide some improvement in disease-related symptoms, including neuropathic pain and vaginal discharge, problems which may be resistant to simpler measures.

Perhaps the most debated data concern the single agent activity of carboplatin and ifosfamide in this disease. Two of the more important carboplatin studies were randomized trials comparing carboplatin and iproplatin, and it is interesting that both showed approximately equivalent activity for the two drugs. The trial with the higher response rate included a greater proportion of patients with extra-pelvic non-irradiated indicator lesions. Whilst the question of whether carboplatin is as active as cisplatin can strictly only be addressed by a prospective randomized trial, the data from these phase II studies are usually interpreted as suggesting that carboplatin may be inferior, citing evidence of cisplatin activity at cross-over. This is arguably unjustified, not only because of the overlapping confidence intervals, but also in view of the carboplatin dosing schemes employed in these trials, which predate the widespread application of pharmacokinetically modelled dose calculations, such as the Calvert formula [79], or those such as the Egorin formula which also include a pharmacodynamic element [80]. Accordingly, the overall incidence of myelosuppression recorded seems low, implying many patients may have been underdosed [81]. In some trials, the carboplatin dose–descalation scheme protocols seem unjustifiably strict (see notes above). The activity of carboplatin therefore remains an important question, given its minimal toxicity, ease of use and facility for out-patient treatment. Several authors have recently taken the view that its activity is probably not significantly different from cisplatin [149].

The second point of frequent contention concerns the activity of ifosfamide. The results of two trials from the UK [36, 43] seem at odds with those of two GOG studies [40, 69]. However, the confidence intervals of all four trials overlap, and in the first GOG trial, most patients had received prior platinum exposure, so arguably a response rate of 11% seems not inconsistent with the UK data, which were largely derived from patients with little prior chemotherapy. The response rate in the second GOG trial was somewhat higher than the first, albeit a very modest 16%. However, whilst it is evident that responses were recorded at both pelvic (irradiated) sites (two patients with CR, three patients with PR) and in extra-pelvic disease (three patients with PR), we are not told about the relative proportion of patients with pelvic and extra-pelvic measurable disease at the outset. However, 90% of patients had received prior radiotherapy, and the poor GOG results might be explained if the majority of indicator lesions were previously irradiated pelvic tumours. Conversely, the better UK data might reflect less radiotherapy, or even selection of patients with longer intervals between radiotherapy and relapse. Both UK studies reported impressive numbers of responses in irradiated lesions.

COMBINATION REGIMENS FOR ADVANCED OR RECURRENT CERVICAL CARCINOMA

Arguably for many uncontrolled trials of combination therapy cited in Tables 23.8 (two drug), 23.9 (three drug) and 23.10 (four drug combinations), patient selection factors may have exerted a more potent influence on treatment outcome than the particular regimen used. Hence, in the absence of firm evidence from large randomized phase III clinical trials, it remains unclear as to whether the apparent trend towards higher response rates in the these combination studies reflects a treatment-specific effect or not. Neither is the clinical utility of these higher response rates certain.

Intuitively, even in the combination studies, the relatively low complete response rates encountered carry little promise for significantly

Table 23.9 Phase II trials of single agents in advanced cervical carcinoma

Drug	Prev. RT/CT	Resp. crit[a]	n eval.	CR (%)	OR (%)	95% CI	Reference	Date
Altretamine	+/+	GOG	26/32	0	0	–	Rose et al. [101]	1996
Carboplatin	47/5	WHO	46/48	4	26	15–41%		
vs iproplatin	41/2	WHO	40/41	5	30	17–47%	Lira-Puerto et al. [103]	1991
Carboplatin	159/0	WHO	175/197	6	15	11–21%		
vs iproplatin	164/0	WHO	177/197	4	11	7–16%	McGuire et al. [104]	1989
Cisplatin								
cisDDP 200 mg/m²	11/1	Stand.	11	0	27		Reichman et al. [105]	1991
cisDDP l00 mg/m²		GOG	166	13	31	24–38%		
vs cisDDP 50 mg/m²		GOG	150	10	21	14–28%	Bonomi et al. [48]	
cisDDP 50 mg/m²			164		16	11–23%	Bonomi et al. [41]	1987
cisDDP 50 mg/m²	/12	GOG	34		38		Thigpen et al. [47]	1981
cisDDP 40 mg/m²/wk	69/0	WHO	67/69	15	27	17–39%	Lele and Piver [106]	1989
Chlorambucil	±		44		25		Wasserman and Cantel [107]	1977
Cyclophosphamide	±		188		15		Wasserman and Cantel [107]	1997
Dibromodulcitol	53/0	GOG	55/60	2	29	19–42%	Stehman et al. [108]	1989
	18/0		18/18		28		Lira-Puerto et al. [109]	1984
Docetaxel	±	XSI	18	–	13	0–32%	Kudelka et al. [91]	1996
Doxorubicin	±	NS	50/?	4	6		Piver et al. [50]	1978
			38		16		Wallace et al. [110]	1978
Epirubicin	23/7	WHO	24/27	–	4	<29%	van der Burg et al. [111]	1993
	38/0	NS	38/41	21	47		Wong et al. [112]	1989
	22/2	WHO	27/30	7	19	8–36%	Calero et al. [113]	1991
Etoposide			31	0	0		Slayton et al. [114]	1979
5-Fluorouracil	184/1	Ns	208	NS	9.6		Malkasian et al. [1]	1977
Gemcitabine	?/–	MRI	45/49	–	18		Goedhals and Bezwoda [115]	1996
Hydroxyurea			14		0		Wasserman and Carter [107]	1977
Ifosfamide	27/3	WHO	30/30	3	33	16–50%	Meanwell et al. [36]	1986
	30/0	WHO	30/30	20	40	22–58%	Coleman [43]	1986
	35/9	WHO	39/41	15	31		Coleman et al. [43]	1986
	28/26	GOG	27/30	–	11		Sutton et al. [40]	1989
	10/0	NS	18/21	17	50		Cervellino et al. [116]	1990
	56/0	GOG	51/56	4	16	7–29%	Sutton et al. [68, 69]	1993
Irinotecan	–/–	XSI	27	5	24	–	Chevallier et al. [59]	1995
	±	XSI	15	–	–	–	Chevallier et al. [59]	1995
Irinotecan	37/42	UICC	42	2	21	10–35%	Verschraegen et al. [117]	1997
Menogaril	±	GOG	22/23	–	–	–	Sutton et al. [118]	1994
Methotrexate			77		16		Wasserman and Carter [107]	1977
Mitomycin-C	±	GOG	52/56	6	12	<21.5%	Thigpen et al. [82]	1995
Mitoxantrone		GOG	25	0	8		Muss et al. [119]	1995
Paclitaxel	±	GOG	55	4	17	8–30%	McGuire et al. [90]	1996
Topotecan	22/20	NA	22	0	18		Noda et al. [89]	1996
Vinblastine			20	0	10		Kavanagh et al. [120]	1985
Vincristine			44		23		Wasserman and Carter [107]	1977
Vindesine			20		30		Rhomberg [121]	1986
Vinorelbine	0/0	UICC	42	5	45	30–60%	Lacava et al. [122]	1997

NA, not available; NS, non-standard, i.e. non-UICC or non-WHO.

Table 23.9 Notes to table

Altretamine – Hexamethylmelamine 260 mg/m^2/day for 21 days q 28 days, 24 patients previously exposed to chemotherapy, of which 22 were cisplatin-containing; 21 patients had prior pelvic radiotherapy: 46% had pelvic and 54% extra-pelvic disease.

Carboplatin – Weiss et al. [81]. Carboplatin 300–400 mg/m^2 q 4/52, depending on whether more than 10% of the bone marrow had been irradiated in the previous 12 months. Fifteen patients commenced at 300 mg/m^2 for this reason. No other details are provided about prior radiotherapy, or location of indicator lesions. WBC nadirs ≤ 1.99 or platelet nadirs ≤ 99 × 10^9/l led to dose reductions. Eighteen patients had performance status (PS) of 2 and 17 patients had a PS of 1. Two septic and one thrombocytopenic death occurred. However, leucopenia (WBC <4.0) occurred in only 38% of cycles (as detected by weekly blood counts), and in only 7% of cycles did the WBC fall to <2.0. In summary this is a disappointingly uninformative study.

Lira-Puerto et al. [103]. Carboplatin vs iproplatin. Carboplatin dose was 400 mg/m^2 every 28 days. Of patients, 79% had recurrence within the pelvis and/or elsewhere, 49% only the pelvis. Disease confined to pelvis responded in 13% of cases, and extra-pelvic disease in 39%.

McGuire et al. [104]. Carboplatin vs iproplatin 300 mg/m^2. Patients treated with prior radiotherapy commenced carboplatin at 340 mg/m^2, and iproplatin 230 mg/m^2. Doses were escalated to a maximum of 450 mg/m^2 in absence of myelotoxicity and decreased in the face of any grade 3/4 toxicity. Grade 3/4 leucopenia occurred in only 10%, and grade 3/4 thrombocytopenia in just 16% of patients treated with carboplatin; 100% of planned doses were administered. Of 22 patients retreated with cisDDP after failing carboplatin or iproplatin, four responded, achieving 2 CRs and 2 PRs, an 18% OR. The results of McGuire et al. [104] on carboplatin vs iproplatin 300 mg/m^2 are given below:

Indicator lesion	Carboplatin n = 175	Iproplatin n = 177
Pelvis only	10/97 (10%)	7/96 (7%)
Nodal	13/24 (54%)	4/25 (16%)
Lung ± others	1/32 (3%)	5/31 (16%)
Liver ± others	0/6 (0%)	0/7 (0%)
Other combinations	5/16 (31%)	3/18 (17%)

Cisplatin – Reichman et al. [105]. Cisplatin 200 mg/m^2 q four out of 52: stopped early after evaluation of the first 11 patients treated revealed only three responses, of short duration, but with significant toxicity. Of responses, two out of three were extra-pelvic.

Lele and Piver [106]. Cisplatin 40 mg/m^2 weekly: responses were seen in: none of six patients with a central recurrence (only), one out of 18 patients with central pelvic tumour and disease at other sites and one out of six patients with pelvic side wall relapse. All patients had prior pelvic radiotherapy, two out of 16 patients with para-aortic lymphadenopathy, four out of ten with supraclavicular nodes, three out of nine with liver metastases and ten out of 21 with pulmonary metastases also responded. Responders included 16 out of 49 squamous cell carcinoma,

two out of 16 adenocarcinoma and none of four adenosquamous cell carcinomas. CAP, cisDDP/5-FU and bleomycin/vincristine/mitomycin-C/cisDDP combination regimens were used to consolidate: only CAP increased response rates, by 21% in those with evaluable disease.

Thigpen et al. [47]. Cisplatin 50 mg/m^2 every 21 days. Patient characteristics; $n = 34$. Median time from diagnosis to recurrence, 13.5 months. PS 0, 12; PS 1, 12; PS 2, 10. Thirty-three out of 34 patients had received prior radiotherapy. Of 22 patients with advanced or recurrent disease with no prior chemotherapy, 40% responded (CR 14%, PR 36%). Of 13 patients previously exposed to chemotherapy, 17% responded (all PR). Thirty-five per cent of patients with only pelvic disease responded; 43% of patients with extra-pelvic disease responded.

Bonomi et al. [48]. A large randomized trial following on from an earlier GOG study [46] designed to define a dose–response relationship and optimal schedule for cisplatin. It compared three cisDDP regimens: (1) 50 mg/m^2 every 21 days in 150 patients with (2) 100 mg/m^2 every 21 days in 166 patients with (3) 20 mg/m^2 days 1–5, every 21 days in 128 patients. Patients were stratified by PS and predominant pelvic vs extra-pelvic disease. Limb (3) was closed early because of inconvenience and lack of obvious schedule-specific benefit (OR 25%). cisDDP 50 mg/m^2 provided a response rate of 20.7% and 100 mg/m^2 31.4%. This difference was significant, $p = 0.015$. There were no significant differences in median duration of response or survival. The prognostic influence of performance status, previous radiotherapy, site of disease, grade and method of tumour measurement were analysed by logistic regression: only performance status proved significant: OR for patients PS 0–1, 29%; OR for patients PS 2–3 was 17% ($p = 0.001$).

Dibromodulcitol – Stehman et al. [105]. Dibromodulcitol 180 mg/m^2 per day, by mouth, for 10 days q 4/52. Of 28 patients with pelvic measurable disease, seven responded (25%), as did nine out of 27 (33%) with disease elsewhere. Thirty-five out of 55 patients had a performance status of 0–1.

Docetaxel – Kudelka et al. [91]. Docetaxel 100 mg/m^2. Sixty-five per cent had previous radiotherapy, no previous chemotherapy, other than as radiation sensitizer (provisional report).

Doxorubicin – Piver et al. [150]. Adriamycin 60 mg/m^2, 75 mg/m^2 or 90 mg/m^2 3–4-weekly. Most patients had received prior pelvic irradiation, but few had received previous chemotherapy. The study describes the activity of six different doxorubicin regimens given to 112 patients, 100 of whom were eligible. Patient characteristics are not specified for each treatment group. Ninety-six patients had recurrent tumour after previous radiation; 87 had received no previous chemotherapy, 13 had progressed on previous chemotherapy.

Epirubicin – van der Burg et al. [111]. Epirubicin 12.5 mg/m^2, weekly. All patients had measurable lesion(s) outside previous radiotherapy field, and no prior anthracyclines. Seven patients had had previous cisplatin-containing chemotherapy: the single responder had no prior chemotherapy.

(continued)

Table 23.9 Notes to table (continued)

Wong *et al.* [112]. Epirubicin 60–120 mg/m^2 every 4 weeks, in an intra-patient dose escalation protocol. Seven out of nine patients with pelvic disease, six out of 12 with distant metastases and five out of 17 patients with multiple sites of disease responded. Median performance status 0 (range 0–1), mean interval of recurrence was 35 months, range 7–120 months.

Calero *et al.* [113]. Epirubicin 80 mg/m^2 every 3 weeks. Patient characteristics: $n = 30$ median KPS 80. Stage III, 14 patients; Stage IV, 16 patients. Median time from diagnosis to relapse was 19 months. Seven patients had no previous treatment. Three were not evaluable, including two early deaths. Toxicity: no grade 4 toxicity. No significant myelosuppression. Three out of six patients not exposed to radiotherapy responded, two out of three achieving CR. Only two out of 21 irradiated patients responded, one in the lungs and one at lymph nodes.

5-Fluorouracil – Malkasian *et al.* [1]. 15 mg/kg body weight (about 600 mg/m^2) days 1–5, repeated every 4 weeks. The group included only patients with disease too widespread for surgery or radiotherapy. No clear trend between time from original treatment to chemotherapy and response rate (≤ 1 year four out of 67, 1–2 years ten out of 46, 2–3 years three out of 34).

Gemcitabine – Goedhals and Bezwoda [87, 115]. Gemcitabine 1250 mg/m^2 days 1, 8 and 15, repeated every 28 days. All patients had either Stage IIIb or IV disease, and were chemotherapy naive; data about radiotherapy are not supplied. Provisional report, 1995, final 1996. Fifty-five per cent had Stage IIb disease and 45% Stage IV. Partial responses were confirmed in five out of 45 (11%). Four patients 'absconded' without objective response measurements. Five patients withdrew once they had obtained symptomatic relief, and two of these were patients with unconfirmed PRs. All nine are included in the denominator, but are not scored as responders.

Ifosfamide – Coleman *et al.* [43]. Ifosfamide 1.5 g/m^2 daily, days 1–5, q 21 days. Patient characteristics: 41 women with advanced or relapsed disease, all FIGO Stages III or IV. Median time from diagnosis was 9 months, range 0–168 months. Twenty-four had prior external beam and intracavitary radiotherapy (RT), eight external beam only and three intracavitary RT only. Six had no prior pelvic RT, but three of these had prior CT. Of a total of nine patients with previous CT, six had previous platinum-containing regimens; none responded to ifosfamide. Responses seen in nine out of 28 patients with previously irradiated pelvic disease, two out of seven with non-irradiated pelvic disease. Time since RT not specified. Overall response was 40% of those with no prior CT, with 20% CR. Four out of 14 patients who received just 41–60% of intended dose responded, three out of seven 61–80%, and five out of 19 81–100%.

Meanwell *et al.* [36, 39]. Ifosfamide 5 g/m^2 q 21 days. No responses seen in three patients previously treated with chemotherapy: all responses seen ≤ 2 cycles and maximal ≤3 Responses seen in four out of 22 previously irradiated lesions and 15 out of 28 non-irradiated lesions. No irradiated lesions achieved CR. Eight patients had received prior brachytherapy alone, six teletherapy alone, and only 13 combined.

Sutton *et al.* [123]. Ifosfamide 1.2 g/m^2 × 5 days q 28 days. Note heavier pretreatment, compared with Meanwell *et al.* [36, 39] and different schedule. Responses were seen in both pelvic and extra-pelvic sites: all three responders had prior platinum drugs. Radiotherapy doses may have been higher in USA, cf. UK.

Cervellino *et al.* [116]. Ifosfamide 3.5 g/m^2 daily, days 1–5, q 28 days; only ten out of 21 patients had received prior radiotherapy and none had previously been exposed to chemotherapy. Sixty-six per cent of all responses occurred in non-irradiated areas. One out of three CRs recorded occurred in a (cobalt 60) irradiated site.

Sutton *et al.* [68, 69]. Ifosfamide 1.2–1.5 g/m^2 per day, days 1–5, q 28 days: both patients achieving pelvic CR had received prior radiotherapy. Of six partial responses, three occurred at previously irradiated pelvic sites and three were extra-pelvic. No data were supplied as to the proportions of patients with pelvic vs extra-pelvic indicator lesions.

Iproplatin – Lira-Puerto *et al.* [103]. Carboplatin vs iproplatin (CHIP). Iproplatin dose was 300 mg/m^2. Ninety-five per cent of patients had disease within the pelvis ± elsewhere, and 73% only in the pelvis. Disease confined to pelvis responded in 37%, and extrapelvic disease in 10%.

Irinotecan – Chevallier *et al.* [88]. Stratified by whether target lesions were in prior radiation field ($n = 15$), or not ($n = 27$).

Verschraegen *et al.* [117]. Irinotecan, 125 mg/m^2 weekly for 4 weeks, every 6 weeks. Patient characteristics: 14% PS 0; 38% PS 1; 48% PS 2. Fifty-two per cent had received one prior chemotherapeutic agent, 38% two drugs, 5% three drugs and 5% four drugs. Toxicity: 77% of patients required ≥ 1 dose reduction, 28% ≥ 2 reductions. Thirty-six per cent of patients suffered grade three out of four neutropenia, seven with fever. Nausea, vomiting (93%) and diarrhoea (76%) were common. Nine per cent suffered allergic reactions. No patients were removed for toxicity. Pelvic disease (including irradiated sites) and lymphadenopathy responded better than other metastatic sites. Lower doses were recommended.

Menogaril – 95% of patients had prior radiotherapy. No prior chemotherapy was permitted, other than radiation chemosensitizers.

Mitomycin-C – No previous chemotherapy other than radiosensitizers; responses in two out of 27 patients with previously irradiated pelvic marker lesions and four out of 25 patients with non-irradiated extra-pelvic disease. All 6 responders had performance status 0.

Paclitaxel – McGuire *et al.* [90]. Paclitaxel l75 mg/m^2 (or 135 mg/m^2 if previous radiotherapy). No prior chemotherapy, except 11 patients given radiation sensitizers, hydroxyurea or cisplatin and 5-FU. Forty-six out of 52 patients had previous pelvic radiation; four responses pelvic, five extra-pelvic.

Topatecan – Noda *et al.* [89]. Topotecan 1.2 mg/m^2 days 1–5 iv. Twenty-six patients were evaluable for toxicity and 22 for activity. The series was unusual for the high proportion (20 patients, 77%) previously treated with radiotherapy and chemotherapy; 18 of these were evaluable with three responses recorded (OR 17%). The table records overall response data for the group as a whole.

Vinorelbine – Lacava *et al.* [122]. Vinorelbine 30 mg/m^2, 20 min infusion, repeated weekly × 12, followed by radical surgery or radiotherapy. Patient characteristics: no prior treatment (neoadjuvant chemotherapy). Eighteen Stage IIb; one Stage IIIa; 19 Stage IIIb; five Stage IVa. PS: 60% 0; 40% 1. Toxicity: 17% grade 3/4 in 17% patients, peripheral neuropathy in 28%, myalgia in 23%. No Stage IV patients responded. Forty-seven per cent of patients with tumours <4 cm responded, and 44% >4 cm.

Table 23.10a Combination chemotherapy in advanced disease: two drug combinations

Drug	Prev. RT/CT	Resp. crit[a]	n eval.	CR (%)	OR (%)	95% CI	Duration resp. (months)	Date	Ref.
Vinc and Bleo 2 × weekly		SWOG	50	16	60				
Vinc and Bleo 1 × weekly			24	4	25				
Bleomycin infusion			41	15	39				
Cisplatin									
and bleomycin			25	32				1988	[124]
Cisplatin									
and methotrexate									
(and folinic acid)			37	14	57		15	1986	[125]
(vs hydroxyurea)			13	0	0		NA	1986	[125]
Cisplatin									
and 5-FU	48/0	GOG	55	13	22	13–34	2	1989	[126]
(non-irrad. indic. lesion)	16/0	WHO	19	21	68		10–18	1990	[126]
(irrad. indic. lesion)	13/0	WHO	13	–	15		4	1990	[127]
	13/1		WHO	18/19	6	61	11	1988	[128]
Cisplatin									
and epirubicin	0/0	WHO	21/22	19	67	47–87	–	1997	[129]
Cisplatin									
and mitomycin-C		SWOG	51	4	25		7	1987	[84]
Cisplatin									
and ifosfamide	37/0	WHO	42	5	39	N/A	7	1990	[44]
	25/0	NS	30/30	3	50		21+	1995	[130]
Carboplatin									
and ifosfamide	0/0	N/A	32	9	59	42–76	N/A	1990	[131]
Cisplatin									
and carboplatin	36/3	UICC	42	2	29	15–42	4.4	1994	[100]
Cisplatin									
and etoposide	28/0	UICC	38	18	39	24–55	12	1997	[132]
Ifostamide									
and 5-FU/leukovorin		WHO	30	33	53		N/A	1994	[83]
Bleomycin									
and mitomycin-C	16/0	NS	15	80	93		NS	1978	[133]
	32/0	NS	33	15	36		12/6	1983	[134]
Methotrexate									
and doxorubicin	45/0	NS	59	22	66		–	1978	[78]

Table 23.10a Notes to table

Cisplatin and methotrexate – Bezwoda et al. [125]. Patients with advanced recurrent or metastatic disease were randomized to cisplatin 20 mg/m^2/day × 3 days, with methotrexate 100 mg/m^2 day 3, plus folinic acid rescue q 3/52 vs oral hydroxyurea 1.5 g/m^2/day × 10 days, then 1 g/m^2/day maintenance after 2 weeks rest. After no responses amongst the first 13 patients randomized to hydroxyurea, the hydroxyurea limb was closed and a further 25 patients treated with cisDDP and methotrexate. The median survival of patients treated with cisDDP and methotrexate was 11 months and for hydroxyurea 4 months, a significant difference.

Methotrexate and doxorubicin – Guthrie and Way [78]. Fifty-nine patients were treated with five similar regimens of doxorubicin 50 mg/m2 and methotrexate 20–40 mg/m^2 day 1 and/or day 8, with or without folinic acid. Overall, 32 had received prior radiotherapy alone, 13 with radiotherapy and surgery, 13 no previous therapy, one with surgery only. Thirteen patients had extra-pelvic tumour. The focus of this paper relates to the impact of minor variations in treatment schedule across five small groups of 11–12 patients on outcome. However, the highest response rates occur in the group with the most extra-pelvic disease, suggesting that patient characteristics are more important. Toxicity differences are more compelling.

Bleomycin and mitomycin-C – Miyamoto et al. [133]. Bleomycin 5 mg daily days 1–7, and mitomycin-C 10 mg day 8, repeated after 7 days rest. The paper implies all patients had received prior pelvic radiotherapy; however, all patients had distant metastases, 12 lung, four liver, eight bone and eight lymph nodes. All patients had symptomatic improvement.

Tropé et al. [134]. Bleomycin 5 mg daily days 1–7 and mitomycin-C 10 mg day 8, q 14 days x4. Twelve patients had pelvic recurrence, ten distant metastases, four recurrence in pelvic nodes, seven recurrence in both pelvis and distal metastases. Four out of five CRs and five out of seven PRs were seen in regional or distal metastases. All lesions achieving CR were ≤3 cm in diameter. Fifty per cent of patients with distant metastases achieved a response, and 25% of those with central recurrences. Of the seven patients with both central and distal recurrences, two responded in their distant metastases.

(continued)

Table 23.10a Notes to table (continued)

Cisplatin and 5-FU – Chiara *et al.* [128]. Cisplatin 20 mg/m^2 and 5-FU 200 mg/m^2 days 1–5, both repeated 3-weekly. Patient characteristics: $n = 19$, one had marker lesion resected. Median PS 2, range 0–4. Twelve had primary disease or pelvic recurrence only, two distant metastases only and five had both. Toxicity: three cases of grade II neurotoxicity, three grade II and one grade II leucopenia. Response by prior site irradiation: eight PRs at 13 irradiated sites; seven objective responses at 12 non-irradiated sites, including two CRs. Five patients had no previous chemotherapy or radiotherapy and locally advanced disease: only two out of five responded but tumour reduction was insufficient to control disease with surgery or radiotherapy.

Bonomi *et al.* [126]. Cisplatin 50 mg/m^2 day 1, 5-FU 1000 mg/m^2 daily, by 24 h infusion, days 1–5, repeated every 21 days. The authors found these results disappointing, not withstanding the broad confidence intervals, and concluded the combination showed no improvement over single agent cisplatin, an opinion bolstered by reference to the patient characteristics of this study, and the similarity of those previously shown as determinant of prognosis, to those of a larger previous cisplatin trial.

Kaern *et al.* [127]. 5-FU 1000 mg/m^2 iv days 1–5, cisplatin 100 mg/m^2 day 1, repeated 3-weekly. Patients were divided into two groups by whether a measurable lesion fell within an irradiated field or not. Toxicity: bone marrow toxicity was dose limiting, 41% of patients received only 50% of the planned dose of 5-FU and 19% of patients were delayed. Ototoxicity and nephrotoxicity required treatment withdrawal in five patients. Overall, 32 out of 37 patients were evaluable for response. Response data by lesion location are tabulated below.

Location	Non-irradiated			Irradiated		
	No.	CR	PR	No.	CR	PR
Central pelvic	0	0	0	5	0	1
Lateral pelvic	4	1	2	8	0	1
Distant	15	1	9	0	0	0
Pulmonary	13	5	3	0	0	0
Total	32	7	14	13	0	2

Sundfør *et al.* [135]. Cisplatin 100 mg/m^2 day 1 and 5-FU 1000 mg/m^2 days 1–5, repeated every 3 weeks for three cycles, as primary chemotherapy (ahead of radiotherapy, in randomized trial). Patient characteristics (randomized to chemotherapy): Stage IIIb, 44 patients; Stage IVa, three patients. KPS 100, 19 patients; KPS 90, 16; KPS 80, 10; KPS 70, 2. Parametrial involvement: unilateral, 14; bilateral, 32. Toxicity: eight patients experienced grade III/IV mucosal toxicity. Reponse after chemotherapy: CR 5%; PR 67%; OR 72%.

Cisplatin and epirubicin – Zanetta *et al.* [129]. Uncontrolled neoadjuvant study (CT/surgery) in cervical/vaginal adenocarcinoma. Cisplatin 50 mg/m^2/week × 9, with epirubicin 70 mg/m^2, at weeks 1, 4 and 7. Patient characteristics: $n = 22$. PS < 2, for all patients. Stage: 16 Stage Ib; one Stage IIa; three Stage IIb; two patients with vaginal adenocarcinoma, one Stage II and one stage III. Nineteen patients completed treatment schedule, 16 with delays 1–5 weeks. Twenty-one were evaluable for response after four cycles. CR 19%, 48% PR, 67% OR.

Cisplatin and etoposide – Al-Saleh *et al.* [132]. Cisplatin 30 mg/m^2/day × 3 days, with etoposide 60 mg/m^2/day × 3 days, followed by oral etoposide 50 mg daily for 7 days.

Patient characteristics: $n = 38$, 28 with recurrent disease (previous radiotherapy 24 patients, radiotherapy and surgery four patients), and ten with newly diagnosed locally advanced disease. Disease sites: pelvis 20; distant seven; both pelvic and distant 11. ECOG PS: PS0, one; PS1, 20; PS2, 12; PS3, four. FIGO stage at diagnosis: Stage I, 17; Stage II, five; Stage III, ten; Stage IV, six. Response: CR 18%; PR 21%; CR 39%. Response rate in non-irradiated sites, 65%. Response rate in irradiated areas, 42%. OR in primarily advanced disease 10%, OR in recurrent disease 50%. Median time to recurrence 2 years; five recurrences occurred after 10 years. Mean survival duration after chemotherapy 9.8 months, range 5–36 months. Symptoms resolved or improved in 28 patients.

Cisplatin and ifosfamide – Coleman *et al.* [44]. Cisplatin 50 mg/m^2 and ifosfamide 1.5 g/m^2 day 1–5, q 21 days. The activity and duration of response of the combination were disappointing to the authors, who contrast both unfavourably with that of single ifosfamide used in the same schedule, and reported earlier by the same group. Thirty-seven out of 44 women had previous pelvic radiotherapy; responses occurred at ten out of 32 (31%) irradiated sites and nine out of 15 (60%) non-irradiated sites. They noted symptomatic improvement in 50%. Response did not seem to be influenced by delivered dose intensity. Other patient characteristics and relation to response are described in detail in a univariate/multivariate analysis in a later paper by Buxton [42].

Cervellino *et al.* [116]. Cisplatin 20 mg/m^2 days 1–5, and ifosfamide 2.5 g/m^2 daily days 1–5, with mesna, all repeated 4-weekly. Patient characteristics: Stage IIIa, six; Stage IIIb, 16; Stage IVa, three; Stage IVb, five. Entry criteria included albumin ≥ 35 g/l, and creatinine clearance >60 ml/min. Of 35 indicator lesions assessed for response, 30 were previously irradiated and 42% responded; only 20% (one out of five) of non-irradiated lesions showed objective response. Response in Stage III patients was 14 out of 22; response in Stage IV patients was one out of eight. Majority of responses occurred at irradiated sites, overall response rate 50% but median duration of response was in excess of 21 months, and median survival of 31+ months; after a median follow-up of 70 months, 11 patients (37%) survive. This is unusual for a series with just one patient achieving CR. Toxicity: grade III anaemia seven patients; nine grade III leucopenia, one grade IV, but no sepsis; one patient suffered encephalopathy 24 h, but survived; two patients, grade III thrombocytopenia; no significant nephrotoxicity. There were no treatment-related deaths.

Bolis *et al.* [58]. Cisplatin 50 mg/m^2 weekly with ifosfamide 3.5 g/m^2 provided an overall response rate of 78%.

Carboplatin and ifosfamide – Kühnle *et al.* [131]. Carboplatin 300 mg/m^2 and ifosfamide 5 g/m^2 day 1 q 28 days. Thirty-two patients presenting with previously untreated Stage IIb to IVb cervical carcinoma.

Ifosfamide, 5-FU, and leukovorin – Stornes and Mejlholm. [83] Ifosfamide 1.2 g/m^2, 5-FU 370 mg/m^2, leukovorin 20 mg/m^2 with mesna, days 1–5, q 28 days. Four out of ten patients achieving CR were alive and disease-free after 19–36 months. Sixty per cent of patients required dose-reduction for toxicity, and the majority of these had received prior radiotherapy. Seventeen patients had prior radiotherapy, 11 relapsed in field, 19 had recurred outside irradiated area. Eight out of ten CRs occurred in out of field lesions, as did five out of six PRs.

Table 23.10b Combination chemotherapy in advanced disease: three drug combinations

Drug	Prev. RT/CT	Resp. crit[a]	n eval.	CR (%)	OR (%)	95% CI	Duration resp.	Date	Reference
Cisplatin, bleomycin									
± methotrexate	0/0	NA	128	15	83		NA	1997	[57]
(Neoadjuvant)	0/0	NS	24	4	67		NA	1991	[28]
(Rec[t]/met[c])	>22/0	NS	36	3	27		NA	1991	[28]
Mitomycin-C									
Vincristine									
Bleomycin		SWOG	14	36	50			1981	[84]
Mitomycin-C									
Cisplatin									
Vincristine									
Cisplatin, vinblastine									
and bleomycin	23/2	NS	23	18	67		18/24 weeks	1983	[136]
	0/0	WHO	92	28	54		–	1995	[137]
Cisplatin, vincristine									
and bleomycin	0/0	UICC	33/36	24	85		–	1993	[64]
5-FU, ifosfamide									
and cisplatin	26/6	NS	30	17	53		12, 6 months	1995	[67]
Cisplatin, ifosfamide									
and etoposide			14	57	0			1991	[138]
Cisplatin, carboplatin									
and ifosfamide	24/0	UICC	31/36	39	74	55–88	23 weeks	1993	[139]
Bleomycin, ifosfamide									
and cisplatin									
Phase II, prev. treated	42/1	UICC	49	20	69	56–82	8 months	1989	[2]
Phase II, prev. treated			12		42			1991	[142]
Phase II neoadjuvant	0/0	NS	20	0	68	58–99	NA	1990	[141]
Randomized neoadj.	0/0	NS	32	6	69	53–85	NA	1990	[141]
Phase II prev. treated	23/0	WHO	20	0	15		9 months	1992	[142]
Phase II prev. treated	10/2	UICC	14	14	29		3 months	1992	[143]
Phase III neoadjuvant	0/0	NS	89	4.5	72		NS	1994	[51]
Phase III neoadjuvant	0/0	NS	36	22	80		NS	1997	[144]
Phase II neoadjuvant	0/0	NS	31	45	90		NS	1997	[144]
Bleomycin, ifosfamide									
and carboplatin	19/21	WHO	35	23	60	44–76	11 months	1994	[145]
Cisplatin, ifosfamide									
and paclitaxel	0/0	NS	15	38	100	NA		1997	[93]

Table 23.10b Notes to table

Cisplatin, (±) methotrexate and bleomycin – Benedetti Panici *et al.* [57]. Drug doses were not stated in this abstract, which appears to be an update of an earlier publication specifying cisplatin 100 mg/m^2 day 1, bleomycin 15 mg day 1 and day 8, and methotrexate 300 mg/m^2 day 8 (with folinic acid rescue), repeated every 21 days for three cycles. One-hundred-and-forty patients with bulky (>4 cm) Stage IB–III tumours were treated with chemotherapy then radical surgery, including aortic and pelvic lymphadenectomy.

Hoskin and Blake [28]. Cisplatin, methotrexate plus folinic acid rescue and bleomycin. Thirty-one patients received cisplatin '60 mg' fractionated over 3 days, with methotrexate '100 mg', and bleomycin '15mg': ten received cisplatin '75 mg', and 19 patients cisplatin '60 mg', both with methotrexate '300' mg' and bleomycin '30 mg'. Note that doses are quoted as 'mg', not mg/m^2. Twenty-four patients received neoadjuvant treatment, 36

patients treatment for recurrent or metastatic disease. Responses are defined by a '>50% reduction in maximum tumour dimension', which may understate UICC partial response rate. No response duration data are provided. Median survival of recurrent/metastatic group was 12 months. Response is detailed by site, rather than patient (see Table 23.10b).

Site	n	OR (%)	CR (%)
Neoadjuvant	24	67	4
Pelvis post-surgery	13	46	8
Pelvis post-radiotherapy	22	9	
Distant metastases	6	50	17

There were 65 sites of disease in 60 patients.

(continued)

Table 23.10b Notes to table (continued)

The overall response rate in previously untreated patients was similar to other combination regimens, albeit with a lowish CR rate. Response rate in previously treated and metastatic patients at 27% overall was comparable with single agent cisplatin.

Cisplatin, vinblastine and bleomycin – Friedlander *et al.* [136]. Cisplatin 60 mg/m^2 day 1, vinblastine 4 mg/m^2 days 1 & 2, and bleomycin 15 mg im, days 1, 8 and 15, every 3 weeks. Patient characteristics: ten patients with advanced inoperable previously untreated tumours; 23 pts with locally recurrent or metastatic disease, including 21 with prior radiotherapy±surgery, and two with prior radiotherapy and chemotherapy. Twenty patients had disease confined to pelvis. Toxicity: 9 patients suffered WBC <2×10^9/l and thrombocytopenia <50×10^9/l, in two patients; three patients suffered neutropenic sepsis, and one died. No neuropathic or nephropathic toxicity recorded. Response data: six out of ten PRs in patients with previously untreated locally advanced disease; 69% of 23 patients with locally recurrent or metastatic disease responded, including 26% CRs.

Namkoong *et al.* [137]. Cisplatin 60 mg/m^2 day 1, vinblastine 4 mg/m^2 days 1 and 2, bleomycin 15 mg/m^2 im days 1, 7 and 14 repeated 3-weekly. Patient characteristics: $n = 92$; 41 patients Stage Ib, 36 Stage IIa, 15 Stage IIb. Tumour size: ≥ 3 cm<4 cm, 41%; ≥ 4 cm<5 cm, 29%; ≥ 5 cm, 29%. Toxicity not described. Response: CR 28%; PR 54%; OR 82%. Increasing number of chemotherapy cycles 1–3 provided progressive increase in complete response rates. After one cycle: OR 71%, CR 0%. After two cycles: OR 85%, CR 26%. After three cycles: OR 83%, CR 42%. Case control study.

Cisplatin, vincristine and bleomycin – Chang *et al.* [64]. Cisplatin 50 mg/m^2, day 1, vincristine 1 mg/m^2 day 2, bleomycin 25 mg/m^2/24 h, days 2–4. Patient characteristics: 36 patients, 21 with Stage Ib disease, 12 Stage IIa. Toxicity: no treatment-related deaths; general malaise in 76%; mild hepatic function disturbance; no pulmonary toxicity; drug fever 64%; grade II/III haematological toxicity in most patients. Response: overall 85%, including 24% CR and 61% PR. A further 14% experienced minor responses.

Cisplatin, ifosfamide and etoposide – Kredentser *et al.* [138]. Cisplatin 25 mg/m^2, ifosfamide 1 g/m^2, and etoposide 75 mg/m^2, days 1–3.

Cisplatin, carboplatin and ifosfamide – Filtenborg *et al.* [139]. Cisplatin 50 mg/m^2 day 1, carboplatin 200 mg/m^2 day 1, ifosfamide 1.5 g/m^2 days 1–3. Patient characteristics: nine patients with PS 0, 18 PS 1, eight PS 2, and one PS 3. Twenty-four had received prior irradiation. Toxicity: one episode of reversible WHO grade 3 encephalopathy; two neutropenic septic deaths. Five patientss suffered non-fatal thrombocytopenic bleeding. Responses: of 19 irradiated lesions, 26% achieved PR and 21% CR, with an overall response of 47%. Of 48 non-irradiated lesions, 23% achieved PR and 48% CR. On an intention to treat analysis of 36 treated patients, 33% achieved PR and 31% CR, with an OR of 64%. Ten responding patients had irradiated lesions, and of ten patients with only irradiated disease, five were responders. All patients with exclusively non-irradiated disease responded. Maximal response was

achieved within the first two cycles of treatment in 14 out of 23 responders, and within four cycles for eight out of 23 responders, and required six cycles in one patient. Median survival was 50 weeks. Subjective palliation of disease-related symptoms was seen in 37% patients within two cycles of treatment.

Mitomycin-C, vincristine and bleomycin – Alberts *et al.* [84]. Small pilot study with 14 patients achieving five CRs, of 8, 20, 23+, 30+ and 34 months duration. Three of the CRs featured complete resolution of pulmonary metastases, resolution of a large pelvic mass, and another disappearance of bony metastases documented at autopsy.

5-FU, ifosfamide and cisplatin – Fanning *et al.* [67]. 5-FU 1500 mg/m^2 (continuous infusion), ifosfamide 3 g/m^2, and mesna 3 g/m^2 in divided doses days 1–3, repeated every 28 days. Just one of 11 (7%) patients with recurrent pelvic disease responded and, interestingly, she was the only patient with recurrent pelvic disease who had not received prior pelvic radiotherapy. Fifteen out of 19 (79%) patients with extra-pelvic recurrence have responded ($p = 0.001$). Five out of 15 (33%) patients relapsing ≤6 months of primary treatment and 11 out of 15 (73%) relapsing >6 months responded ($p = 0.05$). Seven out of nine (78%) patients with adenocarcinoma and nine out of 21 (43%) patients with squamous cell histology responded. Toxicity was minimal. Median number of cycles was six (2–11).

Bleomycin, ifosfamide and cisplatin – Buxton *et al.* [2]. Bleomycin 30 mg/24 h, cisplatin 50 mg/m^2, and ifosfamide 5 g/m^2/24 h, with mesna 6–8 g/m^2/36 h, repeated 3-weekly. Doses were reduced by 30% if creatinine clearance fell to <40 ml/min. Of 49 patients, three received ifosfamide fractionated over 5 days. Patient characteristics: median age 38; median Karnofsky score 90% (32 had a WHO PS of 0; ten of 1, seven of 2). Twenty-three out of 49 patients presented Stage Ib; 42 patients had received prior radiotherapy; 33 patients had pelvic disease, 13 in regional nodes, nine in para-aortic/other nodes and seven lung metastases. Twenty-six out of 36 (72%) previously irradiated lesions responded, and 17 out of 26 (65%) non-irradiated lesions. Only one patient was previously untreated, after presenting Stage 4b. Further patient characteriatics were provided in a later paper [42]. Thirty-four patients relapsed in the pelvis only, nine at both pelvic and distant sites and six at distant sites only. Twenty-eight patients had recurrence only at previously irradiated sites, 14 at non-irradiated, and seven both irradiated and non-irradiated. The later paper, which offered a univariate and multivariate analysis of several phase II trials with ifosfamide (OR 31%, 95% CI 24–44%), ifosfamide/cisplatin (OR 36%, 95% CI 22–50%) and ifosfamide/bleomycin (OR 12%, 95% CI 0–35%) found no significant relationship between time to relapse after 'primary therapy' and response to chemotherapy. Toxicity: six patients developed grade III ifosfamide-related encephalopathy but recovered, two patients died with neutropenic sepsis, one patient had a fatal vaginal bleed. Eligibility for the BIP study was defined using an ifosfamide encephalopathy prediction nomogram, which utilizes serum albumin value, serum creatinine and presence/absence of a pelvic mass [39], plausibly an occult selection factor for patients

(continued)

Table 23.10b   Notes to table (continued)

with a better chance of response. However, the median survival was only 10.2 months, similar to many other platinum-based studies.

Tobias *et al.* [141]. Neoadjuvant bleomycin 30 mg/24 h, cisplatin 50 mg/m^2, and ifosfamide 5 g/m^2/24 h, with mesna 6–8 g/m^2/36 h, repeated 3-weekly, a pilot study in newly presenting advanced disease. Patient characteristics: four stage IIa, five stage IIb, seven stage III, and four Stage IV; median PS 1 (range 0–2). In an interim analysis of the first 66 patients randomized within a larger phase III neoadjuvant trial, 32 were assigned chemotherapy and their data presented. Patient characteristics: five Stage IIa, 18 Stage IIb, 40 Stage III, three Stage IV. Median PS 1, range 0–3. All patients received just two–three cycles of chemotherapy, except in event of definite disease progression. Subjective response or symptomatic improvement was seen in 74% of patients with pelvic pain, 54% with vaginal discharge and 67% with vaginal bleeding at study entry.

Ramm *et al.* [142]. A small series of 24 patients, 22 with pelvic recurrences, and a further two with pelvic plus liver and pelvic plus lung metastases. All 24 patients had received prior external beam radiotherapy to the pelvis (median 46 Gy, range 30–64.8 Gy), and 19 patients intracavitary treatment too. Chemotherapy schedule was the same as Birmingham BIP, except bleomycin was given as a one hour infusion (cf. 24 h). Doses were identical. Median survival was 9 months.

Tay *et al.* [143]. Cisplatin 50 mg/m^2, ifosfamide 1.2 g/m^2 days 1–5 (+ mesna), bleomycin 15 mg iv days 1, 8 and 15, repeated q 28 days. A small detailed series of just 14 patients, of whom ten had pelvic disease, six with concurrent extra-pelvic metastases. Of the remaining patients, all four had cervical lymphadenopathy, two with lung and one with liver metastases. Mean time to recurrence was 18 months, with a range of 5–40 months. There were four responders, whose details are tabulated.

PS	TTP	Histology	Sites of disease	Pelvic DXT	Response
2	30m	Squamous	Cervical nodes and lungs	+	CR (ED)
0	6m	Small cell	Cervical nodes and liver	+	PR (ED)
0	24m	Squamous	Cervical nodes	+	CR
1	0m	Aden/squa.	Pelvis and lungs	+	PR

Eight patients were assessed as showing stable disease and had therapy discontinued after 2–3 cycles.

Four of these were apparently alive at the time of writing, but FU was short (probably ≤ 12 months). Two patients progressed. No responders had irradiated indicator lesions.

Kumar *et al.* [51]. Bleomycin 15 mg iv day 1, ifosfamide with mesna 1 g/m2, iv daily, days 1–5, cisplatin 50 mg/m^2 iv day 1, repeated every three out of 52, for just two cycles

(and responses measured ahead of radiotherapy). Eighty-nine out of 184 patients with Stage IIb-IVa disease were randomized to chemotherapy then radiotherapy: the remainder had radiotherapy alone. Just over half of the patients with partial responses had ≥90% tumour regression.

Kumar *et al.* [144]. In this smaller randomized trial in 71 patients with Stage IIIb disease, 36 patients were treated with three cycles of chemotherapy (cf two cycles in the earlier study) prior to radiotherapy.

Kumar *et al.* [144]. In a follow-on single arm phase II study, again recruiting patients with Stage IIIb disease, chemotherapy doses were increased: bleomycin 15 mg iv day 1, ifosfamide 1.2 g/m^2 days 1–5, cisplatin 20 mg/m^2 iv days 1–5, three cycles ahead of radiotherapy. Complete response rates were higher than the earlier studies.

Bleomycin, ifosfamide and carboplatin – Murad *et al.* [145]. Bleomycin 30 mg bolus day 1; carboplatin 200 mg/m^2 bolus day 1; ifosfamide 2 g/m^2 days 1–3, with iv and oral mesna, in an attempt to reduce the toxicity of BIP, with ifosfamide rescheduled to permit outpatient treatment, albeit with a 20% increased total dose (6 g/m^2, cf 5 g/m^2). The carboplatin dose was calculated using a conversion factor of 4× the cisplatin dose in BIP. Patient characteristics: 27 Stage IVa and eight IVb. Fourteen patients had Karnovsky Performance Status (KPS) 70; 13 KPS 80; eight KPS 90. Toxicity: one death with neutropenic sepsis, and two cases grade 4 thrombocytopenia, little encephalopathy. An unpublished pilot study with 300 mg/m^2 carboplatin with six g/m^2 ifosfamide caused severe throbocytopenia.

Murad *et al.* [145]: response by prior radiotherapy

	n	CR	PR	OR	(%)	Median survival
No radiotherapy	18	7	9	16	(89)	17 months
PrevIOUS radiotherapy	17	1	4	5	(29)	5 months

Patients stable or progressing after two cycles were taken off study.

Cisplatin, ifosfamide and paclitaxel – Colombo *et al.* [93]. Cisplatin 50 mg/m^2 day 1, ifosfamide 5 g/m^2/24 h day 1, with mesna 6 g/m^2/36 h, and paclitaxel 175 mg/m^2, repeated every 3 weeks. Patient characteristics: 22 patients, 11 with bulky Stage Ib (>4 cm dia.), four Stage IIa (>4 cm dia.), three Stage IIb, and four Stage IIIb. Toxicity: 60% grade 3/4 neutropenia; 20% grade 3/4 thrombocytopenia. Thirteen out of 15 evaluable patients responded, eight PR and five CR, and underwent radical surgery. Path CR was found in three out of 13, and Path PR in ten. Four out of ten with Path PR had only microinvasive or *in situ* residual disease.

Table 23.10c Combination chemotherapy in advanced disease: four drug combinations

Drug	Prev. RT/CT	Resp. crit[a]	n eval.	CR (%)	OR (%)	95% CI	Duration resp.	Date	Reference
Carboplatin, vincristine, methotrexate, bleomycin	25/4	UICC	29	10	31	16–52%	30 weeks	1991	[146]
Methotrexate, vinblastine, doxorubicin, cisplatin	23/0	NS	29	21	66		7.5 months	1995	[147]
Bleomycin, vincristine mitomycin-C, cisplatin	11/2	ECOG	13	23	77		4 months	1980	[148]
		SWOG	51	7	22		5 months	1987	[84]
	0/0	WHO	39	25	61		–	1991	[50]
	0/0	NS	28	35	100		–	1991	[62]
Cisplatin, bleomycin and methotrexate/cisplatin and 5-FU									
(Recurrent disease)	18/2	NA	23	26	30		10.5 months		
(Advanced disease)	4/0	N/A	17	18	41		10.5 months	1994	[38]

Table 23.10c Notes to table

Methotrexate, vinblastine, doxorubicin and cisplatin – Long et al. [147]. Methotrexate 30 mg/m^2 days 1, 15, and 22, vinblastine 3 mg/m^2 days 2, 15 and 22, doxorubicin 30 mg/m^2 day 2, cisplatin 70 mg/m^2, with a 28-day cycle. Twenty-nine evaluable patients, eight Stage IVa or b; nine out of 17 (53%) patients with only pelvic disease responded and ten out of 12 (83%) with extrapelvic ± pelvic tumours. Five out of six (83%) patients with no prior irradiation responded and 14 out of 23 (61%) with prior irradiation. Only four patients completed the protocol, but three of these were alive and disease-free at ≥ 46 months. None of these had relapsed in field after standard radical pelvic radiotherapy. Five other patients had potentially 'curative' surgery after MVAC chemotherapy but all relapsed, and none survived long-term.

Bleomycin, vincristine, mitomycin-C and cisplatin – Vogl et al. [148]. A pilot project to modify a three drug combination of cisDDP, bleomycin and high dose methotrexate which had proven highly active, but unacceptably toxic, on account of reduced methotrexate clearance in patients with renal impairment [35]. Treatment comprised: weekly bleomycin 10 u; vincristine 1 mg/m^2 days 1, 8, 22 and 29; mitomycin-C 10 mg/m^2 day 1; cisDDP 50 mg/m^2 days 1 and 22, using 6-weekly cycles. Patient characteristics: only 13 out of 16 patients treated were evaluable for response; four patients PS 0; five patients PS 1; two patients PS 2 and two patients PS 3. Two had prior CT and 11 had progressive tumour after RT. Two patients were previously untreated, Stage IIIb and IVa. Nine patients had pelvic tumour, five pulmonary, four retroperitoneal, three peripheral node, one bone and one liver metastases; which responded is not stated. Symptomatic improvement was usually evident <1 week of starting chemotherapy. Toxicity: one case grade IV thrombocytopenia, five grade II peripheral neuropathies.

Souhami et al. [50]. (Randomized neoadjuvant CT/RT study.) Bleomycin 15 u im 12-h days 1–4 (subsequently days 1–3); vincristine 1 mg/m^2 day 1; mitomycin 10 mg/m^2 iv day 1, and cisDDP 50 mg/m^2 iv day 1 (later day 4 instead), every 21 days for three cycles. Only 39 out of 52 patients randomized to chemotherapy were evaluable. All had Stage IIIb tumour; 24 had unilateral and 15 bilateral parametrial involvement; 24 had pelvic and 13 para-aortic lymph node involvement on lymphangiography. There were four fatal cases of bleomycin pulmonary toxicity.

Dottino et al. [62]. Uncontrolled pilot neoadjuvant study (CT/surgery), cisplatin 50 mg/m^2, mitomycin-C 10 mg/m^2, vincristine 1 mg/m and bleomycin 10 u im repeated every 21 days. Patient characteristics: $n = 28$; four Stage Ib; six Stage IIa; seven Stage IIb; 1 Stage IIIa; 11 Stage IIIb; one Stage IVa. Toxicity not described. Response data: CR 35%, OR 100%. All complete responders had negative pelvic nodes; of 18 PRs, involved nodes were found at subsequent surgery in nine patients. This was a lower than expected incidence of pelvic nodes

Cisplatin, methotrexate and bleomycin, alternating with cisplatin and 5FU – Chambers et al. [38]. An alternating regimen comprising cisplatin 80 mg/m^2 day 1, bleomycin 10 u/day, days 3–6, methotrexate 150 mg/m^2 day 15, day 22 with leukovorin, alternating with cisplatin 100 mg/m^2 day 1, 5-FU 1000 mg/m^2/day, days 2–5, q 28 days. Cisplatin, bleomycin and 5-FU were all given by continuous infusion. The regimen was evaluated separately in two groups of patients: of 17 women with advanced Stage IVa, IVb, or 'non-FIGO' extra-pelvic disease (four with prior radiotherapy), there were seven responders (OR 41%). The two patients who had earlier progressed on radiotherapy did not respond, and seven patients who failed chemotherapy did not respond to subsequent radiotherapy. Of 23 patients with recurrent disease, 21 with prior radiotherapy (75–90 Gy) 12 had relapsed with only pelvic disease, four failed in pelvis and elsewhere, and two with distant metastases alone. Five patients had relapsed after initial surgery. Of seven responses, there were four in patients with pelvic disease alone, and three in patients with both local and distant recurrence. Five responses were seen in 18 patients relapsing after primary radiotherapy, and two responses in the two patients who had received primary surgery alone. No responses were seen in any patient who had previously received treatment of any kind for relapse.

improved survival: amongst uncontrolled single agent studies, complete response rates typically fall in the range 0–10%, and for combinations 10–20%. The bulk of the apparent improvement in overall response rates seen with some combinations typically reflects more of an increase in the partial response rate, and those few combination studies which show higher complete response rates coincidentally report fewer partial responses than might have been anticipated, so their overall response rates remain unremarkable [38, 83]. In Potter's analysis (discussed earlier), all such complete responses occurred at extra-pelvic sites.

However, it seemed plausible to many authors that the numerically increased partial response rates ostensibly provided by some combinations might have the potential to afford improved symptom control. In practice, the usefulness of such an approach would obviously be qualified by any increased toxicity attributable to combination therapy. Ideally, any phase III trial designed to determine the clinical utility of combination therapy in relation to single agent cisplatin in recurrent disease, for example, should therefore include careful assessment of the relative impact of treatment on quality of life, an integral of toxicity on the one hand and symptomatic improvement due to a reduction in disease burden on the other. Very few trials have been designed with this level of stringency, and none have yet been executed satisfactorily.

The CRC phase III trial of single agent cisplatin vs a combination of bleomycin, ifosfamide and cisplatin (BIP) has been closed prematurely because of poor recruitment, with just over 80 patients randomized. Analysis is underway but, as anticipated, there is no significant survival advantage to the combination. Three published studies have made similar though less detailed comparisons. Their results are summarized in Table 23.11 [84–86]. Any advantage in overall response rates for combination chemotherapy is modest and associated with excess toxicity. This suggests the more striking results recorded in some uncontrolled studies reflect patient selection factors. Ironically, these data obviate the necessity for a subtle quality-of-life-based approach. All survival curves appear superimposable [86].

THE FUTURE

Whilst there is a genuine sense of optimism about the application of new classes of cytotoxic drugs (e.g. the taxoids) or scheduling innova-

Table 23.11 Randomized trials of cisplatin-based combinations vs single agent cisplatin

Trial	n	CR	(%)	PR	(%)	OR	(%)	Median survey	Reference
SWOG MmC, Vcr, Blm and cisDDP	54	4	(7)	8	(15)	12	(22)	6.9	
vs MmC and cisDDP	51	2	(4)	11	(21)	13	(25)	7.0	
vs cisDDP	9	1	(11)	2	(22)	3	(33)	17.0	[84]
EORTC Blm, Vind, MmC & cisDDP	143	11	(8)	33	(23)	44	(31)	10	
vs cisDDP	144	8	(6)	20	(14)	28	(19)	9.4	[86]
GOG Mitolactol and cisDDP	147	14	(9)	17	(12)	31	(21)	7.3	
vs Ifosfamide and cisDDP	151	19	(13)	28	(19)	47	(31)	8.3	
vs cisDDP	140	9	(7)	16	(12)	25	(18)	8	[85]

tions (e.g. infusional 5-FU) to the chemotherapy of breast and ovarian carcinoma, data such as those presented here for the activity of new drugs in cervical cancer chemotherapy do not seem to have engendered any such confidence despite response rates of 15–20% following pelvic radiotherapy, which are broadly similar to the activity these drugs have shown in platinum-resistant ovarian carcinoma [87–91]. This dispiriting view seems unjustified.

Perhaps it is time to take a critical look at both the design and direction of chemotherapy trials in cervical carcinoma.

First, there is an obvious requirement that all phase II trial populations should be transparently characterized by performance status and whether the indicator lesions are irradiated or not, and their respective treatment-free interval. Without such information the reporting of bald response rates in phase II trials are uninformative at best or misleading at worst.

Second, since 'false-positive' results are a very real danger in uncontrolled phase II studies, particularly with new platinum combinations, more use should be made of a randomized phase II design (against single agent cisplatin).

Thirdly, whilst the ideal would obviously be a cytotoxic with substantive non-cross-resistance to radiotherapy, in its absence the results of recent neoadjuvant chemotherapy-then-radiotherapy vs radiotherapy studies would hardly seem surprising [10, 51] and until such drugs become available (or unless impending overviews with access to unpublished data such as that being carried out in the UK by the MRC suggest an advantage), it would seem hard to justify more studies of similar design.

Fourth, we need new ways of integrating chemotherapy, surgery and radiotherapy in radical multimodality treatment plans for patients presenting with high-risk disease.

Fifth, we need non-toxic, simple, day case palliative regimens, with an edge over single agent cisplatin.

Sixth, future randomized studies need to be large enough to discriminate plausible improvements in outcome with confidence.

Alternative approaches to the patient with a newly diagnosed bulky tumour include studies of synchronous chemoradiation (discussed elsewhere in this volume; reports on mature GOG and RTOG studies appear imminent: these include GOG 123 and GOG 120, and RTOG 9001 [92]), and also further randomized trials of neoadjuvant chemotherapy ahead of surgery. Considerable work has now been published exploring neoadjuvant chemotherapy-then-surgery in patients presenting with bulky early-stage disease. Despite the majority of these studies being uncontrolled, they have successfully addressed many of the initial safety concerns provoked by their avoiding or delaying the standard treatment modality for this group, namely radical pelvic radiotherapy, and long-term (10 year) follow-up data have recently been reported [57]. Response rates for combination therapy may be very high, and in those patients who respond, it is clear that not only is operability improved, but a significant proportion of patients may obtain either pathological complete remissions, or microscopic residuum [93, 94]. Similarly, there is evidence of a substantial impact on the incidence of pelvic lymph node metastases confirmed at surgery [93]. These findings contrast with the results of studies undertaken in recurrent disease, and suggest a different strategy may be justified in developing new treatments for newly presenting locally advanced disease and that an intensive approach using combination therapy may be more appropriate. Furthermore, the increasing availability of magnetic resonance imaging as an adjunct to examination under anaesthetic might facilitate improved pre-operative assessment of response to chemotherapy, and therefore result in fewer 'open and close' operations than in the past.

The single available randomized trial addressing neoadjuvant chemotherapy-then-surgery in relation to surgery alone has been criticized on a number of counts, not least publication of an interim analysis after just 151 patients had been randomized [53], and the subsequent decision of the authors to halt accrual at 205 patients, on the basis of a subgroup analysis purporting to show an improved survival for patients with tumours 4 cm in diameter [95]. This subgroup included 56 out of 103 controls and 61 out of 102 treated with neoadjuvant chemotherapy, and whilst there was a significant difference in time to progression, there was only a trend towards improved overall survival [96]. A further trial has shown an 81% 3.5-year disease-free survival following neoadjuvant chemotherapy then surgery, compared with 69% with surgery alone.

Larger randomized trials are clearly justified: the GOG-141 protocol was launched in December 1996 using a neoadjuvant cisplatin and vincristine combination in patients with Stage Ib tumours, at least 4 cm in diameter. The target is 340 patients over 3.5 years with an interim analysis after 170 patients have been entered [92]. However, whilst this may provide basic information as to the value of a neoadjuvant surgical strategy, a positive result will prompt more questions. How many cycles of pre-operative chemotherapy are optimal? [65]. What treatment interval should be adopted? Should chemotherapy extend post-operatively? Is it required in all cases, irrespective of resection margins obtained or lymph node status? To what extent might the morbidity of external beam pelvic radiotherapy following surgery be higher than standard radical radiotherapy? Could this be reduced using inflatable devices left in the operative cavity to keep the bowel out of the radiation field? What of those unfortunate patients who fail to respond to primary chemotherapy? As there are ample data to show that such patients do poorly with radiotherapy [51, 54, 55], should they be offered second-line chemotherapy within phase II trials, with an option on surgery if they respond, or might salvage surgery, perhaps followed by radiotherapy, offer more? However, any such approach needs to be viewed in the light of recent suggestions that salvage surgery may represent a better option in the event of a minor response to chemotherapy: Sardi *et al.* have reported that chemotherapy-then-surgery achieved local control in 11 out of 15 such patients (73%), compared with eight out of 21 (38%) with chemotherapy-then-radiation [63, 97].

In contrast to the strategy outlined for the development of more intensive radical treatments for high-risk newly diagnosed disease, the immediate priority for the development of better palliative treatment must be to devise less toxic regimens and algorithms for their safe, successful application. Both these objectives may be within our grasp. Carboplatin, dosed using the Calvert formula, surely deserves another look, and might provide the core of an attractive outpatient regimen [98]. As things stand, ifosfamide is probably the most active of the remaining drugs, and whilst there are few data to suggest that lower delivered dose intensity compromises response [44], lower doses would provide a better tolerated regimen, reduce the risks of toxic encephalopathy and facilitate day case treatment. The ensuing reduction in emesis would also permit the use of oral mesna. The apparent utility of ifosfamide employed at lower doses in combination chemotherapy for non-small cell lung cancer is an encouraging precedent, and the MIC regimen, for example, uses ifosfamide at a dose of just 3 g/m^2 [99].

It has been recognized for 10 years or more that the integration of chemotherapy into multimodality treatment arguably represents our best opportunity to improve the outlook of women with established invasive cervical carcinoma. However, the same period has seen disappointing trial accrual, with at least three randomized neoadjuvant studies closed prematurely. Unfortunately, there can be little doubt that the impact of the change in the presentation of this disease on clinical trial accrual has been potentiated by the reticence of clinicians to approach patients about their potential participation in such studies. The reasons for this are several, and in all probability vary from country to country, trial to trial and time to time, but a general concern about the toxicity of chemotherapy may be less common than sometimes supposed. Arguably the main reason for the failure of the West Midlands Group adjuvant and neoadjuvant studies in the UK, for example, was the development of an early trend towards the increasing use of chemotherapy 'off study', in circumstances where the failure of radiotherapy alone might be anticipated. The practicalities of negotiating 6-week waiting lists for radical radiotherapy may have initially exacerbated the problem, but there emerged an unfortunate but widely held view that primary chemotherapy almost certainly held advantages over standard treatment for high-risk early-stage tumours, a prejudice encouraged by the 60–70% response rates reported for combination regimens in advanced disease, the publication of promising interim analyses [42] and the 'precedent' of demonstrably effective adjuvant chemotherapy for other early-stage solid tumours. With the advent of medical audit and the widespread adoption of 'evidence-based medicine' in the early 1990s, radical pelvic radiotherapy regained its proper status as standard treatment for bulky and inoperable tumours, but diffuse referral patterns, whereby the majority of patients were treated by generic gynaecologists and radiotherapists rather than consultants with subspecialist interest in gynaeoncology, time constraints in busy NHS clinics and, not least, clinician's reluctance to discuss their uncertainty about what might constitute optimal treatment, have all had adverse implications for trial accrual.

Much has changed in the last 5 years and in contrast to the recent past, cautious optimism now appears justified: not only does our increased knowledge of prognostic indices provide the methodology to assay and discard 'fools gold' rapidly in early phase trials, but increasing gynaecological subspecialization amongst radiation and medical oncologists, greater numbers of specialist surgical gynae-oncologists and the advent of clinical nurse specialists in both support and research roles have encouraged the closer teamwork that is crucial to the successful execution of larger randomized studies. Furthermore, there is now a wider recognition of clinical research as intrinsic to the highest standards of individual patient care. These changes in working practice augur well for the recent launch of pan-European collaborative studies to complement those of the Gyneoncology Group and the Southwestern Oncology Group in North America. Importantly, they have the potential to engender and sustain the commitment necessary to complete large trials.

REFERENCES

1. Malkasian, G., Decker, D. and Jorgensen, E. (1977) Chemotherapy for carcinoma of the cervix. *Gynecologic Oncology*, **5**, 109–20.
2. Buxton, E. (1992) Experience with bleomycin, ifosfamide and cisplatin in primary and recurrent cervical cancer. *Seminars in Oncology*, **19**, 9–18.
3. Wingo, P., Tong, T. and Bolden, S. (1995) Cancer Statistics 1995. *CA–A Cancer Journal for Clinicians*, **45**, 8–30.
4. Chung, C., Nahas, W., Stryker, S., *et al.* (1980) Analysis of factors contributing to treatment failures in stage Ib and stage IIa carcinoma of the cervix. *American Journal of Obstetrics and Gynecology*, **138**, 550–6.
5. Fuller, A., Elliot, N., Kosloff, C., *et al.* (1988). Determinants for increased risk of recurrence in patients undergoing radical surgery for stage Ib and IIa carcinoma of the cervix. *Gynecologic Oncology*, **32**, 146.
6. Stehman, F., Blessing, J., McGehee, R., *et al.* (1989) A phase II evaluation of mitolactol in patienrs with advanced squamous cell carcinoma of the cervix: a Gynecologic Oncology Group Study. *Journal of Clinical Oncology*, **17**, 1892–5.
7. Perez, C., Grigsby, P., Nene, S., *et al.* (1992) Effect of tumour size on the prognosis of carcinoma of the uterine cervix treated with irradiation alone. *Cancer*, **69**, 2796–806.
8. Zaino, R., Ward, S., Delgado, G., *et al.* (1992) Histopathologic predictors of the behaviour of surgically treated stage IB squamous cell carcinoma of the cervix: a Gynecologic Oncology Group study. *Cancer*, **69**, 1750–8.

9. Curtin, J., Hoskins, W., Venkatraman, E., *et al.* (1996) Adjuvant chemotherapy versus chemotherapy plus pelvic irradiation for high risk cervical cancer patients after radical hysterectomy and pelvic lymphadenectomy: a randomised phase II trial. *Gynecologic Oncology*, **61**, 3–10.

10. Tattersall, M., Ramirez, C. and Coppleson, M. (1992) A randomised trial comparing platinum-based chemotherapy followed by radiotherapy vs. radiotherapy alone in patients with locally advanced cervical cancer. *International Journal of Gynaecological Cancer*, **2**, 244–251.

11. Cullen, M., Smith, S., Benfield, G., *et al.* (1987) Testing new drugs in untreated small cell lung cancer may prejudice the results of standard treatment: a phase II study of oral idarubicin in extensive disease. *Cancer Treatment Reports*, **71**, 1227–30.

12. Cullen, M. (1989). Evaluating new drugs as first line treatment in patients with small cell lung cancer: guidelines for an ethical approach with implications for other chemotherapy sensitive tumours. *Journal of Clinical Oncology*, **6**, 1356–7.

13. Gehan, E. (1961). The determination of the number of patients required in a preliminary and a follow up trial of a new chemotherapeutic agent. *Journal of Chronic Diseases*, **13**, 346–56.

14. Lee, Y., Catane, R., Rozencweig, M., *et al.* (1979) Analysis and interpretation of response rates for anticancer drugs. *Cancer Treatment Reports*, **63**, 1713–20.

15. Lee, Y., Staquet, M., Simon, R., *et al.* (1979) Two stage plans for patient accrual in phase II clinical trials. *Cancer Treatment Reports*, **63**, 1721–6.

16. Flemming, T. (1982). One-sample multiple testing procedure for phase II clinical trials. *Biometrics*, **38**, 143–51.

17. Moertel, C., Schutt, A., Hahan, R., *et al.* (1974) Effects of patient selection on results of phase II chemotherapy trials in gastrointestinal cancer. *Cancer Chemotherapy Reports*, **59**, 257.

18. Blackledge, G., Lawton, F., Redman, C., *et al.* (1989) Response of patients in phase II studies of chemotherapy in ovarian cancer: implications for patient treatment and design of phase II trials. *British Journal of Cancer*, **59**, 650–3.

19. Gore, M., Fryatt, I., Wiltshaw, E., *et al.* (1990) Treatment of relapsed carcinoma of the ovary with cisplatin or carboplatin following initial treatment with these compounds. *Gynecologic Oncology*, 2071–1.

20. Markman, M., Rothman, R., Hakes, T., *et al.* (1991) Second line platinum chemotherapy in patients with ovarian cancer previously treated with cisplatin. *Journal of Clinical Oncology*, **9**, 389–93.

21. Warwick, J., Kehoe, S. and Earl, H. (1995) Long term follow-up of patients with ovarian cancer involved in clinical trials – important prognostic features. *British Journal of Cancer*.

22. Cutler, S. (1974) Classification of extent of disease in breast cancer. *Seminars in Oncology*, **1**, 91–6.

23. Clark, G., Sledge, G., Osbourne, C., *et al.* (1987) Survival from first recurrence: relative importance of prognostic factors in 1015 breast cancer patients. *Journal of Clinical Oncology*, **5**, 55–61.

24. Brun, B., Benchalal, M., Lebas, C., *et al.* (1997) Response to second-line chemotherapy in patients with metastatic breast carcinoma previously responsive to first-line treatment. *Cancer*, **79**, 2137–46.

25. Peto, R. (1978) Clinical trial methodology. *Biomedicine*, **28**, 24–36.

26. Potter, M., Hatch, K., Potter, M., *et al.* (1989) Factors affecting the response of recurrent squamous cell carc inoma of the cervix to cisplatin. *Cancer*, **63**, 1283–6.

27. Sorbe, B. and Frankendal, B. (1984) Bleomycin–adriamycin–cisplatin combination chemotherapy in the treatment of primary advanced and recurrent cervical carcinoma. *Obstetrics and Gynecology*, **63**, 167–70.

28. Hoskin, P. and Blake, P. (1991) Cisplatin, methotrexate, and bleomycin (PMB) for carcinoma of the cervix: the influence of presentation and previous treatment upon response. *International Journal of Gynecological Cancer*, **1**, 75–80.

29. Picozzi, V., Sikic, B., Carson, R., *et al.* (1985). Bleomycin, mitomycin and cisplatin therapy for advanced squamous cell carcinoma of the uterine cervix: a phase II study of the Northern California Oncology Group. *Cancer Treatment Reports*, **69**, 903–5.

30. Louie, K., Behrens, B., Kinsella, T., *et al.* (1985) Radiation survival parameters of antineoplastic drug-sensitive and -resistant human ovarian cancer cell lines and their modification by buthionine sulfoxamine. *Cancer Research*, **45**, 2110–15.

31. Eastman, A. and Schulte, N. (1988) Enhanced DNA repair as a mechanism of resistance to cis-diamminedichloroplatinum (II). *Biochemistry*, **27**, 4730–4.

32. Masuda, H., Hamilton, T. and Ozols, R. (1988) Increased DNA repair as a mechanism of acquired resistance to L-phenylalanine mustard and diamminedichloroplatinum (II) in human ovarian cancer cell lines. *Cancer Research*, **48**, 5713.

33. Carmichael, J. and Hickson, I. (1991). Keynote address: mechanisms of cellular resistance to cytotoxic drugs and X-irradiation. *International Journal of Radiation Oncology, Biology and Physiology*, **20**, 197–202.

34. Lane, D. (1992). p53, guardian of the genome. *Nature*, **358**, 15–16.

35. Vogl, S., Moukhtar, M. and Kaplan, B. (1979) Chemotherapy of advanced cervical cancer with methotrexate, bleomycin and cis-dichlorodiammineplatinum (II). *Cancer Treatment Reports*, **63**, 1005–6.

36. Meanwell, C., Mould, J., Blackledge, G., *et al.* (1986). Phase II study of ifosfamide in advanced cervical cancer. *Cancer Treatment Reports*, **70**, 727–30.

37. Goren, M., Wright, R., Pratt, C., *et al.* (1986) Dechloroethylation of ifosfamide and neurotoxicity. *Lancet*, **2**, 1219–20.

38. Chambers, S., Lamb, L., Kohorn, E., *et al.* (1994) Chemotherapy of recurrent/advanced cervical cancer: results of the Yale University PBM-PFU Protocol. *Gynecologic Oncology*, **53**, 161–9.

39. Meanwell, C., Blake, E., Kelly, K., *et al.* (1986) Prediction of ifosfamide/mesna associated encephalopathy. *European Journal of Clinical Oncology* **22**, 815–19.

40. Sutton, G., Blessing, J., Adcock, L., *et al.* (1989). Phase II study of ifosfamide and mesna in patients with previously treated carcinoma of the cervix. *Investigational New Drugs*, **7**, 341–3.

41. Bonomi, P., Brady, M., Blessing, J., *et al.* (1988) Prognostic factors related to response and to survival in women with advanced squamous cell carcinoma of the cervix treated with cisplatin. A Gynecologic Oncology Group Study. *Proceedings of the American Association for Cancer Research*, **29**, 207.

42. Buxton, E., Meanwell, C., Hilton, C., *et al.* (1989) Combination bleomycin, ifosfamide, and cisplatin chemotherapy in cervix cancer. *Journal of the National Cancer Institute*, **81**, 359–61.

43. Coleman, R., Harper, P., Gallagher, C., *et al.* (1986) A phase II study of ifosfamide in advanced and relapsed cervical carcinoma of the cervix. *Cancer Chemotherapy and Pharmacology*, **18**, 280–3.

44. Coleman, R., Clark, J., Slevin, M., *et al.* (1990) A phase II study of ifosfamide and cisplatin chemotherapy for metastatic or relapsed carcinoma of the cervix. *Cancer Chemotherapy and Pharmacology*, **27**, 51–4.

45. Zanetta, G., Torri, W., Bocciolone, L., *et al.* (1995) Factors predicting response to chemotherapy and survival in patients with metastatic or recurrent squamous cell cervical carcinoma: a multivariate analysis. *Gynecologic Oncology*, **58**, 58–63.

46. Thigpen, T., Shingleton, H., Homesley, H., *et al.* (1979) Cisdichlorodiaminoplatinum (II) in the treatment of gynaecologic malignancies: phase II trials by the Gynaecologic Oncology Group. *Cancer Treatment Reports*, **63**, 1549–55.

47. Thigpen, T., Shingleton, H., Homesley, H., *et al.* (1981). Cisplatinum in treatment of advanced or recurrent squamous cell carcinoma of the cervix: a phase II study of the Gynaecologic Oncology Group. *Cancer*, 48, 899–903.

48. Bonomi, P., Blessing, J., Stehman, F., *et al.* (1985) Randomised trial of three cisplatin dose schedules in squamous-cell carcinoma of the cervix: a Gynecologic Oncology Group study. *Journal of Clinical Oncology*, **3**, 1079.

49. Piver, M. and Chung, W. (1975) Prognostic significance of cervical lesion size and pelvic node metastases in cervical carcinoma. *Obstetrics and Gynecology*, **46**, 507.

50. Souhami, L., Gil, R., Allan, S., *et al.* (1991) A randomised trial of chemotherapy followed by pelvic radiation therapy in stage IIIB carcinoma of the cervix. *Journal of Clinical Oncology*, **9**, 970–7.

51. Kumar, L., Kaushal, R., Nandy, M., *et al.* (1994) Chemotherapy followed by radiotherapy versus radiotherapy alone in locally advanced cervical cancer: a randomised study. *Gynecologic Oncology*, **54**, 307–15.

52. Tattersall, M., Lorvidhaya, V., Cheirsilpa, A., *et al.* (1995) Randomised trial of epirubicin and cisplatin chemotherapy followed by pelvic radiation in locally advanced cervical cancer. *Journal of Clinical Oncology*, **13**, 444–51.

53. Sardi, J., Sananes, C., Giaroli, A., *et al.* (1993) Results of a prospective randomised trial with neoadjuvant chemotherapy in stage IB bulky squamous carcinoma of the cervix. *Gynecologic Oncology*, **49**, 156–65.

54. Kirsten, F., Atkinson, K., Coppleson, J., *et al.* (1987) Combination chemotherapy followed by surgery or radiotherapy in patients with locally advanced cervical cancer. *British Journal of Obstetrics and Gynecology*, **94**, 583.

55. Symonds, R., Burnett, R., Habeshaw, T., *et al.* (1989) The prognostic value of a response to chemotherapy given before radiotherapy in advanced cancer of the cervix. *British Journal of Cancer*, **59**, 473–5.

56. Benedetti Panici, P., Scambia, G., Baiocchi, G., *et al.* (1991) Neoadjuvant chemotherapy and radical surgery in locally advanced cervical cancer: prognostic features for response and survival. *Cancer*, **67**, 372–9.

57. Benedetti Panici, P., Greggi, S., Scambia, G., *et al.* (1997) Long-term survival following neo-adjuvant chemotherapy and radical surgery in locally-advanced cervical cancer. *ASCO Proceedings* **16**, 1294.

58. Bolis, G., Frigerio, L. and Melpignano, M. (1992) Primary chemotherapy including platinum for bulky and advanced cervical cancer. *Annals of Oncology*, **3**, 106.

59. Bolis, G., Van Zainten-Przybysz, I., Scarfone, G., *et al.* (1996) Determinants of response to a cisplatin-based regimen as neoadjuvant chemotherapy in stage Ib–IIb invasive cervical cancer. *Gynecologic Oncology*, **63**, 62–5.

60. Whitley, N., Brenner, D., Francis, A., *et al.* (1982) Computed tomographic evaluation of carcinoma of the cervix. *Radiology*, **142**, 439–46.

61. Giaroli, A., Sananes, C., Sardi, J., *et al.* (1990) Lymph node metastases in carcinoma of the cervix uteri: response to neoadjucvant chemotherapy and its impact on survival. *Gynecologic Oncology*, **39**, 34–9.

62. Dottino, P., Plaxe, S., Beddoe, A., *et al.* (1991). Induction chemotherapy followed by radical surgery in cervical cancer. *Gynecologic Oncology*, **40**, 7–11.

63. Sardi, J. (1996) To the Editor. *Gynecologic Oncology*, **62**, 321–2.

64. Chang, H.-C., Lai, C.-H., Chou, P., *et al.* (1992). Neoadjuvant chemotherapy with cisplatin, vincristine, and bleomycin and radical surgery in early-stage bulky cervical carcinoma. *Cancer Chemotherapy and Pharmacology*, **30**, 281–5.

65. Kim, D., Moon, H., Hwang, Y., *et al.* (1988) Preoperative adjuvant chemotherapy in the treatment of cervical cancer Stage Ib, IIb with bulky tumor. *Gynecologic Oncology*, **29**, 321–32.

66. Serur, E., Mathews, R., Gates, J., *et al.* (1997) Neoadjuvant chemotherapy in stage Ib2 squamous cell carcinoma of the cervix. *Gynecologic Oncology*, **65**, 348–56.

67. Fanning, J., Ladd, C. and Hilgers, R. (1995) Cisplatin, 5-fluorouracil and ifosfamide in the treatment of recurrent or advanced cervical cancer. *Gynecologic Oncology*, **56**, 235.

68. Sutton, G., Blessing, J., DiSaia, P., *et al.* (1993) Phase II study of ifosfamide and mesna in non-squamous carcinoma of the cervix: a Gynaecologic Oncology Group Study. *Gynecologic Oncology*, **49**, 48–50.

69. Sutton, G., Blessing, J., McGuire, W., *et al.* (1993) Phase II trial of ifosfamide and mesna in patients with advanced or recurrent squamous cell carcinoma of the cervix who have never received chemotherapy: a Gynecologic Oncology Group Study. *American Journal of Obstetrics and Gynaecology*, **168**, 805–7.

70. Field, C., Dockerty, M., and Symmonds, R. (1964) Small cell carcinoma of the cervix. *American Journal of Obstetrics and Gynaecology*, **88**, 4475–1.

71. Mackay, B., Osbourne, B. and Wharton, J. (1979) Small cell tumour of the cervix with neuroepithelial features: ultrastructural observation in two cases. *Cancer*, **43**, 1138–45.

72. Stohler, M., Mills, S., Gersell, D., *et al.* (1991). Small cell neuroendorine carcinoma of the cervix. A human papillomavirus type 18-associated cancer. *American Journal of Surgery and Pathology*, **15**, 28–32.

73. Abeler, V., Holm, R., Nesland, J., *et al.* (1994) Small cell carcinoma of the cervix. A clinicopathologic study of 26 patients. *Cancer*, **73**, 672–7.

74. Sheets, E., Berman, M., Hroutas, C., *et al.* (1988) Surgically-treated, early stage neuroendocrine small cell cervical carcinoma. *Obstetrics and Gynecology*, **71**, 10–14.

75. O'Hanlan, K., Goldberg, G., Jones, J., *et al.* (1991) Adjuvant therapy for neuroendocrine small cell carcinoma of the cervix; review of the literature. *Gynecologic Oncology*, **43**, 167–72.

76. Morris, M., Gershenson, D., Eifel, P., *et al.* (1992) Treatment of small cell carcinoma of the cervix with cisplatin, doxorubicin and etoposide. *Gynecologic Oncology*, **47**, 62–5.

77. Lewandowski, G. and Copeland, L. (1993) A potential role for intensive chemotherapy in the treatment of small cell carcinoma of the cervix. *Gynecologic Oncology*, **48**, 127–31.

78. Guthrie, D. and Way, S. (1978) The use of adriamycin and methotrexate in carcinoma of the cervix. *Obstetrics and Gynaecology*, **52**, 349–54.

79. Calvert, A., Newell, D., Gumbrell, L., *et al.* (1989) Carboplatin dosage: prospective evaluation of a simple formula based on renal function. *Journal of Clinical Oncology*, **17**, 1748–56.

80. Egorin, M., Van Echo, D., Olman, E., *et al.* (1985) Prospective validation of a pharmacologically based dosing scheme for the cis-diamminedichloroplatinum (II) analogue diamminecyclobutanedicarboxylato-platinum. *Cancer Research*, **45**, 6502–6.

81. Weiss, G., Green, S., Hannigan, E., *et al.* (1990) A phase II trial of carboplatin for recurrent or metastatic carcinoma of the uterine cervix: a Southwest Oncology Group Study. *Gynecologic Oncology*, **39**, 332–6.

82. Thigpen, T., Blessing, J., Gallup, D., *et al.* (1995) Phase II trial of mitomycin-C in squamous cell carcinoma of the uterine cervix: a Gynecologic Group study. *Gynecologic Oncology*, **57**, 376–9.

83. Stornes, I. and Mejlholm, J. (1994) A phase II trial of ifosfamide, 5-FU, and leukovorin in recurrent uterine cancer. *Gynecologic Oncology*, **55**, 123–5.

84. Alberts, D., Kronmal, R., Baker, L., *et al.* (1987) Phase II randomised trial of cisplatin chemotherapy regimens in the treatment of recurrent or metastatic squamous cell cancer of the cervix: a Southwest Oncology Group Study. *Journal of Clinical Oncology*, **5**, 1791–5.

85. Omura, G., Blessing, J., Vaccarello, L., *et al.* (1996) A randomised trial of cisplatin versus cisplatin and mitolactol versus cisplatin and ifosfamide in advanced squamous carcinoma of the cervix by the Gynecologic Oncology Group. *Gynecologic Oncology*, **60**, 120.

86. Vermorken, J., Zanetta, G., De Oliveira, C., *et al.* (1996) Cisplatin-based combination chemotherapy (BEMP) versus single agent cisplatin (P) in disseminated squamous cell carcinoma of the uterine cervix: mature data EORTC Protocol 55863. *Annals of Oncology*, **7**C, 67.

87. Goedhals, L. and Bezwoda, W. (1995) A phase II study of gemcitabine in advanced cervical carcinoma. *European Journal of Cancer*, **31**A, S246.

88. Chevallier, B., L'homme, C., Dieras, V., *et al.* (1995) A phase II study of CPT 11 (irinotecan) in chemotherapy naive patients with advanced cancer of the cervix uteri. *European Journal of Cancer*, **31**A, S246.

89. Noda, K., Sasaki, H., Yamamato, K., *et al.* (1996) Phase II trial of topotecan for cervical cancer of the uterus. *ASCO Proceedings*, **15**, 280.

90. McGuire, W., Blessing, J., Moore, D., *et al.* (1996). Paclitaxel has moderate activity in squamous cell cervix cancer: A Gynecologic Oncology Group Study. *Journal of Clinical Oncology*, **14**, 792–5.

91. Kudelka, A., Verschraegen, C., Levy, T., *et al.* (1998). Preliminary report of the activity of docetaxel in advanced or recurrent squamous cell carcinoma of the cervix. *Anticancer Drugs*, **7**, 398–401.

92. PDQ Editorial Board, Ed. (1997) *Current Clinical Trials Oncology (NCI/PDQ)*. Pyros Education Group, Ltd., Green Brook, NJ, USA.

93. Colombo, N., Landoni, F., Pellegrino, A., *et al.* (1997) Phase II study of cisplatin, ifosfamide and paclitaxel (CIP) as neoadjuvant chemotherapy in patients with locally advanced cervical carcinoma. *ASCO Proceedings*, **16**, 367a.

94. Vermorken, J. (1993) The role of chemotherapy in squamous cell carcinoma of the uterine cervix: a review. *International Journal of Gynecological Cancer*, **3**, 129–42.

95. Di Paola, G. and Sardi, J. (1995) To the editor. *Gynecologic Oncology*, **59**, 165.

96. Thomas, G. (1993). Is neoadjuvant chemotherapy a useful strategy for the treatment of stage Ib cervix cancer? *Gynecologic Oncology*, **49**, 153–5.

97. Sardi, J., Sananes, C., Giaroli, A., *et al.* (1996) Results of a phase III trial with neoadjuvant chemotherapy in stage IIIb patients: an unexpected result. *International Journal of Gynecologic Oncology*, **6**, 85–93.

98. Muggia, F. and Murderspach, L. (1994) Platinum compounds in cervical and endometrial cancers: focus on carboplatin. *Seminars in Oncology*, **21**, 35–41.

99. Ferry, D., Chetiyawardana, A., Joshi, R., *et al.* (1992). A randomised trial of Mitomycin, Ifosfamide, and Cisplatin versus best palliative care in extensive stage non-small cell lung cancer: an interim analysis. *British Journal of Cancer*, **65**, 37.

100. Morris, M., Gershenson, D., Burke, T., *et al.* (1994) A phase II study of carboplatin and cisplatin in advanced or recurrent squamous cell carcinoma of the uterine cervix. *Gynaecologic Oncology*, **53**, 234–8.

101. Rose, P., Blessing, J. and Arsenau, J. (1996) Phase II evaluation of altretamine for advanced or recurrent squamous cell carcinoma of the cervix – a Gyneoncology Group Study. *Gynecologic Oncology*, **62**, 100–2.

102. Arseneau, J., Blessing, J., Stehman, F., *et al.* (1986) A phase II study of carboplatin in advanced squamous cell carcinoma of the cervix (a Gynecologic Oncology Group Study). *Investigational New Drugs*, **4**, 187–91.

103. Lira-Puerto, V., Silva, A., Morris, M., *et al.* (1991). Phase II trial of carboplatin or iproplatin in cervical cancer. *Cancer Chemotherapy Pharmacology*, **28**, 391–6.

104. McGuire, W., Arsenau, J., Blessing, J., *et al.* (1989) A randomised trial of carboplatin and iproplatin in advanced squamous carcinoma of the uterine cervix – a Gynecologic Oncology Group study. *Journal of Clinical Oncology*, **7**, 1462–8.

105. Reichman, B., Markman, M., Hakes, T., *et al.* (1991) Phase II trial of high dose cisplatin with sodium thiosulphate nephroprotection in patients with advanced carcinoma of the uterine cervix previously untreated by chemotherapy. *Gynecologic Oncology*, **43**, 159–63.

106. Lele, S. and Piver, S. (1989) Weekly induction chemotherapy in the treatment of recurrent cervical carcinoma. *Gynecologic Oncology*, **33**, 6–8.

107. Wasserman, T. and Carter, S. (1977) The integration of chemotherapy into combined

modality treatment of solid tumours. *Cancer Treatment Reviews*, **4**, 25–46.

108. Stehman, F., Bundy, B., DiSaia, P., *et al.* (1991) Carcinoma of the cervix treated with radiation therapy: a multivariate analysis of prognostic variables in the Gynecologic Oncology Group. *Cancer*, **67**, 2776–85.

109. Lira-Puerto, V., Morales-Canfield, F., Wernz, J., *et al.* (1984) Activity of mitolactol in cancer of the uterine cervix. *Cancer Treatment Reports*, **68**, 669–70.

110. Wallace, H., Hreshchyshyn, M., Wilbanks, G., *et al.* (1978) Comparison of the therapeutic effects of adriamycin alone versus adriamycin plus vincristine versus adriamycin plus cyclophosphamide in the treatment of advanced carcinoma of the cervix. *Cancer Treatment Reports*, **62**, 1435–41.

111. van der Burg, M., Monfardini, S., Guastalla, J., *et al.* (1993) Phase II study of weekly 4'-epidoxorubicin in patients with metastatic squamous cell cancer of the cervix: an EORTC Gynaecological Cancer Cooperative Group Study. *European Journal of Cancer*, **29**A, 147–8.

112. Wong, L.-C., Choy, D., Ngan, H., *et al.* (1989) 4-Epidoxorubicin in recurrent cervical cancer. *Cancer*, **63**, 1279–82.

113. Calero, F., Rodriguez-Escudero, F., Mendana, F., *et al.* (1991) Single agent epirubicin in squamous cell cervical cancer: a phase II trial. *Acta Oncologica*, **30**, 325–7.

114. Slayton, R., Creasman, W., Petty, W., *et al.* (1979) Phase II trial of VP-16–213 in the treatment of advanced squamous cell carcinoma of the cervix and adenocarcinoma of the ovary: a Gynecologic Oncology Group Study. *Cancer Treatment Reports*, **63**, 2089–92.

115. Goedhals, L. and Bezwoda, W. (1996). A phase II study of gemcitabine in advanced cervix carcinoma: final data. *ASCO Proceedings*, **15**, 296.

116. Cervellino, J., Araujo, C., Pirisi, C., *et al.* (1990) Ifosfamide and mesna at high doses for the treatment of cancer of the cervix – a Getlac study. *Cancer Chemotherapy and Pharmacology*, **26**, S1–S3.

117. Verschraegen, C., Levy, T., Kudelka, A., *et al.* (1967) Phase II study of irinotecan in prior chemotherapy-treated squamous cell carcinoma of the cervix. *Journal of Clinical Oncology*, **15**, 625–31.

118. Sutton, G., Blessing, J., Barrett, R., *et al.* (1994) Phase II trial of menogaril in patients with squamous carcinoma of the cervix: a Gyecologic Oncology Group Study. *Gynecologic Oncology*, **52**, 229–31.

119. Muss, H., Sutton, G., Bundy, B., *et al.* (1985) Mitoxantrone (NSC 301739) in patients with advanced cervical carcinoma: a phase II study of the Gynecologic Oncology Group. *American Journal of Clinical Oncology*, **8**, 312–15.

120. Kavanagh, J., Copeland, L., Gershenson, D., *et al.* (1985) Continuous-infusion vinblastine in refractory carcinoma of the cervix: a phase II trial. *Gynecologic Oncology*, **21**, 211–4.

121. Rhomberg, W. (1986) Vindesine for recurrent and metastatic cancer of the uterine cervix. A phase II study. *Cancer Treatment Reports*, **70**, 1455–7.

122. Lacava, J., Leone, B., Machiavelli, M., *et al.* (1997). Vinorelbine as neoadjuvant chemotherapy in advanced cervical carcinoma. *Journal of Clinical Oncology*, **15**, 604–9.

123. Sutton, G., Blessing, J., Photopulos, G., *et al.* (1990) Early phase II Gynecologic Oncology Group experience with ifosfamide/mesna in gynecologic malignancies. *Cancer Chemotherapy Pharmacology*, **26**, S55–8.

124. Edmonson, J., Johnson, P. and Wieand, H.A. (1988) Phase II studies of bleomycin, cyclophosphamide, doxorubicin and cisplatin, and bleomycin and cisplatin in advanced cervical carcinoma. *American Journal of Clinical Oncology (CCT)*, **11**, 149–51.

125. Bezwoda, W., Nissenbaum, M. and Derman, D. (1986) Treatment of metastatic and recurrent cervix cancer with chemotherapy: a randomised trial comparing hydroxyurea with cis-diamminedichloro-platinum plus methotrexate. *Medical and Pediatric Oncology*, **14**, 17–19.

126. Bonomi, P., Blessing, J., Ball, H., *et al.* (1989) A phase II evaluation of cisplatin and 5-FU in patients with advanced squamous cell carcinoma of the cervix: a Gynecologic Oncology Group Study. *Gynecologic Oncology*, **34**, 357–9.

127. Kaern, J., Trope, C., Abeler, V., *et al.* (1990) A phase II study of 5-fluorouracil/cisplatinum in recurrent cervical cancer. *Acta Oncologica*, **29**, 25–8.

128. Chiara, S., Consoli, R. and Falcone, A. (1988). Cisplatin and ifosfamide in advanced and recurrent cervical cancer. *Tumori*, **74**, 471–4.

129. Zanetta, G., Lissoni, A., Gabriele, A., *et al.* (1997) Intense neoadjuvant chmeotherapy with cisplatin and epirubicin for advanced or bulky cervical and vaginal adenocarcinoma. *Gynecologic Oncology*, **64**, 431–5.

130. Cervellino, J., Araujo, C. and Sanchez, O. (1995). Cisplatin and ifosfamide in patients with advanced squamous cell carcinomaof the uterine cervix. *Acta Oncologica*, **34**, 257–9.

131. Kühnle, H., Meerpohl, H.-G., Eiermann, W., *et al. Cancer Chemotherapy Pharmacology* (1990) **26** (Suppl), S33–5.

132. Al-Saleh, E., Hoskins, P., Pike, J., *et al.* (1997) Cisplatin/etoposide chemotherapy for recurrent or primarily advanced cervical carcinoma. *Gynecologic Oncology*, **64**, 468–72.

133. Miyamoto, T., Takabe, Y., Watanabe, M., *et al.* (1978) Effectiveness of a sequential combination of bleomycin and mitomycin-C on an advanced cervical cancer. *Cancer*, **41**, 403–14.

134. Tropé, C., Johnsson, J., Simonsen, E., *et al.* (1983) Bleomycin–mitomycin-C in advanced carcinoma of the cervix, a third look. *Cancer*, **51**, 591–3.

135. Sundfør, K., Tropé, C.G., Hügberg, T. *et al.* (1996) Radiotherapy and neoadjuvant chemotherapy for cervical cancer. *Cancer*, **77**, 2371–8.

136. Friedlander, M., Kaye, S. Sullivan, A., *et al.* (1983) Cervical carcinoma: a drug responsive tumour-experience with combined cisplatin vinblastine and bleomycin. *Gynecologic Oncology*, **16**, 275–81.

137. Namkoong, S., Park, J., Kim, J., *et al.* (1995) Comparative study of patients with locally advanced stages I and II cervical cancer treated by radical surgery with and without preoperative adjuvant chemotherapy. *Gynecologic Oncology*, **59**, 136–42.

138. Kredentser, D. (1991). Etoposide, ifosfamide/mesna, and cisplatin chemotherapy for advanced and recurrent carcinoma of the cervix. *Gynecologic Oncology*, **43**, 145–8.

139. Filtenborg, T., Hansen, H. and Aage, E. (1993) A phase II study of ifosfamide carboplatin and cisplatin in advanced and recurrent squamous cell carcinoma of the uterine cervix. *Annals of Oncology*, **4**, 485–8.

140. Kumar, L. and V. Bhargava (1991). Chemotherapy in recurrent and advanced cervical cancer. *Gynecologic Oncology*, **40**, 107–11.

141. Tobias, J., Buxton, E., Blackledge, G., *et al.* (1990). Neoadjuvant bleomycin, ifosfamide and cisplatin in cervical carcinoma. *Cancer Chemotherapy and Pharmacology*, **26**, S59–62.

142. Ramm, K., Vergote, I., Kaern, J., *et al.* (1992) Bleomycin–ifosfamide–cisplatinum (BIP) in pelvic recurrence of previously irradiated cervical carcinoma: a second look. *Gynecologic Oncology*, **46**, 203–7.

143. Tay, S., Lai, F., Soh, L., *et al.* (1992) Combined chemotherapy using cisplatin, ifosfamide, and bleomycin (PIB) in the treatment of advanced and recurrent cervical carcinoma. *Australia and New Zealand Journal of Obsterics and Gynaecology*, **32**, 263–6.

144. Kumar, L., Pokharel, Y., Grover, R., *et al.* (1997). Neoadjuvant chemotherapy followed by radiotherapy in locally advanced squamous cell cervical cancer: two randomised studies. *Journal of Clinical Oncology*, **16**, 364a.

145. Murad, A., Triginelli, S. and Ribalta, J. (1994) Phase II trial of bleomycin, ifosfamide and carboplatin in metastatic cervical cancer. *Journal of Clinical Oncology*, **12**, 55–9.

146. Junor, E., Davies, J., Habeshaw, T., *et al.* (1991) Carboplatin-based combination chemotherapy for advanced carcinoma of the cervix. *Cancer Chemotherapy and Pharmacology*, **27**, 484–6.

147. Long, H.I., Cross, W., Wieand, H., *et al.* (1995). Phase II trial of methotrexate, vinblastine, doxorubicin, cisplatin in advanced Irecurrent carcinoma of the uterine cervix and vagina. *Gynaecologic Oncology*, **57**, 235–9.

148. Vogl, S., Moukkhtar, M., Calanog, A., *et al.* (1980) Chemotherapy for advanced cervical cancer with bleomycin vincristine mitomycin-C and *cis*-diamminedichloroplatinum (II; BOMP). *Cancer Treatment Reports*, **64**, 1005–7.

149. Thigpen, J., Vance, R., Puneky, L., *et al.* (1995) Chemotherapy as a palliative treatment in carcinoma of the uterine cervix. *Seminars in Oncology*, **22**, 16–24.

150. Piver, M.S., Barlow, J.J. and Xynos, F.P. (1978) Adriamycin alone or in combination in 100 patients with carcinoma of the cervix of vagina. *American Journal of Obstetrics and Gynecology*, **131**, 311–13.

24

INTEGRATED APPROACHES TO THE MANAGEMENT OF CANCER OF THE CERVIX

M.H.N. Tattersall and P.N. Mainwaring

INTRODUCTION

The sequential historical application of surgery, radiotherapy and chemotherapy as methods of treating cancer led in the past to the same sequence of treatments commonly being applied in individual patient management. Over the last decade, the importance of individual prognostic variables and of factors influencing treatment failure has naturally led to a closer integration of treatments in the management of many patients with cervical cancer. Indeed, in some settings new knowledge has caused a change in the order in which these treatment modalities are applied. Combined modality treatment has become the norm in many cancer patients, commonly with improved quality of life and survival.

New technologies are improving pre-treatment staging and helping in the assessment of treatment effects. These techniques better define the extent of disease allowing more rational determination of the treatment goals, and application and evaluation of multimodal treatment. In addition, increasing numbers of randomized trials addressing important questions in better defined patient populations are being undertaken by cooperative groups, sometimes incorporating quality of life measures as well as more traditional end points.

In this chapter the roles of surgery, radiotherapy and chemotherapy in cervical cancer management are reviewed and the issue of how best they may be integrated is considered. Proposals for future research are outlined with the hope that improved treatment outcomes will follow.

GENERAL CONSIDERATIONS

The reasons for integrating different modalities of cancer treatment are either to improve the treatment outcomes in terms of greater disease-control and survival or to reduce treatment morbidity and/or costs.

As outlined in earlier chapters, surgery and radiotherapy are local cancer treatments and their combined use is aimed at improving local control of cervical cancer which it is believed has not spread systemically. In addition, surgery may provide valuable staging information – access to tissue which helps to define the extent of tumour spread and provide evidence of the biological aggressiveness of the tumour. The probability of surgical

Cancer and Pre-cancer of the Cervix
Edited by D.M. Luesley and R. Barrasso. Published in 1998 by Chapman & Hall, London. ISBN 0 412 56600 1.

control of cervical cancer in the pelvis is inversely related to tumour size and disease stage, while the extent of pelvic nodal spread particularly influences the likelihood of distant metastases [1]. The requirement to preserve essential normal tissue and avoid debilitating sequelae necessarily limits the extent of pelvic surgery. Radiotherapy is also an effective treatment of localized cancer and has the advantage over surgery that involved unresectable tissues may be treated within the confines of normal tissue tolerance. Intraoperative radiotherapy has been advocated with the potential advantage of more precise definition of treatment volumes and the opportunity to limit normal tissue toxicity, and encouraging results have been claimed [2]. Both surgery followed by radiotherapy and vice versa have been applied in patients with 'operable' cervical cancer but no comparative randomized trials have been undertaken.

RADIATION THERAPY AND SURGERY

The preferred management of low bulk FIGO Stage Ib cervical cancer is influenced by availability of expertise and equipment, but in most Western countries is effectively a quality of life/referral pattern decision. Surgery in expert hands may preserve ovarian function. Improved sexual function has been claimed after radical hysterectomy compared to pelvic radiotherapy with or without intravaginal irradiation. Treatment costs differ, and radiation therapy which requires capital intensive equipment involves treatment over several weeks. The results of surgery are influenced by surgical expertise and high quality postoperative nursing care is required. No studies have established a survival advantage for one or the other treatment, and there is disagreement about morbidity differences.

Economic evaluation is a recent innovation in the management of patients with cancer. A case has been made for including economic evaluation as an end point in the randomized trials of different cancer treatments [3].

Patients with bulky Stage Ib–IIa disease treated by surgery or radiotherapy have a significant risk of pelvic and systemic relapse and a multi-modality treatment approach has been advocated, particularly when the tumour is 4 cm or more in maximal diameter. However, retrospective data suggest that adjunctive hysterectomy after radiotherapy does not influence survival although pelvic control may be improved [4]. Mendenhall *et al.* reported a retrospective analysis of 150 patients with Stage Ib–IIb tumours equal to or greater than 6 cm treated from October 1964 to June 1983. In the early years patients were treated predominantly with external beam and intracavity radiotherapy, whilst later patients received radiotherapy followed by radical hysterectomy. The local control rate was 74% in patients treated with radiotherapy alone compared to 76% where radiotherapy was followed by surgery. Complications requiring admission to hospital occurred in 5% of patients treated by radiotherapy alone vs 16% with combined treatment [5].

In many units when the resection margins are compromised or pelvic lymph nodes contain metastases surgery is followed by radiotherapy. This strategy was evaluated in 72 of 320 patients with Stage Ib–IIa carcinoma of the cervix [6]. The authors concluded that in the presence of poor prognostic factors, e.g. large tumours, radiation therapy reduced the incidence of pelvic failure. However, these conclusions require confirmation and no effects on survival are known.

The value of para-aortic nodal resection and/or irradiation in women with pelvic node metastases, or with cytologically confirmed para-aortic node metastases, is controversial. The Radiation Therapy Oncology group compared irradiation of the pelvis alone to irradiation of the pelvis plus para-aortic lymph nodes for patients with carcinoma of the cervix bulky ($\geqslant$4 cm) Stage Ib or greater. Patients receiving para-aortic node irradiation had a 65% 5-year survival vs 55% in those treated with pelvic irradiation only ($p < 0.0043$) [7]. By comparison the EORTC reported a reduction in para-aortic nodal failure in a similar trial on 441 patients but pelvic control and overall survival were similar [8]. Concern about the acute and delayed toxicities of para-aortic node radiation and the lack of clear evidence of improved treatment outcome have led to this treatment approach no longer being followed.

Hacker *et al.* (personal communication) have reported a series of patients with para-aortic node metastases that were resected as part of combined modality treatment. An improved outcome was obtained, but this requires confirmation in a prospective randomized trial.

Patients with FIGO Stage IIb–IV disease are usually judged inoperable and commonly treated with curative intent by pelvic ± intravaginal radiotherapy. Reported 5-year survival rates are in the range of 10–40%, and treatment failure is commonly in the pelvis, but also sometimes systemically. Suitably designed randomized trials have investigated multi-modal treatments in these patients only recently.

CHEMOTHERAPY

Chemotherapy was first evaluated in the management of cervical cancer in patients with systemic metastases or with recurrent disease after failed first line treatments [9]. A variety of drugs were reported to sometimes cause tumour shrinkage but responses were usually short-lived. Cisplatin, introduced into clinical trial in 1972, was reported to have activity in cervical cancer in the late 1970s [10, 11]. Alkylating agents had some activity, although with the advent of the urothelial protective agent mesna, ifosfamide has been favoured recently. Anthracyclines have activity and particularly epirubicin [12] in one study was as active as cisplatin.

Combination chemotherapy regimes incorporating cisplatin appear to have higher tumour response rates than cisplatin alone [13, 14] although only one randomized study has reported a survival advantage in advanced disease (Omura *et al.*, personal communication). Treatment with combined 5-fluorouracil, adriamycin, cyclophosphamide, vincristine or ifosfamide has been reported to cause tumour response in patients after failure of first line cisplatin chemotherapy [15]. These results indicate that several drugs have antitumour effects in cervical cancer, but the most effective regime has not been determined. For a detailed review of chemotherapy in cervical cancer see Chapter 23.

The demonstration that cervical cancer commonly shrinks after cisplatin-based chemotherapy lead investigators to explore strategies of integrating chemotherapy into the management of locally advanced cervical cancer, and as a post-surgical adjuvant in patients with earlier stage disease but bad prognostic features. The following types of combined modality treatments incorporating chemotherapy are considered below (Table 24.1):

Table 24.1 Strategies for combining chemotherapy with local cancer treatments

Type	Indication	Rationale
Primary chemotherapy	High local failure rate with local treatment	Chemotherapy → tumour response and improved chance of local control by subsequent local treatment
Postoperative adjuvant chemotherapy	Distant, microscopic metastases, a frequent cause of treatment failure	Micrometastases likely to be more sensitive to chemotherapy and perhaps curable
Radiosensitizing chemotherapy	Radiotherapy alone not likely to achieve local control, e.g. bulky tumours containing hypoxic cells	Enhancement of radiosensitivity by hypoxic sensitizers → ?DNA repair effects, ?cell cycle modulation, ?effect on free radical/glutathione

- neoadjuvant or primary chemotherapy followed by local treatment – surgery and or radiotherapy – in patients with locally advanced disease, i.e. FIGO IIb–IVa;
- 'radiosensitizing' chemotherapy or concurrent chemoradiation where drug treatment is given during radiation with the aim of enhancing the effects of radiotherapy, e.g. sensitizing hypoxic tumour cells without excess toxicity – in patients with locally advanced disease;
- post-operative adjuvant chemotherapy in women with high risk but operable cervical cancer, e.g. with pelvic node metastases Stage Ib–IIa.

NEOADJUVANT OR PRIMARY CHEMOTHERAPY

Some potential advantages of neoadjuvant or primary chemotherapy have been summarized by Alberts *et al.* [16]. These include: enhanced feasibility of tumour resection following tumour regression after chemotherapy, improved control of occult metastatic disease and enhanced probability of tumour control by subsequent radiotherapy. Moreover, drug access to the tumour site may be improved if the vasculature has not been disturbed by surgery or radiotherapy.

Finally, tolerance to chemotherapy may be greater in patients who have received no prior antitumour treatments. Some possible disadvantages include delay in initiating local therapy, prolongation of overall treatment time, possible enhanced toxicity with subsequent radiotherapy/surgery and the possibility of tumour progression during chemotherapy with emergence of drug-resistant and perhaps radioresistant clones.

Primary chemotherapy in cervical cancer has commonly employed two or three cycles of chemotherapy given over 4–9 weeks before surgery or more usually radiotherapy. Tumour response has been reported in about 50% of patients with locally advanced disease and tolerance to subsequent radiotherapy has been unexceptional. The number of treatment cycles prior to local treatment may influence tumour response probability. Which drug combination is most effective has not been investigated in randomized trials. Recently, several randomized trials of neoadjuvant chemotherapy followed by radiotherapy vs radiotherapy alone in patients with bulky Stage Ib or locally advanced disease have confirmed the local tumour response rates, but follow-up has shown no effect on local tumour control and no survival benefit. Table 24.2 presents the published randomized trials

Table 24.2   Randomized trials of neoadjuvant chemotherapy followed by radiotherapy vs radiotherapy alone

Ref.	Disease stage[a]	No. of patients	Type CT No. of cycles	Post CT tumour response[b]	(%CR)	Local CCR CT → RT	RT	Median FU (months)	Survival difference
[18]	IIb–III	138	VMCP × 2–4	35	(1.5)	84.9	88.9	26	No
[19]	IIIb	91	VBMP × 3	61	(25.5)	47.0	32.5	44 and 51	No
[20]	IIIb–IVa	73	PF × 3	NS	(NS)	NS	NS	NS	NS
[21]	IIb–IVa	71	PVB × 3	47	(0)	65	73	42	No
[22]	IIa–IV	66	BIP × 3	69	(6)	75	56	NS	NS
[23]	IIb	25	PEC × 4	73	(9)	64	93	28	NS
[24]	IIb	64	VBP × 3	NS	(NS)	NS	NS	18	NS

CT, chemotherapy; RT, radiotherapy; FU, follow-up; CR, complete response; NS, not stated; [a]FIGO staging; [b]response after chemotherapy; Local CCR, local complete control rate; P, cisplatin, B, bleomycin, V, vinblastine or vincristine in VMCP, VBMP; C, cyclophosphamide, chlorambucil in VMCP; E, epirubicin; F, 5-fluorouracil; I, ifosfamide; M, methotrexate.

of neoadjuvant chemotherapy followed by radiotherapy as summarized by Vermorken [17]. In these trials, most of which had small numbers, patterns of treatment failure have been similar in both treatment arms but in two studies distant relapse rates were lower in the chemotherapy arms [18–24].

A recent large trial from several treatment centres in Asia has evaluated primary chemotherapy with epirubicin 110 mg/m^2 and cisplatin 60 mg/m^2 for two to three cycles followed by radiotherapy vs radiotherapy alone. This study has reported a >50% response to chemotherapy but survival and local control rates in the combined modality treatment patients were significantly inferior [25].

In summary, neoadjuvant chemotherapy followed by radiotherapy has not been shown to improve local control or survival. Indeed the Asian study and that by Souhami *et al.* report a superior outcome for the standard radiotherapy arm [19, 25].

Three factors possibly contributing to these disappointing results can be considered. Withers *et al.* have reported that increased local failure rates following radiotherapy are seen in patients with head and neck cancer whose radiation treatment continues for more than 5–6 weeks [26]. They proposed that prolonged low dose per fraction radiotherapy may select out a population of resistant cells with a high regrowth rate. This mechanism might explain why primary chemotherapy through shrinking the tumour may cause surviving cells to have a high regrowth rate, which reduces the cytotoxic effectiveness of subsequent radiotherapy. The observation that local treatment failure was more common in groups receiving primary chemotherapy in the Asian study is compatible with this proposed mechanism. One possible means of overcoming this effect might be to change radiotherapy fractionation schedules and dose per fraction.

Cross-resistance between drugs and radiotherapy may also explain why tumour response after chemotherapy does not lead to improved patient survival. Reports of primary chemotherapy in cervical cancer and other tumour types have frequently indicated that tumours unresponsive to chemotherapy rarely respond to subsequent radiotherapy [27]. This finding supports the notion that cross-resistance between certain drugs and ionizing radiation may occur, e.g. DNA repair mechanisms.

A final possible factor contributing to the poor outcome of primary chemotherapy is the inevitable delay in starting radiotherapy. However, overt tumour progression during primary chemotherapy is rare.

The strategy of primary chemotherapy followed by surgery rather than radiotherapy does not have the potential disadvantages outlined above. A review of selected trials in which neoadjuvant chemotherapy has been followed by surgery in predominantly early-stage patients reveals encouraging results, summarized in Table 24.3 [28–34]. A low frequency (6–23%) of histologically involved pelvic lymph nodes has also been reported in these patients at the time of radical hysterectomy, suggesting chemotherapy effects in the lymph nodes as well as commonly in the primary tumour.

Benedetti Panici *et al.* reported 75 patients with locally advanced cervical cancer who received three courses of neoadjuvant chemotherapy followed by radical surgery in all responding patients, or by radiotherapy if no change or progressive disease was demonstrated at post-chemotherapy assessment. The authors reported a significantly improved 3-year survival rate in those patients judged to have had a response to chemotherapy compared to those who did not [35].

Recently Sardi *et al.* reported a prospective randomized trial in Stage Ib, bulky, squamous carcinoma of the cervix, in which neoadjuvant chemotherapy followed by surgery proceeding onto radiotherapy was compared with surgery followed by radiotherapy. Patients in the experimental arm were treated with three 'quick' cycles of cisplatin, vincristine and

Table 24.3   Trials utilizing neoadjuvant chemotherapy followed by surgery

Ref.	Disease stage[a]	No. of patients	Type CT No. of cycles	Response following CT (%)	(%CR)	Pathological response CR (%)	LN pos (%)	Median FU (months)
[29]	Ib–IIb	54	PVB × 1–5	94	(44)	13	20	36
[30]	Ib–II	169	VBP × 3	79	(33)	NS	23	24
[31]	Ib–III	75	PBM × 3	83	(15)	13	24	30
[32]	Ib–III	26	PB × 1	88	(19)	19	9.5	18
[33]	Ib–IVa	28	VBMP$_2$ × 1	100	(35)	14	32	24
[34]	Ib–IIb	92	PVB × 2–5	82	(28)	NS	17	42
[35]	Ib–IIb	27	PB × 1	78	(11)	7	15	16.5

CT, chemotherapy; CR, complete response; LNN, lymph node nodal status; FU, follow-up; [a]FIGO staging; Path CR, pathological complete response; P, cisplatin; B, bleomycin; V, vinblastine (vincristine in VBMP); M, methotrexate in PBM; mitomycin in VBMP.

bleomycin prior to definitive local therapy. In both arms patients had a laparotomy at which the surgeon decided if radical hysterectomy could be performed with clear margins. Those patients with tumours considered unresectable received pelvic irradiation according to the Fletcher technique. After 48 months of follow-up, 88% of patients in the neoadjuvant group had not relapsed vs 67% in the control group ($p = 0.03$). Overall survival was not significantly different between the two groups (88% in the neoadjuvant group and 70% in the control group). In the patients treated with chemotherapy prior to definitive local therapy a trend to reduced pelvic recurrences was reported [36].

Results of treatment of patients with Stage Ib–IIb cervical cancer with chemotherapy followed by surgery are promising and justify further study. Large-scale randomized trials that include quality of life measures are now required. These trials require careful planning to address the current issues. The standard treatment arm should be radical hysterectomy, perhaps with pelvic radiotherapy if there are nodal metastases. The experimental arm should be primary chemotherapy followed by surgery with the same criteria for inclusion of pelvic radiotherapy. Eligibility should be restricted to operable patients, and perhaps patients with tumours greater than 4 cm, and/or unusual (not pure squamous) histology.

CONCURRENT CHEMOTHERAPY AND RADIOTHERAPY

Several drugs have been investigated over the last two decades as possible sensitizers to radiation-induced cell damage. The relative radioresistance of the hypoxic cell has been the principle target of radiosensitizer drug development, but conventional chemotherapeutic agents have also been investigated as potentiators of radiation damage, e.g. hydroxyurea, 5-fluorouracil, mitomycin C and cisplatin. Hyperbaric oxygen and hyperthermia have also been studied as potentiators of radiation damage.

During the 1970s small randomized trials of radiosensitizing chemotherapy reported a survival benefit, but toxicity was considerable [37]. More recently the results of larger trials have been reported and the advantageous effects of radiosensitizers are not clear (Table 22.4) [38–45].

In a recently reported large randomized trial patients receiving the investigative radiosensitizer pimonidazole had worse local control and overall survival than those in the radiotherapy alone arm. This result could not be explained by an imbalance in patient prognostic factors between arms of the trial, or inadequate radiotherapy [46].

Current studies are evaluating cisplatin and 5-fluorouracil given with radiotherapy, and

Table 24.4 Randomized trials of concurrent chemotherapy and radiotherapy vs radiotherapy alone

Ref.	Treatment	Stage	No. of patients	Local control	Survival	p value
[38]	RT + HU	IIIb/IVa	51	68%	19.5 months	0.05
	RT		46	49%	10.7 months	
[39]	RT+HU	IIb	20	NS	94% 5 years	0.006
	RT		20	NS	53%	
[40]	RT + HU	IIIb	20	NS	43% 5 years	0.78
	RT		25	NS	50%	
[41]	RT + HU	IIb–IVa	139	NS	48.4 months	
	RT+MIS		157	NS	30.3 months	
[42]	RT + HU	III	14	NS	54% 5 years	
	RT		11	NS	18%	
[43]	RT + CDDP	IIb–III	20	55%	NS	
	RT		25	20%	NS	
[44]	RT + CDDP 2×/week	IIb–III	17	88%		0.83
	RT + CDDP weekly		22	64%	Progression Free interval	
	RT		25	80%		
[45]	RT+PIM		91			
	RT		92			

Local control, local complete control; HU, hydroxyurea; MIS, misonidazole; PIM, pimonidazole; NS, not stated.

unpublished preliminary reports of trials in the USA and Asia are promising.

POST-OPERATIVE ADJUVANT CHEMOTHERAPY

Although radical hysterectomy or pelvic radiotherapy eradicate tumours in most patients with Stage Ib–IIa cervical cancer, in some patients the cancer relapses. The identification of patients at risk for local and or distant recurrence after surgery or radiotherapy is an important research aim. These 'high-risk' women may benefit from adjuvant systemic therapy. Prognostic factors in early-stage cervical cancer patients have been identified, including tumour size and histological type, evidence of vascular invasion and patient age. In addition local surgical pathology findings associated with a poor prognosis include com-

promised surgical margins and pelvic nodal metastases [46–48].

Wertheim *et al.* treated patients with Stage Ib–IIa cervical cancer considered at high risk of relapse, i.e. involving pelvic node metastases, lymphatic or vascular invasion, tumour size >4 cm, grade 3 lesions, adenosquamous histology, parametrial invasion and evidence of locally metastatic (non-contiguous) disease – with post-operative adjuvant chemotherapy comprising two cycles of cisplatin and bleomycin followed by radiotherapy. At 46 months follow-up, 17 of 44 patients were free of relapse and alive and a phase III trial was advocated [49].

Lai *et al.* treated 40 Stage Ib–IIb patients with post-operative adjuvant cisplatin, vinblastine and bleomycin, and reported a 3-year survival rate of 75% for the patients treated with adjuvant chemotherapy compared to

46.8% for those who refused adjuvant therapy ($p < 0.05$) [50].

Ueki *et al.* treated 52 of 114 patients judged to be at increased risk of relapse after primary surgical treatment, with adjuvant tegafur (a 5-fluorouracil derivative). The degree of risk of recurrence was formulated as a discriminant function of risk factors involved in disease recurrence. The rate of recurrence was 25% (13 patients) in patients receiving chemotherapy and 71% (44 patients) without chemotherapy ($p < 0.01$) [51].

Only one randomized trial of post-operative adjuvant chemotherapy has been reported and no significant effects on outcome were found, although the number of patients randomized limited the power of the study [52]. A large-scale multinational trial investigating adjuvant chemotherapy in Stage Ib–IIa cervical cancer treated by radical hysterectomy and pelvic radiotherapy has been commenced.

CONCLUSION

Most patients with early-stage cervical cancer are cured by surgery or pelvic radiation. The identification of high-risk patients who may benefit from combined local and/or systemic treatment is an important future task. Properly defined and executed randomized trials are required to identify the proper place (if any) of systemic therapy in patients at risk of treatment failure or relapse if treated by local measures only [53].

REFERENCES

1. Perez, C.A., Grigsby, P.W., Nene, S.M., Camel, H.M., Galakatos, A., Kao, M.-S. and Lockett, M.A. (1992) Effect of tumour size on the prognosis of carcinoma of the uterine cervix treated with irradiation alone. *Cancer*, **69**, 2796–806.
2. Maruyama, Y., Van Nagel, J.R., Yoneda, J., *et al.* (1990) A review of californium-252 neutron brachytherapy for cervical cancer. *Cancer*, **68**, 1189–97.
3. Bonsl, G.J., Rutten, F.F.H. and Uyl-de Groot, C.A. (1993) Economic evaluation alongside

cancer trials: methodological and practical aspects. *European Journal of Cancer*, **29A**, S10–14.
4. Coleman, D.L., Gallup, D.G., Wolcott, H.D., Otken, L.B. and Stock, R.J. (1992) Patterns of failure of bulky-barrel carcinomas of the cervix. *American Journal of Obstetrics and Gynecology*, **166**, 916–20.
5. Mendenhall, W., McCarthy, P.J., Morgan, L.S., Chafe, W.E. and Millon, R.R. (1991) Stage IB or IIA–B carcinoma of the intact uterine cervix greater than or equal to 6 cm in diameter: is adjuvant extrafascial hysterectomy beneficial? *International Journal of Radiation Oncology, Biology and Physiology*, **21**, 899–904.
6. Soisson, A.P., Soper, J.T., Clarke-Pearson, D.L., Berchuck, A., Montana, G. and Creasman, W.T. (1990) Adjuvant radiotherapy following radical hysterectomy for patients with stage IB–IIA cervical cancer. *Gynecologic Oncology*, **37**, 390–5.
7. Rotman, M., Choi, K., Guze, C., Marcial, V., Hornback, N. and John, M. (1990) Prophylactic irradiation of the para-aortic lymph node chain in stage IIB and bulky IB carcinoma of the cervix: initial treatment results of RTOG 7920. *International Journal of Radiation Oncology, Biology and Physiology*, **19**, 513–21.
8. Haie, C., Pejoric, M.H., Gerbaulet, A., *et al.* (1988) Is prophylactic para-aortic irradiation worthwhile in the treatment of advanced cervical carcinoma? Results of a controlled clinical trial of the EORTC Radiotherapy Group. *Radiotherapy Oncology*, **11**, 101–12.
9. Thigpen, T., Shingleton, H., Homesley, H., Lagasse, L. and Blessing, J. (1981) Cisplatinum in treatment of advanced or recurrent squamous cell carcinoma of the cervix: a phase II study of the Gynecologic Oncology Group. *Cancer*, **48**, 899–903.
10. Cohen, C.J., Castro-Marin, A., Deppe, G., Bruckner, H.W. and Holland, J.F. (1978) Chemotherapy of advanced recurrent cervical cancer with platinum II – preliminary report. *Proceedings of the American Society for Clinical Oncology*, **19**, 401.
11. Thigpen, T., Shingleton, H., Homesleye, H., Lagasse, L. and Blessing, J. (1979) *cis*Dichlorodiammineplatinum(II) in the treatment of gynecologic malignancies: phase II trials by the Gynecologic Oncology Group. *Cancer Treatment Report*, **63**, 1549–55.
12. Wong, L.-C., Choy, D.T.K., Ngan, H.Y.S., Sham, J.S.T. and Ma, H.-K. (1989) 4-Epidoxorubicin in recurrent cervical cancer. *Cancer*, **63**, 1279–82.
13. Guthrie, D. and Way, S. (1974) Treatment of advanced carcinoma of the cervix with adri-

amycin and methotrexate combined. *Obstetrics and Gynecology*, **44**, 586–9.

14. Miyamoto, T., Takabe, Y., Watenabe, M. and Terasima, T. (1978) Effectiveness of a sequential combination of bleomycin and mitomycin-C on an advanced cervical cancer. *Cancer*, **41**, 403–14.

15. Kuehnle, H., Meerpohl, H.-G., Eiermann, W. and Achterrath, W. (1992) Neoadjuvant therapy for cervical cancer. *Seminars in Oncology*, **19**, 94–8.

16. Alberts, D.S., Aristizabal, S, Surwit, E.A., *et al.* (1987) Primary chemotherapy for high-risk recurrence cervical cancer. In *Cervix Cancer* (eds E.A. Surwit and D. Alberts). Martinus Nijhoff, Boston, MA, pp. 161–83.

17. Vermorken, J.B. (1993) The role of chemotherapy in squamous cell carcinoma of the uterine cervix: a review. *International Journal of Gynecology and Cancer*, **3**, 129–42.

18. Chauvergne, J., Rohart, J., Heron, J.F., *et al.* (1988) Randomized phase III trial of neo-adjuvant chemotherapy (CT) + radiotherapy (RT) vs. RT in stage IIB, III carcinoma of the cervix (CACX): a cooperative study of French Oncology Centers. *Proceedings of the American Society for Clinical Oncology*, **7**, 136.

19. Souhami, L., Gil, R.A., Allen, S.E., Canary, P.C.V., Araujo, C.M.M., Pinto, L.H.J. and Silveira, T.R.P. (1991) A randomized trial of chemotherapy followed by pelvic radiation therapy in stage IIIB carcinoma of the cervix. *Journal of Clinical Oncology*, **9**, 970–7.

20. Sundfør, K., Bertelsen, K., Høgberg, T., Onsrud, M., Tropé, C. and Simonsen, E. (1991) A randomized trial of induction chemotherapy in patients with squamous cell carcinoma of the uterine cervix stages IIIB–IVA. *Proceedings of the European Conference on Clinical Oncology*, **6**, S121.

21. Tattersall, M.H.N., Ramirez, C. and Coppleson, M. (1992) A randomised trial comparing platinum based chemotherapy followed by radiotherapy versus radiotherapy alone in patients with locally advanced cervical cancer. *International Journal of Gynecological Oncology*, **2**, 244–51.

22. Tobias, J., Buxton, E.J., Blackledge, G., Mould, J.J., Monaghan, J., Spooner, D. and Chetiyawardana, A. (1990) Neoadjuvant bleomycin, ifosfamide and cisplatin in cervical cancer. *Cancer Chemotherapy and Pharmacology*, **26**, 59–62.

23. Cardenas, J., Olguin, A., Figueroa, F., Pena, J., Becerra, F. and Huizar, R. (1991) Neoadjuvant chemotherapy (CI) in cervical cancer (CC) PEC + RT vs. RT. Preliminary results. *Proceedings of the European Conference on Clinical Oncology*, **6**, S124.

24. Sardi, J. (1991) Early results of a randomized trial with neoadjuvant chemotherapy in squamous carcinoma cervix uteri. *Proceedings of the 3rd Meeting of the International Gynecological Cancer Society*, Blackwell Science, p. 40.

25. Tattersall, M.H.N., Lorvidhaya, M., Vootiprux, V., Cheirsilpa, A., Wong, F., Azhar, I., Lee, H.P., Kang, S.B., Manalo, A., Yen, M.-S., Kampono, N. and Aziz, F. for the Cervical Cancer Study Group of the Asian Oceanian Clinical Oncology Association (1995) Randomised trial of epirubicin and cisplatin chemotherapy followed by pelvic radiation in locally advanced cervical cancer. *Journal of Clinical Oncology*, **13**, 444–51.

26. Withers, H.R., Taylor, J.M.F. and Maciejewski, B. (1988) The hazard of accelerated tumor clonagen repopulation during radiotherapy. *Acta Oncologica*, **27**, 131–46.

27. Ozols, R.F., Masuda, H. and Hamilton, T.C. (1988) Keynote address: Mechanisms of cross resistance between radiation and anti-neoplastic drugs. *NCI Monographs*, **6**, 159–65.

28. Kim, D.S., Moon, H., Kim, K.T., Hwang, Y.Y., Cho, S.H. and Kim, S.R. (1989) Two-year survival: preoperative adjuvant chemotherapy in the treatment of cervical cancer stage Ib and II with bulky tumour. *Gynecologic Oncology*, **33**, 225–30.

29. Giaroli, A., Sananes, C., Sardi, J.E., *et al.* (1990) Lymph nodes metastases in carcinoma of the cervix uteri: response to neoadjuvant chemotherapy and its impact on survival. *Gynecologic Oncology*, **39**, 34–9.

30. Benedetti Panici, P., Scambia, G., Baiocchi, G., *et al.* (1991) Neoadjuvant chemotherapy and radical surgery in locally advanced cervical cancer. Prognostic factors for response and survival. *Cancer*, **67**, 372–9.

31. Benedetti Panici, P., Greggi, S., Scambia, G., *et al.* (1991) High-dose cisplatin and bleomycin neoadjuvant chemotherapy plus radical surgery in locally advanced cervical carcinoma. A preliminary report. *Gynecologic Oncology*, **41**, 212–16.

32. Dottino, P.R., Plaxe, S.C., Beddoe, A., Johnston, C. and Cohen, C.J. (1991) Induction chemotherapy followed by radical surgery in cervical cancer. *Gynecologic Oncology*, **40**, 7–11.

33. Kim, S.J., Namkoong, S.E., Bae, S.N. and Lee, H.Y. (1991) Comparative study of neoadjuvant chemotherapy followed by radical surgery versus simple radical surgery in patients with

locally advanced cervical cancer. *Proceedings of the 3rd Meeting of the International Gynecological Cancer Society,* Blackwell Science, p. 175.

34. Fontanelli, R., Spatti, G., Raspagliesi, F., Zunino, F. and Di Re, F. (1992) A pre-operative single course of high-dose cisplatin and bleomycin with glutathione protection in bulky stage IB/II carcinoma of the cervix. *Annals of Oncology,* **3**, 117–21.

35. Benedetti Panici, P., Scambia, G., Baiocchi, G., Greggi, S., *et al.* (1991) Neoadjuvant chemotherapy and radical surgery in locally advanced cervical cancer: prognostic factors for response and survival. *Cancer,* **67**, 372–9.

36. Sardi, J., Sananes, C., Giaroli, A., *et al.* (1993) Results of a prospective randomized trial with neoadjuvant chemotherapy in stage IB, bulky, squamous carcinoma of the cervix. *Gynecologic Oncology,* **49**, 56–65.

37. Runowicz, C.D., Smith, H.O. and Goldberg, G.L. (1993) Multimodality therapy in locally advanced cervical cancer. *Current Opinion in Obstetrics and Gynecology,* **5**, 92–8.

38. Piver, M.S., Barlow, J.J., Vongtama, V. and Blumenson, L. (1983) Hydroxyurea: a radiation potentiator in carcinoma of the uterine cervix. A randomized double-blind study. *American Journal of Obstetrics and Gynecology,* **147**, 803–8.

39. Piver, M.S., Vongtama, V. and Emrich, L.J. (1987) Hydroxyurea plus pelvic radiation versus placebo plus pelvic radiation in surgically staged stage IIIB cervical cancer. *Journal of Surgical Oncology,* **35**, 129–34.

40. Piver, M.S., Khalil, M. and Emrich, L.J. (1989) Hydroxyurea plus pelvic irradiation in nonsurgically staged stage IIIB cervical cancer. *Journal of Surgical Oncology,* **42**, 120–5.

41. Hreshchyshyn, M.M., Aron, B.S., Boronow, R.C., *et al.* (1979) Hydroxyurea or placebo combined with radiation to treat stages IIB and IV cervical cancer confined to the pelvis. *International Journal of Radiation Oncology, Biology and Physics,* **5**, 317–22.

42. Stehman, F.B., Bundy, B.N., Keys, H., *et al.* (1988) A randomized trial of hydroxyurea versus misonidazole adjunct to radiation therapy in carcinoma of the cervix. *American Journal of Obstetrics and Gynecology,* **159**, 87–94.

43. Choo, Y.C., Choy, T.K., Wong, L.C. and Ma, H.K. (1986) Potentiation of radiotherapy by *cis*-diachiorodiammine platinum (II) in advanced cervical carcinoma. *Gynecologic Oncology,* **23**, 94–100.

44. Wong, L.C., Choo, Y.C., Choy, D., Sham, I.S.T. and Ma, H.K. (1989) Long-term follow-up of potentiation of *cis*-platinum in advanced cervical cancer. *Gynecologic Oncology,* **35**, 159–63.

45. Drescher, C.W., Reid, G.C., Terada, K., Roberts, J.A., Hopkins, M.P., PerezTamayo, C. and Schoeppel, S.L. (1992) Continuous infusion of low-dose 5-fluorouracil and radiation therapy for poor prognosis squamous cell carcinoma of the uterine cervix. *Gynecologic Oncology,* **44**, 227–30.

46. Dische, S., Chassagne, D., Hope-stone, H.F., *et al.* (1993) A trial of Ro 03-8799 (Pinomidazole) in carcinoma of the uterine cervix: an interim report from the Medical Research Council Working Party on Advanced Carcinoma of the Cervix. *Radiotherapy Oncology,* **26**, 93–103.

47. Van Bommel, P.F.J., Van Lindert, A.C.M., Kock, H.C.L.K., Leers, H.W. and Neijt, J.P. (1987) A review of prognostic factors in early-stage carcinoma of the cervix (FIGO IB and IIA) and implications for treatment strategy. *European Journal of Obstetrics Gynecology and Reproductive Biology,* **26**, 69–84.

48. Stockler, M., Russell, P., McGahan, S., Elliott, P.M., Dalrymple, C. and Tattersall, M.H.N. (1996) Prognosis and prognostic factors in node-negative cervical cancer. *International Journal of Gynecological Cancer,* **6**, 477–82.

49. Wertheim, M.S., Hakes, T.B., Daghestani, A.N., *et al.* (1985) A pilot study of adjuvant therapy in patients with cervical cancer at high risk or recurrence after radical hysterectomy and pelvic lymphadenectomy. *Journal of Clinical Oncology,* **3**, 912–16.

50. Lai, C.H., Lin, T.S., Soong, Y.K. and Chen, H.F. (1989) Adjuvant chemotherapy after radical hysterectomy for cervical carcinoma. *Gynecologic Oncology,* **35**, 193–8.

51. Ueki, M., Ilamura, S. and Maeda, T. (1987) Individualization of patients for adjuvant chemotherapy after surgical treatment of cervical cancer. *British Journal of Obstetrics and Gynaecology,* **94**, 985–90.

52. Tattersall, M.H.N., Ramirez, C. and Coppleson, M. (1992) A randomized trial of adjuvant chemotherapy after radical hysterectomy in stage IB–IIA cervical cancer patients with pelvic lymph node metastases. *Gynecologic Oncology,* **46**, 176–81.

53. Tattersall, M.H.N. and Rose, B.R. (1996) (Editorial) Prognostic factors for survival in cervical cancer – warts and all. *Journal of the National Cancer Institute,* **88**, 1331–2.

M.A. van Eijkeren and A.P.M. Heintz

INTRODUCTION

Cervical cancer can be divided into early cervical cancer with a low risk of recurrence after treatment (International Federation of Gynaecology and Obstetrics (FIGO) Stages Ib and IIa) and cancers with a high risk of recurrence (FIGO Stages IIb, III and IV). Five-year survival rates decrease from 85% in Stage I disease, 60% in Stage II disease, 35% in Stage III disease to 15% in Stage IV disease [1]. Apart from cancer stage, other cervical cancer characteristics, such as high-grade, large tumour volume, lymph-vascular space invasion and metastases to lymph nodes are also poor prognostic factors [2].

Approximately 75% of relapsed and/or persistent cervical cancer is diagnosed within the first 2 years following initial treatment [3, 4]. The diagnosis and management of relapsed cervical cancer will be discussed in this chapter. Firstly, various aspects on diagnosing suspected relapsed cervical cancer will be addressed. Thereafter we will discuss the treatment modalities, such as radiotherapy, surgery and chemotherapy. Special emphasis will be put on the different aspects of the pelvic exenteration procedure. Finally, new developments, i.e. combined operative and radiotherapeutic treatment procedures, hyperthermia in combination with chemotherapy and immunotherapy will be highlighted.

DIAGNOSIS

SIGNS AND SYMPTOMS

Cervical cancer tends to recur predominantly locally, i.e. centrally in the pelvis or retroperitoneally at the pelvic side wall. Lung and bone metastases are rare. During follow-up visits a history should be taken, which focuses on complaints related to persistent or relapsed cervical cancer. General symptoms of cancer relapse are loss of appetite, weight loss and general fatigue. Central pelvic relapses of cervical cancer can cause deep pelvic and lower back pain and difficulties in voiding and/or defecation, whereas pelvic side wall relapses usually cause referred pain in the buttocks, legs or groins due to tumour growth into the sacral plexus. Pelvic side wall relapses can also cause oedematous leg or legs through compression of veins and/or lymph circulation. An ominous sign for cervical cancer relapse is the combination of weight loss, unilateral leg oedema and pain in the pelvis, lower back or legs, which usually reflects unilateral cancer recurrence, compressing lymph and/or venous flow.

If the uterus is still *in situ*, a central relapse will usually cause vaginal discharge. Regarding complaints due to distant metastases; lung metastases can cause dyspnoea, whereas bone metastases cause localized pain.

Cancer and Pre-cancer of the Cervix
Edited by D.M. Luesley and R. Barrasso. Published in 1998 by Chapman & Hall, London. ISBN 0 412 56600 1.

General physical examination should focus on peripheral enlarged lymph nodes, especially in the scalene region and in the groins. Also, the abdomen should be palpated to search for enlarged peri-aortal lymph nodes and an enlarged liver. Oedematous legs should be noted. The purpose of the gynaecological examination is to assess any evidence of pelvic disease. After radiation therapy this can be difficult, especially if the uterus is still *in situ*, for radiation can cause severe fibrotic changes in the pelvis, which make it difficult to distinguish between relapsed cervical cancer and post-radiation fibrosis. However, post-radiation fibrosis usually feels smooth on palpation, while relapsed cervical cancer mostly shows up with some nodularity.

DIAGNOSIS DURING FOLLOW-UP

It is commonly accepted good practice to see patients for follow-up on a regular basis after they have finished their initial treatment for cancer. Usually, patients are seen every 2 months for half a year, every 3 months for the year thereafter, then every 6 months until 5 years after the initial treatment. From there on, it varies widely if patients are still seen on a yearly basis or not at all. In different gynaecological oncology centres there is also no consensus as to what additional examinations should be performed. These vary from no smear, annual cytological smear only, to annual cytological smears, chest X-rays and intravenous pyelograms, sometimes even combined with annual computerized axial tomography (CT) scans.

The rationale of follow-up visits has been questioned for endometrial carcinoma [5] and recently also a retrospective study on 674 patients treated with radical hysterectomy and lymph node dissection for low-risk cervical cancer showed that 112 patients (17%) developed relapses of their cancer and only 29 (26%) of these relapses were found during follow-up visits [6]. Nearly half of the patients with relapsed cervical cancer were referred by

their general practitioner between scheduled visits. Of the women who were diagnosed in the course of their follow-up visits, only three benefited from this: they are alive and free of disease after treatment. It seems worthwhile to investigate prospectively if, after treatment for low-risk cervical cancer, patients can be instructed carefully as to which symptoms should make them contact their gynaecological oncologist. Another question that needs to be addressed in this respect is the need for psychological rather than medical support.

Recurrence rates for high-risk cervical cancers do not warrant such a restrained follow-up policy.

Apart from the search for the best care for cancer patients, these issues are also important for future health care policies, since downregulation of follow-up visits saves on health care budgets.

ADDITIONAL EXAMINATIONS

Imaging techniques

If cervical cancer relapse is suspected, the following imaging techniques are of value: transvaginal and abdominal ultrasound, intravenous pyelogram, CT scan (see Figure 25.1) and magnetic resonance imaging (MRI). CT scans and abdominal ultrasound are very specific in determining para-aortic lymph nodes [7], whereas MRI is superior in the evaluation of the local, i.e. pelvic tumour size and extension [8]. A chest X-ray needs to be performed routinely if relapsed cervical cancer is suspected.

Pathology

In order to confirm the diagnosis of relapsed cervical cancer, tissue should be obtained for histology. This can be done by endocervical curettage in the case where the uterus is still present, or needle biopsies of palpatory suspicious pelvic areas. If the suspect lesion is deeper in the pelvis or abdomen, ultrasound-, CT-

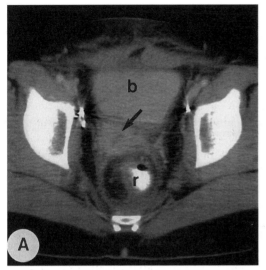 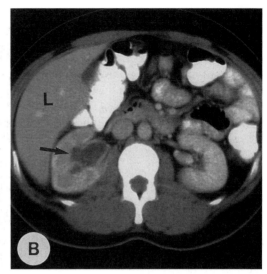

Figure 25.1   CT images of a patient who developed a central recurrence of an early-stage cervical cancer, initially treated with a radical hysterectomy and bilateral pelvic lymphadenectomy, followed by external radiotherapy. A shows the recurrence on the vaginal top (arrow); b = bladder; r = recto-sigmoid. The recurrence does not seem to extend to the pelvic side walls. B depicts the upper abdomen and shows a hydronephrosis on the right (arrow), due to a ureter obstructed by the tumour recurrence; L = liver.

or MRI-directed biopsies are an elegant possibility. In rare cases an exploratory laparotomy is necessary to distinguish between cervical cancer recurrence and fibrosis due to radiotherapy in a patient with bilateral ureteral obstruction, although approximately 90% of these cases are caused by cancer recurrence. On the other hand, if fibrosis is the cause of ureteral obstruction, an exploratory laparotomy combined with a urinary diversion procedure is life-saving.

Another reason for stressing the importance of histological confirmation is the experience that patients who have been treated with radiotherapy as the initial treatment can have pain and functional complaints of their bladder and bowels, which can be similar to complaints due to cervical cancer relapse. Cramping bowel pains coupled with diarrhoea due to radiation gastroenteritis can cause loss of appetite and general wasting and thus mimic a relapse, while in such a case proper dietary measures or bowel surgery can cure the patient.

Relapsed and/or persistent cervical cancer

In the high-risk cervical cancer patient, it can be difficult or even impossible to distinguish between relapsed and persistent cervical cancer, for instance in patients who are wasting away, have pain complaints or vaginal discharge, while physical and gynaecological examinations as well as imaging techniques cannot locate the suspected cancer and/or no abnormalities can be found in curettings. These patients need to be followed carefully and usually at a certain stage in time a tumour localization can be found and the diagnosis can be histologically confirmed.

TREATMENT

In general, it must be emphasized that the majority of patients with persistent or relapsed cervical cancer have a very poor prognosis. The 5-year survival rate is less than 5% and only 10–15% of patients will survive the first year [3, 4], so treatment will generally have a palliative character. Traditionally, only those patients who present with a central pelvic relapse of cervical cancer are candidates for treatment with a curative intention. Both treatment possibilities, with a palliative and with a curative intention, will be further discussed, with an emphasis on the latter.

We will discuss radiotherapy, surgery and chemotherapy in relapsed cervical cancer and new developments in these treatments, leaving aside the more general palliative modalities such as pain medication and medication against nausea and vomiting.

RADIOTHERAPY (SEE TABLE 25.1)

Relapsed cervical cancer following radical hysterectomy is treated with radiotherapy, if possible. If the recurrence is only located in the pelvis and is not too large to be completely sterilized by radiotherapy (diameter 1–2 cm), radiotherapy should be set up in a curative fashion, comparable to the primary treatment of more extended cervical cancers or of primary vaginal cancers (see also Chapter 22, Management of cervical cancer by radio-

Table 25.1 Radiotherapy in relapsed cervical cancer

A. External beam radiotherapy
   First choice therapy for
   1. Local recurrence if patient has not been radiated before and
      (a) recurrence <2 cm diameter or
      (b) if local recurrence extends to pelvic side wall
   2. Distant metastases.

B. Re-irradiation
   Only described with old-fashioned radiotherapy techniques. Nowadays considered to cause too many major complications.

C. Interstitial radiotherapy
   Poor results if local recurrence is extended to pelvic side wall.
   Small central recurrences might benefit, but only four patients have been described.

---

therapy). If disseminated disease is diagnosed on CT scan, MRI or chest X-ray, palliative radiotherapy is an option. However, the timing of this palliative radiotherapy depends on the complaints these recurrences cause the patient, the likelihood they will cause complications in the near future and the patient's wish to be treated. For instance, bone metastases causing pain should be treated with radiotherapy without delay. Also pelvic locations should be radiated, since they will usually cause vaginal discharge, pain and bowel and/or bladder dysfunction in the near future, whereas lung metastases and para-aortal lymph node metastases can await treatment until they cause dyspnoea or pain, respectively.

Re-irradiation

If, in the course of the initial treatment, radiotherapy has already been given, re-irradiation with external beam radiation should not be given. In older reports re-irradiation was regularly described with salvage percentages between 10% and 25% [9, 10]. However, it is now thought these successes reflect those patients who initially had an early stage cervical cancer, which recurred predominantly centrally in the pelvis and did not initially receive the full radiotherapy dose, a situation which does not occur any more with modern, highly sophisticated radiotherapy methods. Pelvic re-irradiation would cause unacceptable complication rates in most cases, as is illustrated by the study of Russell *et al.* [11] who re-irradiated with external beam a selected group of patients with relapsed cervical cancer, who were medically or technically unsuitable for exenteration, but of the 14 patients in whom the tumour was controlled after re-irradiation (of a total of 25 patients), seven developed major complications.

Interstitial radiotherapy

Other researchers have described interstitial radiotherapy for relapsed cervical cancer [12–15]. These results appear to be dependent on the location of the recurrence: if the recurrence is extended to the pelvic side wall, results are poor (10% of 70 patients were alive after 15 months [12], in another study all five patients with pelvic side wall relapses died [15]), whereas central pelvic recurrences may be treated well with this modality, although only four patients have been described [14]. Nori *et al.* [13] did not give detailed information on the localization of the recurrences in 33 patients treated with interstitial radiotherapy. In their report the 5-year survival rate was 7%.

SURGERY (SEE TABLE 25.2)

As stated above, surgery with curative intention, up until now has been traditionally reserved for central pelvic relapses of cervical cancer after initial treatment including radiotherapy and without any evidence of distant metastases.

Radical hysterectomy

There has been some debate whether or not a radical hysterectomy might be indicated in

Table 25.2 Surgery in relapsed cervical cancer

Traditionally only for central pelvic recurrences in:
1. Previously radiated patients.
2. Central recurrences >2 cm diameter.

A. Pelvic exenteration
   First choice treatment.
      Mortality <10%.
      5-year survival rates: 30–60%.
   1. Total pelvic exenteration: removing recurrence, uterus if *in situ*, recto-sigmoid and bladder.
   2. Posterior pelvic exenteration: leaving bladder *in situ*.
   3. Anterior pelvic exenteration: leaving recto-sigmoid *in situ*.

B. Radical hysterectomy
   Not widely used because of high mortality and morbidity rate (mainly fistulas).

relapsed cervical cancer after radiotherapy. The main reason for this is the possibility of preserving the surrounding pelvic organs, i.e. bladder and distal large bowel. Rubin *et al.* [16] described their experience in 21 patients with relapsed cervical cancer after radiation therapy in whom they performed a radical hysterectomy. The mortality and morbidity rate were high: two patients died of sepsis and 10 patients developed a total of 17 fistulas, both urinary and intestinal. Thirteen patients survived. These survivors were patients with only an initial low-risk stage of cervical cancer presenting with a central recurrence smaller than 2 cm. Of these, eight required an urinary diversion, two of them combined with a colostomy as well. More recently Rutledge *et al.* [17] published comparable results. In their study the best results were also obtained in 13 patients with an initial low-risk cervical cancer and a small central recurrence (a 5-year survival rate of 84% and a 31% chance of fistula formation). These data suggest that radical hysterectomy should only be considered in small central recurrences of low-risk cervical cancers, at the cost of a substantial percentage of major complications, mainly fistulas.

In very small central recurrences an even more limited surgical procedure, such as a simple abdominal hysterectomy, might be sufficient, but this procedure is also correlated with a high percentage of complications, mostly fistulae, requiring further surgery. For this reason most gynaecological oncologists treat centrally relapsed cervical cancer after radiotherapy with ultra-radical surgery via a pelvic exenteration.

Pelvic exenteration

Pelvic exenteration was first introduced in 1948 by Brunschwig [18], originally as a palliative procedure, but later it became evident that pelvic exenteration might be curative. At first this procedure was heavily criticized for being a too aggressive treatment for recurrent cancer, but since it showed limited mortality and morbidity, it gradually became accepted and is currently considered the procedure of choice for relapsed cervical cancer after initial radiotherapy, located centrally in the pelvis, and occasionally after initial surgical treatment only, i.e. if the centrally recurred cancer is considered too large for curative radiotherapy. (Pelvic exenteration is also indicated in centrally recurrent endometrial, vaginal and vulvar cancers and, although infrequently, in recurrent ovarian cancers.)

It is also generally agreed that this procedure should only be planned if distant metastases are excluded as completely as possible. Also, palliative pelvic exenterations should be avoided, since following palliative exenteration quality of life is usually poor [19]. For these reasons likely candidates should undergo thoracic and abdominal CT scans to detect lung metastases and extrapelvic metastases, respectively. If necessary, a fine needle biopsy should be undertaken to prove or exclude the likelihood of distant metastases. If distant metastases are found, pelvic exenteration should not be undertaken. In the pelvis, however, CT scan abnormalities are difficult to interpret after radiation therapy. Recently, CT scan findings of women with a central relapse of cervical cancer were retrospectively reviewed by radiologists blinded to the outcome of the operation, in order to ascertain whether CT scan can predict operability in these patients. Of the 31 reviewed patients, 21 eventually underwent a pelvic exenteration. Using classic CT criteria such as extension of the pelvic mass to the pelvic side wall, encasement of adjacent vessels, ureteric dilation, lymph node enlargement and ascites, sensitivity was 93%, but specificity was poor (47%). If ascites and lymphadenopathy were excluded as signs of peritoneal and lymph node metastases, respectively, specificity rose to 82% [20]. This reflects the impossibility of differentiating by CT scan whether enlarged lymph nodes are due to metastases, radiation or inflammation. Factors positively influencing specificity were tumour extending to the pelvic side

walls, encasement of adjacent vessels and ureteric dilation. We agree with Crawford *et al.* that a specificity of 82% to predict inoperability is too low to cancel the procedure. However, patients with pelvic CT scan findings suggestive for inoperability should be counselled carefully that during the laparotomy the procedure may be abandoned.

The same argument holds true for assessing operability on palpation. A fixed pelvic mass may be caused by radiation fibrosis as well as by cancer spread to the pelvic side wall, and true assessment of operability is only possible during laparotomy.

The foregoing makes it clear that it is a very difficult task for the gynaecological oncologist to distinguish between patients who are suitable for this procedure and for whom it might be life-saving and those for whom it is not, where further palliative care can attempt to provide good quality of life, but no survival.

Apart from assessing operability and excluding distant metastases, as best as can be done with imaging techniques and biopsies, patients planned to undergo a pelvic exenteration also need a thorough general medical examination, in which operation risks should be weighed against the patient's overall health, especially cardiopulmonary status. General physicians may need several days or weeks to optimize the patient's cardiopulmonary situation. It has been demonstrated in 63 patients of 65 years or older that age alone is no contraindication for pelvic exenteration [21].

Last but not least, careful counselling of the patient is vital for the outcome of the operation. It takes several visits to explain to the patient and her family the full extent of the procedure. Not only the technical, but also the psychological and sexual impact of the procedure needs to be fully understood and reflected upon, before a patient and her family can decide whether or not to go ahead with the offered procedure. Psychological counselling in the sexually active patient as well as her partner is mandatory. A sexualist is of great value in the decision-making about vaginal reconstruction. This decision depends, among other factors, on the sexual life the patient wants after her operation and on her willingness to use dilators or vibrators to keep the reconstructed vagina open. Also, the stoma therapist should be consulted before the operation to counsel the patient and to pre-operatively mark the best stoma site on the abdominal wall.

With all of these above-mentioned aspects taken into account, some patients decline pelvic exenteration because they feel the mutilation and anxieties of a pelvic exenteration are too high a cost for the possibility of survival, but the majority of patients will accept the proposed procedure.

Before the operation patients receive bowel cleansing. The actual procedure should be planned as a major one, which implies precautions, such as the insertion of a central line before the procedure starts, minor heparinization combined with the application of elastic stockings or calf compressors and antibiotics intravenously during and 5 days after operation.

At laparotomy, first the abdominal cavity is inspected and palpated for evidence of extra-pelvic tumour spread. If suspected areas are encountered, frozen sections should be obtained, and if proven positive, the procedure should be abandoned. Frozen sections should also be obtained of pelvic and low para-aortic lymph nodes. There has been some dispute as to whether or not pelvic exenteration should be accomplished if positive pelvic nodes are encountered [22, 23] but the overall opinion based on the survival rates of such patients (approximately 9%) is not to continue once positive pelvic nodes have been indicated on frozen section.

Thereafter the pelvic retroperitoneal spaces, the pararectal and paravesical space, should be carefully opened in order to assess the possibility of extirpating the central tumour and adjacent bladder and/or rectum with margins free of tumour (see Figure 25.2). If tumour

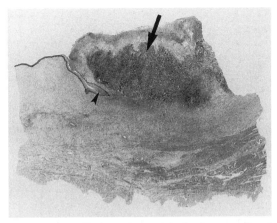

Figure 25.2 Light microscopical photograph of the total exenteration specimen of the patient whose CT images are described in Figure 25.1. This photograph shows the tumour recurrence (arrow) in the vaginal top in relation to the vaginal wall (arrowhead).

spreading to the pelvic side wall is encountered during this part of the procedure, again the operation should be abandoned. Miller *et al.* [24] reviewed why explorative laparotomies in order to perform a pelvic exenteration for relapsed cervical cancer were abandoned in the experience of the M.D. Anderson Cancer Center for a total of 394 patients. One-hundred-and-eleven were abandoned, predominantly for peritoneal spread (44%) and nodal disease (40%). Other reasons were parametrial fixation (13%) and hepatic or bowel involvement (5%).

If all the precautions to exclude extra-pelvic spread or positive surgical margins are taken, the ureters and/or the recto-sigmoid are divided, the pelvic web is dissected to the pelvic floor, a perineal incision is made, the paravaginal tissue is dissected free and the central tumour with the adjacent bladder and/or recto-sigmoid is separated from its surroundings and taken out *en bloc* (see Figure 25.3). In cases of small, predominantly anterior recurrences, it is occasionally possible to extirpate the bladder with the tumour and leave the large bowel intact, a so-called anterior exenteration (see Figure 25.3a). A posterior

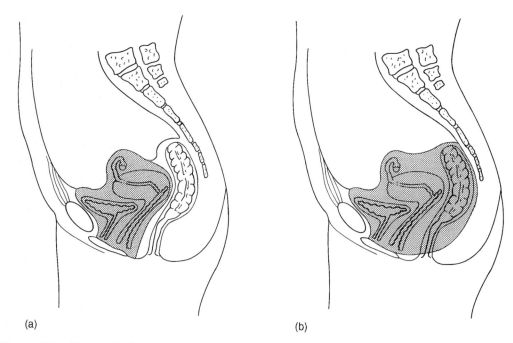

(a)   (b)

Figure 25.3 The marked area in (a) depicts the organs extirpated in the case of an anterior pelvic exenteration, i.e. bladder, uterus, vagina and ovaries; whereas the marked area in (b) reflects the total pelvic exenteration, also including an extirpation of the recto-sigmoid.

exenteration, leaving the bladder *in situ*, however, is seldom indicated in the treatment of relapsed cervical cancer, since it is usually anatomically impossible to have a tumour-free margin if the bladder is left *in situ* and secondly, leaving the bladder *in situ* causes a large percentage of complications, mainly fistulae and voiding impossibilities, since most patients have already had pelvic radiation.

When the specimen is taken out, all that remains in the pelvis is a large denuded area and various techniques have been described to cover this area, predominantly to prevent the small bowels from adhering and to diminish the infection rate. Vicryl mesh [25], pelvic peritoneum [26], and dura mater allografts [27] have all been described for this purpose, but mostly the gastro-colic omentum is used, a technique described by Rutledge *et al.* [28]. Lately a sodium hyaluronate-based bioresorbable membrane (Sepra film R), which

has to be laid on denuded areas was shown to be effective in preventing post-operative adhesions [29]. This membrane may also be of use in covering the denuded pelvic floor after exenterations.

Directly connected with covering the denuded pelvic floor are the various diversions and reconstructions that need to be performed, since they also fill the pelvic cavity.

In the early days of exenterative surgery, urinary diversion was accomplished by transplanting the ureters into the colon proximal to the colostomy [18]. This caused severe urinary tract infections and electrolyte disturbances. Since the distal ileum has been introduced as a conduit for urinary diversion [30], these problems have diminished greatly. Although the Bricker diversion is very satisfactory, currently more and more continent urostomies with a valve construction are performed. Instead of wearing urostomy bags, patients empty the urinary conduit with a disposable catheter three to four times per day. These continent conduits also reduce the incidence of pyelonephritis due to reflux of contaminated urine. Various continent urostomies have been described, of which the most used initially was the ileal continent urostomy or Kock's pouch [31], in which 80 cm of the distal ileum is used to construct the continent pouch. Because of the risk of metabolic disturbances when large amounts of distal ileum are used for the pouch, currently most surgeons prefer the Miami/Indiana pouch [32] or the Mainz pouch [33] in which a small part of the terminal ileum and the ascending and proximal transverse colon are used to construct the continent pouch.

The handling of the dissected recto-sigmoid has also been modified recently. In the early days of pelvic exenteration a colostomy usually had to be constructed (since the distal margin was too low) less than 6 cm from the anus, to perform a hand-sewn end-to-end anastomosis of the rectum and the sigmoid. However, an automatic stapler device for very low colpoproctostomies is now available, thus avoiding the traditional colostomy [34]. Because these very low anastomoses can lead to tenesmus and unacceptable faecal frequency, some authors also recommend creating a rectal J-pouch reservoir at the same time [35].

At the end of the reconstructive phase of the operation a vaginal reconstruction, if required, needs to be performed. Various techniques for vaginal reconstructions have been described. Frequently used are the sigmoid neovagina [36] and the omental J flap neovagina with a split-thickness skin graft [37]. The sigmoid neovagina has the slight disadvantage of mucous secretions, which tend to decrease in time, whereas the split-thickness skin graft vagina produces no discharge. The decision if and which vaginal reconstruction can be performed may depend on the amount of omentum and/or viable sigmoid available after urinary and recto-sigmoid reconstruction.

If it is possible to construct a continent urinary conduit, an end-to-end recto-sigmoid anastomosis and a vaginal reconstruction, everyday life will have deteriorated less and be more acceptable to the patient after pelvic exenteration.

For the specific surgical pelvic techniques needed to perform these operations readers are referred to gynaecological oncological surgical textbooks.

Patients will be monitored post-operatively in an intensive care unit for the first few days, because of the various imbalances such as electrolyte disturbances and haemodynamic fluid shifts, which can be expected after major surgery.

As to the results of the pelvic exenteration procedure for relapsed cervical cancer, up until now several thousand of such procedures have been described [27, 28, 38–42] with 5-year survival rates ranging from 30% to 60%. Five-year survival has increased during the past two decades, reflecting better peri-operative medical care, such as intensive care monitoring and improved anaesthetic techniques, as well as better patient selection, which is possible with CT and MRI imaging and directed fine needle biopsies. Mortality rates have also dropped to below 10%. It is clear that unknown distant micrometastases at the time of operation have a great influence on the 5-year survival rate.

CHEMOTHERAPY

Chemotherapy plays a modest role in the treatment of cervical cancer in general (see Chapter 23, Chemotherapy in the management of cervical cancer). In the treatment of persistent or relapsed cervical cancer its role is even more limited, since many recurrences will be localized in previously radiated locations with diminished blood flow thus jeopardizing the effects of chemotherapy. Despite numerous efforts with all kinds of chemotherapeutics in various combinations in persistent and relapsed cervical cancer, single-agent cisplatin has been proven the most effective with response rates of around 30%, complete responses around 10% and a median response duration from 3 to 5 months [43, 44].

Intra-arterial infusion with various combinations of chemotherapeutic agents in the treatment of relapsed cervical cancer has not proven to prolong survival and hence is of little value [45, 46].

NEW DEVELOPMENTS (SEE TABLE 25.3)

RADICAL SURGERY IN COMBINATION WITH RADIOTHERAPY

Extension of the relapsed cervical cancer to the pelvic side wall is traditionally a contra-indication to perform a pelvic exenteration. In the search for a curative treatment for these patients and based on the finding that 50% of women who died of a pelvic relapse had no distant metastases [3, 4], some gynaecological oncologists have developed techniques combining extensive debulking surgery, comparable with pelvic exenteration, with radiotherapy at the site of the greatest residual tumour load [47–51]. Hoeckel *et al.* [47, 48] use brachytherapy with intra-operatively placed guiding tubes, in which other organs are protected by a muscle and omentum flap. They describe disease-free survival in seven of 18 patients

treated as such with a median follow-up of 24 months [48]. If the relapses have a diameter greater than 5 cm, all patients recurred. Other authors [49–51] favour extensive debulking surgery in combination with intra-operative radiation. Results of these studies are variable: Garton *et al.* describe 19 patients, of which 33% survive 5 years; Stelzer *et al.* describe comparable results: 43% of 22 patients survive 5 years, but Mahe *et al.* report far worse results in 70 patients with recurrent cervical cancer, of which 59 patients with pelvic side wall extension: 8% survive 3 years and the median survival in his study is 11 months. Reported treatment-related complications include predominantly peripheral neuropathy. In our view the development of this combination of debulking surgery and radiotherapy is praiseworthy in its intention to find a curative treatment for this group of patients, but in view of the results we agree with Abe and Shibamoto [52] that previously radiated patients with unresectable tumours are not appropriate candidates for intra-operative radiotherapy (or brachytherapy) following debulking surgery.

HYPERTHERMIA IN COMBINATION WITH CHEMOTHERAPY

Recently a 52% response rate was observed in 23 patients with a pelvic relapse of cervical cancer after initial radiotherapy treated with weekly loco-regional hyperthermia and simultaneous cisplatin 50 mg/m^2 for 12 cycles [53] with an acceptable toxicity. The median response duration was 9 months and 1-year survival was 42%. Hyperthermia improves the cytotoxicity of cisplatin and hence this combination is promising in patients with relapsed cervical cancer after radiotherapy who are not suitable for surgical treatment.

IMMUNOTHERAPY

Immunotherapy is still promising but is as yet unavailable in the therapy of cervical cancer.

Table 25.3 Experimental therapies in relapsed cervical cancer

---

A. Radical surgery combined with radiotherapy
   If pelvic recurrence extends to pelvic side wall: radical surgery followed by:
   1. Brachytherapy pelvic side wall, or
   2. Intra-operative external beam radiotherapy.
   Until now only small series.
   Results vary from survival rates of 8–40%.

B. Hyperthermia with chemotherapy
   50% response is described in previously radiated patients.

C. Immunotherapy
   Promising, but not clinically available as yet.
   Various vaccination trials focusing on the elimination of HPV are underway.

Cervical cancer, however, has evident connections with the immune system, since one of the causal factors is an infection with human papillomavirus (HPV), predominantly 16 or 18 [54]. These viruses probably deliver strategies to evade the immune system, thus promoting further tumour growth [55]. However, the exact subcellular mechanisms involved are not fully understood as yet, but treatment with vaccines against HPV has been successful in the mouse model [56] and currently several studies are proceeding to test various vaccines against HPV in recurrent cervical cancer in humans. Further results are required before any conclusions can be drawn for future strategies.

CONCLUSION

Relapsed cervical cancer still has a very bad prognosis and only a select group of patients can be treated with curative intention. These treatments are often tedious, both for the patient and for the doctor, and only half of these patients can eventually be cured. Before any progress can be achieved, our knowledge of the origin and behaviour of cancer needs to increase.

REFERENCES

1. Sondik, E.J., Young, J.L., Horm J.W. and Gloeckler, L.A. (1985) *Annual Cancer Statistics Review*. NIH publication, 2786–9.
2. Bommel P.F. van, Lindert, A.C. van, Kock, H.C., Leers, W.H. and Neijt, J.P. (1987) A review of prognostic factors in early-stage carcinoma of the cervix (FIGO IB and IIA) and implications for treatment strategy. *European Journal of Obstetrics and Gynaecology and Reproductive Biology*, **26**, 69–84.
3. Burke, T.W., Hoskins, W.J., Heller, P.B., *et al.* (1987) Clinical patterns of tumor recurrence after radical hysterectomy in stage IB cervical carcinoma. *Obstetrics and Gynecology*, **69**, 382–5.
4. Sommers, G.M., Grigsby, P.W., Perez, C.A., *et al.* (1989) Outcome of recurrent cervical carcinoma following definitive irradiation. *Gynecologic Oncology*, **35**, 150–5.
5. Owen, P. and Duncan, I.D. (1996) Is there any value in the long-term follow-up of women treated for endometrial cancer? *British Journal Of Obstetrics and Gynaecology*, **103**, 710–13.
6. Ansink, A., Barros Lopez, A. de., Naik, R. and Monaghan, J.M. (1996) Recurrent stage Ib cervical carcinoma: evaluation of the effectiveness of routine follow up surveillance. *British Journal of Obstetrics and Gynaecology*, **103**, 1156–8.
7. Heller, P.B., Malfetano, J.H., Bundy, B.N., *et al.* (1990) Clinical-pathologic study of stage IIB, III and IVA carcinoma of the cervix: extended diagnostic evaluation for paraaortic node metastasis – a gynecologic oncology group study. *Gynecologic Oncology*, **38**, 425–30.
8. Kim, S.H., Choi, B.L., Lee, H.P., *et al.* (1990) Uterine cervical carcinoma: comparison of CT and MR findings. *Radiology*, **175**, 45–51.
9. Murphy, W.T. and Schmitz, A. (1956) The results of re-irradiation in cancer of the cervix. *Radiology*, **67**, 378.
10. Nolan, J.F., Vidal, J.A. and Anson, J.H. (1957) Treatment of recurrent carcinoma of the uterine cervix with cobalt 60. *Western Journal of Surgery*, **65**, 358.
11. Russell, A.H., Koh, W.J., Markette, K., *et al.* (1987). Radical reirradiation for recurrent or second primary carcinoma of the female reproductive tract. *Gynaecological Oncology*, **27**, 226–32.
12. Evans, S.R., Hilaris, B.S. and Barber, H.R.K. (1971) External vs interstitial irradiation in unresectable recurrent cancer of the cervix. *Cancer*, **28**, 1284–8.
13. Nori, D., Hilaris, B.S., Kim, H.S., *et al.* (1981) Interstitial irridiation in recurrent gynaecological cancer. *International Journal of Radiation Oncology, Biology and Physiology*, **7**, 1513–17.
14. Randall, M.E., Evans, L., Greven, K., *et al.* (1991) Interstitial reirradiation for gynecologic malignancies: results and analysis of prognostic factors. *Gynecologic Oncology*, **48**, 23–31.
15. Monk, B.J., Walker, J.L., Tewari, K., *et al.* (1994) Open interstitial brachytherapy for the treatment of local–regional recurrences of uterine corpus and cervix cancer after primary surgery. *Gynecologic Oncology*, **52**, 222–8.
16. Rubin, S.C., Hoskins, W.J. and Lewis, J. (1987) Radical hysterectomy for recurrent cervical cancer following radiation therapy. *Gynecologic Oncology*, **27**, 316–22.
17. Rutledge, S., Carey, M.S., Prichard, H., *et al.* (1994) Conservative surgery for recurrent or persistent carcinoma of the cervix following irradiation: is exenteration always necessary? *Gynecologic Oncology*, **52**, 353–9.

18. Brunschwig, A. (1948) Complete excision of pelvic viscera for advanced carcinoma. *Cancer,* **1**, 177–83.

19. McCullough, W.M. and Nahhas, W. (1985) Palliative pelvic exenteration – futility revisited. *Gynecologic Oncology,* **27**, 97–103.

20. Crawford, R.A.F., Richards, P.J. and Reznek, R.H. (1996) The role of CT in predicting the surgical feasibility of exenteration in recurrent carcinoma of the cervix. *International Journal of Gynaecological Cancer,* **6**, 231–4.

21. Matthews, C.M., Morris, M., *et al.* (1992) Pelvic exenteration in the elderly patient. *Obstetrics and Gynecology,* **79**, 773–7.

22. Creasman, W.T. and Rutledge, F. (1974) Is positive pelvic lymphadenopathy a contradiction to radical surgery in recurrent cervical carcinoma? *Gynecological Oncology,* **2**, 482–5.

23. Stanhope, C.R. and Symmonds, R.E. (1985) Palliative exenteration – what, when, and why? *American Journal of Obstetrics and Gynaecology,* **152**, 12–16.

24. Miller, B., Morris, M., Rutledge, F., *et al.* (1993) Aborted exenterative prosedures in recurrent cervical cancer. *Gynecologic Oncology,* **50**, 94–9.

25. Buchsbaum, H.J., Christoferson, W. and Lifshitz, S. (1985) Vicryl mesh in pelvic floor reconstruction. *Archives of Surgery,* **120**, 1389–91.

26. Perticucci, S. (1976) Pelvic floor reconstruction following gynaecologic exenterative surgery. *Obstetrics and Gynecology,* **50**, 31–4.

27. Averette, H.E., Lichtinger, M. and Sevin, B. (1984) Pelvic exenteration: a 15-year experience in a general metropolitan hospital. *American Journal of Obstetrics and Gynecology,* **150**, 179–84.

28. Rutledge, F.N., Smith, J.P. and Wharton, J.T. (1977) Pelvic exenteration: analysis of 296 patients. *American Journal of Obstetrics and Gynecology,* **129**, 881–92.

29. Becker, J.M. and Dayton, M.T. (1996) Prevention of postoperative adhesions by a sodium hyaluronate-based bioresorbable membrane: a prospective randomized, double-blind multicenter study. *Journal of the American College of Surgeons,* **183**, 297–306.

30. Bricker, E.M. (1950) Symposium on clinical surgery: bladder substitution after pelvic evisceration. *Surgical Clinics of North America,* **30**, 1511–21.

31. Kock, N.G., Nilson, A.E., *et al.* (1982) Urinary diversion via continent ileal reservoir: clinical results in 12 patients. *Journal of Urology,* **128**, 469–75.

32. Rowland, R.G., Mitchell, M.E., *et al.* (1987) Indiana continent urinary reservoir. *Journal of Urology,* **137**, 1136–9.

33. Thueroff, J.W., Alken, P. and Engelmann, U. (1986) The Mainz pouch (mixed augmentation ileum and cecum) for bladder augmentation and continent diversion. *Journal of Urology,* **136**, 17–26.

34. Harris, W.J. and Wheeless, C.R. (1986) Use of the end-to-end anastomosis stapling device in low colorectal anastomosis associated with radical gynecological surgery. *Gynecological Oncology,* **23**, 350–7.

35. Wheeless, C.R. and Hempling, R.E. (1989) Rectal J pouch reservoir to decrease the frequency of tenesmus and defaecation in low coloproctostomy. *Gynecologic Oncology,* **35**, 136–8.

36. Kindermann, G. (1987) The sigmoid vagina: experiences in the treatment of congenital absence or later loss of vagina. *Zeitschrift für Geburtshilfe und Frauenheilkunde,* **47**, 650–9.

37. Berek, J.S., Hacker, N.F. and Lagasse, L.D. (1984) Vaginal reconstruction performed simultanious with pelvic exenteration. *Obstetrics and Gynecology,* **63**, 318–23.

38. Barber, H.R.K. (1969) Relative prognostic significance of preoperative and operative findings in pelvic exenteration. *Surgical Clinics of North America,* **49**, 431–47.

39. Symmonds, R.E., Pratt, J.H. and Webb, M.J. (1975) Exenterative operations: experience with 198 patients. *American Journal of Obstetrics and Gynecology,* **121**, 907–18.

40. Jones, W.B. (1987) Surgical approaches for advanced or recurrent cancer of the cervix. *Cancer,* **60**, 2094–103.

41. Lowhead, R.A., Clark, D.G.C. and Smith, D.H. (1989) Pelvic exenteration for recurrent or persistent gynaecologic malignancies: a 10-year review of the Memorial Sloan Kettering Cancer Center experience (1972–1981). *Gynecologic Oncology,* **33**, 279–82.

42. Morley, G.W., Hopkins, M.P., Lindenauer, S.M., *et al.* (1989) Pelvic exenteration, University of Michigan, 100 patients at 5 years. *Obstetrics and Gynecology,* **74**, 934–43.

43. Park, R.C. and Tate Thigpen, J. (1993) Chemotherapy in advanced and recurrent cervical cancer. *Cancer,* **71**, 1446–50.

44. Vermorken, J.B. (1993) The role of chemotherapy in squamous cell carcinoma of the uterine cervix: a review. *International Journal of Gynecological Cancer,* **3**, 129–42.

45. Scarabelli, C., Tumolo, S. and De Paoli, A. (1987) Intermittent pelvic aterial infusion with peptichemio, doxorubicin and cisplatin for locally advanced and recurrent carcinoma of the uterine cervix. *Cancer,* **60**, 25–30.

46. Rettenmaier, M.A., Moran, M.F. and Ramsinghani, N.F. (1988) Treatment of advanced and recurrent squamous carcinoma of the uterine cervix with constant intraarterial infusion of cisplatin. *Cancer*, **61**, 1301–3.

47. Hoeckel, M. and Knapstein, P.G. (1992) The Combined Operative and Radiotherapeutic Treatment (CORT) of recurrent tumors infiltrating the pelvic wall: first experience with 18 patients. *Gynecologic Oncology*, **46**, 20–8.

48. Hoeckel, M., Baussmann, E., Mitze, M., *et al.* (1994) Are pelvic side-wall recurrences of cervical cancer biologically different from central relapses? *Cancer*, **74**, 648–55.

49. Garton, G.R., Gunderson, L.L., Webb M.J., *et al.* (1993) Intraoperative radiation therapy in gynecologic cancer: the Mayo Clinic experience. *Gynecologic Oncology*, **48**, 328–32.

50. Stelzer, K.J., Koh, W.J., Greer, B., *et al.* (1995) The use of intraoperative radiation therapy in radical salvage for recurrent cervical cancer: outcome and toxicity. *American Journal of Obstetrics and Gynecology*, **172**, 1881–8.

51. Mahe, M.A., Gerard, J.P., *et al.* (1996) Intraoperative radiation therapy in recurrent carcinoma of the uterine cervix: report of the French Intraoperative Group on 70 patients. *International Journal of Radiation Oncology, Biology and Physiology*, **34**, 21–6.

52. Abe, M. and Shibamoto, Y. (1996) The usefulness of intraoperative radiation therapy in the treament of pelvic recurrence of cervical cancer. *International Journal of Radiation Oncology, Biology and Physiology*, **34**, 513–14.

53. Rietbroek, R.C., Schilthuis, M.S., Bakker, P.J.M., *et al.* (1996) Phase II trial of weekly regional hyperthermia and cisplatin in patients with a previously irradiated recurrent carcinoma of the cervix. Thesis, University of Amsterdam, October 1996.

54. Hausen, H. zur (1991) Human Papilloma Viruses in the pathogenesis of anogenital cancer. *Virology*, **184**, 9–13.

55. Natali, P.G., Nicotra, M.R., *et al.* (1989) Selective changes in expression of HLA class I polymorphic determinants in human solid tumors. *Proceedings of the National Academy of Science of the USA*, **86**, 6719–23.

56. Feltkamp, M.C.W., Smits, H.L., *et al.* (1993) Vaccination with cytotoxic T lymphocyte epitope-containing peptide protects against a tumor induced by human papillomavirus type 16-transformed cells. *European Journal of Immunology*, **23**, 2242–9.

QUALITY OF LIFE AND PSYCHOLOGICAL ISSUES IN WOMEN WITH CERVICAL CANCER

Teresa Beynon, Charles D.A. Wolfe and K. Shanti Raju

INTRODUCTION

A report to the American Cancer Institute [1] stated that the physical, psychological and social loss that results from cancer treatment is a major determinant of cancer patients' quality of life after treatment. The physical effect of surgical treatment commonly means the loss of the uterus, ovaries, upper vagina, and at times the bladder and rectum. Surgical treatment almost always causes functional losses, e.g. the cessation of menstruation, loss of fertility and hormonal disequilibrium. These changes, in turn, affect sexual function, resulting in loss of sensory perception, dyspareunia, vaginal atrophy and stenosis [2]. From the psychological perspective, radical surgical procedures can be experienced as a serious assault on body image and sexuality of the woman and in turn damage her self-esteem [3, 4].

QUALITY OF LIFE IN CANCER

Maximal individual quality of life is intuitively felt to be a worthy goal by all modern societies [5]. However, quality of lfe is difficult to define. Many attempts have been made to address this issue, but there is, as yet, no agreed definition. Kenneth Calman, the Chief Medical Officer in England, has suggested it could be defined as the gap, at any point in time, between a person's hopes, expectations and present experience [6]. This allows for each individual to define their own quality of life. As one might expect, due to the difficulty of definition, there is as yet no universally accepted tool for measuring quality of life.

Quality of life is a multidimensional concept, it includes many conditions and experiences of life which are common to us all such as physical well-being, perceived health, emotional well-being, activities, social and sexual functioning.

The importance of this topic for medicine lies principally in assessment of treatment; any treatment given should add some form of quality to life. The term Quality Adjusted Life Years (QALY) has been adopted in an attempt to measure quantity of life, time gained and cost of any health intervention, in order to compare the benefits of different treatments.

In clinical oncology it has become accepted within the last 20 years that a quality of life measure should be used in assessing the benefit of treatments such that, not only should life be prolonged but also the quality of that life should be improved. Only with this information can a person make fully informed

Cancer and Pre-cancer of the Cervix
Edited by D.M. Luesley and R. Barrasso. Published in 1998 by Chapman & Hall, London. ISBN 0 412 56600 1.

decisions about treatment; for instance, a considerable reduction in quality of life may be tolerated for a period if the outcome of treatment is likely to be positive; conversely, if the personal risk outweighs the benefit a treatment option may be declined. This has to be the individual's decision. Slevin showed in his study that many patients with recurrent disease would opt for intensive chemotherapy even if there was only a 1% chance of cure [7].

Many of the early quality of life tools were developed for the assessment of chemotherapy treatments, for example, the Rotterdam Symptom Check List (RSCL) and the European Organization for Research and Treatment of Cancer core questionnaire (EORTC-QLQ30). This has tended to mean they are unsuitable for use in other areas, as they do not cover all the dimensions that make up quality of life, such as relationships, work, finances and more nebulous concepts such as the meaning of life.

Individual perceptions of quality of life are modified because of individual attributes, although in general, patients with terminal illness report lower quality of life compared to those termed cancer survivors [5].

QUALITY OF LIFE ASSESSMENT

Many tools have been used, adapted and developed in an attempt to serve this purpose, for example:

- Karnofsky Performance Status
- Spitzer Quality of Life Index
- Support Team Assessment Schedule (STAS); Palliative Care Assessment Tool (PACA), SEIQOL.

The number of tools available suggest that there is currently no gold standard which is both universally acceptable for the clinical setting and reliable enough to make it possible for intermittent ratings, in what is inevitably a changing situation. Ideally such a tool would be simple to complete, comprehensive, repeatable, sensitive across the course of an illness and useful for both daily management and research.

Such a tool has not and (due to the inherent difficulties) probably will not, be developed.

As a quality of life tool reflects an individual's life, in general, a quality of life tool should be self-rated. In severely ill people this may not be possible. It may be necessary to use health care professionals, relatives or carers as a proxy. There is some evidence to support a good correlation between relatives and patients in assessing their quality of life, using STAS; there is also evidence suggesting different health care professionals can agree on a patient's quality of life, using the Spitzer Index.

A detailed description of assessment tools is felt to be beyond the scope of this book. Interested readers are referred to other texts on this subject, although a summary of the more common quality of life indices, along with some of their advantages and disadvantages, are included in Tables 26.1 and 26.2.

QUALITY OF LIFE IN ADVANCED CANCER

In palliative care, quality of life is of paramount importance because of the aim of 'adding life to years' when time is limited due to an incurable illness. Within the specialty there is currently much interest in how to measure quality of life, in part to assess the benefits of palliative interventions.

Gynaecological cancers present a challenge to palliative care physicians. The natural histories are long compared to many other cancers, and such patients often have prolonged periods of increasing physical morbidity, which can impact severely on their quality of life.

Cervical cancer, in particular, causes physical challenges due to:

1. **Pain:** Involvement of the lumbosacral nerves may cause neuropathic pain which is difficult to control, either due to a pre-sacral mass intrapelvic disease or local vertebral invasion. As always in the management of cancer pain, a stepwise approach to pain control, using the World Health Organization

Table 26.1 Quality of life assessment in oncology

Tools	Year	Design	Rater	Population	Domains	Psychometric evaluation
Spitzer Quality of Life Index	1981	Five items ranked 0, 1, 2, Total score 0–10	Observer <1 min Subject rated also	General oncology and chronic illness	Physical Functional Social Outlook Daily living Health	Good inter-rater reliability Valid Accurate Precise Responsive
Functional Living Index Cancer	1984	22 questions Numerical scale (1–7) Verbal anchors	Subject <10 min	General oncology (ambulatory)	Physical Functional Psychological Social	Reliable Valid
Rotterdam Symptom Checklist	1986	30 questions Numerical scale	Subject	Breast cancer	Physical Functional Ambulatory	Reliable Validity – face, construct, concurrent
Suffering in terminal illness	1987	20 multiple choice + open-ended global questions	Observer 30 min	Advanced cancer	Physical Psychological Social Spiritual Knowledge of disease	Internationally tested Validity – discriminant, construct Internally consistent
European Organization for Research and Treatment of Cancer	1988	30 questions + 1 global Disease, protocol specific modules	Subject 10 min	General oncology (chemotherapy)	Physical Functional Psychological Social	Reliable Valid Acceptable Sensitive
Functional Assessment of Cancer Therapy	1988 1993	33 questions + 1 global Numeral response scale (0–4)	Subject 5 min	General oncology (included advanced cancer)	Physical Functional Psychological Social	Reliability – internal consistency Validity – content, concurrent, discriminant
Cancer Rehabilitation Education System	1988	59 questions Numerical response scale	Subject 20 min	General oncology	Physical Functional Psychological Social	Reliability – test–retest Validity – convergent, discriminant

Source: Donnelly, S. and Walsh, D. (1996) *Palliative Medicine*, **10**, 4, 279, published by Arnold Publishers, London (reproduced with permission).

Table 26.2   Quality of life assessment instruments: applicability to patients with advanced cancer

Instruments	Advantages	Disadvantages
Spitzer Quality of Life Index	Brevity Simplicity	Excluding advanced cancer; questions on spiritual, emotional needs Overemphasis on physical symptoms Weak support domain
Functional Living Index Cancer (FLIC)	Early trials (excluding very ill) showed good reliability and validity Questions answered in <10 min	Procedural and administrative problems led to low completion rate Fails to distinguish between individuals with advanced cancer Functional and social domains inadequate
Rotterdam Symptom Checklist	High sensitivity, specifically on psychological dimensions Simplicity	Length of questionnaire Absence of questions on family Support, spiritual issues Some relevant questions
Suffering in Terminal Illness	Questions relevant to 95% of 259 patients with advanced cancer	Vague wording Excluded weakness, fatigue Few near death
European Organization for Research and Treatment of Cancer (FACT-G)	Simplicity Brevity Validated in several languages	Not validated in advanced cancer Study population mainly male Too many questions for ill patients
Functional Assessment of Cancer Therapy (FACT-G)	Brevity Simplicity Physical, functional, emotional domains sensitive to change	Not tested in advanced cancer
Cancer Rehabilitation Education System (CARES)	Comprehensive Self-administered Aim is to identify patient's problems	Focuses on clinical problems amenable to rehabilitation planning Inability to complete 59 questions Omission of relevant and inclusion of inappropriate questions

Source: Donnelly, S. and Walsh, D. (1996) *Palliative Medicine*, **10**, 4, 280, published by Arnold Publishers, London (reproduced with permission).

(WHO) Analgesic Ladder [8] should be followed. Pains not quickly relieved following this strategy should be referred to a pain team as local anaesthetic nerve blocks are often needed for optimal pain relief.

2. **Weakness**: Invasion of lumbosacral nerves may also cause motor weakness and at times incontinence, both of which are of course distressing. Advice may be needed about continence and walking aids. Radiotherapy and chemotherapy may be helpful in delaying progression of such symptoms, although on occasions surgical diversion of the ureters or an end colostomy may be needed to be performed.

3. **Fistulae**: The development of fistulae, either due to disease progression or to radiation necrosis, is naturally disruptive to body

image and self-esteem as well as causing considerable physical inconvenience and embarrassment. Palliative surgery may well be justified in a woman who has a reasonable prognosis.

4. **Haemorrhage and discharge**: One or other of these symptoms may well be the initial symptom where disease is advanced at presentation. However, despite surgery and radiotherapy both may recur as the disease progresses; the palliative care physician may be left with fewer 'treatment' options than are available at an earlier stage. Fibrinolytic agents, such as tranexamic acid, may prevent capillary bleeds; metronidazole is useful in reducing odour. Massive haemorrhage is rarely the terminal event, and if it is deemed to be so, should be appropriately treated with sedation. Selective embolization of the bleeding vessel can be useful in some selected cases.

5. **Renal failure**: This is probably the commonest cause of death in this group. If renal failure occurs due to ureteric obstruction at an early stage in the patient's disease, or there are treatment options still available, or the patient deems her life to be of sufficient quality to merit it, ureteric stenting is indicated.

QUALITY OF LIFE IN CARCINOMA OF THE CERVIX

Quality of life in patients with carcinoma of the cervix has rarely been evaluated independently.

Speca *et al.* [9] attempted to establish the validity of the EORTC-QLQ30 in a group of women participating in a phase II chemotherapy trial evaluating the addition of cisplatin to a standard radiotherapy regime, for treatment of locally advanced carcinoma of the cervix. The EORTC-QLQ30 has site-specific modules which can be appended for different cancer sites; the 'most appropriate' site-specific module was used although there is as yet no module for carcinoma of the cervix. Those with recurrent or progressive disease were excluded.

There were difficulties with recruitment, with the final sample consisting of only five women, who formed a focus group.

The consensus view of the women was that the EORTC-QLQ30 could not adequately reflect their quality of life in the context of cancer treatment. They felt it overemphasized physical symptoms to the exclusion of emotional and psychological symptoms. The issues that these women felt to be of paramount importance related to concepts such as lack of control, the desire to retain a sense of 'normalcy', the feeling of having been violated and the fear and anxiety associated with the unpredictability of having had a cancer diagnosis. The authors readily acknowledge the deficiencies of a small retrospective study of this nature. It should also be acknowledged that this tool was not developed for the purpose for which it was used here.

However, it does illustrate that the factors which this group felt contributed most to their quality of life were less likely to be physical than emotional and psychological, and that changes in function, and treatment side-effects, were of less importance to overall quality of life.

The patients did not actually complete the questionnaire in this study so we do not know how they would have scored. It is of course possible that had they scored highly on physical symptoms this would have overridden any other concerns.

PSYCHOSOCIAL NEEDS

The literature in this area referring particularly to carcinoma of the cervix is scant. It tends to focus on early adjustment rather than later disease and there is far more written on sexual dysfunction than any other area, even though general texts on psycho-oncology [10] would suggest that up to 50% of people with cancer suffer either from an adjustment reaction or a formal psychiatric diagnosis such as depression or anxiety, and some 30% with a cancer diagnosis have some degree of sexual dysfunction.

PSYCHOLOGICAL REACTIONS TO A CANCER DIAGNOSIS

Up to 50% of patients are reported to have depressive symptoms at some time during the course of their illness. Some 7–20%, depending on the study, are regarded as having full-blown depression. Depression in advanced cancer is often perceived to be difficult to treat, at times unrewarding and possibly inappropriate to treat. It can be confused with either physical symptoms, concurrent treatment or probably best sadness. Advice from experts suggests that if symptoms of depression persist for more than a few weeks treatment should be considered, an assessment which is probably best made by a psychiatrist.

Anxiety, which is considered normal following initial diagnosis, if it persists or becomes disabling can be termed an anxiety disorder. Explanation, behavioural techniques and/or medication may be appropriate in this situation.

A normal adaptive process such as denial, taking the form of avoidance or suppression of information, protects patients to a degree from the more maladaptive symptoms of anxiety and depression. However, if denial becomes extreme, communication is blocked, compliance with treatment may be poor and necessary practical arrangements may be avoided. To prevent maladaptive behaviour, each patient should be allowed ample opportunity to ask questions wherever possible. All information need not be given at once. Information can and should be given with empathy, open communication and mutual respect.

Anger is frequently seen in the emotionally charged situations that inevitably surround a cancer diagnosis. Sometimes anger is justified, in which case it must be dealt with honestly and undefensively; however, it may be displaced and need redirecting.

Contrary to expectations, there is some evidence that suicidal thoughts diminish as cancer progresses. The incidence of suicide is thought to be approximately twice that of the general population, highest in the year after cancer diagnosis and possibly underestimated. Suicidal ideation may represent unrelieved distress accumulating and should always be explored, psychiatric help being sought where necessary. Voluntary euthanasia is currently against the law in the United Kingdom.

The importance of good communication skills to deal with such difficult issues cannot be overemphasized. Communicating information honestly, sensitively, undefensively and at the patient's pace can prevent much psychological morbidity [11]. A partner or friend, a nurse who can stay with the patient to answer any unresolved questions after a difficult interview and a quiet room are all helpful. 'Bad' news may need to be given more than once for the patients to absorb it.

PSYCHOLOGICAL DYSFUNCTION IN GYNAECOLOGICAL CANCER

Cain *et al.* [12] compared 60 patients with gynaecological cancers 1 month after diagnosis with a group of acutely depressed women and a control group. They used standard validated psychological assessment tools and found that of the group with gynaecological malignancy, 60% were mildly depressed, 33% moderately so and 3% severely depressed at 1 month. There was a tendency for depression to increase with the stage of disease, although this was not statistically significant. Compared to the group who were acutely depressed the cancer group had less interference with their social role but greater impairment in social adjustment, particularly at work, home and with relationships; compared to the 'normal' group they were significantly worse in most respects.

Less tangible adaptations to the diagnosis need to take place. A study of 20 patients with cervical intra-epithelial neoplasia showed that although fear of cancer overrode all other concerns at the time of diagnosis, this concern decreased with time and definitive treatment. Lack of attractiveness then became of prime importance [13]. This study also demonstrated that these women suffered from poor self-esteem.

Body image is more likely to be disrupted with gynaecological cancers than breast cancer. Even after hysterectomy for benign conditions, women have a disturbed body image [14]. The authors of this study demonstrated that there is a unique pattern to the emotional distress accompanying a cancer diagnosis. Depression and confusion were significantly higher in newly diagnosed gynaecological cancer patients than in women with benign conditions at the same body site, or in healthy women.

Femininity was also affected; both fertility and youth are lost due to treatment, which may have an impact on such feelings.

Relationships may be disrupted; three out of five women divorced after a pelvic exenteration in one study, though this could have been affected by the pre-existing relationship; longer relationships endured the stress of treatment better [15]. Pre-treatment counselling has been advocated by some in order to reduce psychosocial morbidity [16].

PSYCHOSOCIAL NEEDS IN CERVICAL CANCER

Areas of psychosocial need in cervical, as in any other gynaecological cancer patients, include the need for emotional and social support, information concerning the natural history of the disease and the treatment, advice about interaction with the medical environment and care-givers, advice concerning financial matters and information on other issues including sexuality and performing daily tasks [17].

Following the diagnosis of cancer of the cervix, it is important to discuss topics such as the cessation of menses, the end of fertility, the possible need for hormone replacement therapy and surgical reconstruction of the vagina prior to taking informed consent for treatment. Not only should this be undertaken prior to initiation of treatment, but opportunities for further discussion should be made

available during treatment and subsequent visits. Both the woman and her husband or partner should be encouraged to participate in these discussions. The quality of sexual experiences and general emotional health before treatment is the greatest determinant of satisfactory adjustment after therapy.

Reality issues need to be confronted at this stage. A common misapprehension of the male partner is that vaginal intercourse will be painful for the woman or impede her treatment. This fear is usually groundless unless there is a large cervical lesion present. Many men also feel guilt at having 'caused' the cancer. Occasionally poorly informed partners believe that a malignancy can be contagious or that the treatment modality, radium or drugs may remain in the woman to damage her close contacts. Such fears should be looked for, as they are not often revealed to the medical attendant during initial interviews.

An Australian study published in 1990 [17] looked at an unselected group of cancer patients attending an out-patient clinical oncology department. Areas of likely concern were reduced to eight. Patients were asked to rank these concerns according to their importance to them. One-hundred-and-eighty people were included, 129 of whom were female. There was little variation between the sexes. Of the total the majority (38%) had breast cancer and 7% of the study sample had cervical cancer. This study is of interest in that of the issues raised, 96% ranked family, dealing with emotional stress and getting information as their top three concerns. Other issues included were financial matters, social life, work, sex and dealing with hospital staff.

SEXUAL DYSFUNCTION WITH A GYNAECOLOGICAL CANCER DIAGNOSIS

With the diagnosis of cancer, the woman's life, relationships, responsibilities and activities are invaded. Her relationship and responsibilities to her partner, children, family and society may be disrupted both emotionally and physically. Previously successful adaptive behaviour may fail, anxiety is generated and self-esteem may be lost [18]. Very little attention has been paid to the psychosocial issues consequent on the diagnosis of cervical cancer. It is well-recognized that cancer influences the psychological and social functioning of patients, as well as their physical state [19].

There are many studies looking at the effect of treatment on sexual functioning following diagnosis and treatment of gynaecological malignancies. These have not all been in advanced malignancy. As suggested previously the importance of this particular issue may decrease with time as other issues gain more importance.

Surgery, which is felt to be associated with the least morbidity, has nevertheless been shown to affect sexual function. Corney *et al.* [20] demonstrated that following radical pelvic surgery for gynaecological malignancy, 50% of women felt their sexual relationship had worsened, though only 16% felt their relationship had deteriorated. Twenty-eight per cent of the women would have liked more information. Younger women appeared more likely to attribute personal and mental problems to sexual dysfunction. Whilst the median age of death for cervical cancer is 51 years [21], increasing age should not be assumed to mean lack of interest in sexual activity. Sexual concerns should be sought regardless of age. One study of 250 women with a mean age of 70 (range 60–93) suggested that 54% of the married women were sexually active [22].

Stomas influence psychology adversely at 1 year independently of the diagnosis. Those with neurotic or obsessional traits are particularly at risk [23].

Radiotherapy causes more sexual dysfunction than surgery. Dyspareunia is more common due to precipitation of menopause, reduced oestrogen and radiation changes causing dryness of the vagina. In Corney's study [20] 80% of those who had had radiotherapy developed sexual dysfunction. A small study by Rollinson and Strang [24] showed intrauterine brachytherapy to be at least moderately painful for most and caused great anxiety prior to admission.

The effect of chemotherapy on sexual dysfunction appears to have been little studied, the assumption possibly being that sex is of little relevance at this stage in a person's illness. This cannot be assumed.

PSYCHOSEXUAL DYSFUNCTION IN WOMEN WITH CANCER OF THE CERVIX

The physical mechanisms underlying the female sexual response may be impaired by surgery, pelvic or vaginal radiation and chemotherapy, or indirectly by ovarian failure. These treatment modalities cause vaginal stenosis, atrophy and irritation of the mucosa and inadequate lubrication, which may result in painful intercourse. The depth and calibre of the vagina itself may be altered following surgery and radiation treatment. The female sexual response may be impaired by psychological mechanisms and by changes in corticol sensory input. Psychological reactions to genital cancer that may result in sexual dysfunction have been described by Anderson and Wolf [25].

Sensitivity to cultural and religious beliefs is of paramount importance in sexual rehabilitation. These should be respected during counselling of these patients regarding sexual alternatives.

Cultural pressure on those who perceive a loss of physical female identity compounds irrevocable physiologic problems of altered anatomical structure and loss of sexual reproductive organs. Frequently it is hardly possible to distinguish between the physiological/ anatomical and psychological sources of sexual dysfunction. According to the sexual script theory and social learning theory [26, 27] sexuality can only be evaluated in subjective terms because there is no such thing as 'objective sexuality'. Therefore, evaluation of sexual functioning after treatment for cancer of the cervix has to be operationalized in terms of personal subjective experiences, closely related to the extent of similarity in ideal experiences and peak experiences at an individual level. Whether the individual succeeds in realizing her desire depends on a psychological level, on the presence or absence of psychologic capacities to perform sexually in such a way that it gives access to these ideal experiences. The subjective evaluation of sexual functioning can be seen as the outcome of a cognitive process in which ideal sexual experiences and real sexual experiences are weighed against each other.

Women with cancer of the cervix are the group of women amongst all other gynaecological cancers who have received the greatest amount of interest from investigators of sexual functioning following treatment. This is probably related to the fact that this disease affects younger women and has a comparatively more favourable prognosis, and the controversy about which treatment modality is superior, radical hysterectomy with lymph node dissection or radiotherapy. In terms of survival, the two modes of treatments are equally effective for early-stage disease. Therefore, the studies have focused on determining whether the treatment modalities produce different rates of sexual disruption. All the retrospective studies [28–32] except for that of Tamburini *et al.* [32] found a post-treatment difference in sexual disruption to the detriment of radiation therapy. Tamburini *et al.* [32] found no significant differences in sexual functioning between women treated with radical hysterectomy, radiotherapy or combined treatment.

Krumm and Lamberti [33] showed decreased frequency of masturbation and intercourse, decreased frequency of orgasm, through non-coital sexual activities, less satisfaction with sex and less enjoyment of intercourse following radiation treatment for cervical cancer. In their study the majority believed that changes were due to radiation therapy and reflected decreased feelings of self-esteem, sex desirability and attractiveness. Women who did not follow advice regarding the use of vaginal dilators

and did not resume their premorbid level of sex developed more physical and sexual changes. Non-compliance with the use of vaginal dilators might be related to underlying psychosexual concerns which could be dispelled with more information and counselling.

The timing for these discussions is relevant. Discussions regarding the illness and hazards of treatment should begin just prior to treatment. Researchers have focused on determining the predictors of post-treatment sexual dysfunction. Schover *et al.* [34] and Weijman Schultz *et al.* [35] advocated a stronger focus on relationship status and prior sexual function for predictive post-treatment sexual dysfunction. Andersen [36] generally supported magnitude of treatment as the more powerful determinant. After radiotherapy has begun conversations and suggestions about sexuality should be broached. At the completion of the treatment, the effects of illness and treatment on the patient and her partner should be discussed as well as the need for regular vaginal dilation coitus and other sexual activity.

Careful interviewing can elicit fears about the impact of the disease and its treatment on the individual's body image, self-identity and self-esteem. The fact that such an important organ as the cervix has been affected by disease, and the vagina by treatment, can bring out regressive fantasies and fears about defectiveness, desirability and attractiveness. These anxieties may promote a pre-occupation with sexual function, fear of loss of love and possible rejection by the partner. Airing these concerns in the context of understanding, medical personnel and a supportive partner can help assuage the woman's perception of her physical changes and enhance her self-esteem and self-acceptance.

Provision of information and education of the patient and her partner in order to help them understand and deal with any changes in sexual behaviour following treatment either by surgery, radiotherapy or combination of the two should minimize and prevent the psychological and physical trauma in these women.

PSYCHOSEXUAL ISSUES AFTER EXENTERATIVE SURGERY

Exenterative surgery results in the loss of a variable amount of tissue including possibly the bladder, anus and vagina with resulting ileal conduit and/or colostomy. It is often observed in clinical practice that the patient feels mutilated or crippled by her physical loss and suffers guilt, shame and self-disgust anticipating that her partner will see her unattractive and crippled. Loss of bladder and bowel control to many women means loss of autonomy and may produce anger and resentment.

Following exenteration operations, many women have difficulty in accepting their altered body image. Special difficulties in sexual rehabilitation may occur as a result of the altered body's external or internal configurations. These are further exacerbated by simultaneous reactive depression and loss of libido when the use of dilators and early intercourse is most important in maintaining a neovagina. Exenterative surgery may arouse strong negative emotions in the partner as well as the patient. He is forced to confront his own emotions and to evaluate his feelings for the woman and their relationship. He may have difficulty accepting a woman with an ileal conduit, colostomy or neovagina. To some extent these psychological problems in both the patient and her partner may be alleviated by psychosexual assessment prior to the surgery and providing general information similar to that conveyed in any sexual counselling or therapy situation shared with the patient and her partner. Sometimes meeting and discussing the personal adjustment necessary following exenterative surgery with a patient who has previously undergone similar treatment may be very helpful.

Prior to discharge from the hospital the patient and her partner should be given the opportunity to discuss any questions and concerns regarding the care of the ostia and neovagina and should be assured continuing support from the hospital staff if need arises.

REHABILITATION

Vincent *et al.* [37] found that 75% of gynaecological cancer patients have received no information regarding sexual adjustment before, during or after treatment. However, 80% of these women desired this kind of information and also stated that they would not bring up the subject themselves, but preferred the discussion to be initiated by the medical team. A needs assessment by Bullard *et al.* [38] among cancer patients revealed that 63% of the participants would have liked to have received more information about sexual functioning after treatment and that 64% would participate in a specific counselling programme on the topic if this should become possible. Wiel *et al.* [39] showed that only 23% of patients had received any information about the possible consequences of treatment on their sexual functioning and that, of this informed group, only 55% had regarded this information as adequate. The inadequate information about the effect of disease and therapeutic interventions on people's sexual functioning is also reflected in the scarce amount of research available on this issue.

Treatments for cervical cancer range from local tissue destruction to total pelvic exenteration. The ensuing coital and reproductive alterations produce a spectrum of damage from vaginal discharge to premature menopause and even extensive mutilation with exenterative procedures to achieve the goal of a cure. There is much that needs to be achieved by sexual rehabilitation of such patients. The ability to provide the support for this rehabilitation is available to all the carers of a woman with cervical cancer, namely the attitudes of interest, openness, support and the ability to provide simple information patiently and repetitively. The expertise of professional sex counsellors is rarely required. Simple drawings of the pelvic organs help in teaching the woman about her malignancy and the anatomic changes which may follow her treatment.

Knowledge of the woman's customary sexual practices as well as awareness of her social class and religious background are most important in planning emotional and sexual rehabilitation.

Discussion of the individual's mechanics of sexual intercourse and physiologic changes of sexual arousal provide a foundation for the description of any changes expected after treatment. The alternative methods possible after the treatment need to be fully addressed. These discussions are best initiated as soon as the diagnosis of cervical cancer has been made. Whether sexual dissatisfaction occurs will also depend on personal and social factors and the context in which these 'negative' changes occur. Sexual dysfunction represents the final common pathology of biological, social and psychological factors in the patient's personal and interpersonal world.

PREVENTING MORBIDITY

Most patients will benefit from a full discussion about their disease and the treatment being planned. This has been shown to reduce psychological morbidity [11].

In addition to help given by health professionals, patients often find it of benefit to be involved in self-help groups. There is controversial evidence that these may prolong the prognosis by engendering a fighting spirit, possibly improving immunity [40]. Other organizations, such as BACUP (British Association of Cancer United Patients) [41], produce patient information leaflets which describe the disease, treatments and potential complications in layman's terms. They also provide counselling.

FOLLOW-UP

During the years of follow-up after the treatment, it is not uncommon for women to experience 'check-up anxiety' with each follow-up visit. Feelings or anxiety are described by women as early as 2 weeks before the appointment; the anxiety persists, increases and reaches its peak in the last 2–3 days before the hospital visit. Insomnia, moodiness, irritability, tearfulness and problems with concentration and memory are common stress symptoms during this time. Even after many years following their treatment women are painfully aware of the risk of recurrence at the time of their hospital check-up. The anxiety and stress dissipate rapidly when the checkup is completed and the woman is reassured by the medical attendant. Occurrence of unexplained symptoms causes a state of anxiety until a recurrence of the disease has been ruled out. Women who have recurrence diagnosed often experience greater anxiety than they did at the time of the initial diagnosis. The fear of dying is intensified as fears about successful treatment for the recurrence are experienced. Psychological support is invaluable at this time from the medical attendant and the supportive staff, as the woman's fears of dying are naturally stronger at the time of diagnosis of recurrence.

REFERENCES

1. Sugarbaker, P.H., Barefsky, I., Rosenberg, S.A. and Gionala, F.J. (1982) Quality of life assessment of patients in extremity sarcoma clinical trials. *Surgery*, **91**, 17–23.
2. Andersen, B.L. (1987) Sexual functioning complications in women with gynaecologic cancer outcomes and directions for prevention. *Cancer*, **60**, 2123–8.
3. Holland, C.B.J. (1976) Coping with cancer. A challenge to the behavioural sciences. In *Cancer. The Behavioural Dimensions* (eds J.W. Cullen, B.H. Fox and R.N. Isam). Raven Press, New York.
4. Vettese, J.M. Problems of the patients confronting the diagnosis of cancer. In *Cancer. The Behavioural Dimensions* (eds J.W. Cullen, B.H. Fox and R.N. Isam). Raven Press, New York.
5. McCartney, C. and Larson, D. (1997) Quality of life in patients with gynaecologic cancer. *Cancer*, **60**, 2129–36.
6. Calman, K.C. (1984) Quality of life in cancer patients – an hypothesis. *Journal of Medical Ethics*, **10**, 124–7.

7. Slevin, M., Stubbs, L., Plant, H.J., *et al.* (1990) Attitudes to chemotherapy: comparing views of patients with cancer with those doctors, nurses, and general public. *British Medical Journal*, **300**, 1458–60.

8. World Health Organization (1986) *Cancer Pain Relief*, WHO, Geneva.

9. Speca, M., Robinson, J., Goodey, E. and Frizzell, B. (1994) Patients evaluate a quality of life scale: whose life is it anyway? *Cancer Practice*, **2**, 365–70.

10. Barraclough, J. (1995) *Cancer and Emotion: A Practical Guide to Psychooncology*. John Wiley, Chichester.

11. Rainey, L.C. (1985) Effects of preparatory patient education for radiation oncology patients. *Cancer*, **56**, 1056–61.

12. Cain, E., Kohorn, E., Quinlan, D., *et al.* (1983) Psychological reactions to the diagnosis of gynaecologic cancer. *Obstetrics and Gynecology*, **62**, 635–41.

13. McDonald, T.W., Neutens, J.J., Fischer, L.M., *et al.* (1989) Impact of CIN diagnosis and treatment on self esteem and body image. *Gynaecologic Oncology*, **84**, 345–9.

14. Andersen, B. and Jochinsen, P. (1985) Sexual functioning among breast cancer, gynaecologic cancer and healthy women. *Journal of Consulting Clinical Psychologists*, **53**, 25–32.

15. Andersen, B. and Macker, N. (1983) Psychosexual adjustment following pelvic exenteration. *Obstetrics and Gynecology*, **61**, 331–8.

16. Auchincloss, S.S. (1995) After treatment: psychosocial issues in gynaecologic cancer survivorship. *Cancer*, **76** (Suppl.), 2117–24.

17. Liang, L.P., Dunn, S.M., Gorman, A., *et al.* (1990) Identifying priorities of psychosocial need in cancer patients. *British Journal of Cancer*, **62**, 1000–3.

18. Corney, R.H., Everest, H., Howells, A., *et al.* (1992) Psychological adjustment following major gynaecological surgery for carcinoma of the cervix and vulva. *Journal of Psychosomatic Research*, **36**, 561–8.

19. Stam, H.J., Bultz, B.D. and Pittman, C.A. (1986) Psychological problems and interventions in a referred sample of cancer patients. *Psychosomatic Medicine*, **48**, 539–44.

20. Corney, R.H., Crowther, M.E., Howells, A., *et al.* (1993) Psychosexual dysfunction in women with gynaecological cancer following radical pelvic surgery. *British Journal of Obstetrics and Gynaecology*, **100**, 73–8.

21. Manetta, A., Benman, M.C. and Disaria, P.J. (1992) Advanced and recurrent carcinoma of the cervix. In *Gynaecological Oncology.*

Fundamental Principles and Clinical Practice, Vol. 1 (ed. M. Coppleson). Churchill Livingstone, Edinburgh.

22. Newman, G. and Nichols, C.R. (1960) Sexual activities and attitudes in older persons. *Journal of the American Medical Association*, **173**, 33–5.

23. Thomas, C., Madden, F. and Jehu, D. (1987) Psychological effects of stomas – II factors influencing outcome. *Journal of Psychosomatic Research*, **31**, 317–23.

24. Rollinson, B. and Strang, P. (1995) Pain, nausea and anxiety during intra-uterine bracytherapy or cervical carcinoma. *Support Care Cancer*, **3**, 205–7.

25. Anderson, B.J. and Wolf, F.M. (1986) Chronic physical illness and sexual behaviour: psychological issues. *Journal of Consulting Clinical Psychologists*, **54**, 168–75.

26. Simon, W. and Gagnon, J.H. (1986) Sexual scripts: permanence and change. *Archives of Sexual Behaviour*, **15**, 97–120.

27. Rolter, J.B. (1954) *Social Learning and Clinical Psychology*. Prentice Hall, New York.

28. Vasicka, A., Popovich, N.R. and Brausch, C.C. (1958) Post medication course of patients with cervical carcinoma. A clinical study of psychic, sexual and physical well being of sixteen patients. *Obstetrics and Gynecology*, **11**, 403–14.

29. Decker, W.H. and Schwartzman, E. (1962) Sexual function following treatment for carcinoma of the cervix. *American Journal of Obstetrics and Gynecology*, **83**, 401–5.

30. Abitol, M. and Davenport, J.H. (1974) The irradiated vagina. *Obstetrics and Gynecology*, **44**, 249–56.

31. Bertelsen, K. (1983) Sexual function after treatment of cervical cancer. *Danish Medical Bulletin*, **30** (Suppl.), 31–34.

32. Tamburini, M., Filibert, A., Ventapridda, V., *et al.* (1985) Emotional status, sexuality and quality of life in patients treated for carcinoma of the uterine cervix. *Cervix*, **2**, 261–7.

33. Krumm, S. and Lamberti, J. (1993) Changes in sexual behaviour following radiation therapy for cervical cancer. *Journal of Psychosomatic Obstetrics and Gynaecology*, **14**, 51–63.

34. Schover, L.R., Fife, N. and Gershewen, D. (1989) Sexual dysfunction and treatment for early stage cervical cancer. *Cancer*, **63**, 204–12.

35. Weijman Schultz, W.C.M., Van de Wiel, H.B.M., Bonma, J., *et al.* (1990) Psychological functioning after the treatment of cancer of the vulva: a longitudinal study. *Cancer*, **66**, 402–7.

36. Andersen, B.L. (1993) Predicting sexual and psychologic morbidity and improving the quality of life for women with gynaecological cancer. *Cancer*, **71** (Suppl. 4), 1678–90.

37. Vincent, C.E., Vincent, B., Greiss, F.C., *et al.* (1975) Some marital concomitants of carcinoma of the cervix. *Southern Medical Journal*, **68**, 552–8.

38. Bullard, D.G., Cansey, G.G., Newman, A.B., *et al.* (1980) Sexual health care and cancer: a needs assessment. In *Frontiers of Radiation Therapy and Oncology* (ed. J.M. Vaeth). Karger, Basel, pp. 55–8.

39. Wiel van de, H.B.M., Weijmar Schultz, W.C.M., Hengeveld, M.W., *et al.* (1991) Sexual functioning after ostomy surgery. *Sex and Marital Therapy*, **2**, 177–94.

40. Spiegal, D., Bloom, J.R., Kraevel, M.C., *et al.* (1989) Effect of psychosocial treatment on survival of patients with metastatic breast cancer. *Lancet*, **2**, 888–91.

41. BACUP, 3 Bath Place, Rivington Street, London EC2A 3JR, UK.

INDEX